PRINCIPLES OF
THE THEORY OF SOLIDS

TO ROSEMARY

*

How much better is it to get wisdom than gold. And to get understanding rather to be chosen than silver.

Proverbs xvi. 16

PRINCIPLES OF THE THEORY OF SOLIDS

BY

J. M. ZIMAN, F.R.S.

Melville Wills Professor of Physics in the University of Bristol

SECOND EDITION

CAMBRIDGE UNIVERSITY PRESS

Published by the Press Syndicate of the University of Cambridge
The Pitt Building, Trumpington Street, Cambridge CB2 1RP
40 West 20th Street, New York, NY 10011-4211 USA
10 Stamford Road, Oakleigh, Melbourne 3166, Australia

Library of Congress catalogue card number: 72-80250

First published 1964
Reprinted 1965, 1969
Second edition 1972
First paperback edition 1979
Reprinted 1986, 1988, 1989, 1992, 1995

ISBN 0 521 29733 8 paperback

Transferred to digital printing 1999

PREFACE

The Frontiers of Knowledge (to coin a phrase) are always on the move. Today's discovery will tomorrow be part of the mental furniture of every research worker. By the end of next week it will be in every course of graduate lectures. Within the month there will be a clamour to have it in the undergraduate curriculum. Next year, I do believe, it will seem so commonplace that it may be assumed to be known by every schoolboy.

The process of advancing the line of settlements, and cultivating and civilizing the new territory, takes place in stages. The original papers are published, to the delight of their authors, and to the critical eyes of their readers. Review articles then provide crude sketch plans, elementary guides through the forests of the literature. Then come the monographs, exact surveys, mapping out the ground that has been won, adjusting claims for priority, putting each fact or theory into its place.

Finally we need textbooks. There is a profound distinction between a treatise and a textbook. A treatise expounds; a textbook explains. It has never been supposed that a student could get into his head the whole of physics, nor even the whole of any branch of physics. He does not need to remember what he can easily discover by reference to monographs, review articles and original papers. But he must learn to *read* those references: he must learn the language in which they are written: he must know the basic experimental facts, and general theoretical principles, upon which his science is founded.

This book aims to present, as simply as possible, the elements of the theory of the physics of perfect crystalline solids. It is a book full of *ideas*, not facts. It is an exposition of the principles, not a description of the phenomena.

A theory is an analysis of the properties of a hypothetical model. In physics, which may almost be defined as the intellectual exercise of subsuming the universe to mathematics, our models are mathematical. The theories discussed in this book are mathematical theories; the most important concepts in the field, such as 'the Fermi Surface', are abstract mathematical constructions, which cannot be explained or understood properly without formal analysis.

What I have tried to do is to give a self-contained mathematical treatment of the simplest model that will demonstrate each principle. If most of the interesting properties of superconductors can be derived from a model of free electrons with a curtailed attractive interaction, then that is the framework of the calculation. At this stage, it is better to appreciate the conditions that are essential to the appearance of the phenomenon at all, than it is to try to wield far heavier equations, based upon more realistic but much more complex physical specifications, in order to anticipate the observed deviations from the simple formulae. The reader must go to the original papers and treatises for these elaborations of the elementary models.

On the other hand, having defined the model, one must not shirk the mathematical analysis. It is my experience that the direct derivation of many simple, well-known formulae from first principles is not easy to find in print. The original papers do not follow the easiest path, the authors of reviews find the necessary exposition too difficult—or beneath their dignity—and the treatises are too self-conscious about completeness and rigour. I have tried to make the mathematical argument complete in itself—or at least intelligible in principle—without frequent appeal to that *deus ex machina* of the tired author 'it can be shown that...'. An advantage of trying to cover such a wide field is that one can invoke general principles, such as Bloch's theorem and the theory of zones, to unify many branches of the subject and save much mental effort.

How much is the reader expected to know already? He should be acquainted with the elementary descriptive facts about solids—for example, the free-electron theory of metals—as taught in undergraduate courses. I also assume the elements of quantum mechanics—especially the Schrödinger equation, perturbation theory, and the theory of scattering—such as graduate students of experimental physics are now expected to acquire. I have tried to keep the mathematical techniques to that level; whenever the algebra threatened to get difficult, I have stopped. There are no density matrices, bubble diagrams, branch points, character tables, or other bits and pieces of the apparatus of advanced theory; professional 'theoreticians' must look elsewhere for their fodder.

For the benefit of those reviewers who judge a book by what is absent from it, let me admit that there is no serious discussion of alloys, dislocations, F-centres, impurity centres, etc. There is nothing about magnetic resonances associated with single atoms or nuclei, and no attempt to interpret essentially macroscopic phenomena such as ferromagnetic domains, p–n junctions or the intermediate state of a superconductor. The exclusion of all except simple, perfect, crystalline solids is artificial, but it is convenient, for it gives some unity to the discourse, which is centred on the mathematical consequences of lattice periodicity. Also, friends, life is short.

No novelty is claimed for this account of an active branch of physics. These are the theories that are currently used, and accepted as well established, by those who work in this field. There is no attempt to criticize the theories, to discuss their validity as interpretations of natural phenomena, to derive them with full mathematical rigour, or to demonstrate the full flowering of their applications. The effort has been focused upon clarity of exposition. The reader is being asked to grasp new concepts; let him suspend his critical faculties until he has understood what the new ideas mean; if, then, he is rightly sceptical, let him turn to the enormous literature upon which this science is built, and help his unbelief from those copious, if muddy, sources. I have deliberately refrained from making any direct reference to the original papers; for such information the student should consult the monographs and review articles listed at the end of the book.

This book began as a course of lectures for graduate students of Theoretical and Experimental Physics in the University of Cambridge: it was written in my last two terms, and completed in my last few weeks, of active membership of the staff of the Cavendish Laboratory. It is a great privilege to have belonged, for nine years, to that peerless institution. I am only too conscious of the impossibility of living up to the unique standard that it has set, and continues to set, in the world of physics.

But the Cavendish is more than a famous laboratory; it is an abode of humanity and friendship. May I offer thanks to those friends—especially to Nevill Mott, to Brian Pippard, and to Volker Heine—who brought me to Cambridge, who welcomed me, taught me, wisely controverted me, abundantly assisted me, and generally made life here agreeable, interesting and exciting. They have heard the music of the spheres; and yet they know that science is made for man, not man for science.

J. M. Z.

Cavendish Laboratory, Cambridge
June 1963

PREFACE TO THE SECOND EDITION

This new edition is meant still to conform to the principles expounded in the above Preface. But eight years is about the doubling time of modern scientific knowledge and solid state physicists have not been idle in the interval. Most of the original text still stands, but several new sections have been added, to cover topics that have come into greater prominence lately or where there has been a significant shift of understanding or emphasis. I have also attempted to make reference in passing to a number of phenomena or fields of study that are relevant to the basic theory, even if they cannot be discussed in detail. In this way, the general scope of the book has been widened, to include, for example, something about magnetic and non-magnetic impurities, F-centres, surfaces, tunnelling, junctions, and type II superconductivity. But the general level of mathematical sophistication has not been raised, even though the technical formalism of advanced quantum theory is now becoming more commonplace in this field.

I am most grateful to many colleagues—especially to Bob Chambers here in Bristol and to Federico Garcia-Moliner in Madrid—for a number of detailed comments of which I have tried to take account in the new text. Bob Evans helped greatly by preparing a new index. And lest the reader may feel that absence from Cambridge has been a long period of exile, may I simply add that Bristol, too, is just as good a 'ole to go to.

J. M. Z.

H. H. Wills Physics Laboratory, Bristol
March 1971

CONTENTS

CHAPTER 1

PERIODIC STRUCTURES

Again! again! again! THOMAS CAMPBELL

1.1 Translational symmetry

A theory of the physical properties of solids would be practically impossible if the most stable structure for most solids were not a regular crystal lattice. The N-body problem is reduced to manageable proportions by the existence of *translational symmetry*. This means that *there exist basic vectors,* $\mathbf{a}_1$, $\mathbf{a}_2$, $\mathbf{a}_3$ *such that the atomic structure remains invariant under translation through any vector which is the sum of integral multiples of these vectors.*

In practice, this is only an ideal. A laboratory specimen is necessarily finite in size, so that we must not carry our structure through the boundary. But the only regions where this matters are the layers of atoms near the surface, and in a block of N atoms these constitute only about $N^{\frac{2}{3}}$ atoms—say 1 atom in 10^8 in a macroscopic specimen. Most crystalline solids are also structurally imperfect, with defects, impurities and dislocations to disturb the regularity of arrangement of the atoms. Such imperfections give rise to many interesting physical phenomena, but we shall ignore them, except incidentally, in the present discussion. We are mainly concerned here with the perfect ideal solid, and with the properties it shows; the phenomena which are associated with the solid as the matrix, vehicle, or background for little bits of dirt, or tiny cracks and structural flaws, belong to a different realm of discourse.

We represent the translational group by a *space lattice* or *Bravais net*. Start from some point and then construct all points reached from it by the basic translations. These are the *lattice sites*, defined by the set

$$\boldsymbol{l} = l_1 \mathbf{a}_1 + l_2 \mathbf{a}_2 + l_3 \mathbf{a}_3, \tag{1.1}$$

where l_1, l_2, l_3 are integers (Fig. 1).

But a solid is a physical structure—not a set of mathematical points. Suppose that there are some atoms, etc., in the neighbourhood of our origin O. The translational invariance insists that there must be exactly similar atoms, placed similarly, about each lattice site (Fig. 2).

It is obvious that we can define the physical arrangement of the whole crystal if we specify the contents of a single *unit cell*—for example, the parallelepiped subtended by the basic vectors $\mathbf{a}_1$, $\mathbf{a}_2$, $\mathbf{a}_3$. The whole crystal is made up of endless repetitions of this object stacked like bricks in a wall. But the actual definition of a unit cell is to some extent arbitrary. It is obvious enough that any parallelepiped of the right size, shape and orientation would do—as we see in

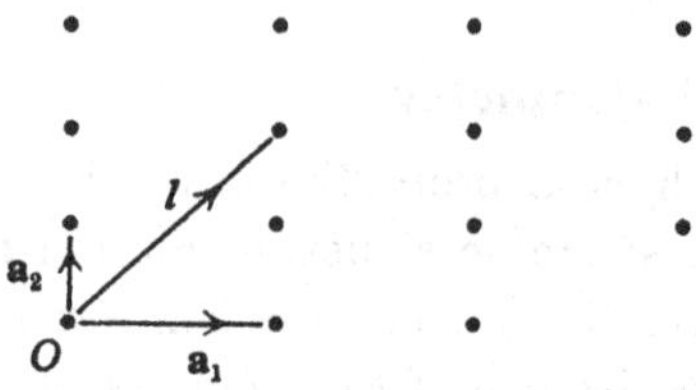

Fig. 1

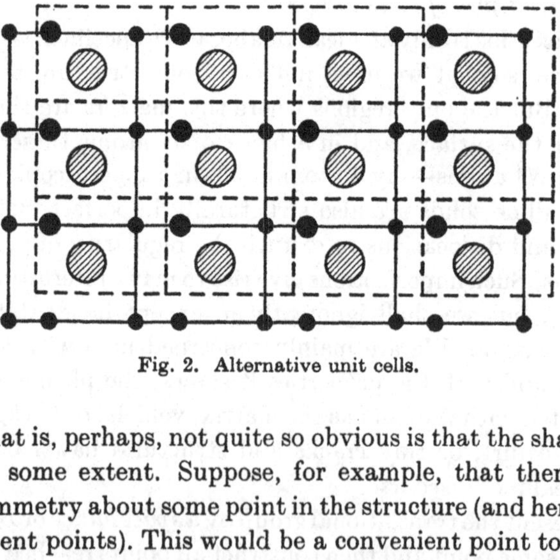

Fig. 2. Alternative unit cells.

Fig. 2. What is, perhaps, not quite so obvious is that the shape can be altered to some extent. Suppose, for example, that there is some central symmetry about some point in the structure (and hence, about all equivalent points). This would be a convenient point to choose as the centre of a cell, itself with central symmetry. One can do this systematically by constructing a *Wigner–Seitz* cell, that is, by drawing the perpendicular bisector planes of the translation vectors from the chosen centre to the nearest equivalent lattice sites. The volume inside all the bisector planes is obviously a unit cell—it is the region whose elements lie nearer to the chosen centre than to any other lattice site.

The unit cell can contain one or more atoms. Naturally, if it contains only one atom, we centre that on the lattice site, and say that we have a *Bravais lattice*. If there are several atoms per unit cell, then we have a *lattice with a basis*. In most of what follows, we shall assume, without special notice, that the structure is a Bravais lattice. This is for simplicity; in reality only a few elementary solids, such as the alkali metals, have this structure.

The science of *crystallography* is concerned with the enumeration and classification of all possible types of crystal structure, and the determination of the actual structure of actual crystalline solids.

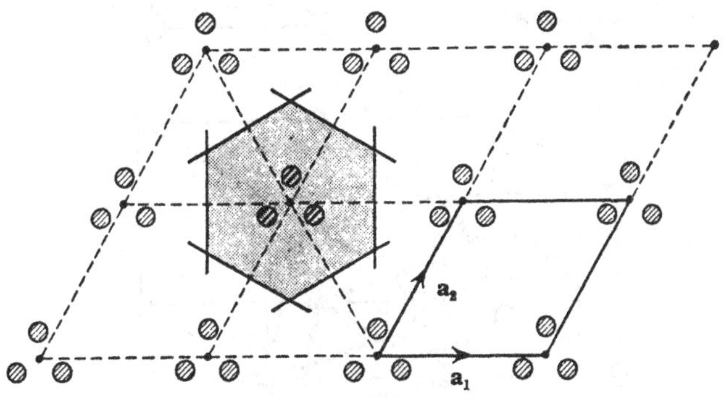

Fig. 3. Wigner–Seitz cell.

Structures are classified according to their symmetry properties, such as invariance under rotation about an axis, reflection in a plane, etc. These symmetries are often of great importance in the simplification of theoretical computations, and can be used with great power in the discussion of the numbers of parameters that are necessary to define the macroscopic properties of solids. However, to take full advantage of this theory, one needs the mathematics of *group theory*, which would take us too far away from our main topic. If we restrict ourselves mainly to very simple solids, then most of the symmetry properties can be discovered by inspection without formal algebraic analysis. In any case, there are many excellent books on crystallography and on group theory and its applications to the theory of solids.

In these books, the various types of Bravais lattice, etc., are set out in detail. We shall consider here only one case, which exemplifies many of the principles of the subject, and which is also of great importance as a structure which is actually assumed by some elements. This is the *body-centred cubic* (B.C.C.) structure illustrated in Fig. 4.

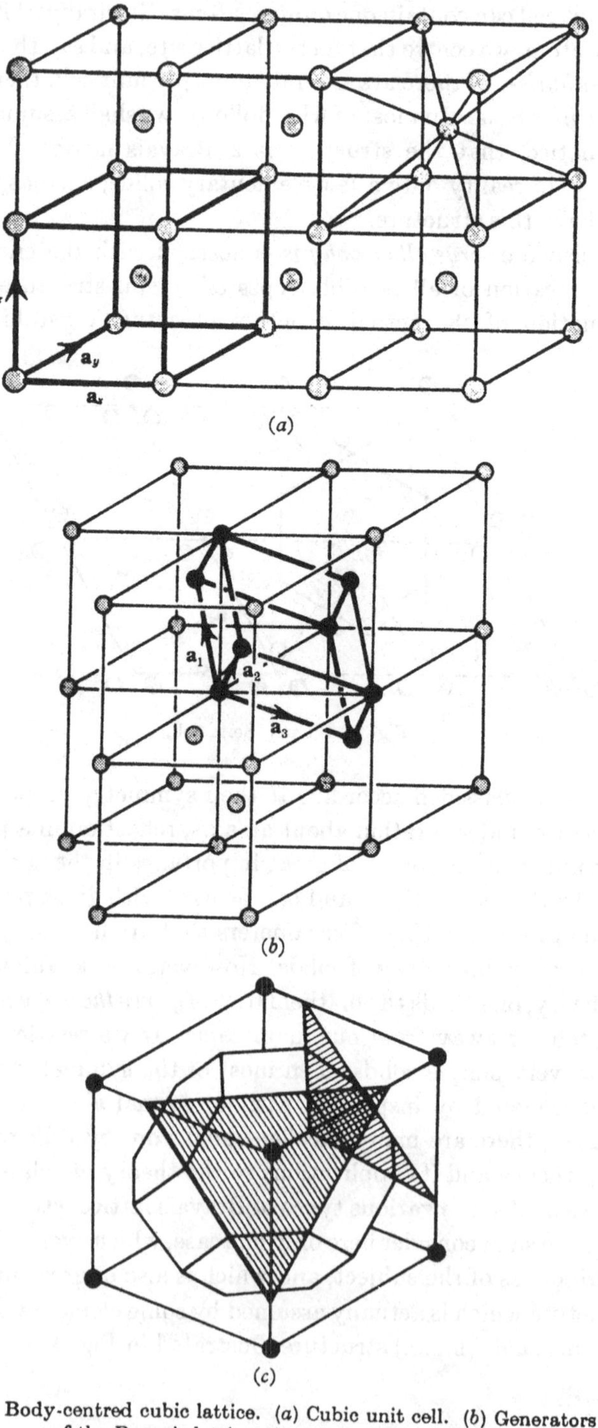

Fig. 4. Body-centred cubic lattice. (*a*) Cubic unit cell. (*b*) Generators of the Bravais lattice. (*c*) Wigner–Seitz cell.

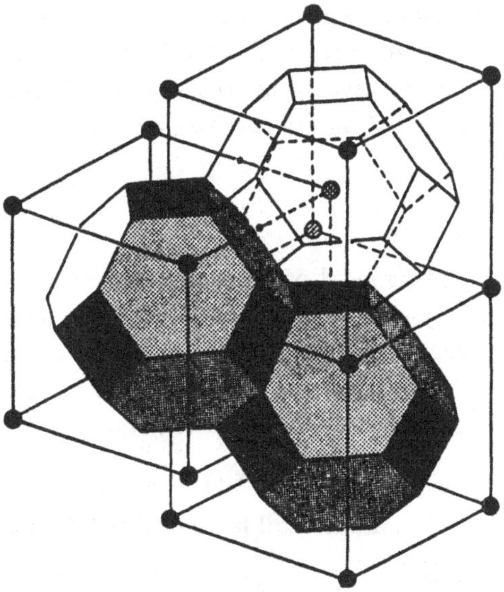

Fig. 5. Stacking Wigner–Seitz cells of the B.C.C. lattice.

At first sight, this is a cubic lattice with two atoms per unit cell, or two interpenetrating simple cubic sublattices defined by

$$\left.\begin{aligned} \mathbf{l} &= l_x\,\mathbf{a}_x + l_y\,\mathbf{a}_y + l_z\,\mathbf{a}_z, \\ \mathbf{l}' &= (l_x+\tfrac{1}{2})\,\mathbf{a}_x + (l_y+\tfrac{1}{2})\,\mathbf{a}_y + (l_z+\tfrac{1}{2})\,\mathbf{a}_z, \end{aligned}\right\} \tag{1.2}$$

where l_x, l_y, l_z are all integers. But if we write

$$\left.\begin{aligned} \mathbf{a}_1 &= \tfrac{1}{2}(-\mathbf{a}_x+\mathbf{a}_y+\mathbf{a}_z), \\ \mathbf{a}_2 &= \tfrac{1}{2}(\mathbf{a}_x-\mathbf{a}_y+\mathbf{a}_z), \\ \mathbf{a}_3 &= \tfrac{1}{2}(\mathbf{a}_x+\mathbf{a}_y-\mathbf{a}_z), \end{aligned}\right\} \tag{1.3}$$

we can generate all the points of both sublattices by

$$\mathbf{l} = l_1\mathbf{a}_1 + l_2\,\mathbf{a}_2 + l_3\,\mathbf{a}_3 \tag{1.4}$$

with l_1, l_2, l_3 integers. We shall find ourselves at a cube centre, or corner, according as $(l_1+l_2+l_3)$ is odd, or even.

Thus (Fig. 4(*b*)) this is really a Bravais lattice. Instead of using a cubic unit cell we may use the Wigner–Seitz cell (Fig. 4(*c*)), which is constructed by chopping off all the corners of a cube half way along a diagonal from the centre to a corner point. This figure obviously has the same symmetry as a cube—for example, the original vectors, $\mathbf{a}_x, \mathbf{a}_y, \mathbf{a}_z$ are axes of four-fold symmetry. It also shows, more clearly perhaps than the original cube, that the vectors $\mathbf{a}_1$, $\mathbf{a}_2$, $\mathbf{a}_3$, i.e. the diagonals of the cube, are axes of three-fold rotational symmetry, for they pass through the hexagonal faces of the cell. It is a good exercise to visualize how these polyhedra can be packed to make the original lattice (Fig. 5).

1.2 Periodic functions

To define a physical model of a crystal structure we need to give values to some function $f(\mathbf{r})$ in space—local electron density, electrostatic potential, etc., which can be recognized as an arrangement of atoms (e.g. Fig. 6). Our assertion of translational symmetry means that this must be a *multiply periodic function*

$$f(\mathbf{r}+\mathbf{l}) = f(\mathbf{r}) \tag{1.5}$$

for all points $\mathbf{r}$ in space, and for all lattice translations.

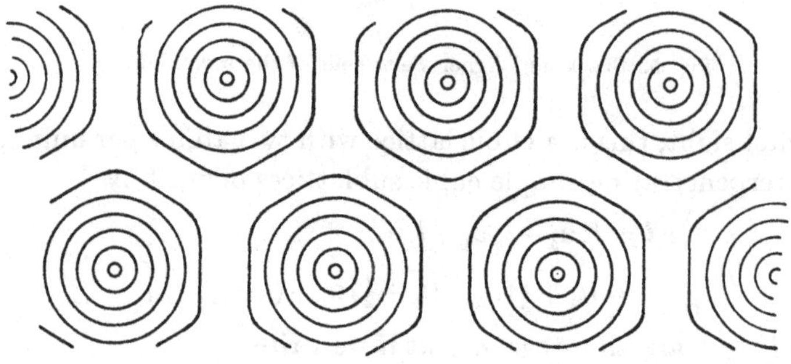

Fig. 6. Multiply periodic function.

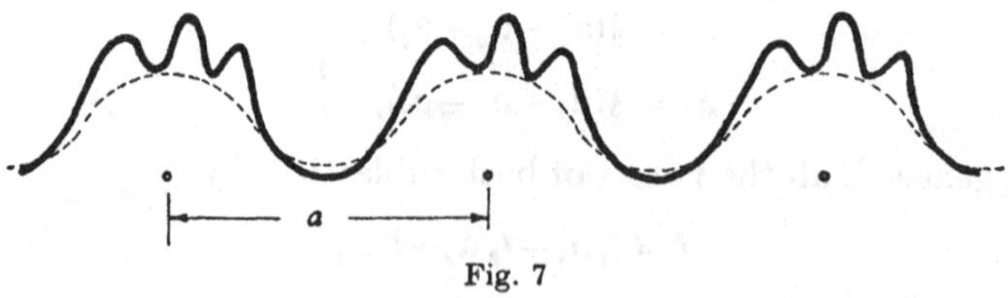

Fig. 7

In one dimension we are perfectly familiar with periodic functions. Thus, as in Fig. 7, we may have

$$f(x+l) = f(x), \tag{1.6}$$

where l is of the form $l_1 a$, with l_1 an integer and a the period of the function. We also know that any such function can be expressed as a *Fourier series*,

$$f(x) = \sum_n A_n e^{2\pi inx/a}, \tag{1.7}$$

where n is an integer. Let us write this in the form

$$f(x) = \sum_g A_g e^{igx}, \tag{1.8}$$

where the quantities g belong to the set of *reciprocal lattice lengths*,†

$$g_n = n\frac{2\pi}{a}. \tag{1.9}$$

The coefficients in (1.8) are well known to be defined by

$$A_g = \frac{1}{a}\int_{\text{cell}} f(x)\, e^{-igx}\, dx, \tag{1.10}$$

where in this case the range of integration is, say $0 < x \leqslant a$—it need only be one cell of the lattice. The elementary proof that (1.8) implies (1.6) can be derived from the condition that, for any value of g,

$$e^{igl} = 1 \tag{1.11}$$

for all translations l. This means no more than that

$$gl = n\frac{2\pi}{a} l_1 a = nl_1 2\pi = \text{integer} \times 2\pi \tag{1.12}$$

when g takes one of its allowed values g_n and $l = l_1 a$. The derivation of (1.10) also follows from this same relation. We are not concerned here with mathematically pathological functions, and may use naïve Fourier theory quite freely.

Extension of this theorem to a structure with three rectangular axes is easy enough. If these axes are $\mathbf{a}_x$, $\mathbf{a}_y$, $\mathbf{a}_z$, then, if

$$f(\mathbf{r}+\mathbf{l}) \equiv f(\mathbf{r}+l_x\mathbf{a}_x+l_y\mathbf{a}_y+l_z\mathbf{a}_z) = f(\mathbf{r}), \tag{1.13}$$

we have

$$f(\mathbf{r}) = \sum_{g_x, g_y, g_z} A(g_x, g_y, g_z) \exp\{i(g_x x + g_y y + g_z z)\}, \tag{1.14}$$

where each of g_x, etc., is a reciprocal length out of the set $2\pi n/a_x$, etc. One can prove this by first analysing $f(\mathbf{r})$ as a Fourier series in the x direction, then showing that each coefficient in this series is a periodic function of y, and can be analysed into a further series—and similarly for z.

Let us rewrite (1.14) in the form

$$f(\mathbf{r}) = \sum_{\mathbf{g}} A_{\mathbf{g}} e^{i\mathbf{g}\cdot\mathbf{r}}, \tag{1.15}$$

† It is convenient to include the factor 2π in this definition, although in crystallography the convention is to leave 2π explicitly in the exponent.

where $\mathbf{g}$ is the vector with components (g_x, g_y, g_z). This vector has the property expressed in (1.11), i.e. for any such vector

$$\begin{aligned}\mathbf{g}\cdot\boldsymbol{l} &= (g_x l_x a_x + g_y l_y a_y + g_z l_z a_z) \\ &= \frac{2\pi n_x}{a_x} l_x a_x + \frac{2\pi n_y}{a_y} l_y a_y + \frac{2\pi n_z}{a_z} l_z a \\ &= 2\pi \times \text{integer},\end{aligned} \tag{1.16}$$

whatever the value of $\boldsymbol{l}$. Thus,

$$e^{i\mathbf{g}\cdot\boldsymbol{l}} = 1 \tag{1.17}$$

for all lattice vectors $\boldsymbol{l}$, and for all *reciprocal lattice vectors* $\mathbf{g}$.

It is obvious that this condition is sufficient to make the series (1.15) represent a function like (1.5) with the periodicity of the lattice:

$$\begin{aligned}f(\mathbf{r}+\boldsymbol{l}) &= \sum_{\mathbf{g}} A_{\mathbf{g}} e^{i\mathbf{g}\cdot(\mathbf{r}+\boldsymbol{l})} = \sum_{\mathbf{g}} A_{\mathbf{g}} e^{i\mathbf{g}\cdot\mathbf{r}} e^{i\mathbf{g}\cdot\boldsymbol{l}} \\ &= \sum_{\mathbf{g}} A_{\mathbf{g}} e^{i\mathbf{g}\cdot\mathbf{r}} = f(\mathbf{r}).\end{aligned} \tag{1.18}$$

One can readily devise a proof that it is a necessary condition, i.e. that the sum (1.15) may only contain terms corresponding to values of $\mathbf{g}$ that satisfy the condition (1.17).

It only remains to construct the reciprocal lattice vectors for a non-rectangular lattice. This is easily done as follows: take the *reciprocal triad* of the basic vectors, i.e.

$$\mathbf{b}_1 = \frac{\mathbf{a}_2 \wedge \mathbf{a}_3}{\mathbf{a}_1 \cdot \mathbf{a}_2 \wedge \mathbf{a}_3}, \quad \mathbf{b}_2 = \frac{\mathbf{a}_3 \wedge \mathbf{a}_1}{\mathbf{a}_1 \cdot \mathbf{a}_2 \wedge \mathbf{a}_3}, \quad \mathbf{b}_3 = \frac{\mathbf{a}_1 \wedge \mathbf{a}_2}{\mathbf{a}_1 \cdot \mathbf{a}_2 \wedge \mathbf{a}_3}, \tag{1.19}$$

and write

$$\begin{aligned}\mathbf{g} &= g_1 \mathbf{b}_1 + g_2 \mathbf{b}_2 + g_3 \mathbf{b}_3 \\ &= 2\pi n_1 \mathbf{b}_1 + 2\pi n_2 \mathbf{b}_2 + 2\pi n_3 \mathbf{b}_3,\end{aligned} \tag{1.20}$$

where n_1, etc., are integers.

By elementary vector analysis, we verify that $\mathbf{b}_1 \cdot \mathbf{a}_1 = 1$, etc., and $\mathbf{b}_1 \cdot \mathbf{a}_2 = 0$, etc., so that

$$\begin{aligned}\mathbf{g}\cdot\boldsymbol{l} &= (g_1 \mathbf{b}_1 + g_2 \mathbf{b}_2 + g_3 \mathbf{b}_3)\cdot(l_1 \mathbf{a}_1 + l_2 \mathbf{a}_2 + l_3 \mathbf{a}_3) \\ &= g_1 l_1 + g_2 l_2 + g_3 l_3 \\ &= 2\pi \times \text{integer}.\end{aligned} \tag{1.21}$$

Thus, every vector of the set (1.20) satisfies the condition (1.17).

We can also generalize the formula (1.10) for each coefficient in the series (1.15). It must be a volume integral, which we write

$$A_{\mathbf{g}} = \frac{1}{v_{\text{cell}}} \int_{\text{cell}} f(\mathbf{r})\, e^{-i\mathbf{g}\cdot\mathbf{r}} d\mathbf{r}. \tag{1.22}$$

The proof follows by multiplying the series through by $\exp(-i\mathbf{g}\cdot\mathbf{r})$, and integrating. An integral of $\exp(i\mathbf{g}\cdot\mathbf{r})$ through a unit cell is obviously zero, if $\mathbf{g}$ is of the form (1.20), unless $\mathbf{g} = 0$.

1.3 Properties of the reciprocal lattice

The vectors defined by (1.20), i.e.

$$\mathbf{g} = n_1 . 2\pi\mathbf{b}_1 + n_2 . 2\pi\mathbf{b}_2 + n_3 . 2\pi\mathbf{b}_3, \tag{1.23}$$

generate a lattice with basic cell spanned by the vectors $2\pi\mathbf{b}_1$, $2\pi\mathbf{b}_2$, $2\pi\mathbf{b}_3$. This is the *reciprocal lattice* of our original direct lattice. It is an invariant geometrical object, whose properties are fundamental in the theory of solids. Some of the elementary geometrical properties are easily deduced.

(i) *Each vector of the reciprocal lattice is normal to a set of lattice planes of the direct lattice.*

Choose a particular reciprocal lattice vector $\mathbf{g}$, and a lattice vector $\boldsymbol{l}$ and consider the relation (1.21)

$$\begin{aligned} \mathbf{g}\cdot\boldsymbol{l} &= 2\pi(n_1 l_1 + n_2 l_2 + n_3 l_3) \\ &= 2\pi N, \end{aligned} \tag{1.24}$$

where N is an integer. This tells us that the projection of the vector $\boldsymbol{l}$ on the direction of $\mathbf{g}$ has the length

$$d = \frac{2\pi N}{|\mathbf{g}|}. \tag{1.25}$$

But there are infinitely many points of the lattice with this property. For example, suppose $\boldsymbol{l}'$ is a lattice point represented by the integers

$$l_1' = l_1 - mn_3, \quad l_2' = l_2 - mn_3, \quad l_3' = l_1 + m(n_1 + n_2), \tag{1.26}$$

where m is an integer. It is obvious that

$$\mathbf{g}\cdot\boldsymbol{l}' = \mathbf{g}\cdot\boldsymbol{l} = 2\pi N. \tag{1.27}$$

Thus $\boldsymbol{l}'$ has the same projection on $\mathbf{g}$ and must therefore also be on the plane normal to $\mathbf{g}$, at distance d from the origin. Thus, if there is one lattice point on this plane, there is an infinity of such points; we have constructed one of the *lattice planes*. This relationship between the direct and reciprocal lattices is, of course, a special case of the familiar duality between points and planes in three-dimensional geometry.

(ii) *If the components of* $\mathbf{g}$ *have no common factor, then* $|\mathbf{g}|$ *is inversely proportional to the spacing of the lattice planes normal to* $\mathbf{g}$.

This follows from (1.25) (see Fig. 8). If (n_1, n_2, n_3) have no common factor, then we can always find a lattice vector $\boldsymbol{l}''$ with components such that

$$\mathbf{g}\cdot\boldsymbol{l}'' = 2\pi(N+1). \tag{1.28}$$

This means that the lattice plane containing $\boldsymbol{l}''$ is at a distance

$$d'' = \frac{2\pi(N+1)}{|\mathbf{g}|} \tag{1.29}$$

from the origin—i.e. is spaced $2\pi/|\mathbf{g}|$ from the plane containing $\boldsymbol{l}$.

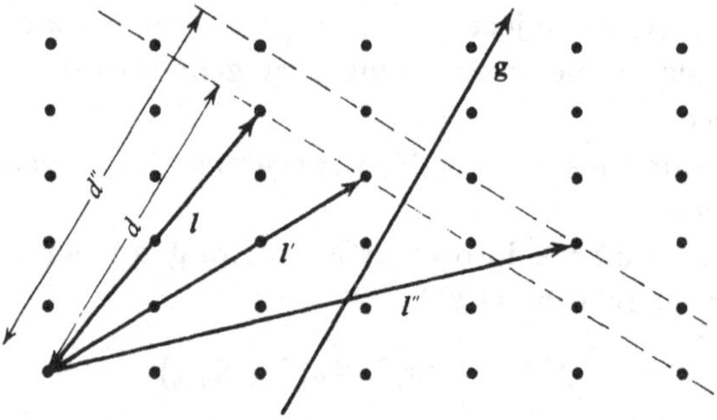

Fig. 8

From these two elementary geometrical results we see that the simplest way of characterizing the planes of a lattice is by their normals, expressed as vectors of the reciprocal lattice. The most prominent planes in the direct lattice are those which are most densely populated with lattice sites. Since the density of lattice sites is constant in space the most prominent planes must be those which are most widely separated, i.e. those with the smallest reciprocal lattice vectors.

The labelling of lattice planes by their corresponding reciprocal lattice vectors is equivalent to the *Miller indices* used in classical crystallography. Suppose that we have a lattice plane, with normal $\mathbf{g}$, such that (1.24) is satisfied for all points $\boldsymbol{l}$ upon it. Then if we take a point such that $l_2 = l_3 = 0$, we have $l_1 = N/n_1$, so that the intercept of this plane along the $\mathbf{a}_1$ axis has length

$$d_1 = \left(\frac{N}{n_1}\right) a_1. \tag{1.30}$$

Similarly this plane will cut the $\mathbf{a}_2$ axis at a distance

$$d_2 = \left(\frac{N}{n_2}\right) a_2 \tag{1.31}$$

from the origin. The intercepts of this plane along the axes, measured in units of the lengths of the corresponding basic vectors, are inversely proportional to the integers n_1, n_2, n_3. These integers, after removal of any common factors, are the Miller indices of the plane, expressed in the form (n_1, n_2, n_3).

It is obvious that densely populated planes—i.e. ones with small Miller indices—are those which are most likely to show up in natural crystals, either in growth or after cleavage. The study of the geometry of such faces was of the essence of classical crystallography, and the

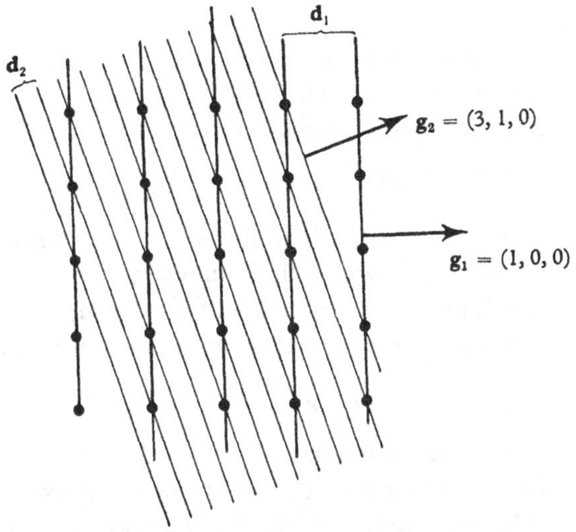

Fig. 9. The spacing of planes with large Miller indices is closer than the spacing of the principal symmetry planes.

key discovery was that a single set of vectors $\mathbf{a}_1, \mathbf{a}_2, \mathbf{a}_3$ could be found such that all the observed faces of a macroscopic crystal could be represented by small Miller indices.

Two conventions of notation should be mentioned. The symbol $(1\bar{1}\bar{1})$ represents the planes $(1, -1, -1)$; the minus sign is put above the number for conciseness. The symbol $\{110\}$, with curly brackets, represents all the different sets of planes that are equivalent by symmetry to the set (110). Thus, for a cubic structure it might include (101), (011), $(\bar{1}\bar{1}0)$, $(1\bar{1}0)$, etc.

(iii) *The volume of a unit cell of the reciprocal lattice is inversely proportional to the volume of a unit cell of the direct lattice.*

This follows by elementary vector analysis. The reciprocal lattice cell is spanned by $2\pi\mathbf{b}_1$, $2\pi\mathbf{b}_2$, $2\pi\mathbf{b}_3$. Using (1.19), its volume would be

$$\begin{aligned}(2\pi)^3(\mathbf{b}_1\cdot\mathbf{b}_2\wedge\mathbf{b}_3) &= (2\pi)^3\frac{(\mathbf{a}_2\wedge\mathbf{a}_3)\cdot\{(\mathbf{a}_3\wedge\mathbf{a}_1)\wedge(\mathbf{a}_1\wedge\mathbf{a}_2)\}}{(\mathbf{a}_1\cdot\mathbf{a}_2\wedge\mathbf{a}_3)^3}\\ &= (2\pi)^3\frac{(\mathbf{a}_2\wedge\mathbf{a}_3)\cdot[\{\mathbf{a}_3\cdot(\mathbf{a}_1\wedge\mathbf{a}_2)\}\,\mathbf{a}_1-\{\mathbf{a}_1\cdot(\mathbf{a}_1\wedge\mathbf{a}_2)\}\,\mathbf{a}_3]}{(\mathbf{a}_1\cdot\mathbf{a}_2\wedge\mathbf{a}_3)^3}\\ &= \frac{(2\pi)^3}{(\mathbf{a}_1\cdot\mathbf{a}_2\wedge\mathbf{a}_3)}\\ &= \frac{8\pi^3}{v_c},\end{aligned} \tag{1.32}$$

where v_c is the volume of the unit cell spanned by $\mathbf{a}_1$, $\mathbf{a}_2$, $\mathbf{a}_3$.

The factor $8\pi^3$ comes in here because of the way we have defined the reciprocal lattice. A more common convention is to write

$$e^{i\mathbf{g}\cdot\mathbf{l}} \equiv \exp(2\pi i\mathbf{K}_g\cdot\mathbf{R}_l), \tag{1.33}$$

where $\mathbf{R}_l$ is a lattice vector, and the vectors $\mathbf{K}_g$ are defined as vectors of a reciprocal lattice with unit cell volume $1/v_c$. The disadvantage of this notation is that the factor 2π keeps turning up in the exponent; it is also incommensurable with the conventional symbolism

$$\psi = e^{i\mathbf{k}\cdot\mathbf{r}} \tag{1.34}$$

for a free-electron wave-function.

(iv) *The direct lattice is the reciprocal of its own reciprocal lattice.*

This is implicit in the name, and can be verified by constructing, say, the vector $(\mathbf{b}_1\wedge\mathbf{b}_2)/(\mathbf{b}_1\cdot\mathbf{b}_2\wedge\mathbf{b}_3)$, and showing that it is identical with $\mathbf{a}_3$. One can see this by inspection of the derivation of (1.32).

(v) *The unit cell of the reciprocal lattice need not be a parallelepiped.* In fact, we almost always deal with the Wigner–Seitz cell of the reciprocal lattice. This is called a *Brillouin zone.*

As an example of the properties of the reciprocal lattice, consider an important and commonly observed structure—the *face-centred cubic* lattice (F.C.C. structure). This is built up out of four interpenetrating simple cubic lattices, arranged so that if we look at any one of them, we see an atom at the centre of each face of its unit cell, as well as on the sites at the corners of the cube (Fig. 10 (*a*)).

This looks at first, like a lattice with a basis, with four atoms in the cubic unit cell generated by $(\mathbf{a}_x, \mathbf{a}_y, \mathbf{a}_z)$.

But if we choose the half diagonals of the faces of the cube,

$$\left.\begin{aligned} \mathbf{a}_1 = (0, \tfrac{1}{2}a, \tfrac{1}{2}a), \quad \mathbf{a}_2 = (\tfrac{1}{2}a, 0, \tfrac{1}{2}a), \\ \mathbf{a}_3 = (\tfrac{1}{2}a, \tfrac{1}{2}a, 0), \end{aligned}\right\} \tag{1.35}$$

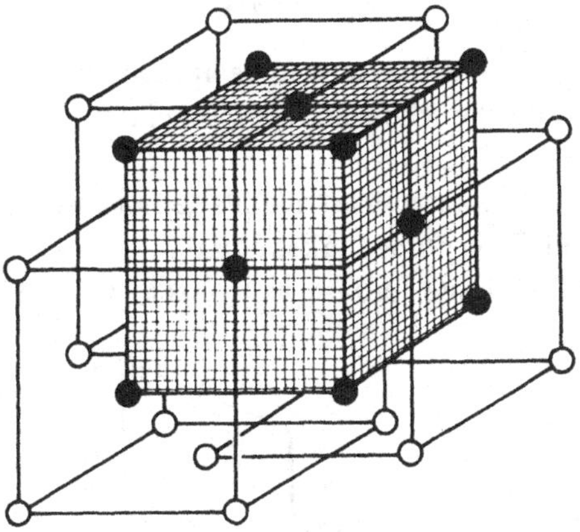

(a)

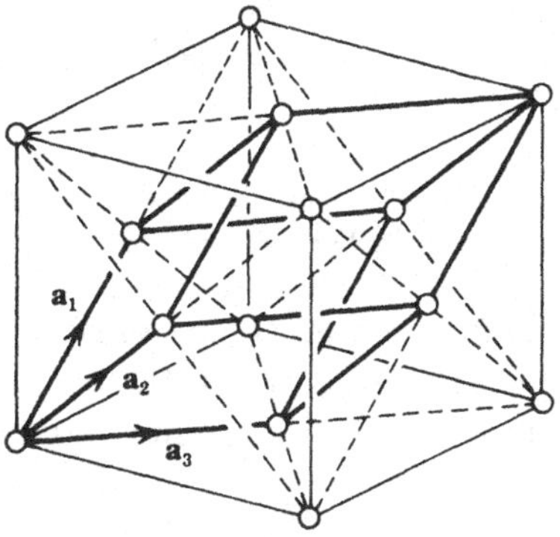

(b)

Fig. 10. Face-centred cubic lattice. (a) As four interpenetrating sublattices. (b) As Bravais lattice.

then we can reach any site by a combination of multiples of these (Fig. 10 (b)). Thus, the F.C.C. lattice is truly a Bravais lattice with $\mathbf{a}_1, \mathbf{a}_2, \mathbf{a}_3$ as generators.

We could now construct the reciprocal lattice by algebra. But it is easy to use the geometrical properties noted above. It is obvious, for example, that there are important lattice planes normal to the edges of the cube. Relative to ordinary Cartesian axes along $\mathbf{a}_x$, $\mathbf{a}_y$, $\mathbf{a}_z$, these planes must have Miller indices (100) (010) (001), etc., i.e. they are the set {100}. We can see, also, that they are spaced a distance $\frac{1}{2}a$ apart. Thus, they correspond to reciprocal lattice vectors of length

$$|\mathbf{g}| = 2\pi/\tfrac{1}{2}a$$

along rectangular Cartesian axes in reciprocal space. Our reciprocal lattice must include the whole simple cubic lattice generated by these vectors.

There is also an important lattice plane normal to the diagonal of the cubic unit cell. This plane makes equal intercepts on all three axes, and must therefore have indices in the set {111}. These planes are

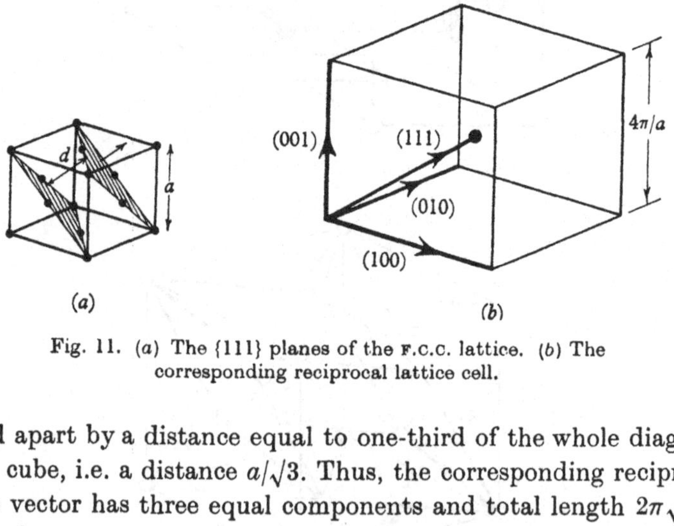

Fig. 11. (*a*) The {111} planes of the F.C.C. lattice. (*b*) The corresponding reciprocal lattice cell.

spaced apart by a distance equal to one-third of the whole diagonal of the cube, i.e. a distance $a/\sqrt{3}$. Thus, the corresponding reciprocal lattice vector has three equal components and total length $2\pi\sqrt{3}/a$; it must be

$$\mathbf{g}' = \frac{2\pi}{a}(1, 1, 1). \tag{1.36}$$

In reciprocal space, it is obviously directed along the diagonal of the cubic cell—but is only half its length. Thus, it generates the body centres of the simple cubic lattice.

One can now study all other lattice planes—there are, for example, the {110} sets, which are normal to the diagonals of the faces of the basic cube—but the reciprocal lattice vectors for these are all contained already in the points generated by the {100} and {111} planes. Thus, the reciprocal of the F.C.C. lattice is B.C.C. (and, of course, *vice versa*).

The Brillouin zone of the F.C.C. lattice is the Wigner–Seitz cell of this reciprocal lattice. We have already seen this figure (Fig. 4 (*c*))—a polyhedron with square and hexagonal faces. The centres of the square faces correspond to the directions of the normals of the cube planes of the direct lattice. The centres of the hexagonal faces are the normals of the diagonal planes.

It is interesting to note, further, that the F.C.C. structure is 'close-packed'. It can be built up by superposing successive (111) planes, each of which is a hexagonal network, as of rigid spheres. This is one of the reasons why it is often adopted by metals, whose cohesion is not associated with strongly directional bonds but is largely a volume effect.

1.4 Bloch's theorem

We have learnt how to represent functions with the periodicity of the lattice. But these are not enough for a physical theory; we need to consider various excitations of the structure, which will destroy the exact translational symmetry. Of these there are several types, of which the most important are *lattice waves*, i.e. vibrations of the atoms about their equilibrium positions, and *electron states*, i.e. motion of electrons in the field of the static lattice. In Chapter 10 we shall also consider *spin waves*, which are excitations of the spins localized on the atoms of the crystal.

All these excitations are characterized by a dynamical equation, or a Schrödinger equation, or a spin-exchange Hamiltonian, which is *invariant under lattice translation*. Thus, if $\mathscr{V}(\mathbf{r})$ is the potential seen by an electron at $\mathbf{r}$, then $\mathscr{V}(\mathbf{r}+\boldsymbol{l}) = \mathscr{V}(\mathbf{r})$ for all $\boldsymbol{l}$. The electron wave-function $\psi(\mathbf{r})$ must satisfy a Schrödinger equation

$$\left(-\frac{\hbar^2}{2m}\nabla^2+\mathscr{V}(\mathbf{r})-\mathscr{E}\right)\psi = 0, \tag{1.37}$$

which remains the same after we have substituted $\mathbf{r}+\boldsymbol{l}$ for $\mathbf{r}$ in the operator that acts on ψ.

In the case of lattice waves and spin waves, the formalism becomes more complicated but the principle is the same. Suppose that

$$(u_1, u_2, \ldots, u_{\mathbf{n}}, \ldots)$$

are the displacements (or spin displacements, or spin deviation operators, or something) of the atoms at sites $1, 2, \ldots, \mathbf{n}, \ldots$. The equations of motion (or the Hamiltonian) will contain these variables (or operators) in such a way that if we make a lattice translation $\boldsymbol{l}$—that is, if we put $u_{\mathbf{n}+\boldsymbol{l}}$ in the formulae wherever $u_{\mathbf{n}}$ occurred previously—we shall not be able to tell the difference. In other words, all cells of the lattice are equivalent, and indistinguishable. It is rather like the 'cosmological principle' which says that the universe looks essentially the same from whatever point we view it.

To formalize this property, let us use the following notation. Let $\mathcal{H}(0)$ and $|0\rangle$ represent the Hamiltonian operator and wave-function before a lattice translation; let $\mathcal{H}(\boldsymbol{l})$ and $|\boldsymbol{l}\rangle$ represent the same mathematical objects after the translation $\boldsymbol{l}$. Thus, for electrons $|0\rangle \equiv \psi(\mathbf{r})$, and $|\boldsymbol{l}\rangle \equiv \psi(\mathbf{r}+\boldsymbol{l})$. For lattice waves, if $|0\rangle$ represents the state in which the atom at $\mathbf{n}$ is going through some particular motion, then $|\boldsymbol{l}\rangle$ represents the state in which the atom at $(\mathbf{n}+\boldsymbol{l})$ is performing that same motion.

The statement of translational invariance is then

$$\mathcal{H}(\boldsymbol{l}) = \mathcal{H}(0). \tag{1.38}$$

Now the eigenstates must satisfy equations of the form†

$$\mathcal{H}(0)\,|0\rangle = \mathscr{E}\,|0\rangle. \tag{1.39}$$

This equation is identical with

$$\mathcal{H}(\boldsymbol{l})\,|\boldsymbol{l}\rangle = \mathscr{E}\,|\boldsymbol{l}\rangle, \tag{1.40}$$

which comes merely by a formal relabelling of all the variables in both operators and state functions. But, by (1.38), this implies

$$\mathcal{H}(0)\,|\boldsymbol{l}\rangle = \mathscr{E}\,|\boldsymbol{l}\rangle, \tag{1.41}$$

i.e. the state function $|\boldsymbol{l}\rangle$ is also a solution of the equation satisfied by $|0\rangle$. Since $|\boldsymbol{l}\rangle$ is not necessarily identical with $|0\rangle$, it looks as if we can generate an immense number of solutions of the equations of motion just by translations of any one solution that we have happened to find. All these solutions are, of course, degenerate in energy.

This is obviously ridiculous; the new solutions must all be equivalent, in some way, to our original solution. There are two cases to consider:

(i) Suppose $|0\rangle$ is really *non-degenerate* (which is never actually the case, because all lattices have reflection symmetries, as well as translational invariance). Then the only possibility is that each new state

† In the case of the classical equations of motion of lattice waves, the square of the frequency ω plays the role of the energy, and the 'Hamiltonian' is the matrix of the linked linear differential equations derived upon the assumption that we are dealing with small deviations from equilibrium. See §2.1.

$|\boldsymbol{l}\rangle$ is a multiple of $|0\rangle$, and hence physically indistinguishable from it. There must be some number λ_1 such that

$$|\mathbf{a}_1\rangle = \lambda_1 |0\rangle \tag{1.42}$$

for the result of a single step in the direction $\mathbf{a}_1$. Normalization requires that

$$|\lambda_1|^2 = 1, \tag{1.43}$$

so that we might have

$$\lambda_1 = e^{ik_1}, \tag{1.44}$$

where k_1 is a real number. Similarly, we might have

$$|\mathbf{a}_2\rangle = e^{ik_2}|0\rangle; \quad |\mathbf{a}_3\rangle = e^{ik_3}|0\rangle, \tag{1.45}$$

for unit translations in the other basic directions. Then a general translation must lead to the following relation

$$\begin{aligned} |\boldsymbol{l}\rangle &\equiv |l_1\,\mathbf{a}_1 + l_2\,\mathbf{a}_2 + l_3\,\mathbf{a}_3\rangle \\ &= e^{ik_1}|(l_1-1)\,\mathbf{a}_1 + l_2\,\mathbf{a}_2 + l_3\,\mathbf{a}_3\rangle \\ &= e^{ik_1 l_1}|0 + l_2\,\mathbf{a}_2 + l_3\,\mathbf{a}_3\rangle \\ &= e^{i(k_1 l_1 + k_2 l_2 + k_3 l_3)}|0\rangle \end{aligned} \tag{1.46}$$

as we make l_1 steps in direction $\mathbf{a}_1$, l_2 steps in direction $\mathbf{a}_2$, etc., and multiply by the appropriate factor for each step.

Let us define a vector $\mathbf{k}$, in the form

$$\mathbf{k} = k_1\,\mathbf{b}_1 + k_2\,\mathbf{b}_2 + k_3\,\mathbf{b}_3, \tag{1.47}$$

where $\mathbf{b}_1, \mathbf{b}_2, \mathbf{b}_3$ are the reciprocal triad of $\mathbf{a}_1, \mathbf{a}_2, \mathbf{a}_3$ as in (1.19). The relation (1.46) becomes

$$|\boldsymbol{l}\rangle = e^{i\mathbf{k}\cdot\boldsymbol{l}}|0\rangle, \tag{1.48}$$

which is the result we are seeking to prove:

For any wave-function/state function that satisfies the Schrödinger equation (or its classical or quantal equivalent) there exists a vector $\mathbf{k}$ *such that translation by a lattice vector* $\boldsymbol{l}$ *is equivalent to multiplying by the phase factor* $\exp(i\mathbf{k}\cdot\boldsymbol{l})$.

Each different state function may have a different *wave-vector* $\mathbf{k}$, but each one must satisfy this condition. It is a strong condition imposed on the form of an elementary excitation by the strong initial circumstance that the lattice is translationally invariant.

But we need the proof in a more general case.

(ii) Suppose $|0\rangle$ is degenerate. For simplicity, suppose that it is doubly degenerate, so that there are certainly two distinct states $|0\rangle_1$

and $|0\rangle_2$ of the same energy. Then the lattice translation $\mathbf{a}_1$ can produce, at most, linear combinations of these same two states. Thus

$$\left.\begin{aligned} |\mathbf{a}_1\rangle_1 &= T_1^{11}\,|0\rangle_1 + T_1^{12}\,|0\rangle_2, \\ |\mathbf{a}_1\rangle_2 &= T_1^{21}\,|0\rangle_1 + T_1^{22}\,|0\rangle_2. \end{aligned}\right\} \tag{1.49}$$

The numbers T_1^{11}, etc., are written like this because they are the elements of a matrix $\mathbf{T}_1$, which must be unitary to preserve normalization. For conciseness, let us write this pair of equations in the form

$$(|\mathbf{a}_1\rangle) = \mathbf{T}_1(|0\rangle). \tag{1.50}$$

But $|0\rangle_1$ and $|0\rangle_2$, being degenerate, are not uniquely defined. We could have started with any two other states which were linear combinations of these. For example, we might have started with

$$\left.\begin{aligned} |0\}_1 &= S^{11}\,|0\rangle_1 + S^{12}\,|0\rangle_2, \\ |0\}_2 &= S^{21}\,|0\rangle_1 + S^{22}\,|0\rangle_2, \end{aligned}\right\} \tag{1.51}$$

where the numbers S^{11}, etc., are elements of another (arbitrary) unitary matrix $\mathbf{S}$.

Now let us choose $\mathbf{S}$ so that the matrix $\mathbf{S}\mathbf{T}_1\mathbf{S}^{-1}$ is a diagonal matrix. By standard algebraic theory, this is always possible. Suppose, for example, that it reduces to the form

$$\mathbf{S}\mathbf{T}_1\mathbf{S}^{-1} = \begin{pmatrix} e^{ik_1} & 0 \\ 0 & e^{ik_1'} \end{pmatrix}. \tag{1.52}$$

It can easily be verified that the states $|0\}_1$ and $|0\}_2$ will now transform under translation exactly as if they were non-degenerate, i.e. as in (1.42–1.45),

$$|\mathbf{a}_1\}_1 = e^{ik_1}\,|0\}_1, \quad |\mathbf{a}_1\}_2 = e^{ik_1'}\,|0\}_2. \tag{1.53}$$

The state $|0\}_1$ has a wave-vector with component k_1 for the direction $\mathbf{a}_1$; the state $|0\}_2$ has wave-vector with component k_1' (in general, different) for the same translation direction.

But now what about translations in other directions. Consider $\mathbf{a}_2$. For the initial states $|0\rangle_1$ and $|0\rangle_2$ there will be another matrix $\mathbf{T}_2$ defining the result of this translation, i.e. symbolically as in (1.49) and (1.50)

$$(|\mathbf{a}_2\rangle) = \mathbf{T}_2(|0\rangle). \tag{1.54}$$

We could now reduce this matrix $\mathbf{T}_2$ to diagonal form, just as in (1.52), by choosing a suitable set of starting functions. At first sight, this does not seem compatible with the set we have used to make $\mathbf{T}_1$ diagonal. But consider the following alternative reductions:

$$(|\mathbf{a}_1+\mathbf{a}_2\rangle) = \begin{Bmatrix} \mathbf{T}_2(|\mathbf{a}_1\rangle) = \mathbf{T}_2\mathbf{T}_1(|0\rangle), \\ \mathbf{T}_1(|\mathbf{a}_2\rangle) = \mathbf{T}_1\mathbf{T}_2(|0\rangle). \end{Bmatrix} \tag{1.55}$$

The matrices $\mathbf{T}_1$ and $\mathbf{T}_2$ thus commute with one another. There is a theorem of matrix algebra which tells us that there then exists a unitary matrix $\mathbf{S}$ such that *both* $\mathbf{T}_1$ and $\mathbf{T}_2$ can be reduced *simultaneously* to diagonal form. Thus, the states $|0\}_1$ and $|0\}_2$ will not be mixed by the translation $\mathbf{a}_2$, but will simply be multiplied by phase factors

$$|\mathbf{a}_2\}_1 = e^{ik_2}|0\}_1, \quad |\mathbf{a}_2\}_2 = e^{ik_2'}|0\}_2. \tag{1.56}$$

A similar result holds for translation by $\mathbf{a}_3$, so that we are back at essentially the same as (1.45). For each of $|0\}_1$ and $|0\}_2$ there exists a wave-vector such that

$$|\boldsymbol{l}\} = e^{i\mathbf{k}\cdot\boldsymbol{l}}|0\}. \tag{1.57}$$

The extension to more than two states degenerate in energy is obviously trivial.

Thus, the theorem is proved in general, in the sense that any degenerate solution of the wave-function can be represented as a linear combination of solutions of the same energy each of which must satisfy a condition of this sort, though not all with the same value of $\mathbf{k}$.

This is *Bloch's theorem*. Actually, there is a general group-theoretical proof, which follows as a corollary from the theorem that 'in the field of complex numbers any representation of an Abelian group can be reduced to a sum of one-dimensional representations'. The group of translations of the crystal is Abelian—that is to say, all the elements of the group commute with one another. This follows from the obvious circumstance that the translation $(\mathbf{a}_1+\mathbf{a}_2)$, say, is identical with $(\mathbf{a}_2+\mathbf{a}_1)$—we can get to a lattice point by many different paths, but the result is just the same. The matrices $\mathbf{T}_1$, $\mathbf{T}_2$, etc., are representations of these translations, and so can be reduced to diagonal form.

1.5 Reduction to a Brillouin zone

Bloch's theorem is of such generality that it is hard to grasp it at this stage. In the case of electron waves, it means that we can label every wave-function by its wave-vector $\mathbf{k}$, and write

$$\psi_{\mathbf{k}}(\mathbf{r}+\boldsymbol{l}) = e^{i\mathbf{k}\cdot\boldsymbol{l}}\psi_{\mathbf{k}}(\mathbf{r}). \tag{1.58}$$

We notice that an ordinary free-electron wave satisfies this condition;

$$\psi_{\mathbf{k}}(\mathbf{r}) = e^{i\mathbf{k}\cdot\mathbf{r}}. \tag{1.59}$$

This is to be expected, for we are still dealing with the solution of the Schrödinger equation in a periodic potential; but this potential happens to be zero everywhere. The *empty-lattice test* is often very valuable in the theory of solids.

It is sometimes convenient to try to make an electron wave-function of given value of **k** look as much as possible like a free-electron wave. We put

$$\psi_{\mathbf{k}}(\mathbf{r}) = e^{i\mathbf{k}\cdot\mathbf{r}}u_{\mathbf{k}}(\mathbf{r}) \tag{1.60}$$

and hope that $u_{\mathbf{k}}$ will be nearly constant. In fact, as may be proved at once from (1.58), the function $u_{\mathbf{k}}(\mathbf{r})$ must be periodic, i.e.

$$u_{\mathbf{k}}(\mathbf{r}+\boldsymbol{l}) = u_{\mathbf{k}}(\mathbf{r}). \tag{1.61}$$

Bloch's theorem is sometimes stated in this form.

The factor $\exp(i\mathbf{k}\cdot\boldsymbol{l})$ that appears in the theorem is similar to the factor $\exp(i\mathbf{g}\cdot\boldsymbol{l})$ that appeared when we studied periodic functions. A wave-vector **k** evidently has the same dimensions as a reciprocal lattice vector **g**; it belongs in reciprocal space. If some state happened to have wave-vector **g**, then it would be a periodic function:

$$\begin{aligned}\psi_{\mathbf{g}}(\mathbf{r}+\boldsymbol{l}) &= e^{i\mathbf{g}\cdot\boldsymbol{l}}\psi_{\mathbf{g}}(\mathbf{r}) \\ &= \psi_{\mathbf{g}}(\mathbf{r}),\end{aligned} \tag{1.62}$$

because

$$e^{i\mathbf{g}\cdot\boldsymbol{l}} = 1 \tag{1.63}$$

for all $\boldsymbol{l}$.

Again, suppose the state $\psi_{\mathbf{k}}$ has wave-vector **k** such that

$$\mathbf{k} = \mathbf{g}+\mathbf{k}', \tag{1.64}$$

where **g** is some reciprocal-lattice vector, and **k**′ is another vector. Then by (1.58) and (1.63)

$$\begin{aligned}\psi_{\mathbf{k}}(\mathbf{r}+\boldsymbol{l}) &= e^{i(\mathbf{g}+\mathbf{k}')\cdot\boldsymbol{l}}\psi_{\mathbf{k}}(\mathbf{r}) \\ &= e^{i\mathbf{g}\cdot\boldsymbol{l}}e^{i\mathbf{k}'\cdot\boldsymbol{l}}\psi_{\mathbf{k}}(\mathbf{r}) = e^{i\mathbf{k}'\cdot\boldsymbol{l}}\psi_{\mathbf{k}}(\mathbf{r}).\end{aligned} \tag{1.65}$$

That is to say, the state $\psi_{\mathbf{k}}$ satisfies Bloch's theorem as if it had the wave-vector **k**′. The original label **k** is not unique; every state has a whole host of possible wave-vectors, differing from one another by the vectors of the reciprocal lattice. This does not, of course, contradict the theorem, which merely asserted that each state must have at least *one* wave-vector.

We are thus faced with a problem; how are we to define uniquely the wave-vector of a given state? One could use (1.60) as a guide, and try to choose $\mathbf{k}$ so that $u_{\mathbf{k}}(\mathbf{r})$ is as constant as possible—but this is arbitrary, and, as we shall see later, even misleading. The proper procedure is as follows:

Consider what happens in one dimension (cf. 1.8–1.12): the analogue of the reciprocal lattice is the set of reciprocal lattice lengths

$$g_n = n\frac{2\pi}{a}. \tag{1.66}$$

A state may be assigned any wave-number in the set

$$k = n\frac{2\pi}{a} + k', \tag{1.67}$$

that is to say, $\mathbf{k}$ is only defined *modulo* $(2\pi/a)$. All the points $\mathbf{k}$ in Fig. 12 are equivalent.

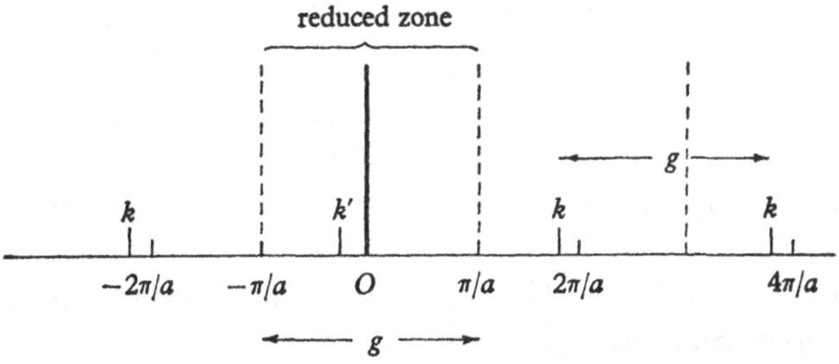

Fig. 12. The points k all reduce to k' in the one-dimensional reciprocal lattice.

It is natural to take k' as the representative of all of them, with $|k'|$ as small as possible. In other words, we always choose for a wave-number the value that lies in the range

$$-\frac{\pi}{a} < k \leqslant \frac{\pi}{a}. \tag{1.68}$$

It is evident that this range is equivalent to a Brillouin zone for our one-dimensional system; it is a unit cell of the reciprocal lattice. In three dimensions we do the same thing—we choose our wave-vector to lie in the first Brillouin zone in reciprocal space. That this is always possible follows by elementary geometry. We choose the value of $\mathbf{g}$ in (1.64) to make $|\mathbf{k}'|$ as small as possible—that is, to lie as near to the origin of the reciprocal lattice as it can. This means that $|\mathbf{k}'|$ is to lie nearer to the origin than to any other sites of the reciprocal lattice—which amounts to saying that it lies in the Wigner–Seitz cell of that lattice, i.e. in the Brillouin zone. It is evident that we can *reduce* any point **k** in reciprocal space to a point in the Brillouin zone, so that any state can be given a label in the *reduced zone scheme.*

Any state may thus be characterized by its *reduced wave-vector.* But there may be many states with the same reduced wave-vector, and different energies. The use of the reduced zone scheme thus prevents us from assigning a distinct value of **k** to each state.

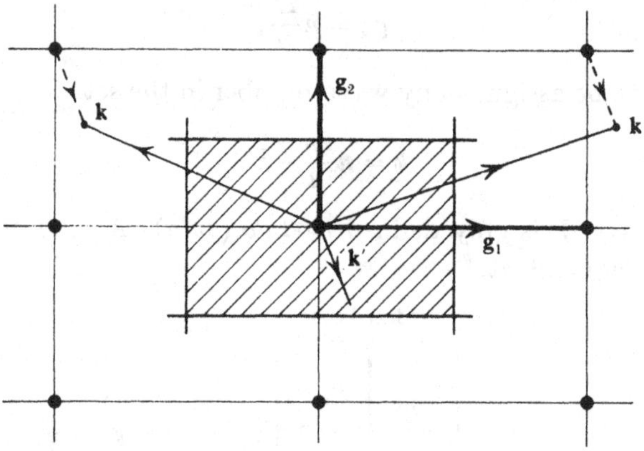

Fig. 13. The wave-vectors **k** all reduce to **k**′, which lies in the Brillouin zone.

It is interesting to see what happens to free-electron states in an empty lattice. Suppose

$$\begin{aligned}\psi &= e^{i\mathbf{k}\cdot\mathbf{r}}\\ &= e^{i(\mathbf{k}-\mathbf{g})\cdot\mathbf{r}}\,e^{i\mathbf{g}\cdot\mathbf{r}}\\ &= e^{i\mathbf{k}'\cdot\mathbf{r}}\,(e^{i\mathbf{g}\cdot\mathbf{r}}),\end{aligned} \tag{1.69}$$

where **k**′ is the reduced value of the actual wave-vector, **k**. This is in the form (1.60), since $\exp(i\mathbf{g}\cdot\mathbf{r})$ is a periodic function in the lattice: but it is an extremely artificial representation of a plane wave. If the energy is given, in the one-dimensional case, by

$$\mathscr{E}(\mathbf{k}) = \frac{\hbar^2 k^2}{2m}, \tag{1.70}$$

then it will appear as a multivalued function of **k**′ in the reduced zone. In Fig. 14, the section AB of the parabola is moved to $A'B'$ by a translation of a reciprocal lattice vector and so on. It is obvious that the *extended zone scheme*, in which each state is represented by its 'true' wave-vector, **k**, would be more natural in this case.

1.6 Boundary conditions: counting states

All this assumes complete translational symmetry, which would require an infinite lattice. This is tiresome mathematically, because there would be an infinite number of atoms, and of states to deal with. We can only deal with these by treating a system of a finite number of atoms, and then going to the limit as this number tends to infinity. We might introduce boundaries as if they were real surfaces, at which, say, the wave-functions must vanish. But this also is inconvenient, because the exact stationary states of the system would then have to take account of all the reflections from the boundaries, and would turn out to be standing waves. To use such functions to represent the states of the conduction electrons, which are usually scattered incoherently by impurities or thermal vibrations within a very short distance, is mathematically clumsy and essentially unphysical.

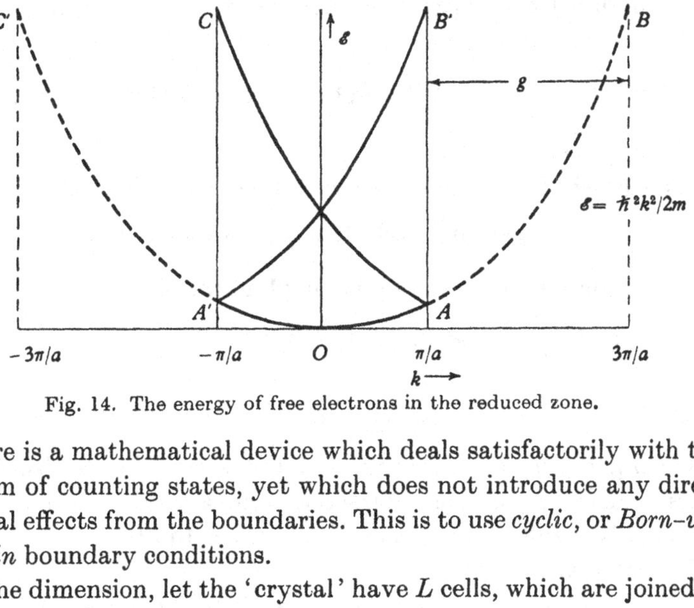

Fig. 14. The energy of free electrons in the reduced zone.

There is a mathematical device which deals satisfactorily with the problem of counting states, yet which does not introduce any direct physical effects from the boundaries. This is to use *cyclic*, or *Born–von Kármán* boundary conditions.

In one dimension, let the 'crystal' have L cells, which are joined in a circle. The consequence of this for an electron wave-function might be expressed by the condition

$$\psi(x+La) = \psi(x), \tag{1.71}$$

ensuring that there is no discontinuity at the junction point.

But our Bloch condition in one dimension says

$$\psi_k(x+La) = e^{ikLa}\psi_k(x), \tag{1.72}$$

so that we must have

$$e^{ikLa} = 1, \tag{1.73}$$

i.e. $$k = \frac{2\pi m}{La}, \tag{1.74}$$

where m is an integer.

In a reduced zone in one dimension we have $-\pi/a < k \leqslant \pi/a$, so that the integers in the range

$$-\tfrac{1}{2}L < m \leqslant \tfrac{1}{2}L \tag{1.75}$$

will give all essentially different values of reduced wave-number. There are exactly L such values (when one has taken care of the different cases where L is odd and L is even—a trivial point) and they are spaced $(1/L)(2\pi/a)$ apart. Since L is supposed to be very large, this distribution is effectively continuous, and of constant density in reciprocal space.

In three dimensions we extend this argument by asserting that our macroscopic system is cyclic in three dimensions. We take it to be a crystal of dimensions $L_1\mathbf{a}_1$, $L_2\mathbf{a}_2$, $L_3\mathbf{a}_3$ in the three basic directions of the lattice, and we write

$$\psi(\mathbf{r}+L_1\mathbf{a}_1) = \psi(\mathbf{r}), \quad \psi(\mathbf{r}+L_2\mathbf{a}_2) = \psi(\mathbf{r}), \quad \psi(\mathbf{r}+L_3\mathbf{a}_3) = \psi(\mathbf{r}), \tag{1.76}$$

just as in (1.71).

For a Bloch state of wave-vector $\mathbf{k}$, these conditions imply that

$$e^{i\mathbf{k}\cdot(L_1\mathbf{a}_1)} = e^{i\mathbf{k}\cdot(L_2\mathbf{a}_2)} = e^{i\mathbf{k}\cdot(L_3\mathbf{a}_3)} = 1, \tag{1.77}$$

which can only be achieved by having $\mathbf{k}$ in the form

$$\begin{aligned}\mathbf{k} &= k_1\mathbf{b}_1 + k_2\mathbf{b}_2 + k_3\mathbf{b}_3 \\ &= \frac{2\pi m_1}{L_1}\mathbf{b}_1 + \frac{2\pi m_2}{L_2}\mathbf{b}_2 + \frac{2\pi m_3}{L_3}\mathbf{b}_3,\end{aligned} \tag{1.78}$$

where m_1, m_2, m_3 are integers, and the vectors $\mathbf{b}_1$, $\mathbf{b}_2$, $\mathbf{b}_3$, are, of course, the reciprocal triad of the set $(\mathbf{a}_1, \mathbf{a}_2, \mathbf{a}_3)$ as in (1.19).

But this result is obviously comparable to the definition of a reciprocal lattice vector

$$\mathbf{g} = n_1 . 2\pi\mathbf{b}_1 + n_2 . 2\pi\mathbf{b}_2 + n_3 . 2\pi\mathbf{b}_3, \tag{1.79}$$

where n_1, n_2, n_3 are integers. The *allowed values* of $\mathbf{k}$ in (1.78) are obtained by dividing the generators of the cell of the reciprocal lattice into L_1 parts in direction $\mathbf{b}_1$, L_2 parts in direction $\mathbf{b}_2$ and L_3 parts in direction $\mathbf{b}_3$. Thus, a fine rash of points is distributed evenly through reciprocal space, as in Fig. 15.

To calculate the density of these points, it is only necessary to notice that we can cover a unit cell of the reciprocal lattice by allowing the integers m_1, m_2, m_3 to run through the values

$$0 \leqslant m_1 < L_1, \quad 0 \leqslant m_2 < L_2, \quad 0 \leqslant m_3 < L_3. \tag{1.80}$$

But this is not the cell we should naturally use as a basic zone for **k**-vectors. We would do better, perhaps, with the range

$$-\tfrac{1}{2}L_1 < m_1 < \tfrac{1}{2}L_1, \quad -\tfrac{1}{2}L_2 < m_2 < \tfrac{1}{2}L_2, \quad -\tfrac{1}{2}L_3 < m_3 < \tfrac{1}{2}L_3, \tag{1.81}$$

by analogy with (1.75). This would give a parallelepiped unit cell centred on the origin. The best choice of unit cell, however, is the Wigner–Seitz cell of the reciprocal lattice, i.e. our old friend the Brillouin zone.

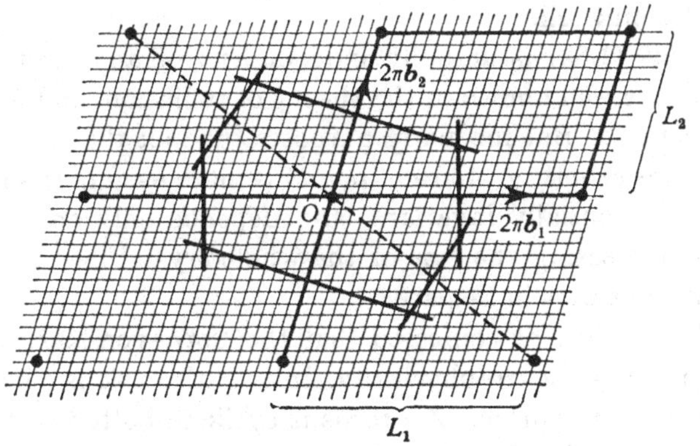

Fig. 15. The Brillouin zone covers the same area as the parallelepiped unit cell of the reciprocal lattice, and contains just N 'allowed **k**-vectors'.

Now the volume of the Brillouin zone is the same as the parallelepiped unit cell, so that it must contain exactly the same number of 'allowed values' of **k**. From (1.80) or (1.81) this number is obviously $L_1 \times L_2 \times L_3 = N$ which is exactly the number of unit cells in our whole macroscopic crystal. This is a most important theorem: *there are exactly as many allowed wave-vectors in a Brillouin zone as there are unit cells in the block of crystal.*

We have already shown (in (1.32)) that the volume of a zone is $8\pi^3/v_c$. If there are N unit cells in the volume V of crystal, then

$$v_c = V/N. \tag{1.82}$$

The volume in **k**-space per allowed **k**-vector is thus

$$\frac{1}{N}\frac{8\pi^3}{v_c} = \frac{8\pi^3}{V}. \tag{1.83}$$

Thus, *there are* $V/8\pi^3$ *allowed* **k**-*vectors per unit volume of reciprocal space.*

In practice, N is very large, so that this distribution is treated as continuous. We often express a sum over **k**-vectors as an integral

$$\sum_{\mathbf{k}} \rightarrow \int d\mathbf{k} \equiv \frac{V}{8\pi^3} \iiint d^3k, \tag{1.84}$$

using the single integral as a concise notation for the limit of the sum. It is important to remember, however, that when we actually come to a definite quadrature in **k**-space we must include the weight factor $V/8\pi^3$. We shall usually assume, for simplicity, that $V = 1$, so that N is the number of cells per unit volume of crystal and $1/N$ is the volume of a unit cell.

The result (1.83) is actually independent of any assumed zone structure; it is well known in the theory of radiation, and in the case of free electrons. But for many purposes the fact that the Brillouin zone contains exactly N allowed points is more useful. It shows that the zone is essentially invariant, and depends only on the crystal structure; increasing the size of the whole crystal merely increases the density of states in **k**-space.

Actually, it must be confessed that the Born–von Kármán conditions cannot be achieved physically. In two dimensions a network of cells drawn on the surface of a torus is cyclic in both directions, but for a three-dimensional lattice all three conditions cannot be satisfied simultaneously by any topological contortions. However, it works admirably as a mathematical device, and it can be justified by some messy mathematics. The fact is that the density of states in the asymptotic limit of large quantum numbers is very insensitive to the precise form of the boundary conditions. So long as we do not propose to discuss boundary effects themselves, the cyclic conditions are far and away the most convenient.

CHAPTER 2

LATTICE WAVES

For those well ordered motions, and regular paces, though they give no sound unto the ear, yet to the understanding they strike a note most full of harmony. SIR THOMAS BROWNE

2.1 Lattice dynamics

The simplest solid is probably solid argon, which is a regular array of neutral atoms, with tightly bound closed shells of electrons, held together by van der Waals forces, which act mainly between nearest neighbours in the lattice. The physics of such a crystal is the thermal motion of the atoms about their idealized equilibrium positions.

To discuss this motion the most elementary idea is the *Einstein model*, in which each atom vibrates like a simple harmonic oscillator in the potential well of the force fields of its neighbours. This field can never be precisely a quadratic well with spherical symmetry, but this is a reasonable approximation on the average. The excitation spectrum of the crystal then consists of levels spaced a distance $\hbar\nu_E$ apart, where ν_E is the *Einstein frequency*, i.e. the frequency of oscillation of each atom in its potential well.

This model is useful in some contexts where a very crude account of thermal vibrations is sufficient—especially at relatively high temperatures, when the assumption that the various atoms vibrate independently is justified. But we can see at once that if two or more atoms move in unison the forces between them, tending to restore each of them to its equilibrium position, will be reduced, so that the energy required to excite a quantum may be rather less. There is a tendency for the motion of the adjacent atoms to be correlated.

To get the spectrum of the whole lattice, we must put in the local forces and describe the motion completely. This would be an impossible problem if it were not for the translational invariance of the lattice. The excitations must satisfy Bloch's theorem, as proved in § 1.4. The theory of *lattice dynamics* provides the most elementary example of the mathematical machinery built up in Chapter 1.

Let us first define *lattice displacements*: $\mathbf{u}_{sl}$ is the vector defining the displacement of the sth atom, in the lth unit cell, from its equilibrium position. Suppose this atom has mass M_s. Then the kinetic energy of the system is defined by

$$\text{K.E.} = \sum_{sl} \tfrac{1}{2} M_s |\dot{\mathbf{u}}_{sl}|^2. \tag{2.1}$$

We assume here, quite generally, a lattice with a basis, so that the label s runs through the one or more atoms in the unit cell.

For the potential energy term we should need to define the interatomic forces. But we can assume, quite generally, that there is a function $\mathscr{V}(\mathbf{u}_{sl})$ which expresses the potential energy of the whole crystal in terms of the positions of all the atoms, that is, in terms of their actual displacements, at some moment, from the sites of the equilibrium lattice. We may assume, further, that this function has a minimum when all the $\mathbf{u}_{sl}$ are zero, for the perfect lattice is presumably a configuration of stable equilibrium. We use the theory of small vibrations, and expand $\mathscr{V}$ as a power series in the cartesian components, u_{sl}^j, of $\mathbf{u}_{sl}$ around this point

$$\mathscr{V} = \mathscr{V}_0 + \sum_{slj} u_{sl}^j \left[\frac{\partial \mathscr{V}}{\partial u_{sl}^j}\right]_0 + \tfrac{1}{2} \sum_{ss',ll',jj'} u_{sl}^j u_{s'l'}^{j'} \left[\frac{\partial^2 \mathscr{V}}{\partial u_{sl}^j \, \partial u_{s'l'}^{j'}}\right]_0 + \dots \tag{2.2}$$

The constant term is unimportant in the present context and the coefficient of the linear term must vanish because we are near equilibrium; the first term of importance must be the one that is quadratic in the displacements—the so-called *harmonic term.*

From these two functions (2.1) and (2.2), using the ordinary Lagrangian procedure of classical mechanics, we get equations of motion for the components

$$M_s \ddot{u}_{sl}^j = -\sum_{s'l'j'} \left[\frac{\partial^2 \mathscr{V}}{\partial u_{sl}^j \, \partial u_{s'l'}^{j'}}\right]_0 u_{s'l'}^{j'} \tag{2.3}$$

for all sites s in the unit cell, for all unit cells l in the lattice, and for each Cartesian component $j = x, y, z$ of the displacement vector.

These represent, of course, an immense number of coupled differential equations. But let us think about the coefficients on the right-hand side. Let us write

$$\left[\frac{\partial^2 \mathscr{V}}{\partial u_{sl}^j \, \partial u_{s'l'}^{j'}}\right]_0 \equiv G_{sl;s'l'}^{jj'}. \tag{2.4}$$

These coefficients form the components of a Cartesian tensor, so that we can write (2.3) somewhat more simply in vector form

$$M_s \ddot{\mathbf{u}}_{sl} = -\sum_{s'l'} \mathbf{G}_{sl;s'l'} \cdot \mathbf{u}_{s'l'}. \tag{2.5}$$

This equation may be interpreted as follows: each term in the sum on the right is the *force acting on the sth atom in the lth cell due to the displacement* $\mathbf{u}_{s'l'}$ *of the atom on the s'th site of the l'th cell.* We could write this down directly in terms of interatomic force constants if we knew that the whole lattice potential energy could be built up out of forces between pairs of atoms.

But whatever may be the physical interpretation of the forces, one thing is necessary: they cannot depend on the absolute positions of $\mathbf{l}$ and $\mathbf{l}'$ in the lattice. The tensor $\mathbf{G}$ must be a function only of their relative positions;

$$\mathbf{G}_{sl;s'l'} = \mathbf{G}_{ss'}(\mathbf{h}), \quad \text{where} \quad \mathbf{h} = \mathbf{l}' - \mathbf{l}. \tag{2.6}$$

Our equations of motion now read

$$M_s \ddot{\mathbf{u}}_{sl} = -\sum_{s'\mathbf{h}} \mathbf{G}_{ss'}(\mathbf{h}) \cdot \mathbf{u}_{s,\,l+\mathbf{h}}. \tag{2.7}$$

These equations are of the translationally invariant form required to satisfy Bloch's theorem; changing the label from $\mathbf{l}$ to $\mathbf{l}''$, say, merely repeats exactly the same set. Suppose now that we have found a solution—a set of functions of the time, describing $\mathbf{u}_{sl}$ for each value of $\mathbf{l}$. These functions must satisfy the Bloch condition (1.48) or (1.57). There exists a wave-vector† $\mathbf{q}$ such that

$$\mathbf{u}_{sl}(t) = e^{i\mathbf{q}\cdot\mathbf{l}}\mathbf{u}_{s,o}(t), \tag{2.8}$$

where $\mathbf{u}_{s,o}(t)$ is the displacement at whatever cell happens to have been chosen as origin for the lattice vectors $\mathbf{l}$. Notice that in every cell the atom on site s moves in the same direction with the same amplitude; only the phase varies from cell to cell.

For every solution of the dynamical equations, there exists a value of $\mathbf{q}$ such that (2.8) is satisfied. To find out which solution has which wave-vector, we must go back to (2.7). Substituting from (2.8)

$$M_s \ddot{\mathbf{u}}_{s,o} e^{i\mathbf{q}\cdot\mathbf{l}} = -\sum_{s'\mathbf{h}} \mathbf{G}_{ss'}(\mathbf{h}) \cdot \mathbf{u}_{s',o} e^{i\mathbf{q}\cdot\mathbf{h}} e^{i\mathbf{q}\cdot\mathbf{l}}. \tag{2.9}$$

The factor $\exp i\mathbf{q}\cdot\mathbf{l}$ can be cancelled. To show that the origin was quite arbitrary, but that we are treating a solution with a definite value of $\mathbf{q}$, let us write

$$\mathbf{u}_{s,o} = \mathbf{U}_{s,\mathbf{q}}. \tag{2.10}$$

† We use this symbol as the wave-vector for lattice waves, and $\mathbf{k}$ for electron waves. These are, of course, both vectors in the same reciprocal space.

We are left with

$$M_s \ddot{\mathbf{U}}_{s,\mathbf{q}} = -\sum_{s'}\{\sum_{\mathbf{h}} \mathbf{G}_{ss'}(\mathbf{h})\, e^{i\mathbf{q}\cdot\mathbf{h}}\}\cdot \mathbf{U}_{s',\mathbf{q}}$$

$$= -\sum_{s'} \mathbf{G}_{ss'}(\mathbf{q})\cdot \mathbf{U}_{s',\mathbf{q}}, \tag{2.11}$$

where

$$\mathbf{G}_{ss'}(\mathbf{q}) \equiv \sum_{\mathbf{h}} \mathbf{G}_{ss'}(\mathbf{h})\, e^{i\mathbf{q}\cdot\mathbf{h}} \tag{2.12}$$

—a Fourier transform of the force tensor **G**.

The new equations (2.11) are relatively easy to solve. The main point is that there are a limited number of them. Suppose that there are n atoms per unit cell, and N cells in the lattice. The original set of equations (2.3) or (2.5) would then be $3nN$ in number, allowing separately for the three Cartesian components of the vectors. The new set, (2.11), are only $3n$ equations—3 component equations for each of the n values of s.

This is a powerful example of translational invariance. The essential structural identity of all unit cells means that we only need information about one of them to calculate the dynamics of the whole assembly. It is true that we have to make a separate calculation if we want a solution for each value of $\mathbf{q}$ (of which the boundary conditions allow exactly N different values), but in practice we are only interested in a finite sample of such solutions and can interpolate for the rest.

The procedure now is elementary. As in the ordinary theory of vibrations we assume that $\mathbf{U}_{s,\mathbf{q}}$ contains a time factor $\exp(i\nu t)$. We must then solve the set of $3n$ equations

$$\sum_{s'j'}\{\mathbf{G}_{ss'}^{jj'}(\mathbf{q}) - \nu^2 M_s \delta_{ss'}\delta_{jj'}\}\, U_{s',\mathbf{q}}^{j'} = 0, \tag{2.13}$$

for the components of $\mathbf{U}_{s,\mathbf{q}}$. This is an eigenvalue equation, whose $3n$ solutions are obtained by finding the roots of the equation in ν^2 resulting when the determinant of the matrix { } is made to vanish. In fact, all that we are doing is finding the $3n$ normal modes of vibration of the n atoms in a unit cell, supposing them to interact via the force tensor $\mathbf{G}_{ss'}(\mathbf{q})$. This tensor differs from the original interatomic force tensor $\mathbf{G}_{ss'}(\mathbf{h})$, and is different for each value of $\mathbf{q}$; for given values of s and s' it sums up the interaction of *all* s-type atoms—those on site s of each unit cell of the lattice—with *all* the atoms on site s', making allowance for their relative phases.

2.2 Properties of lattice waves

To understand the physical properties of lattice waves, it is useful to study a few simple cases. Of these, the *linear chain* is the simplest.

Suppose we have a one-dimensional 'lattice' with one atom per unit cell and only nearest-neighbour forces. The equations of motion are easily written down. The potential energy (cf. (2.2)) is of the form

$$\mathscr{V} = \sum_l \tfrac{1}{2}\alpha(u_l - u_{l+a})^2, \tag{2.14}$$

where α is the force constant. The equations of motion (cf. (2.5)) become

$$M\ddot{u}_l = -\alpha(2u_l - u_{l+a} - u_{l-a}) \tag{2.15}$$

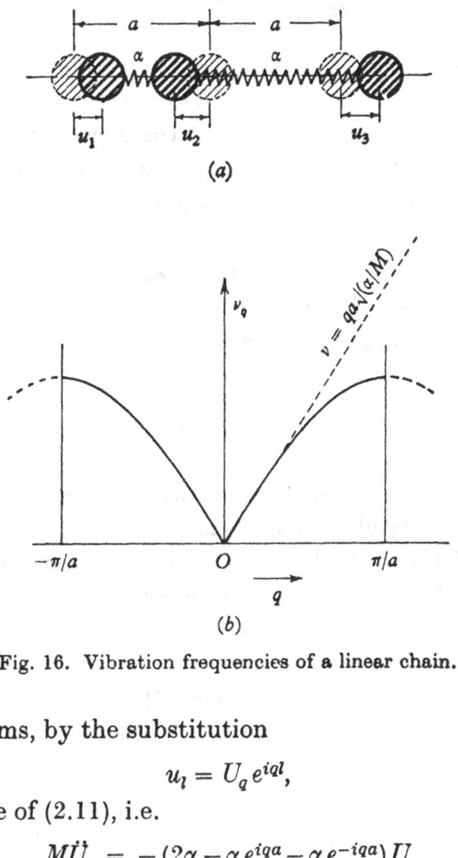

Fig. 16. Vibration frequencies of a linear chain.

which transforms, by the substitution

$$u_l = U_q e^{iql}, \tag{2.16}$$

to the analogue of (2.11), i.e.

$$\begin{aligned} M\ddot{U}_q &= -(2\alpha - \alpha e^{iqa} - \alpha e^{-iqa})\,U_q \\ &= -2\alpha(1-\cos qa)\,U_q, \end{aligned} \tag{2.17}$$

if a is the spacing of atoms in the chain. This is the equation of a simple-harmonic oscillator, having the frequency

$$\nu_q = \sqrt{\left(\frac{\alpha}{M}\right)}\, 2\sin\left(\frac{qa}{2}\right). \tag{2.18}$$

This well-known result exemplifies many points of the theory:

(i) All possible vibrations are given by values of q in the range

$$-\frac{\pi}{a} < q \leqslant \frac{\pi}{a}. \tag{2.19}$$

This is exactly our Brillouin zone. Any value of q lying outside this range simply repeats precisely the same motion;

$$u_l = U\, e^{i(q+g)l} \equiv U\, e^{iql}. \tag{2.20}$$

It would be quite unnatural (though not meaningless!) to assign wave-vectors outside the Brillouin zone. Yet if we had used values of $|\mathbf{q}|$ greater than π/a we should have got the correct answer. In (2.18), ν_q is a periodic function of q, so that all values of q that would reduce to the same point in the Brillouin zone have the same frequency.

There are exactly N different solutions, corresponding to the N allowed values of q in the Brillouin zone. This is consistent with the N degrees of freedom of the original lattice.

(ii) For small values of q (i.e. $qa \ll 1$)

$$\nu_q \sim \sqrt{\left(\frac{\alpha}{M}\right)}\, aq. \tag{2.21}$$

The proportionality of frequency to wave-number corresponds to the well-known property of ordinary elastic waves in a continuum. If the wavelength of the disturbance is much greater than the lattice constant, then the chain of atoms behaves like a 'heavy elastic string' in classical mechanics.

At large values of q, however, the velocity of the wave is not constant, and at $q = \pi/a$, i.e. when the wavelength equals $2a$, the function ν_q bends over to a horizontal tangent. This shows the property of *dispersion*.

Consider now a more complicated case: a linear chain of atoms with the same spacing, and the same force constants as before, but of two different masses, M_1 and M_2, placed alternately. We now have a unit cell containing two atoms. The analogue of (2.11) is more complicated than (2.17):

$$\left.\begin{aligned} M_1 \ddot{U}_1 &= -2\alpha U_1 + 2\alpha \cos qa\,.\,U_2,\\ M_2 \ddot{U}_2 &= -2\alpha U_2 + 2\alpha \cos qa\,.\,U_1.\end{aligned}\right\} \tag{2.22}$$

To find the frequency ν, we must solve the determinantal equation

$$\begin{vmatrix} 2\alpha - M_1\nu^2 & -2\alpha\cos qa \\ -2\alpha\cos qa & 2\alpha - M_2\nu^2 \end{vmatrix} = 0, \tag{2.23}$$

which has the two roots

$$\nu_{\pm}^2 = \alpha\left(\frac{1}{M_1}+\frac{1}{M_2}\right) \pm \alpha\sqrt{\left\{\left(\frac{1}{M_1}+\frac{1}{M_2}\right)^2 - \frac{4\sin^2 qa}{M_1M_2}\right\}}. \tag{2.24}$$

These two roots, when plotted as functions of q, are as shown in Fig. 18.

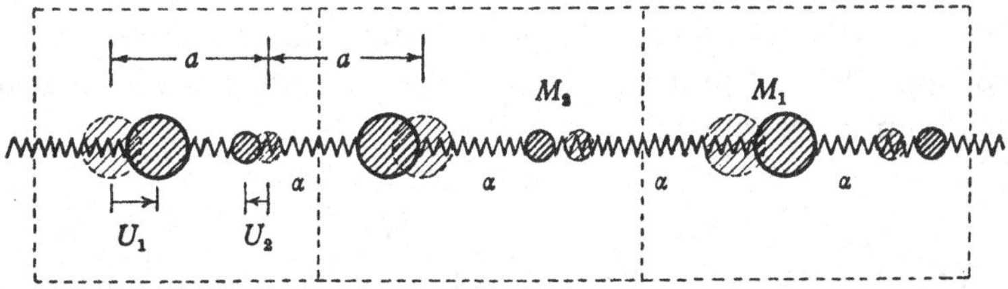

Fig. 17. Diatomic linear chain.

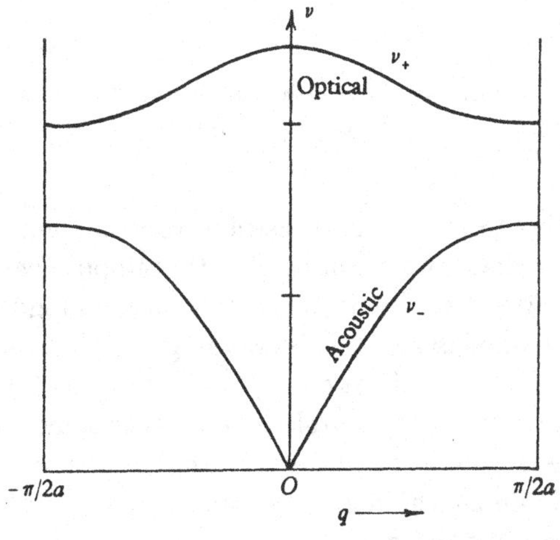

Fig. 18. Vibration frequencies of diatomic chain.

As in the monatomic chain, there is a root, ν_-, which tends to be proportional to q near $q = 0$. We call this the *acoustic mode*, because it is the analogue of a long-wavelength vibration of the chain, conceived as an elastic continuum. We leave it as an exercise for the reader to show that the normal value for the velocity of sound in such a medium is reproduced for this case.

But there is another branch, ν_+, such that

$$\nu_+^2 \sim 2\alpha\left(\frac{1}{M_1}+\frac{1}{M_2}\right) \tag{2.25}$$

near $q = 0$. This branch is separated considerably from the acoustic mode, but the two branches tend to approach in frequency as q increases. This is called an *optical mode*. We can easily check that in the case $q = 0$ the two sublattices of light and heavy atoms are each moving rigidly in opposition to one another, or, as one might say, the diatomic molecule in each cell is vibrating as if independent of its neighbours. If, as is the case in ionic crystals, the two types of atom are of opposite electric charge, this gives an oscillating dipole moment which is optically active.

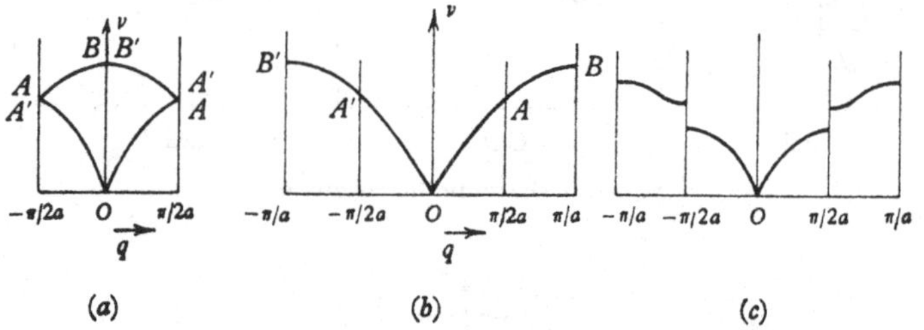

Fig. 19. Vibration frequencies of: (*a*) monatomic linear chain treated as diatomic; (*b*) monatomic linear chain, showing zone boundaries; (*c*) diatomic linear chain, treated as monatomic.

It is instructive to study the transition from the monatomic chain of Fig. 16 to the diatomic chain of Fig. 18. Suppose we had started with a unit cell with two atoms in the first case, and ignored the fact that they are identical in mass. It is easy to prove that the optical and acoustic modes will then join at $q = \pi/2a$, as in Fig. 19(*a*). Comparing with Fig. 16, we seem to have doubled the number of modes. However, by doubling the length of the unit cell, we have apparently halved the size of the Brillouin zones. Fig. 19(*a*) can be opened out, by translations through π/a (i.e. by a reciprocal lattice vector for our double cell), to give Fig. 19(*b*). The 'acoustic' and 'optical' branches then join continuously. The reduced zone of Fig. 19(*a*) is spuriously half what it should be.

We can put this another way. The effect of changing the masses of alternate atoms of the chain is to introduce new zone boundaries at $\pm\pi/2a$. The frequency is no longer continuous across this boundary; a gap appears. We could represent this by Fig. 19(*c*)—another way of doing the diatomic chain. We shall encounter this phenomenon frequently in the next chapter.

The next case to consider would be a two- or three-dimensional Bravais lattice with nearest-neighbour interactions. Already, however, the equations become too complicated for an easy general solution. Each dimension introduces an extra Cartesian component of the displacement vector, so for three dimensions we have to solve a cubic equation in ν^2.

The three roots, in that case, can be identified physically if we go to the long-wavelength limit. We are dealing, in effect, with an elastic continuum, whose structure is too fine to be significant in the dynamics of long waves. It is well known that in such a continuum one can propagate three different types of acoustic wave of different velocities.

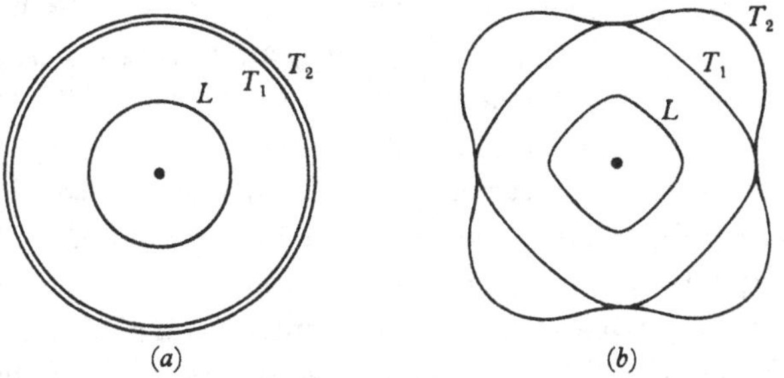

Fig. 20. Constant frequency surfaces: (*a*) for isotropic medium; (*b*) for real crystal.

The essential difference between these three *acoustic modes* is in the nature of their *polarization*. In an *isotropic* continuum, for example, one mode is *longitudinally* polarized—the displacement vector of each atom is along the direction of propagation of the wave. The other two modes are of the same velocity, and *transversely* polarized—the atoms move in the planes normal to the wave-vector. Generally speaking, the longitudinal mode has higher velocity than the transverse mode, so that the surfaces of constant frequency in **q**-space would look like Fig. 20(*a*).

A crystalline solid is not usually isotropic in its macroscopic elastic properties, so that this picture is misleading. The velocity of a wave of given type will depend on the direction of propagation. Thus (Fig. 20(*b*)) the surface of constant frequency for the transverse branch may be very far from a sphere. The transverse modes will then no longer be degenerate, except in special symmetry directions, so that their frequency surfaces may intersect in a complicated pattern. Indeed, the formal classification into transverse and longitudinal modes is somewhat misleading, since the actual polarization vector (i.e. the direction of the vector $\mathbf{U}_\mathbf{q}$ that solves (2.11) or (2.13)) need not be strictly along or normal to $\mathbf{q}$ in such a medium.

This fixes the long-wavelength behaviour of the lattice waves. For $\mathbf{q}$ in a given direction, we must have a relation between ν and q, of the form sketched in Fig. 21—a linear relation near $q = 0$, with slope equal to the corresponding macroscopic sound velocity. But as q increases, we approach the boundary of the Brillouin zone, and there must be dispersion.

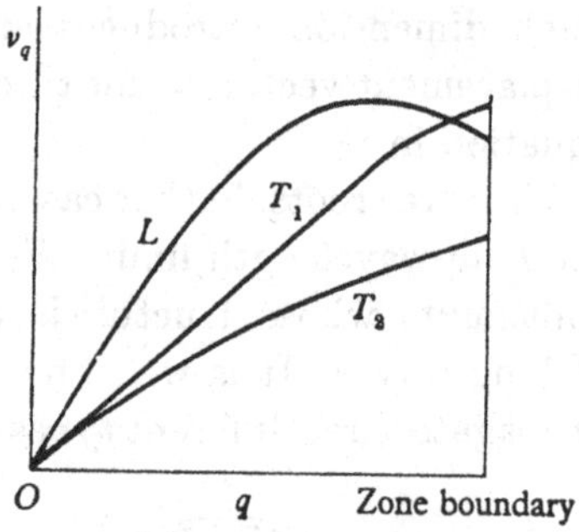

Fig. 21. Dispersion of lattice waves in a given direction.

The exact behaviour then will be rather complicated. In general, $\nu_\mathbf{q}$ bends over, as in the linear case, but not necessarily to a horizontal tangent. The only rule is that the whole pattern of roots of $\nu_\mathbf{q}$, as a function of $\mathbf{q}$ in reciprocal space, is continuous, and has the periodicity of the reciprocal lattice. This follows by inspection of (2.8) and (2.12). As in the linear case, if $\mathbf{g}$ is a reciprocal lattice vector, the solutions for wave-vector $\mathbf{q}' = \mathbf{q} + \mathbf{g}$ are identical to the solutions for wave-vector $\mathbf{q}$, so that $\nu_\mathbf{q}$ is periodic with period $\mathbf{g}$. It is a continuous function in $\mathbf{q}$-space because it is the solution of an eigenvalue equation, whose matrix is symmetric and whose coefficients are, themselves, continuous periodic functions in $\mathbf{q}$-space.

This means that when we extend a figure like Fig. 20(*b*) out to the whole zone it must form part of such a pattern, as sketched in Fig. 22. Such diagrams can be calculated out, if one knows the force constants—although in practice it is more usual to determine the dispersion curves in various directions by neutron or X-ray diffraction (see § 2.8) and then try to find atomic force constants to fit.

In the more complicated case of a three-dimensional lattice with a basis we find, in addition, transverse and longitudinal optic modes, with their own dispersion curves and energy surfaces. All these cases are covered in principle by our general equations (2.13).

Fig. 22. A branch of the lattice frequency function, $\nu_{\mathbf{q}}$ in the repeated zone scheme.

2.3 Lattice sums

The equivalence, in the long-wave limit, of the solution of the equations for the normal modes of a lattice to the equations for the elastic waves of a macroscopic continuum, provides a convenient method for correlating atomic force constants with observed elastic constants of solids. In principle, this is a relatively simple programme, though in practice it may involve laborious calculations.

There is one major point of difficulty. In one common case we know the interatomic forces very well; in an ionic solid the major part of the cohesion of the solid comes from the Coulomb forces between the oppositely charged ions. For an assembly of ions at points $\mathbf{R}_i$, for example, the total electrostatic energy must be given by

$$\mathscr{E} = \tfrac{1}{2}\sum_{ij} \frac{\pm e^2}{|\mathbf{R}_i - \mathbf{R}_j|}, \tag{2.26}$$

the sign depending on the relative sign of the charges at $\mathbf{R}_i$ and $\mathbf{R}_j$.

This may be rewritten in the form

$$\mathscr{E} = -\tfrac{1}{2}N\alpha\frac{e^2}{a}, \tag{2.27}$$

where a is a lattice constant and the coefficient α, now dimensionless, depends upon the arrangement of positive and negative ions in the lattice. The *Madelung constant*, α, is thus of some interest as a parameter characterizing a particular type of crystal structure, regardless of its scale.

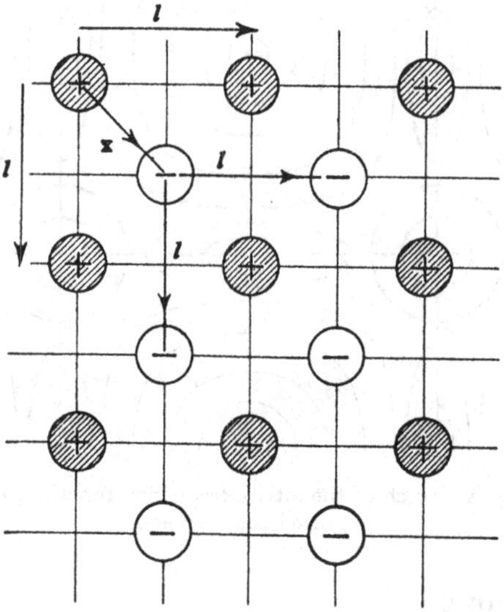

Fig. 23

Suppose, for example, that our ionic crystal consists of two sub-lattices—positive ions at a set of sites $\mathbf{l}$ and negative ions at $\mathbf{l}+\mathbf{x}$. We need to calculate

$$\alpha = a\sum_{l}\left\{\frac{1}{|\mathbf{l}|} - \frac{1}{|\mathbf{l}+\mathbf{x}|}\right\} \tag{2.28}$$

(excluding $\mathbf{l} = 0$ in the first term). The trouble is that the two separate terms in this sum diverge; at a distance l there are of the order of l^2 points to account for.

The series is conditionally convergent, but even then only slowly. The problem of computing such a sum becomes very serious. The best direct method is that of Evjen, where one treats successive shells, going outward from the origin, each one being exactly neutral in total electrostatic charge.

This difficulty is not confined to the calculation of the cohesive energy. It arises in the evaluation of such expressions as (2.12), the Fourier transform of the force tensor between the atoms. There again, we have to sum over all lattice points, and the convergence is slow for Coulomb forces. Any attempt to compute the lattice dynamics of an ionic crystal must run into this difficulty.

There is a very elegant procedure, known as *Ewald's method*, by which this problem may be solved. Perhaps it is only a trick rather than a basic principle of the theory of solids, but it is so neat that it is worth a brief digression. It also illustrates some of the theory of periodic functions in a lattice.

First consider the function

$$\frac{2}{\sqrt{\pi}}\sum_{l} e^{-|\boldsymbol{l}-\mathbf{r}|^2\rho^2} = F(\mathbf{r},\rho). \tag{2.29}$$

This function is periodic in $\mathbf{r}$, with the periodicity of the lattice. Therefore, by (1.15), it may be expanded in a Fourier series

$$F(\mathbf{r},\rho) = \sum_{\mathbf{g}} F_{\mathbf{g}} e^{i\mathbf{g}\cdot\mathbf{r}}, \tag{2.30}$$

where $$F_{\mathbf{g}} = \frac{1}{V}\int \frac{2}{\sqrt{\pi}}\sum e^{-|\boldsymbol{l}-\mathbf{r}|^2\rho^2} e^{-i\mathbf{g}\cdot\mathbf{r}}\,d\mathbf{r}, \tag{2.31}$$

as in (1.22) (the integral being over the whole crystal).

Now let us introduce a factor $\exp(i\mathbf{g}\cdot\boldsymbol{l}) = 1$ in each term of the series. In that case

$$\begin{aligned} F_{\mathbf{g}} &= \frac{1}{V}\frac{2}{\sqrt{\pi}}\sum_{l}\int e^{-|\boldsymbol{l}-\mathbf{r}|^2\rho^2} e^{-i\mathbf{g}\cdot(\mathbf{r}-\boldsymbol{l})}\,d\mathbf{r} \\ &= \frac{N}{V}\frac{2}{\sqrt{\pi}}\int e^{-r^2\rho^2-i\mathbf{g}\cdot\mathbf{r}}\,d\mathbf{r}, \end{aligned} \tag{2.32}$$

because each term in the sum is now the same, except for being measured from a different origin $\boldsymbol{l}$ in the crystal. Since the integral converges rapidly, and all sites are equivalent, this follows easily.

The integral in (2.32) can easily be evaluated. It gives

$$F_{\mathbf{g}} = \frac{2\pi}{v_c}\frac{1}{\rho^3}e^{-g^2/4\rho^2}. \tag{2.33}$$

The substitution of (2.33) in (2.29) and (2.30) constitutes a relation known as the *theta-function transformation* in pure mathematics

(where it is usually proved only for a simple cubic lattice; our result is more general)

$$\frac{2}{\sqrt{\pi}}\sum_{l} e^{-|l-\mathbf{r}|^2\rho^2} = \frac{2\pi}{v_c}\sum_{\mathbf{g}}\frac{1}{\rho^3}e^{-g^2/4\rho^2}e^{i\mathbf{g}\cdot\mathbf{r}}. \tag{2.34}$$

But consider the identity

$$\frac{2}{\sqrt{\pi}}\int_0^\infty e^{-z^2\rho^2}\,d\rho = \frac{1}{|z|}. \tag{2.35}$$

Applying this to the left-hand side of (2.34), we have

$$\sum_{l}\frac{1}{|l-\mathbf{r}|} = \sum_{l}\frac{2}{\sqrt{\pi}}\int_0^\infty e^{-|l-\mathbf{r}|^2\rho^2}\,d\rho. \tag{2.36}$$

We split the integration over the dummy variable ρ into two parts at some arbitrary point G, and substitute for the integrand from (2.34) in the first part

$$\sum_{l}\frac{1}{|l-\mathbf{r}|} = \frac{2\pi}{v_c}\int_0^G\sum_{\mathbf{g}}\frac{1}{\rho^3}e^{-g^2/4\rho^2}e^{i\mathbf{g}\cdot\mathbf{r}}\,d\rho + \sum_{l}\frac{2}{\sqrt{\pi}}\int_G^\infty e^{-|l-\mathbf{r}|^2\rho^2}\,d\rho$$

$$= \frac{\pi}{v_c}\frac{1}{G^2}\sum_{\mathbf{g}} e^{i\mathbf{g}\cdot\mathbf{r}}\frac{e^{-g^2/4G^2}}{g^2/4G^2} + \sum_{l}\frac{1}{|l-\mathbf{r}|}\operatorname{erfc}\{G\,|l-\mathbf{r}|\}, \tag{2.37}$$

where the second term contains the *complementary error function*, which tends rapidly to zero at large values of its argument.

The procedure is to choose a value of G which makes both the series in l and the series in $\mathbf{g}$ rapidly convergent; it is easy to see that both these series converge very much better than the original series (2.36). 'Physically' speaking, it is as if we had built up the charge at $\mathbf{r}$ in two stages. First we make a spherically symmetrical Gaussian distribution, with charge density at radius z proportional to $\exp(-G^2z^2)$. The lattice of point charges will interact electrostatically with this cloud over long distances, to give the first term in (2.37). Then we correct our calculation by putting a point charge at $\mathbf{r}$ and subtracting the Gaussian distribution. Being neutral, this object has negligible long-range interaction with the lattice, and contributes a rapidly convergent sum over neighbouring sites.

This is not quite the whole story, since the above sum has all positive signs, and when $\mathbf{r} = 0$ it contains the effect of the charge at the origin on itself. This would be seen as a divergent contribution near $\mathbf{r} = 0$ in the term for $l = 0$ in the second summation. We subtract $1/r$ from the series on the left, which is equivalent to removing this first term in the sum over $\boldsymbol{l}$ on the right, and replacing it, in the limit $r \to 0$, by

$$\begin{aligned}\frac{1}{r}\operatorname{erf} Gr - \frac{1}{r} &= \frac{2}{\sqrt{\pi}}\frac{1}{r}\left\{\int_{Gr}^{\infty} e^{-z^2}\,dz - \frac{\sqrt{\pi}}{2}\right\} \\ &= -\frac{2}{\sqrt{\pi}}\frac{1}{r}\int_0^{Gr} e^{-z^2}\,dz \\ &\to -2G/\sqrt{\pi}. \end{aligned} \tag{2.38}$$

Thus, $$\sum_{l\neq 0}\frac{1}{|\boldsymbol{l}|} = \frac{\pi}{v_c}\frac{1}{G^2}\sum_{\mathbf{g}}\frac{e^{-g^2/4G^2}}{g^2/4G^2} + \sum_{l\neq 0}\frac{1}{|\boldsymbol{l}|}\operatorname{erf}\{G|\boldsymbol{l}|\} - \frac{2G}{\sqrt{\pi}}. \tag{2.39}$$

Both (2.37) and (2.39) still contain a singularity; the term for $\mathbf{g} = 0$ diverges. But when we come to calculate the Madelung constant, as in (2.28), we subtract one series from another, and the two singularities cancel out. This means, simply, that the average potential due to each sublattice, if calculated separately, would be infinite, but that the over-all charge of the crystal is zero.

We see here some physical significance in the transformation. Consider a more complicated case. Suppose that we have a lattice wave creating a dipole at each site of the lattice,

$$\mathbf{p}(\boldsymbol{l}) = \mathbf{p}\,e^{i\mathbf{q}\cdot\boldsymbol{l}}. \tag{2.40}$$

The Coulomb field at $\mathbf{r}$ due to this distribution of charge is given by elementary electrostatics: it is the gradient of the potential created by the dipoles,

$$\begin{aligned}\mathbf{E}(\mathbf{r}) &= \nabla\left\{\sum_{l}\mathbf{p}(\boldsymbol{l})\cdot\nabla\left(\frac{1}{|\boldsymbol{l}-\mathbf{r}|}\right)\right\} \\ &= \nabla(\mathbf{p}\cdot\nabla)\left\{\sum_{l}\frac{e^{i\mathbf{q}\cdot\boldsymbol{l}}}{|\boldsymbol{l}-\mathbf{r}|}\right\}.\end{aligned} \tag{2.41}$$

The sum over lattice sites is a slightly more complicated version of (2.36). It can be evaluated by exactly the arguments that we have used above for the special case $\mathbf{q} = 0$; the result is

$$\sum_{l}\frac{e^{i\mathbf{q}\cdot\boldsymbol{l}}}{|\boldsymbol{l}-\mathbf{r}|} = \frac{\pi}{v_c}\frac{1}{G^2}\sum_{\mathbf{g}}\frac{e^{i(\mathbf{q}+\mathbf{g})\cdot\mathbf{r}}\,e^{-|\mathbf{q}+\mathbf{g}|^2/4G^2}}{|\mathbf{q}+\mathbf{g}|^2/4G^2} + \sum_{l}\frac{e^{i\mathbf{q}\cdot\boldsymbol{l}}}{|\boldsymbol{l}-\mathbf{r}|}\operatorname{erf}\{G|\boldsymbol{l}-\mathbf{r}|\}, \tag{2.42}$$

which can be differentiated term by term, as required by (2.41). In fact, we have here the sort of formula needed to calculate the coefficients of the force tensor, as in (2.12), when there are Coulomb forces between the ions.

But consider the contribution to the electric field from the term for $\mathbf{g} = 0$ in the first sum:

$$\mathbf{E}_0(\mathbf{r}) = \frac{-4\pi}{v_c} e^{-q^2/4G^2} \frac{(\mathbf{p}\cdot\mathbf{q})}{q^2}\mathbf{q}. \qquad (2.43)$$

Apart from the factor $\exp(-q^2/4G^2)$, this is the macroscopic electric field, $\bar{\mathbf{E}}(\mathbf{r})$, associated with a macroscopic wave of polarization

$$\mathbf{p}(\mathbf{r}) = \mathbf{p}\, e^{i\mathbf{q}\cdot\mathbf{r}} \qquad (2.44)$$

in the crystal treated as if it were a continuum. Thus we can express (2.41) in the form

$$\mathbf{E}(\mathbf{r}) = \bar{\mathbf{E}}(\mathbf{r}) + \frac{4\pi}{v_c}(1 - e^{-q^2/4G^2})\frac{(\mathbf{p}\cdot\mathbf{q})}{q^2}\mathbf{q} + \sum_{\mathbf{g}\neq 0}\{\,\} + \sum_{\mathbf{l}}\{\,\}, \qquad (2.45)$$

where we shall not bother to write out all the terms in the sums over $\mathbf{g}$ and $\mathbf{l}$, but simply note that they all converge rapidly (except in the neighbourhood of $\mathbf{r} = 0$ when the device used in (2.39) can be employed to eliminate the self-energy of the dipole at the origin). In this expression we have separated out the macroscopic field $\bar{\mathbf{E}}(\mathbf{r})$, which behaves badly near $\mathbf{q} = 0$, from the remaining terms, which converge rapidly, for all values of $\mathbf{q}$, to a well-defined local field (see § 8.2).

This local field is important in the lattice dynamics of an ionic crystal, for it may polarize the ions themselves. The resulting distribution of induced dipoles contributes, in turn, to the forces acting on the ions: and must therefore be included in the effective force tensor (2.4): central pair-wise forces between point charges are not adequate to explain the observations. Starting from this end of the theory, the formulae look rather elaborate, but they can be derived from a simple physical model in which each ion is replaced by a massive 'core', of charge e^+, say, connected to a weightless 'shell' of valence electrons of charge e^-. The various charges and spring constants in this *shell model* not only provide sufficient adjustable parameters to fit the lattice spectrum to experiment but also represent analytically the real *non-central* and many-body interatomic forces in the crystal.

In a covalent or metallic crystal the convergence of lattice sums is easily arranged because the long-range Coulomb fields of the positive ions are rapidly screened out by the valence electrons (see Chapter 5). But calculation of the lattice spectrum is made complicated by other types of many-body force, such as resistance to the bending of covalent bonds (§ 4.2), or to the compression of the gas of conduction electrons (§ 6.11). The dynamical coupling of the electromagnetic field with polar lattice vibrations in *polariton* modes (§ 8.3) must also be taken into account in non-conducting crystals.

2.4 Lattice specific heat

A major consequence of the existence of lattice waves is their thermal excitation, observable as a contribution to the specific heat of the solid. To calculate this, we really ought to use quantum-mechanical operators in place of the classical co-ordinates $\mathbf{u}_{sl}$, and their corresponding momenta. All that we have done in § 2.1 is to show that these co-ordinates may be transformed canonically to a new set in which the equations of motion are those of an assembly of independent simple harmonic oscillators. But the excitations must, therefore, be of the Bose–Einstein type; a normal mode may be given any number of quanta, of energy $\hbar\nu_q$, where ν_q is the corresponding classical frequency.

The theory of statistical mechanics then tells us that there will, on the average, be

$$\bar{n}_{\mathbf{q}} = \frac{1}{e^{\hbar\nu_{\mathbf{q}}/kT}-1} \tag{2.46}$$

quanta in the $\mathbf{q}$th mode, and these will contribute an energy

$$\bar{\mathscr{E}}_{\mathbf{q}} = (\bar{n}_{\mathbf{q}}+\tfrac{1}{2})\,\hbar\nu_{\mathbf{q}}, \tag{2.47}$$

counting the zero-point energy which we shall for the moment drop. Thus, the average total energy of the system is given by

$$\bar{\mathscr{E}} = \sum_{\mathbf{q}} \frac{\hbar\nu_{\mathbf{q}}}{e^{\hbar\nu_{\mathbf{q}}/kT}-1}, \tag{2.48}$$

where the summation is over all modes (i.e. over all polarizations, as well as all different wave-vectors).

Because $\mathbf{q}$-vectors are distributed with density $V/8\pi^3$ in reciprocal space, we can write this sum as an integral; we can also calculate the specific heat by differentiating with respect to temperature;

$$C_V = \frac{1}{V}\frac{\partial\bar{\mathscr{E}}}{\partial T} = \frac{1}{kT^2}\frac{1}{8\pi^3}\sum_{\text{polarizations}}\iiint \frac{(\hbar\nu_{\mathbf{q}})^2\,e^{\hbar\nu_{\mathbf{q}}/kT}}{(e^{\hbar\nu_{\mathbf{q}}/kT}-1)^2}\,d^3q. \tag{2.49}$$

In a formal way, we can simplify this by introducing a *lattice spectrum*. Let us define $\mathscr{D}(\nu)$ such that $\mathscr{D}(\nu)\,d\nu$ is the fraction of modes with frequency in the range $\nu \to \nu+d\nu$. For a structure with n atoms per unit cell, there are $3nN$ modes altogether. Thus

$$C_V = 3nNk\int \frac{(\hbar\nu/kT)^2\,e^{\hbar\nu/kT}}{(e^{\hbar\nu/kT}-1)^2}\,\mathscr{D}(\nu)\,d\nu. \tag{2.50}$$

But this only pushes the problem back on to a computation of $\mathscr{D}(\nu)$, which may require a complete solution of the dynamical equations for the lattice modes. We can get a crude formula by the following assumptions

(i) Only acoustic modes need be considered, and all have the same constant sound velocity, i.e.

$$\nu_{\mathbf{q}} = sq. \tag{2.51}$$

(ii) The Brillouin zone, which limits the range of allowed values of **q**, can be replaced by a sphere of the same volume in reciprocal space.

The second assumption implies that there is a maximum wave-number for lattice waves, the *Debye wave-number* q_D. The *Debye sphere* is an approximation to the zone; if it is to contain exactly N points, at a density $V/8\pi^3$ in **q**-space, we must have

$$N = \frac{V}{8\pi^3}\tfrac{4}{3}\pi q_D^3, \tag{2.52}$$

i.e.

$$q_D = \left(\frac{6\pi^2}{v_c}\right)^{\frac{1}{3}}. \tag{2.53}$$

We often make a similar approximation in the direct lattice, and replace the Wigner–Seitz cell by a *Wigner–Seitz sphere* of radius r_s and the same volume

$$\tfrac{4}{3}\pi r_s^3 = v_c. \tag{2.54}$$

Hence,

$$q_D = \left(\frac{9\pi}{2}\right)^{\frac{1}{3}}\frac{1}{r_s}, \quad \lambda_D = \frac{2\pi}{q_D} \sim 2{\cdot}6 r_s. \tag{2.55}$$

In other words, the *cut-off wavelength* for an acoustic mode is a bit more than the average diameter of a unit cell. The lattice will not propagate waves of shorter wavelength.

From these two assumptions the lattice spectrum takes the form

$$\begin{aligned}\mathscr{D}(\nu)\,d\nu &= 4\pi q^2\,dq/\tfrac{4}{3}\pi q_D^3 \\ &= \frac{3\nu^2}{\nu_D^3}\,d\nu,\end{aligned} \tag{2.56}$$

where ν_D is the *Debye frequency*, sq_D. The formula for the specific heat becomes

$$\begin{aligned}C_V &= 3Nk\int_0^{\nu_D}\frac{(\hbar\nu/kT)^2\,e^{\hbar\nu/kT}}{(e^{\hbar\nu/kT}-1)^2}\,\frac{3\nu^2}{\nu_D^3}\,d\nu \\ &= 3Nk\left(\frac{T}{\Theta}\right)^3 3\int_0^{\Theta/T}\frac{z^4\,e^z}{(e^z-1)^2}\,dz,\end{aligned} \tag{2.57}$$

where we have put $z = \hbar\nu/kT$, and where

$$k\Theta = \hbar\nu_D \tag{2.58}$$

defines the *Debye temperature* Θ.

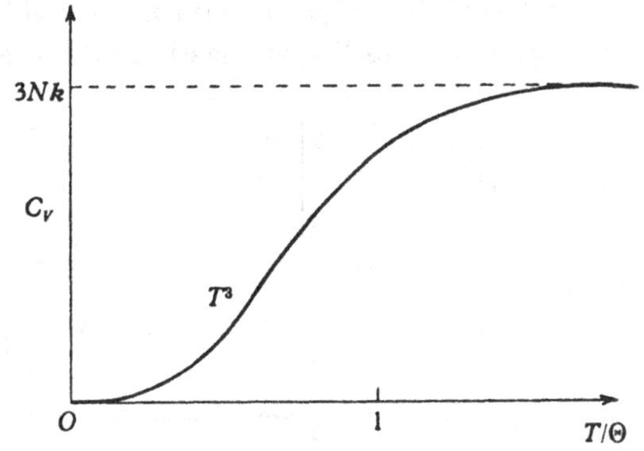

Fig. 24. The Debye law of specific heat.

This is the famous *Debye formula for specific heats.* It has a very simple structure.

(i) When T/Θ is large, the upper limit in the integral is small, and the integrand can be expanded, to give

$$\int_0^{\Theta/T} \frac{z^4}{z^2}\,dz = \frac{1}{3}\left(\frac{\Theta}{T}\right)^3. \tag{2.59}$$

In that case

$$C_V = 3N\boldsymbol{k}, \tag{2.60}$$

which is the *Dulong and Petit Law.*

(ii) At low temperatures, the upper limit, Θ/T, of the integral may be taken to be infinite for all practical purposes. The integral then tends to a constant, $(4\pi^4/15)$, and

$$C_V \sim \frac{12\pi^4}{5} N\boldsymbol{k}\left(\frac{T}{\Theta}\right)^3, \tag{2·61}$$

which is the well-known *T^3-law of specific heats,* valid at low temperatures.

The Debye formula works very well for most solids, and the Debye temperature is tabulated as a physical parameter of the solid. It is the most convenient scale temperature for the dynamical motion of the lattice: $\boldsymbol{k}\Theta$ represents the energy of the maximum quantum that can be excited amongst the lattice modes: Θ ought also to be related to the average velocity of sound in the crystal, by the relation

$$\boldsymbol{k}\Theta = \hbar q_D s. \tag{2.62}$$

Nevertheless, we have made the most drastic assumptions in deriving this formula. We have assumed a very special form for the lattice spectrum (Fig. 25 (*a*)). The relation $\mathscr{D}(\nu) \propto \nu^2$ should hold near $\nu = 0$, when the material behaves like an elastic continuum, but the

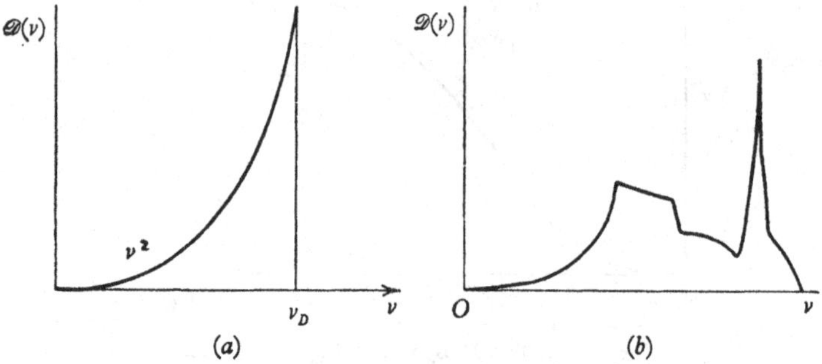

Fig. 25. (*a*) The Debye spectrum. (*b*) A true lattice spectrum.

sharp cut-off at ν_D is not justified. Exact calculations (e.g. Fig. 25 (*b*)) show that there is a long spread, with several peaks, corresponding to the modes of different polarization having very different velocities. There also tends to be a considerable peak at high frequencies, arising from the strong dispersion near the zone boundary. The curves for $\nu_{\mathbf{q}}$ as a function of **q** flatten out (cf. Fig. 22, § 2.2) so that larger volumes of **q**-space are included between surfaces differing in frequency by $d\nu$.

If we have a lattice with a basis, we ought also to allow for the contributions of optical modes. The frequency of these is more nearly independent of wave-number, so that an *Einstein model*, where each mode has the same frequency, might be more appropriate, i.e.

$$\begin{aligned} C_V &= 3N\boldsymbol{k}\frac{(\hbar\nu_E/\boldsymbol{k}T)^2\, e^{\hbar\nu_E/\boldsymbol{k}T}}{(e^{\hbar\nu_E/\boldsymbol{k}T}-1)^2} \\ &= 3N\boldsymbol{k}\frac{(\Theta_E/T)^2\, e^{\Theta_E/T}}{(e^{\Theta_E/T}-1)^2}, \end{aligned} \tag{2.63}$$

where Θ_E is an *Einstein temperature*, defined by

$$\boldsymbol{k}\Theta_E = \hbar\nu_E. \tag{2.64}$$

It should be remarked, however, that we could still represent the specific heat of a complex crystal by a single Debye term if we took N in (2.57) to be the total number of atoms in the crystal, instead of the number of unit cells. This is what we should get if we ignored the differences in position and mass between one atom and another in the unit cell. The Debye formula contains no hint of the actual crystal structure, so this might not be a bad approximation. In effect, we are using an artificially enlarged Brillouin zone—an extended zone scheme (Fig. 26 (*a*))—and ignoring the gaps that are produced, as in Fig. 19 (*c*) [p. 34] by the differences in mass, and in dynamical interaction, of the different atoms in the unit cell.

2.5 Lattice spectrum

The actual calculation of the frequency distribution of the normal modes is a problem in numerical computing. There is no way to do it in the end except to solve the equations (2.13) for a closely spaced net of values of $\mathbf{q}$, and find how many values of $\nu_{\mathbf{q}}$ fall into each range $d\nu$.

But there is an important principle, *van Hove's theorem*, which governs the function $\mathscr{D}(\nu)$, and the nature of its singularities. The full proof of this theorem is beyond us, but the general argument is of interest and importance.

Let us consider, for simplicity, a single branch of the spectrum. The proportion of modes with frequency in the range $d\nu$ is equal to

$$\mathscr{D}(\nu)\,d\nu = \frac{v_c}{8\pi^3}\iiint d^3q, \tag{2.65}$$

where the integration is through the volume of the shell in $\mathbf{q}$-space where $\nu \leqslant \nu_{\mathbf{q}} \leqslant \nu + d\nu$.

Let us introduce the vector

$$\mathbf{v}_{\mathbf{q}} \equiv \nabla_{\mathbf{q}}\,\nu_{\mathbf{q}} \equiv \left(\frac{\partial \nu_q}{\partial q_x}, \frac{\partial \nu_q}{\partial q_y}, \frac{\partial \nu_q}{\partial q_z}\right), \tag{2.66}$$

the gradient of the frequency function in $\mathbf{q}$-space. This vector has the dimensions of a velocity (in the Debye model it is the velocity of sound) and plays an important role in all problems where there is propagation of energy by the wave. It is, in fact, the *group velocity* of a wave packet of wave-vector $\mathbf{q}$ in the dispersive medium of the lattice.

We can use (2.66) to effect a formal simplification of (2.65). Let our element of volume, in the shell in $\mathbf{q}$-space through which we integrate, be a cylinder, of area dS_ν on the frequency surface $\nu_{\mathbf{q}} = \nu$ and of height

$$dq_{\perp} = \frac{d\nu}{|\partial \nu_{\mathbf{q}}/\partial q|} \tag{2.67}$$

measured normal to the surface. This direction is the direction of $\mathbf{v}_{\mathbf{q}}$ so that

$$\mathscr{D}(\nu)\,d\nu = \frac{1}{8\pi^3 N}\iint dS_\nu\,dq_{\perp} \tag{2.68}$$

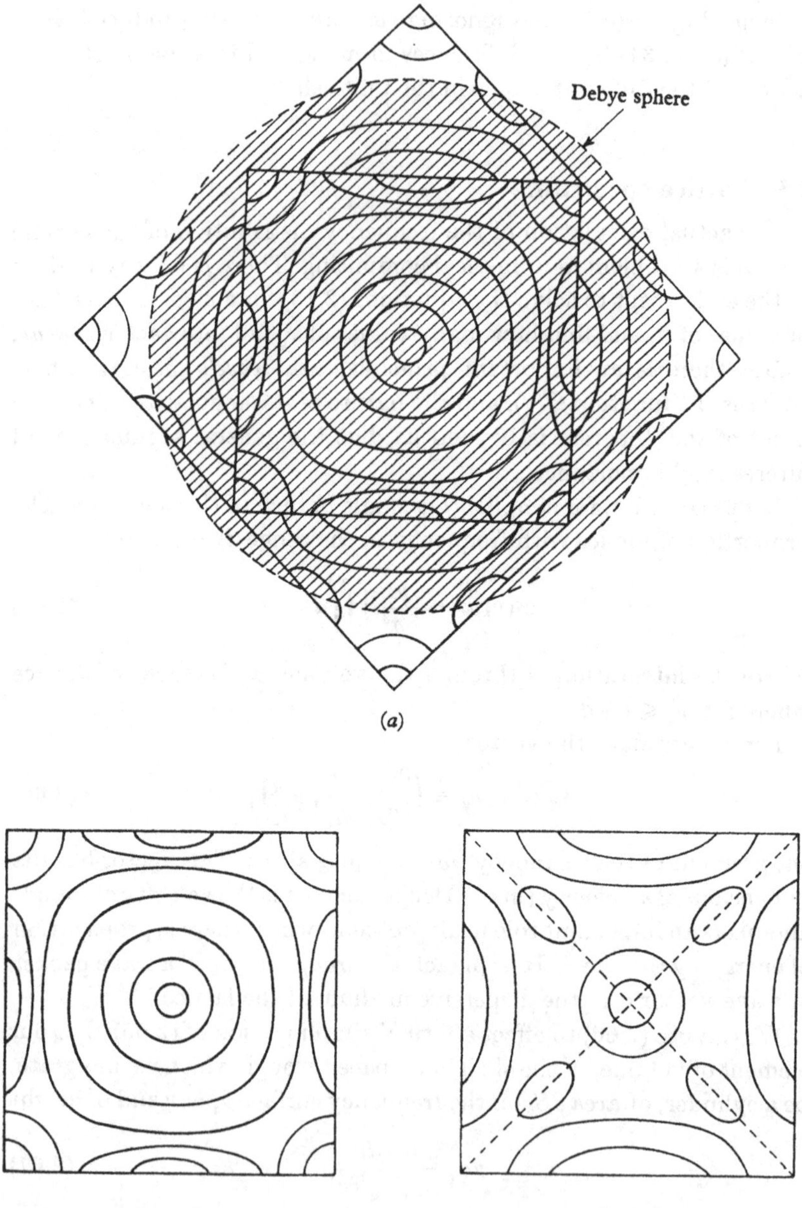

Fig. 26. Acoustic and optical frequency surfaces: (a) in an extended zone scheme; (b) in two separate zones.

gives
$$\mathscr{D}(\nu) = \frac{1}{8\pi^3 N}\int \frac{dS_\nu}{v_\mathbf{q}}, \tag{2.69}$$

where $v_\mathbf{q}$ is just the modulus of the vector $\mathbf{v}_\mathbf{q}$.

We often use this formula in the theory of the spectrum of lattice modes and of electron states. It does not tell us much in itself, except concerning the nature of the singularities of $\mathscr{D}(\nu)$. These must come at *critical points*, wherever $v_\mathbf{q}$ vanishes, i.e. wherever there is a locally 'flat' region of the function $\nu_\mathbf{q}$.

Let us suppose that $\mathbf{q}_c$ is a critical point. Since $\nu_\mathbf{q}$ is a continuous function of $\mathbf{q}$, it can be expanded as a Taylor series around that point. The linear terms vanish, because $\mathbf{v}_\mathbf{q} = 0$, and the quadratic terms can be reduced to a sum of squares by a principal axes transformation. Thus, we can write
$$\nu_\mathbf{q} = \nu_c + \alpha_1 \xi_1^2 + \alpha_2 \xi_2^2 + \alpha_3 \xi_3^2 + \ldots, \tag{2.70}$$

where $\boldsymbol{\xi} = \mathbf{q} - \mathbf{q}_c$ is the vector distance from the critical point, referred to the local principal axes, and the coefficients $\alpha_1, \alpha_2, \alpha_3$ depend on local second derivatives of $\nu_\mathbf{q}$ with respect to $\mathbf{q}$.

For example, suppose that $\alpha_1, \alpha_2, \alpha_3$ are all negative. Then ν is near a local maximum. The constant frequency surfaces (2.70) are ellipsoids; by elementary analytical geometry, the volume enclosed by the surface ν, around $\mathbf{q}_c$, is given by

$$\tfrac{4}{3}\pi \frac{(\nu_c - \nu)^{\frac{3}{2}}}{|\alpha_1 \alpha_2 \alpha_3|^{\frac{1}{2}}}, \tag{2.71}$$

whence we find, from (2.65), after a differentiation with respect to ν,

$$\mathscr{D}(\nu) = \frac{1}{4\pi^2 N |\alpha_1 \alpha_2 \alpha_3|^{\frac{1}{2}}}(\nu_c - \nu)^{\frac{1}{2}}. \tag{2.72}$$

This holds for $\nu < \nu_c$; when $\nu > \nu_c$, there is no contribution to $\mathscr{D}(\nu)$ from the neighbourhood of q_c. Thus, this singularity does not spoil the continuity of $\mathscr{D}(\nu)$, but its slope, $\partial\mathscr{D}(\nu)/\partial\nu$, is discontinuous and tends to $-\infty$ as $\nu \to \nu_c$ from below.

There are other possibilities for the coefficients α_1, α_2, α_3. Thus, if one is positive and the other two negative, we have a *saddle-point of index* 1. In that case the form of the spectrum in the neighbourhood of ν_c becomes, by exactly the same sort of analytical geometry,

$$\mathscr{D}(\nu) = \begin{cases} C + O(\nu - \nu_c) & (\nu < \nu_c), \\ C - \dfrac{1}{4\pi^2 N |\alpha_1 \alpha_2 \alpha_3|^{\frac{1}{2}}}(\nu - \nu_c)^{\frac{1}{2}} + O(\nu - \nu_c) & (\nu > \nu_c), \end{cases} \tag{2.73}$$

again, only a discontinuity of slope, with a vertical tangent on one side, imposed on the smooth function $C+O(\nu-\nu_c)$ generated by other regions of the zone.

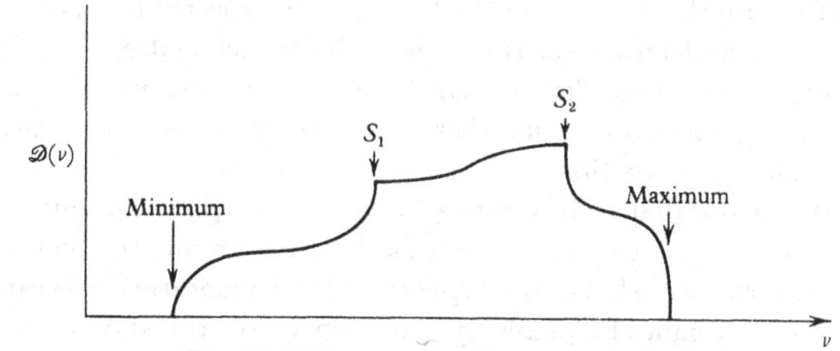

Fig. 27. Different types of van Hove singularity.

There are, in fact, four types of critical points, as indicated in Fig. 27, where S_1 and S_2 refer to saddle-points of index 1 and index 2, respectively (the latter has two of the α's positive, the other negative, and shows behaviour like (2.73) reversed). The case of a minimum is obviously similar to that of a maximum.

These singularities are not very serious—but they are essential. Van Hove's theorem states, in effect, that *the spectrum must contain at least one critical point of each of the types S_1 and S_2, and the slope of $\mathscr{D}(\nu)$ must tend to $-\infty$ at the upper end.* The proof of this requires some theorems of functional topology, which are beyond our present scope. But the argument may be illustrated in two dimensions (where the saddle-points produce logarithmic singularities and the extrema produce finite discontinuities of $\mathscr{D}(\nu)$).

The point is that the function $\nu_{\mathbf{q}}$ is a continuous periodic function of $\mathbf{q}$ in reciprocal space. Suppose that it has a minimum at some point m in the zone. Then it has similar minima at each corresponding point of the reciprocal lattice. Again, if it has a maximum value at some point M in the zone, then this is a critical point corresponding to the peak of a hill surrounding M in the repeated zone.† There will thus be a distribution of such peaks throughout reciprocal space.

Now consider a path MM joining two maxima. The function $\nu_{\mathbf{q}}$

† If the maximum is at a corner of the zone we might feel that we should only include in the integration the part that lies within the zone. But there will be, automatically, contributions from the other corners to make up a complete term like (2.72). We could ensure this by shifting our region of integration so as to include one whole peak. *Any* unit cell of the reciprocal lattice is a possible region of integration; the Brillouin zone is only a conventional choice.

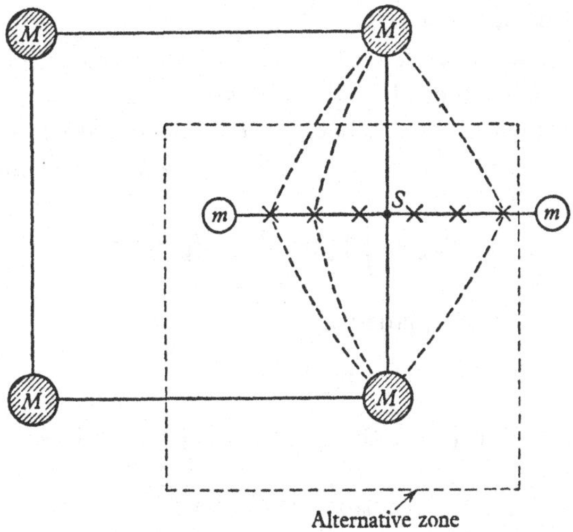

Fig. 28. Finding a saddle-point.

must vary along this path; at some point X it will have its minimum value for the path. Now move this path around. A series of points, X, X, X will be generated. These form another continuous path; because m and m are absolute minima of $\nu_{\mathbf{q}}$, this path must pass through these points.

But now we have a path mm passing between two minima. On this path there must be at least one maximum, S. This point is a saddle-point—it is a minimum for paths like MM and a maximum for paths like mm. Thus, our function $\nu_{\mathbf{q}}$ has at least one critical point of type S.

In the general three-dimensional problem, with three acoustic branches of the spectrum the argument is more complicated. The origin of the zone $\mathbf{q} = 0$ is an absolute minimum of all three branches—but the function $\nu_{\mathbf{q}}$ cannot be expanded in a Taylor series there, and the velocity does not vanish. The transverse branches are also not distinct so that $\nu_{\mathbf{q}}$ must be treated as a two-valued function of $\mathbf{q}$. However, the theorem as stated holds for the total spectrum. There may, of course, be many more critical points than this minimum number.

2.6 Diffraction by an ideal crystal

If all atoms in a crystal were exactly at their official lattice sites, then all the measurable physical quantities associated with the crystal would be exactly periodic functions. For example, the potential energy of an electron would satisfy the condition

$$\mathscr{V}(\mathbf{r}+\mathbf{l}) = \mathscr{V}(\mathbf{r}). \tag{2.74}$$

Suppose that a beam of fast electrons is directed into the crystal. There will be transitions from one electron state to another, because of scattering by this potential. In the Born approximation of perturbation theory, the rate of transition between the initial state $\Psi_{\mathbf{k}}$ and the final state $\Psi_{\mathbf{k}'}$, is proportional to the square of the matrix element

$$\mathscr{M}_{\mathbf{k}'\mathbf{k}} \equiv \int \Psi^*_{\mathbf{k}'}(\mathbf{r})\mathscr{V}(\mathbf{r})\Psi_{\mathbf{k}}(\mathbf{r})\,d\mathbf{r}. \tag{2.75}$$

We take these to be plane-wave states

$$\Psi_{\mathbf{k}} = e^{i\mathbf{k}\cdot\mathbf{r}}. \tag{2.76}$$

We also know from (1.15) that the potential may be expanded in a Fourier series

$$\mathscr{V}(\mathbf{r}) = \sum_{\mathbf{g}} \mathscr{V}_{\mathbf{g}} e^{i\mathbf{g}\cdot\mathbf{r}}. \tag{2.77}$$

Thus

$$\begin{aligned} \mathscr{M}_{\mathbf{k}'\mathbf{k}} &= \int e^{-i\mathbf{k}'\cdot\mathbf{r}} \sum_{\mathbf{g}} \mathscr{V}_{\mathbf{g}} e^{i\mathbf{g}\cdot\mathbf{r}} e^{i\mathbf{k}\cdot\mathbf{r}}\,d\mathbf{r} \\ &= \sum_{\mathbf{g}} \mathscr{V}_{\mathbf{g}} \int e^{i(\mathbf{k}+\mathbf{g}-\mathbf{k}')\cdot\mathbf{r}}\,d\mathbf{r} \\ &= \begin{cases} \mathscr{V}_{\mathbf{g}} & \text{if} \quad \mathbf{k}+\mathbf{g}-\mathbf{k}' = 0, \\ 0 & \text{otherwise.} \end{cases} \end{aligned} \tag{2.78}$$

If $\mathbf{k}$ is fixed, i.e. if the incident beam of electrons is monochromatic and well defined in direction—then we can observe diffracted beams only in directions corresponding to wave-vectors satisfying

$$\mathbf{k}' = \mathbf{k} + \mathbf{g}, \tag{2.79}$$

where $\mathbf{g}$ is one of the reciprocal lattice vectors of the crystal.

We must also insist in this case that the energy of the diffracted beam be the same as the incident energy, i.e. the wave-vectors have the same length. This means a condition on the angle of scatter. If 2θ is the angle between $\mathbf{k}$ and $\mathbf{k}'$ then

$$|\mathbf{g}| = 2\,|\mathbf{k}|\sin\theta. \tag{2.80}$$

To satisfy these geometrical conditions in reciprocal space we construct the *Ewald sphere* (Fig. 29 (*b*)), with radius OP equal to the incident wave-vector. If the origin of the reciprocal lattice is located at P, then the vector $\mathbf{g}$ must carry us to Q, where OQ is the wave-vector $\mathbf{k}'$. Diffraction occurs, therefore, whenever the orientation of the crystal relative to the incident beam puts a point of the reciprocal lattice on the sphere.

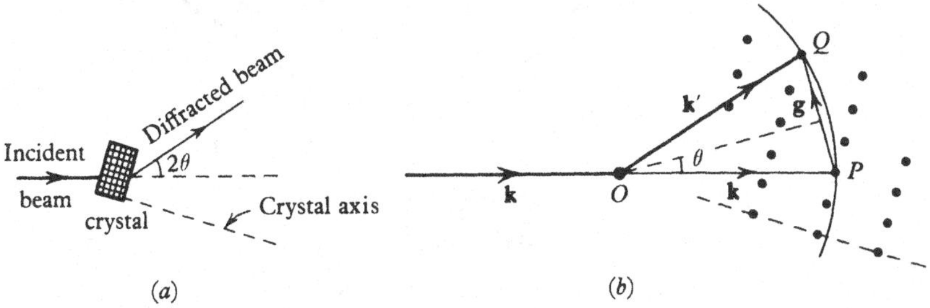

Fig. 29. (*a*) X-ray diffraction; (*b*) Ewald's construction.

We can express this another way. As shown in § 1.3, the length of a reciprocal lattice vector is proportional to the reciprocal of the spacing, d, of the planes to which it is normal; i.e.

$$|\mathbf{g}| = \frac{2\pi n}{d}, \tag{2.81}$$

where n is an integer (the H.C.F. of the components of $\mathbf{g}$ relative to the reciprocal triad axes). Thus, if λ is the wavelength of the incident electrons, (2.80) and (2.81) give

$$n\lambda = 2d \sin\theta. \tag{2.82}$$

This is known as the *Bragg reflection law.* It can easily be deduced by consideration of the phase relation between beams reflected off successive lattice planes. For coherent diffraction, the extra path ABC must be an integral number of wavelengths (Fig. 30).

The magnitude of the scattering matrix element depends on $\mathscr{V}_{\mathbf{g}}$. For electrons this would seem to be the Fourier components of the local potential—but this is misleading when we deal with very slow electrons. We return to this point in § 3.6. The same theory holds also for diffraction of X-rays, except that these would respond to the local electron density in the crystal. Since this also is a periodic function like (2.74), the diffraction pattern will be qualitatively the same. The theory of this section leads on, of course, to the marvellous complexities and successes of the X-ray analysis of crystal structure.

In the case of neutrons, the scattering arises mainly from the nucleus, which behaves like a delta-function potential in space, but which varies considerably from element to element. Indeed, because different isotopes of the same element may have very different neutron

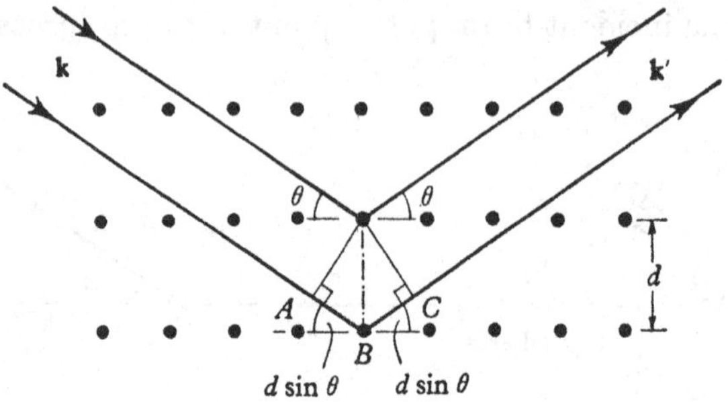

Fig. 30. Bragg diffraction.

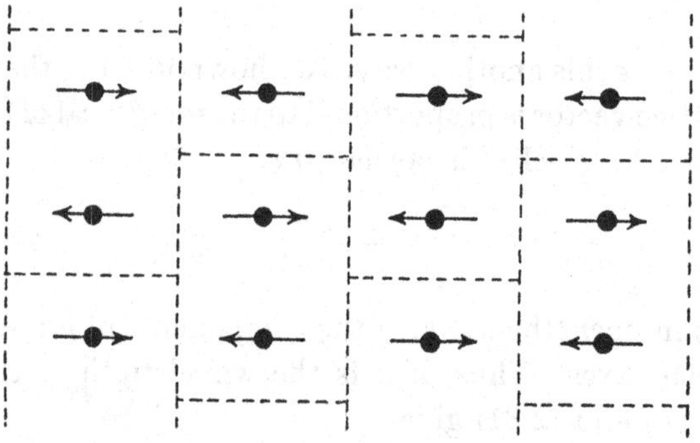

Fig. 31. Antiferromagnetic array, showing doubled unit cell.

scattering cross-sections, and will be mixed randomly in the crystal, there is always some *incoherent* scattering which produces a diffuse background to the diffraction pattern. This can even occur when only one isotope is present, because neutrons are sensitive to the spin-orientation of the nucleus.

Neutron diffraction is of particular interest in the analysis of *antiferromagnetic* crystals. In these the atoms have permanent atomic moments which are polarized and ordered in such a way that all the atoms on one sublattice have moments pointing one way, and all atoms on the other sublattice have moments pointing in the opposite direction. We shall consider the magnetic properties of such a system in §§ 10.6, 10.11.

Here we simply note that a neutron, having a magnetic moment, is sensitive to the local magnetic field in each atom, and can distinguish between atoms in which the magnetic moments are differently oriented. The result is that the unit of periodicity of the structure is increased, which means that the size of the effective Brillouin zone is reduced. New reciprocal lattice vectors are introduced (cf. Fig. 19) so that new lines appear in the diffraction pattern. These are called *magnetic superlattice* lines.

This elementary *geometrical theory* of crystal diffraction ignores several practical complications, such as the effect of thermal vibrations discussed in the next section. We must also allow for the continued interaction of the incident and diffracted beams with successive atomic layers as they pass through a relatively thick specimen. This gives rise to several complicated effects such as *primary extinction* and the *pendulum effect* ('Pendellösung'), whose explanation calls on the *dynamical theory* of X-ray diffraction. The essence of this is that X-rays (or electrons, or neutrons) suffer velocity dispersion as they propagate through the crystal, especially when close to the conditions of coherent diffraction (see § 3.3). Two or more beams on different dispersion surfaces may be excited simultaneously and then interfere with one another as they go through the specimen. In the case of *low energy electron diffraction* (L.E.E.D.) such effects completely dominate the experimental phenomena (see § 6.9).

2.7 Diffraction by crystal with lattice vibrations

Consider now a more general case, where the potential is not necessarily periodic. Suppose that it is the result of the superposition of atomic potentials

$$\mathscr{V}(\mathbf{r}) = \sum_l \mathscr{V}_a(\mathbf{r} - \mathbf{R}_l), \tag{2.83}$$

where $\mathbf{R}_l$ is the actual position of the atom (or ion) that ought to have been at the lattice site $\boldsymbol{l}$, and where $\mathscr{V}_a$ is the potential field of an individual atom. For simplicity of notation, we assume a Bravais lattice.

We then may calculate a matrix element for scattering from state $\Psi_\mathbf{k}$ to state $\Psi_{\mathbf{k}'}$ as in (2.75):

$$\begin{aligned}\mathscr{M}_{\mathbf{k}'\mathbf{k}} &= \int e^{-i\mathbf{k}'\cdot\mathbf{r}} \sum_l \mathscr{V}_a(\mathbf{r} - \mathbf{R}_l)\, e^{i\mathbf{k}\cdot\mathbf{r}}\, d\mathbf{r} \\ &= \sum_l \int e^{i(\mathbf{k}-\mathbf{k}')\cdot\mathbf{r}}\, \mathscr{V}_a(\mathbf{r} - \mathbf{R}_l)\, d\mathbf{r} \\ &= \sum_l e^{i(\mathbf{k}-\mathbf{k}')\cdot\mathbf{R}_l} \int e^{i(\mathbf{k}-\mathbf{k}')\cdot(\mathbf{r}-\mathbf{R}_l)}\, \mathscr{V}_a(\mathbf{r} - \mathbf{R}_l)\, d\mathbf{r}.\end{aligned} \tag{2.84}$$

Each of the integrals in the sum is nominally over all space—or at least over the whole crystal. But we expect an atomic potential $\mathscr{V}_a(\mathbf{r})$ to be of very limited range, so that the result of the integration can scarcely depend upon the centre, $\mathbf{R}_l$, from which it is measured. Let us introduce the *scattering vector*

$$\mathbf{K} = \mathbf{k}' - \mathbf{k}; \tag{2.85}$$

then our matrix element becomes

$$\mathscr{M}_{\mathbf{k}'\mathbf{k}} = \mathscr{V}_a(\mathbf{K}) \cdot \frac{1}{N} \sum_l e^{-i\mathbf{K}\cdot\mathbf{R}_l}, \tag{2.86}$$

where we define the Fourier transform of the atomic potential by†

$$\mathscr{V}_a(\mathbf{K}) \equiv \frac{1}{v_c} \int \mathscr{V}_a(\mathbf{r})\, e^{-i\mathbf{K}\cdot\mathbf{r}}\, d\mathbf{r}. \tag{2.87}$$

The separation of (2.86) into an *atomic factor* and a *structure factor* is of the greatest utility. Let us first consider the structure factor for an ideal lattice, i.e. let

$$\mathbf{R}_l = \boldsymbol{l} = l_1 \mathbf{a}_1 + l_2 \mathbf{a}_2 + l_3 \mathbf{a}_3 \tag{2.88}$$

for all lattice sites.

Let us write the scattering vector in the form

$$\mathbf{K} = K_1 \mathbf{b}_1 + K_2 \mathbf{b}_2 + K_3 \mathbf{b}_3, \tag{2.89}$$

where $\mathbf{b}_1, \mathbf{b}_2, \mathbf{b}_3$ are the reciprocal triad (1.19). Then the structure factor becomes

$$\begin{aligned} \sum_l e^{-i\mathbf{K}\cdot\boldsymbol{l}} &= \sum_{l_1, l_2, l_3} e^{-i(K_1 l_1 + K_2 l_2 + K_3 l_3)} \\ &= \Big(\sum_{0 \leqslant l_1 < L_1} e^{-iK_1 l_1}\Big)\Big(\sum_{0 \leqslant l_2 < L_2} e^{-iK_2 l_2}\Big)\Big(\sum_{0 \leqslant l_3 < L_3} e^{-iK_3 l_3}\Big) \\ &= \left(\frac{1 - e^{-iK_1 L_1}}{1 - e^{-iK_1}}\right)\left(\frac{1 - e^{-iK_2 L_2}}{1 - e^{-iK_2}}\right)\left(\frac{1 - e^{-iK_3 L_3}}{1 - e^{-iK_3}}\right), \end{aligned} \tag{2.90}$$

where, as in (1.76), L_1, L_2, L_3 are the numbers of unit cells along the sides of the block of crystal.

Each factor in (2.90) is only of the order of unity, and fluctuates wildly as $\mathbf{K}$ varies. Any averaging process over a range of values of $\mathbf{K}$ will give zero—except when it includes a zero of all the denominators. In that case

$$e^{-iK_1} = e^{-iK_2} = e^{-iK_3} = 1, \tag{2.91}$$

† We introduce $N = 1/v_c$, the number of unit cells in a macroscopic unit volume of crystal, to leave $\mathscr{V}_a(\mathbf{K})$ with the dimensions of energy. The plane wave states (2.76) used in (2.84) are obviously also normalized to this macroscopic unit volume.

which is only true if each of K_1, K_2, K_3 is an integral multiple of 2π. But according to (1.20), this is the condition that **K** should coincide with one of the reciprocal lattice vectors $\mathfrak{g}$.

When **K** is actually equal to a reciprocal lattice vector, both numerator and denominator of each term in (2.90) vanish. But we then return to the original sum, and note that each term, being of the form $\exp(-i\mathfrak{g}\,.\,\boldsymbol{l})$ is exactly unity and the sum is just N. We thus have the important rule

$$\frac{1}{N}\sum_{l} e^{-i\mathbf{K}.\boldsymbol{l}} = \delta_{\mathbf{K}\,\mathfrak{g}}, \tag{2.92}$$

where the δ-symbol vanishes unless **K** is one of the vectors of the set $\mathfrak{g}$.

Putting (2.92) back into (2.86), we get

$$\mathscr{M}_{\mathbf{k}'\mathbf{k}} = \mathscr{V}_a(\mathfrak{g})\,\delta_{\mathbf{k}'-\mathbf{k},\mathfrak{g}}, \tag{2.93}$$

which is precisely our previous result (2.78).

But now suppose that the lattice modes of our crystal are excited. Each atom is displaced from its ideal lattice site

$$\begin{aligned}\mathbf{R}_l &= \boldsymbol{l}+\mathbf{u}_l\\ &= \boldsymbol{l}+\sum_{\mathbf{q}>}(\mathbf{U}_\mathbf{q}e^{i\mathbf{q}.\boldsymbol{l}}+\mathbf{U}_\mathbf{q}^{*}e^{-i\mathbf{q}.\boldsymbol{l}}),\end{aligned} \tag{2.94}$$

where $\mathbf{U}_\mathbf{q}$ is, as in (2.8) and (2.10), the vector amplitude of the mode of wave-vector **q**. The sum is over only half the range of **q**, so that we may display the complex conjugate term $\mathbf{U}_\mathbf{q}^{*} = \mathbf{U}_{-\mathbf{q}}$ making the displacements real.

We substitute (2.94) in the structure factor of (2.86)

$$\begin{aligned}\frac{1}{N}\sum_{l} e^{-i\mathbf{K}\cdot\mathbf{R}_l} &= \frac{1}{N}\sum_{l}\exp\,[-i\mathbf{K}\cdot\{\boldsymbol{l}+\sum_{\mathbf{q}>}(\mathbf{U}_\mathbf{q}e^{i\mathbf{q}.\boldsymbol{l}}+\mathbf{U}_\mathbf{q}^{*}e^{-i\mathbf{q}.\boldsymbol{l}})\}]\\ &= \frac{1}{N}\sum_{l}[e^{-i\mathbf{K}.\boldsymbol{l}}\prod_{\mathbf{q}>}\exp\{-i\mathbf{K}\cdot(\mathbf{U}_\mathbf{q}e^{i\mathbf{q}.\boldsymbol{l}}+\mathbf{U}_\mathbf{q}^{*}e^{-i\mathbf{q}.\boldsymbol{l}})\}].\end{aligned} \tag{2.95}$$

This is exact. Let us expand the exponential function whose argument is proportional to $\mathbf{U}_q$—supposedly a small displacement:

$$\begin{aligned}&\exp\{-i\mathbf{K}\cdot(\mathbf{U}_\mathbf{q}e^{i\mathbf{q}.\boldsymbol{l}}+\mathbf{U}_\mathbf{q}^{*}e^{-i\mathbf{q}.\boldsymbol{l}})\}\\ &\qquad = 1-i\mathbf{K}\cdot(\mathbf{U}_\mathbf{q}e^{i\mathbf{q}.\boldsymbol{l}}+\mathbf{U}_\mathbf{q}^{*}e^{-i\mathbf{q}.\boldsymbol{l}})-|\mathbf{K}\cdot\mathbf{U}_\mathbf{q}|^2\ldots.\end{aligned} \tag{2.96}$$

Substituting in (2.95), it is evident that, when the product is multiplied out, there will be a term like (2.90), giving the diffraction pattern of the ideal crystal structure, followed by a sum of all the terms that are linear in $\mathbf{K}\cdot\mathbf{U}_\mathbf{q}$, i.e. terms like

$$\frac{1}{N}\sum_l e^{-i\mathbf{K}\cdot l}\sum_\mathbf{q}(-i\mathbf{K}\cdot\mathbf{U}_\mathbf{q})\,e^{i\mathbf{q}\cdot l} = \frac{1}{N}\sum_\mathbf{q}(-i\mathbf{K}\cdot\mathbf{U}_\mathbf{q})\sum_l e^{i(\mathbf{q}-\mathbf{K})\cdot l}$$
$$= \sum_\mathbf{q}(-i\mathbf{K}\cdot\mathbf{U}_\mathbf{q})\,\delta_{\mathbf{K}-\mathbf{q},\mathbf{g}} \qquad (2.97)$$

by (2.92). As we shall see in § 2.9, the modulus-squared term in (2.96) is also important, because of its invariant sign.

To understand this result let us consider two cases. In Fig. 32 (*a*), **K** lies in the Brillouin zone. That is, we look for the scattered electron (or X-ray, or neutron) in the region of states where **k′** does not differ by more than half a reciprocal lattice vector from the initial state **k**. By convention the wave-vectors **q** are restricted to a Brillouin zone, so that the only term left in (2.97) is the one for which $\mathbf{q} = \mathbf{K}$. In other words, the matrix element for the transition is

$$\mathscr{M}_{\mathbf{k}'\mathbf{k}} = -i\{(\mathbf{k}'-\mathbf{k})\cdot\mathbf{U}_\mathbf{q}\}\mathscr{V}_a(\mathbf{k}'-\mathbf{k}), \qquad (2.98)$$

where
$$\mathbf{q} = \mathbf{k}'-\mathbf{k}. \qquad (2.99)$$

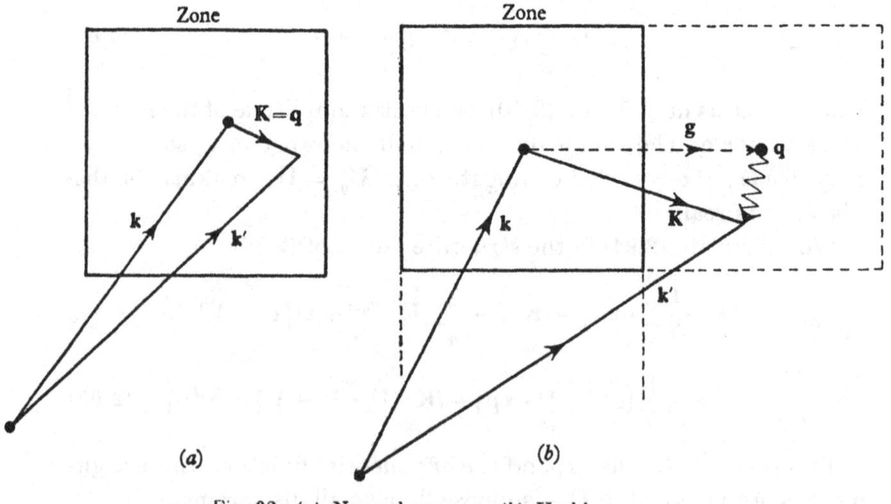

Fig. 32. (*a*) Normal process. (*b*) Umklapp process.

If **K** does *not* lie in the first Brillouin zone (Fig. 32 (*b*)), there is still one term left in the sum (2.97). That is, there is still a value of **q** such that $\mathbf{K}-\mathbf{q} \doteq \mathbf{g}$, i.e. such that

$$\mathbf{q} = \mathbf{k}'-\mathbf{k}-\mathbf{g}. \qquad (2.100)$$

The Brillouin zone, being a unit cell of the reciprocal lattice, can be repeated so as to cover exactly the whole of the space of the vector **K**; any value of **K** can be constructed uniquely out of a value of **g**—a translational vector of the reciprocal lattice—and a value of **q**, a vector constrained to lie in one unit cell of that lattice. In the language of § 1.5, **q** is the value of **K**, *reduced* to the first Brillouin zone.

In both cases, therefore, the formula (2.98) holds, with the value of **q** defined by (2.99) or (2.100), whichever is appropriate; if we include $\mathbf{g} = 0$ in the complete set of reciprocal lattice vectors, the relation (2.100) is always valid. Whatever direction we choose, there will be some scattering, only its amplitude will depend on the amplitude of the lattice vibrations that happen to have the correct wave-vector to satisfy these conditions.

2.8 Phonons

The possibility of a scattering process is not determined solely by the non-vanishing of the matrix element. There are also energy conditions to be satisfied. In the case of the ideal lattice, the whole system is static, so that the initial and final electron (X-ray, neutron) states are of the same energy.

In truth, the lattice is in constant motion. The vector $\mathbf{U}_\mathbf{q}$ in (2.94) should contain a time factor like $\exp(i\nu_\mathbf{q} t)$, where $\nu_\mathbf{q}$ is the frequency of the normal mode of wave-vector **q**. Indeed, $\mathbf{U}_\mathbf{q}$ is the sum of three such vectors, of different amplitudes, directions, and frequencies, corresponding to the three different acoustic modes of that wave-vector. But take one of these at a time, and remember that the states $\Psi_\mathbf{k}$ and $\Psi_{\mathbf{k}'}$, of the incident and scattered beams, must have energies $\mathscr{E}(\mathbf{k})$ and $\mathscr{E}(\mathbf{k}')$, giving them time factors like $\exp\{i\mathscr{E}(\mathbf{k})t/\hbar\}$ and $\exp\{i\mathscr{E}(\mathbf{k}')t/\hbar\}$. When we average over time, we shall find integrals like

$$\int \exp[i\{\mathscr{E}(\mathbf{k}) - \mathscr{E}(\mathbf{k}') + \hbar\nu_\mathbf{q}\}t/\hbar]\,dt,$$

which vanish unless $$\mathscr{E}(\mathbf{k}) - \mathscr{E}(\mathbf{k}') + \hbar\nu_\mathbf{q} = 0. \tag{2.101}$$

In other words the diffraction is *inelastic*; the beam has gained one quantum of the energy of the lattice mode with which it interacts.

Actually this is not complete, since the conjugate time-dependence $\exp(-i\nu_\mathbf{q} t)$ will also occur in the lattice vibrations, giving a diffraction in which the energy $\hbar\nu_\mathbf{q}$ is lost by the electron. In both cases, the argument that we have sketched can be made perfectly precise by an appeal to the standard theory of time-dependent perturbations.

The two conditions (2.100) and (2.101) can be given a very graphic interpretation. We say that the incident electron (or X-ray, or neutron) has interacted with the lattice to destroy (or create) a *phonon*, of wave-vector $\mathbf{q}$ and energy $\hbar\nu_{\mathbf{q}}$. We represent these processes by a diagram such as Fig. 33 (*a*) or (*b*). The condition (2.101) is the rule for energy conservation in that process.

If we look at the condition (2.99), which holds for all of values of $(\mathbf{k}'-\mathbf{k})$ in the first zone, we see that it looks like a law of conservation of momentum. Multiply by $\hbar$ and we have

$$\hbar\mathbf{k}' = \hbar\mathbf{k} + \hbar\mathbf{q}. \tag{2.102}$$

The momentum of the electron has taken up the *crystal momentum*, $\hbar\mathbf{q}$ of the phonon which was destroyed. This is the argument that justifies the name *phonon*, as a quantized acoustic excitation with 'particle-like' properties, by analogy with 'photon'.

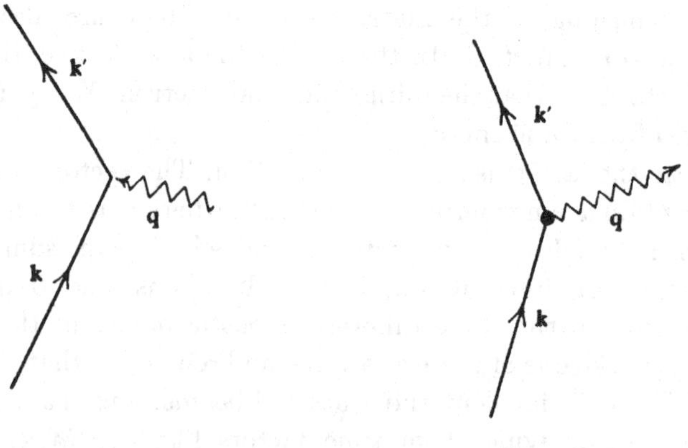

(*a*) (*b*)

Fig. 33. Electron scattering processes: (*a*) phonon absorption; (*b*) phonon emission.

But in the general case the 'rule of conservation of momentum' does not hold; as we see in (2.100) the electron can gain, or lose, the momentum $\hbar\mathbf{g}$ in addition to the momentum of the phonon. We call such a process an *Umklapp process*, or *U-process*. The special case (2.102) may be referred to as a *Normal process*, or *N-process*.

There is nothing really surprising about this result. The extra momentum $\hbar\mathbf{g}$ is simply transferred to the crystal as a whole. In constructing the lattice vibrational states we ignored the motion of the centre of mass of the ideal lattice, against which the lattice vibrations were measured. One can check that the fundamental law of conservation of momentum is obeyed over-all, when the motion associated with this degree of freedom is included. An Umklapp process can be thought of as the creation (or destruction) of a phonon with, simultaneously, a Bragg reflection. The momentum in the latter process is obviously transferred to the crystal as a whole.

The phenomenon of inelastic diffraction is a valuable tool for the study of the lattice dynamics of crystals. The beam diffracted in a particular direction is associated with lattice modes having a definite wave-vector $\mathbf{q}$. One can look at the change of energy of the diffracted particles, and hence measure $\hbar\nu_{\mathbf{q}}$. By looking in different directions, and moving the crystal into different orientations, one can plot out the whole function $\nu_{\mathbf{q}}$. Of course, this has several branches, of different polarization, but these may be separated by systematic analysis. The sensitivity of neutrons to local magnetic moments (§ 2.6) implies a similar theory for the inelastic diffraction of polarized neutron beams by *magnons* (see § 10.10), yielding similar information about their dispersion properties.

However, this experiment is only practicable with 'thermal' neutrons, whose wavelength is of the order of the lattice spacing at energies of the order of 0·1 eV. The shift due to the phonon energy, which is of the order of $k\Theta$ or less—perhaps 0·01 eV.—can easily be observed. For electrons and X-rays the beam energy must be much higher—tens or hundreds of electron volts—so that the small change in energy in the diffraction process cannot be detected.

The selection rule (2.102) for crystal momentum depends on the underlying long-range order of the lattice. What do we observe when this order is absent, as in a liquid or glass? Elastic diffraction would depend on the square of the matrix element (2.86), which is proportional to the square of the structure factor:

$$S(\mathbf{K}) \equiv \left|\frac{1}{N}\sum_{l} e^{-i\mathbf{K}\cdot\mathbf{R}_l}\right|^2$$

$$= \frac{1}{N^2}\sum_{ll'} e^{-i\mathbf{K}\cdot(\mathbf{R}_l-\mathbf{R}_{l'})}$$

$$= \frac{1}{N}\int P(\mathbf{R})\, e^{-i\mathbf{K}\cdot\mathbf{R}}\, d\mathbf{R}. \qquad (2.103)$$

The diffraction pattern depends on the Fourier transform of the pair correlation function $P(\mathbf{R})$—the probability of finding two atoms at a distance R apart. In a liquid this is just the familiar *radial distribution function* of the system.

By a simple generalization of this argument, we may show that the amplitude of *inelastic* diffraction, $S(\mathbf{K}, \nu)$, must be the double Fourier transform of a *time-dependent correlation function*, $P(\mathbf{R}, t)$ which measures the probability of finding an atom at point $\mathbf{R}$ at time t, given that an atom (possibly the same one) was at the point 0 at time 0. Neutron diffraction, for example, may measure the frequency spectrum of density fluctuations of wave vector $\mathbf{K}$—a physical quantity that happens to be quite sharp only for relatively perfect lattices. Similarly, polarized neutrons are diffracted by fluctuations of *spin density* defined by analogous correlation functions or their Fourier transforms in space or momentum, time or energy.

2.9 The Debye–Waller factor

The terms in (2.97), corresponding to scattering processes involving one phonon, are not all that may occur in the structure factor (2.95). It is easy to see that the product of factors like (2.96), and also higher terms in the expansion of the exponential, give rise to contributions containing various products of factors like $\mathbf{U}_\mathbf{q} \exp(i\mathbf{q}\cdot\mathbf{l})$. For each such factor, one says that the corresponding phonon has been created or destroyed, so that these terms refer to *multiphonon processes*.

Generally speaking, these processes fall off rapidly in rate as we go to higher order, and do not contribute very heavily to the background of inelastic diffraction. There is, however, an important class of terms, arising from the squared term $-|\mathbf{K}\cdot\mathbf{U}_\mathbf{q}|^2$ in the expansion (2.96), which do not average to a small contribution. If we look at (2.95) and (2.96), we find that the square of the matrix element for both elastic and inelastic scattering ought to be multiplied by

$$e^{-2W} = \prod_{\mathbf{q}} \{1 - |\mathbf{K}\cdot\mathbf{U}_\mathbf{q}|^2\}, \tag{2.104}$$

returning now to the whole zone for $\mathbf{q}$.

This is called the *Debye–Waller factor*. It is written in this form because we can use a standard theorem of algebra,

$$\lim_{N\to\infty} \prod_{n=1}^{N} \left(1 - \frac{1}{N} a_n\right) = \exp\left\{-\lim_{N\to\infty} \frac{1}{N} \sum_{n=1}^{N} a_n\right\}, \tag{2.105}$$

to transform the product into a sum. Thus

$$e^{-2W} = \exp\{-\sum_{\mathbf{q}} \tfrac{1}{2}|\mathbf{K}\cdot\mathbf{U}_{\mathbf{q}}|^2\}, \tag{2.106}$$

i.e.

$$W = \tfrac{1}{2}\sum_{\mathbf{q}} |\mathbf{K}\cdot\mathbf{U}_{\mathbf{q}}|^2. \tag{2.107}$$

To evaluate this sum, we need to know the amplitude, $\mathbf{U}_{\mathbf{q}}$, of the $\mathbf{q}$th lattice mode. This will be a function of the temperature. We know that the average energy in this mode is given by (2.47);

$$\bar{\mathscr{E}}_{\mathbf{q}} = (\bar{n}_q + \tfrac{1}{2})\hbar\nu_{\mathbf{q}},$$

where $\bar{n}_{\mathbf{q}}$ is the average number of phonons in the mode, as given by the Bose–Einstein formula (2.46).

Classically, we can calculate the energy of each simple-harmonic oscillator mode as the sum of its kinetic and potential energies. These are known to be equal—so we have, from (2.1) and (2.8),

$$\begin{aligned}\bar{\mathscr{E}} &= \sum_{sl} M_s |\dot{\mathbf{u}}_{sl}|^2 \\ &= \sum_{s\mathbf{q}} NM_s |\dot{\mathbf{U}}_{s\mathbf{q}}|^2 \\ &= \sum_{s\mathbf{q}} NM_s \nu_{\mathbf{q}}^2 |\mathbf{U}_{s\mathbf{q}}|^2.\end{aligned} \tag{2.108}$$

If there is only one atom per unit cell, of mass M, then

$$\begin{aligned}|\mathbf{U}_{\mathbf{q}}|^2 &= \bar{\mathscr{E}}_{\mathbf{q}}/NM\nu_{\mathbf{q}}^2 \\ &= (\bar{n}_{\mathbf{q}} + \tfrac{1}{2})\hbar/NM\nu_{\mathbf{q}}.\end{aligned} \tag{2.109}$$

We know the polarization of $\mathbf{U}_{\mathbf{q}}$ for each branch of the lattice spectrum, so that we can, in principle, calculate the Debye–Waller factor exactly.

To see how it should behave let us assume a Debye model with all three modes having the same velocity. For any one polarization we should find, on the average

$$|\mathbf{K}\cdot\mathbf{U}_{\mathbf{q}}|^2 = \tfrac{1}{3}K^2|\mathbf{U}_{\mathbf{q}}|^2, \tag{2.110}$$

but with three different polarizations the factor $\frac{1}{3}$ is removed.†

† There is a useful rule: in this model, the three modes are degenerate at each value of $\mathbf{q}$. Therefore one can choose the polarization vectors at will, provided they are orthogonal. One therefore chooses one mode to have $\mathbf{U}_{\mathbf{q}}$ parallel to $\mathbf{K}$, the other two normal to $\mathbf{K}$. This yields the result we need.

From (2.46), (2.56) and (2.109) we have

$$\begin{aligned} W &= \tfrac{1}{2}K^2 \frac{\hbar}{NM}\frac{1}{8\pi^3}\iiint \frac{\bar{n}_{\mathbf{q}}+\frac{1}{2}}{\nu_{\mathbf{q}}}\, d^3q \\ &= \frac{1}{2}\frac{\hbar^2 K^2}{M}\int_0^{\nu_D}\left\{\frac{1}{e^{\hbar\nu/kT}-1}+\frac{1}{2}\right\}\frac{3\nu^2}{\hbar\nu\nu_D^3}\, d\nu \\ &= \frac{3}{2}\frac{\hbar^2 K^2 T^2}{Mk\Theta^3}\int_0^{\Theta/T}\left\{\frac{1}{e^z-1}+\frac{1}{2}\right\} z\, dz, \end{aligned} \tag{2.111}$$

as in (2.57).

At high temperatures the upper limit of the integral is small, and the exponential factor in the integrand can be expanded in powers of z. The result is

$$W \to \frac{3}{2}\frac{\hbar^2 K^2 T}{Mk\Theta^2}. \tag{2.112}$$

Thus, the X-ray diffraction pattern, which is proportional to the square of the matrix element $\mathscr{M}_{\mathbf{k}'\mathbf{k}}$, is reduced in intensity by a factor

$$e^{-2W} \sim \exp(-3\hbar^2 K^2 T/Mk\Theta^2). \tag{2.113}$$

This factor depends quite strongly on the temperature, and also on the magnitude of the scattering vector. This result would be obtained if we used classical statistics for the average energy of each mode, i.e. $\bar{\mathscr{E}}_q = kT$.

At temperatures below the Debye temperature the formula is more complicated, but we note that W will tend to a constant at the lowest temperatures. This is due to the term $\frac{1}{2}$ in the integral—a term arising from the *zero-point motion* of the lattice. This is not a negligible effect:

$$W \to \frac{3}{8}\frac{\hbar^2 K^2}{Mk\Theta} \quad \text{as} \quad T \to 0, \tag{2.114}$$

which is $\frac{1}{4}$ of the value of W at $T \sim \Theta$. Zero-point *energy* may be physically irrelevant, but the *motion* associated with it can be observed directly.

Very similar arguments explain the *Mössbauer effect*. The spectrum of nuclear γ-emission may contain a very narrow line which does not appear to have been broadened by the thermal motion of the atom about its equilibrium site in the crystal. This is really a simple classical phenomenon. Radiation of frequency ω emitted from a source moving with uniform velocity v is Doppler shifted by an amount $\omega v/c$. But suppose this source is carried by a bound atom, oscillating back and forth with frequency $\nu_{\mathbf{q}}$ and amplitude $\mathbf{U}_{\mathbf{q}}$. Since the velocity now varies with time, we observe frequency modulation of the emitted radiation. As in the elementary theory of FM broadcasting, the spectrum now splits into lines at frequencies ω, $\omega \pm \nu$, $\omega \pm 2\nu$ etc. For small amplitudes of vibration, the fraction of the spectrum in the unmodified line is approximately $1 - \frac{1}{2}(\mathbf{k}\cdot\mathbf{U}_\mathbf{q})^2$, where $\mathbf{k}$ is the wave-vector of the γ-photon. Combining the contributions of all possible modes of vibration, as in (2.104–7), we find that the relative strength of this line is a finite quantity, given by a formula just like the Debye–Waller factor.

In the standard language of solid state theory, we may say that the impulse given to the atom by the emission of the γ-ray is partly taken up by the creation of phonons, whose energy is accounted for by broadening of the line. But a finite proportion of the momentum is transferred to the crystal as a whole, without creation (or absorption) of phonons, and hence with infinitesimal loss (or gain) of energy. As in our analysis of X-ray diffraction, the Debye–Waller factor merely measures the proportion of such elastic or *recoilless* events. This extremely sharp line is, of course, immensely useful as a probe for magnetic fields in molecules and solids, in tests of relativity theory, etc.

The derivation of (2.111) provides us, incidentally, with another interesting result. It is clear from (2.1) and (2.8) that the mean square amplitude of the vibration of each atom about its lattice site is given by

$$\begin{aligned}\frac{1}{N}\sum_l |\mathbf{u}_l|^2 &= \sum_\mathbf{q} |\mathbf{U}_\mathbf{q}|^2 \\ &\approx \frac{9\hbar^2 T}{Mk\Theta^2}\end{aligned} \tag{2.115}$$

above the Debye temperature. At the temperature T the root-mean-square displacement of each atom from its equilibrium site will thus be a fraction x of, say, the mean radius r_s of a unit cell, where

$$x = \sqrt{\frac{9\hbar^2 T}{Mk\Theta^2 r_s^2}}. \tag{2.116}$$

The *Lindemann melting formula* is based upon this idea. It is suggested that a solid must melt when x attains some standard value, x_m. Thus, the melting temperature T_m is related to the other atomic constants of the solid by

$$T_m = \frac{x_m^2}{9\hbar^2} Mk\Theta^2 r_s^2. \tag{2.117}$$

It seems that x_m is in the range 0·2–0·25 in most solids.

This rule may be used to estimate the Debye temperature approximately from knowledge of T_m. It also provides a convenient short cut for estimating values of quantities, like W, which depend on the amplitude of the atomic vibrations. For example, we can show that (2.112) can be written

$$W \approx x_m^2 \frac{T}{T_m} \frac{K^2}{q_D^2}, \tag{2.118}$$

where q_D is the Debye wave-number (2.53).

2.10 Anharmonicity and thermal expansion

At the beginning of this chapter, we expanded the potential energy of the crystal as a Taylor series in the lattice displacements. But this series, (2.2), was curtailed at the second-order term. There will be further terms, such as

$$\mathscr{V}^{(3)} = \frac{1}{3!} \sum_{\substack{ss's'' \\ ll'l'' \\ jj'j''}} u_{sl}^{j} u_{s'l'}^{j'} u_{s''l''}^{j''} \left[\frac{\partial^3 \mathscr{V}}{\partial u_{sl}^{j} \partial u_{s'l'}^{j'} \partial u_{s''l''}^{j''}} \right]_0, \tag{2.119}$$

and so on. The actual calculation of the coefficients is a very complicated problem since they involve geometrical factors as well as third derivatives of the interatomic potentials.

Several important physical phenomena are associated with the *anharmonic terms*. Of these the most familiar is *thermal expansion*. It is not easy to derive this directly from expressions like (2.119), but the general physical idea is easy enough. As the temperature rises the amplitude of the lattice vibrations increases, so that the average R.M.S. values of the displacements $\mathbf{u}_{sl}$, etc., increase. The anharmonic terms contribute to the free energy of the crystal, which is now no longer necessarily a minimum for vibrations around the assumed 'equilibrium' configuration in which each $\mathbf{u}_{sl}$ vanishes. The whole crystal then expands (or contracts) until it finds the volume where the total free energy is a minimum.

In default of adequate information about the anharmonic terms for a 'first principle' calculation, we may assume phenomenologically that the frequency of the lattice modes is a function of volume. For simplicity suppose that a change of volume ΔV gives rise to the same relative change of frequency of every mode;

$$\frac{\Delta \nu}{\nu} = -\gamma \frac{\Delta V}{V}. \tag{2.120}$$

The total free energy of the crystal, as a function of volume, may then be written

$$F = \frac{1}{2}\frac{1}{\kappa}\left(\frac{\Delta V}{V}\right)^2 + kT\sum_{\mathbf{q}} \ln\left(2\sinh\frac{\hbar\nu_{\mathbf{q}}}{2kT}\right). \tag{2.121}$$

The first term is the potential energy associated with the compressibility κ of the solid as an elastic continuum. The second term is the sum of the free energies in the lattice modes, as given by the conventional statistical mechanics of Bose–Einstein oscillators.

Differentiating with respect to volume, and using (2.120), we find the condition for minimum free energy,

$$\begin{aligned}\frac{1}{\kappa}\left(\frac{\Delta V}{V}\right) &= \sum_{\mathbf{q}} \gamma\hbar\nu_{\mathbf{q}}\tfrac{1}{2}\coth\frac{\hbar\nu_{\mathbf{q}}}{2kT} \\ &= \gamma\bar{\mathscr{E}}(T),\end{aligned} \tag{2.122}$$

where $\bar{\mathscr{E}}$ is the energy in the lattice modes at temperature, T, as in (2.48). We thus arrive at the *Grüneisen formula*: the dilatation at temperature T is proportional to the mean thermal energy density, i.e.

$$\frac{\Delta V}{V} = \kappa\gamma\bar{\mathscr{E}}(T). \tag{2.123}$$

The *thermal expansion coefficient*, being the derivative of the dilatation with temperature, is proportional to the specific heat, C_V.

This formula may be compared with experiment to yield the *Grüneisen constant*, γ, which is usually about 2. It is a convenient dimensionless parameter for the effects of anharmonicity. In the Debye model, we can write (2.120) in the form

$$\gamma = -\frac{\partial \ln \Theta}{\partial \ln V} \tag{2.124}$$

showing the effect of volume on the Debye temperature. In truth, this is much too simplified as a model of thermal expansion. Dilatation affects different modes in different ways; the value of γ for longitudinal modes is usually much larger than for transverse modes, so these must be counted as separate contributions in (2.122)

Another effect of anharmonicity is on the elastic constants, which vary with volume and with temperature. These are complicated phenomena, for which there is no elementary theory; they depend on a number of different parameters.

2.11 Phonon–phonon interaction

From a formal point of view, the most significant effect of the anharmonic terms is that it destroys the dynamical independence of the different lattice modes. Suppose, for example, that we make the substitutions (2.8) and (2.10) in (2.119). For a Bravais lattice we may write

$$\mathscr{V}^{(3)} = \sum_{ll'l''} \mathbf{A}_{ll'l''} : \mathbf{u}_l \mathbf{u}_{l'} \mathbf{u}_{l''}$$

$$= \sum_{ll'l''} \sum_{\mathbf{q}\mathbf{q}'\mathbf{q}''} \mathbf{A}_{ll'l''} : \mathbf{U}_{\mathbf{q}} \mathbf{U}_{\mathbf{q}'} \mathbf{U}_{\mathbf{q}''} \exp\{i(\mathbf{q}\cdot\boldsymbol{l} + \mathbf{q}'\cdot\boldsymbol{l}' + \mathbf{q}''\cdot\boldsymbol{l}'')\}, \quad (2.125)$$

where we use a very condensed notation for the tensor of the third-order coefficients, and for its scalar product with each of the lattice displacement vectors.

This term would have to be counted in the equations of the motion. We may illustrate the nature of such terms by applying, as in (2.6), the principle of translational invariance. The coefficients of the tensor in (2.125) must depend only on the relative positions of sites $\boldsymbol{l}$, $\boldsymbol{l}'$ and $\boldsymbol{l}''$. Thus

$$\mathbf{A}_{ll'l''} = \mathbf{A}(\mathbf{h}', \mathbf{h}'') \quad \text{where} \quad \mathbf{h}' = \boldsymbol{l}' - \boldsymbol{l},\ \mathbf{h}'' = \boldsymbol{l}'' - \boldsymbol{l}: \quad (2.126)$$

substituting in (2.125), we get

$$\mathscr{V}^{(3)} = \sum_{l\mathbf{h}'\mathbf{h}''} \sum_{\mathbf{q}\mathbf{q}'\mathbf{q}''} \mathbf{A}(\mathbf{h}', \mathbf{h}'') : \mathbf{U}_{\mathbf{q}} \mathbf{U}_{\mathbf{q}'} \mathbf{U}_{\mathbf{q}''} e^{i(\mathbf{q}+\mathbf{q}'+\mathbf{q}'')\cdot \boldsymbol{l}} e^{i\mathbf{q}'\cdot\mathbf{h}'} e^{i\mathbf{q}''\cdot\mathbf{h}''}$$

$$= \sum_{\mathbf{q}\mathbf{q}'\mathbf{q}''} \Big\{\sum_{\mathbf{h}'\mathbf{h}''} \mathbf{A}(\mathbf{h}', \mathbf{h}'') e^{i\mathbf{q}'\cdot\mathbf{h}'} e^{i\mathbf{q}''\cdot\mathbf{h}''}\Big\} : \mathbf{U}_{\mathbf{q}} \mathbf{U}_{\mathbf{q}'} \mathbf{U}_{\mathbf{q}''} \sum_{l} e^{i(\mathbf{q}+\mathbf{q}'+\mathbf{q}'')\cdot\boldsymbol{l}}$$

$$= \sum_{\mathbf{q}\mathbf{q}'\mathbf{q}''} N\delta_{\mathbf{q}+\mathbf{q}'+\mathbf{q}'',\mathbf{g}} \mathbf{A}(\mathbf{q}', \mathbf{q}'') : \mathbf{U}_{\mathbf{q}} \mathbf{U}_{\mathbf{q}'} \mathbf{U}_{\mathbf{q}''}. \quad (2.127)$$

We see that the tensor of the coupling coefficient between the modes is a Fourier transform of the tensor of the anharmonic terms. More particularly, we notice that the sum over $\boldsymbol{l}$ introduces a factor of the form (2.92), which vanishes unless

$$\mathbf{q} + \mathbf{q}' + \mathbf{q}'' = \mathbf{g}, \quad (2.128)$$

where, as usual, $\mathbf{g}$ is a reciprocal lattice vector or zero. Thus, there is a condition on the crystal momenta of the processes that are coupled.

We can interpret this as follows. Suppose that we treat the anharmonic terms as a perturbation on the motion of the lattice modes, governed by the equations (2.11). These can be reduced, by a canonical transformation, to the equations of motion of a set of independent simple harmonic oscillators. It is well known that such equations are represented quantum mechanically by a Hamiltonian that can again be transformed into an expression involving the creation and annihilation operators of quanta in each mode.

In fact, we find we can write

$$\mathbf{U}_{\mathbf{q}} = -i\left(\frac{\hbar}{2NM\nu_{\mathbf{q}}}\right)^{\frac{1}{2}} \mathbf{e}_{\mathbf{q}}(a_{\mathbf{q}} - a^*_{-\mathbf{q}}), \tag{2.129}$$

where $\mathbf{e}_{\mathbf{q}}$ is a unit vector in the direction of polarization of the mode of wave-vector $\mathbf{q}$ in some branch of the spectrum, and $a_{\mathbf{q}}, a^*_{\mathbf{q}}$ are annihilation and creation operators for a phonon in that mode. The polarization vector is determined, of course, by the solution of the eigenvalue equation (2.13) for the frequency $\nu_{\mathbf{q}}$.

In terms of these operators, the Hamiltonian derived from the harmonic terms is reduced to the very simple form

$$\mathscr{H}_0 = \sum_{\text{polarizations}} \sum_{\mathbf{q}} \hbar\nu_{\mathbf{q}}(a_{\mathbf{q}}\, a^*_{\mathbf{q}} + a^*_{\mathbf{q}}\, a_{\mathbf{q}}), \tag{2.130}$$

whose eigenstates are the phonon states, $|n_{\mathbf{q}}\rangle$ say, with $n_{\mathbf{q}}$ quanta in the $\mathbf{q}$th mode. Substituting from (2.129) into the anharmonic term (2.127), we get a very complicated expression, containing products of the operators $a_{\mathbf{q}}$, $a_{\mathbf{q}'}$, etc. For example, there is a term containing (after shuffling some signs)

$$\delta_{\mathbf{q}-\mathbf{q}'-\mathbf{q}'',\mathbf{g}}\, a^*_{\mathbf{q}}\, a_{\mathbf{q}'}\, a_{\mathbf{q}''}. \tag{2.131}$$

This term, when applied as a perturbation to the eigenstates of $\mathscr{H}_0$, gives rise to transitions in which one phonon is created in the state $\mathbf{q}$, and one is destroyed in each of the states $\mathbf{q}'$ and $\mathbf{q}''$. The selection rule for this is

$$\mathbf{q}' + \mathbf{q}'' = \mathbf{q} - \mathbf{g}. \tag{2.132}$$

This is akin to (2.100). If $\mathbf{g}$ were zero we should say that crystal momentum had been conserved: the momenta $\hbar\mathbf{q}'$ and $\hbar\mathbf{q}''$ have been destroyed to make the momentum $\hbar\mathbf{q}$. When $\mathbf{g}$ is not zero, crystal momentum is no longer conserved—but it can only change by $\hbar\mathbf{g}$. Thus, the *phonon–phonon interaction* can be described by N-processes and by U-processes, just as in the scattering of an external beam of X-rays. There are obviously other processes in which one phonon is destroyed and two are created—the conjugate of (2.130).

One can show that energy must be conserved for real processes; in the case (2.132) we must have

$$\hbar\nu_{\mathbf{q}'} + \hbar\nu_{\mathbf{q}''} = \hbar\nu_{\mathbf{q}}. \tag{2.133}$$

The argument follows from the time-dependence of the lattice modes.

It is obvious that the actual coefficients for the phonon–phonon interaction are very complicated, being derived from the anharmonic coefficients and mixed in with polarization vectors, etc. The rules for the real transitions between the modes of different polarization are also complex, since they depend on the possibility of simultaneously satisfying both energy and crystal-momentum conditions. It is also clear that there can be higher-order phonon–phonon interactions, where, say, n phonons are destroyed and n' created. These would arise from terms of order $n+n'$ in the Taylor expansion of the potential energy of the lattice or from higher-order terms in the perturbation series. They must satisfy a selection rule such as (2.128)—the total change of crystal momentum must be $\hbar\mathbf{g}$, where $\mathbf{g}$ is zero or a reciprocal lattice vector. Phonon–phonon interaction processes are of importance in the theory of lattice conduction. We shall return to this point in § 7.10.

This approach to lattice dynamics needs modification for *quantum crystals*—especially solid helium—where the zero-point motion of the atoms is comparable with the lattice spacing. The time-dependent perturbation series, defined by a succession of coefficients of higher and higher order, describing successively more complex phonon interaction processes, no longer converges absolutely: we may even find *negative* values for the second-order harmonic force constants (2.4) calculated as derivatives of the potential energy at the equilibrium lattice spacing. To patch up the algebra we must apply one of the standard techniques of *many-body theory*. We may, for example, follow the path of *diagrams* or *cluster expansions*, and *renormalize* all the interaction coefficients after summing various infinite sets of terms in the perturbation series. Alternatively, with more immediate physical intuition, we set up trial wave-functions for the atoms, in the spirit of the *Hartree method*, and look for variational conditions on the parameters. The results are much the same in either case. The system behaves quite normally, with a phonon spectrum, phonon–phonon interactions, etc., but the effective force constants and higher coefficients must be determined *self-consistently*. Thus, in place of (2.4) we should write something like

$$\mathbf{G}_{ll'} = \left\langle \left| \frac{\partial^2 \mathscr{V}}{\partial \mathbf{u}_l \partial \mathbf{u}_{l'}} \right| \right\rangle, \tag{2.134}$$

where the derivatives of the potential energy with respect to atomic displacement are evaluated as *expectation values* in a phonon state $|\ \rangle$ consistent with these values of the force constants.

2.12 Vibrations of Imperfect Lattices

A real crystal can never be perfect, or infinite in extent; we must give thought to the effects of imperfections and surfaces on the lattice vibrations. This subject is open-ended leading to such intractable problems as the dynamics of completely disordered systems such as liquids and glasses. Nevertheless, if we confine ourselves to isolated defects—for example, substitutional atoms of different mass—we can now understand a number of interesting physical phenomena.

The essential steps in the argument are usually clouded in a mass of trivial detail. To reduce the apparent complexity of the algebra, let us use an abstract matrix notation, in which, for example, the general equations for the vibrations of a *perfect* (Bravais) lattice take the form

$$\mathsf{L}(\nu^2)\, u = 0 \tag{2.135}$$

representing say

$$M\nu^2\mathbf{u}_l - \sum_{l'} \mathbf{G}_{ll'} \cdot \mathbf{u}_{l'} = 0. \tag{2.136}$$

In these equations which are essentially the same as (2.5) the column matrix u stands for the set of displacements $\mathbf{u}_l$ in a mode of frequency ν, under the action of interatomic forces $\mathbf{G}_{ll'}$. The *dynamical matrix*, L, has the translational symmetry of the lattice. By the argument of § 2.1, there is a root of (2.135) at each phonon frequency $\nu = \nu_\mathbf{q}$, for we may then find a (normalized) polarization vector $\mathbf{U}_\mathbf{q}$ for the corresponding phonon state

$$\mathbf{u}_l = \mathbf{U}_\mathbf{q} e^{i\mathbf{q}\cdot\boldsymbol{l}} \tag{2.137}$$

by solving the eigenvector equation (2.13).

Now suppose that we have made a change δM in the mass of the atom at $\boldsymbol{l} = 0$ (an arbitrary site, really) and/or some change $\boldsymbol{\delta}\mathbf{G}_{ll'}$ in the interactions of this atom with its neighbours in the crystal. Our dynamical equations (2.136) would now read

$$M\nu^2\mathbf{u}_l - \sum_{l'} \mathbf{G}_{ll'} \cdot \mathbf{u}_{l'} + \delta M \nu^2 \mathbf{u}_0 + \sum_{l'} \boldsymbol{\delta}\mathbf{G}_{ll'} \cdot \mathbf{u}_{l'} = 0, \tag{2.138}$$

which we symbolize as

$$\mathbf{L}u + \delta\mathbf{L}u = 0. \tag{2.139}$$

The changes in the physical system can thus be represented as a perturbation of the dynamical matrix of the originally perfect crystal.

In the special case of a linear chain (cf. (2.15)) these equations can be solved more or less by inspection. But the *Green function method* is more generally applicable, and is scarcely more complicated. The trick is to rewrite (2.138) in the algebraically equivalent form

$$(1 + \mathbf{R}\,\delta\mathbf{L})\,u = 0, \tag{2.140}$$

where the *resolvent* or *Green function*, **R**, is the matrix inverse of the dynamical matrix **L**. If **R** exists and can be calculated, we are on the way to solving the linear equations symbolized by (2.140).

The resolvent must be a function of ν^2, satisfying the abstract formula

$$\mathbf{R}(\nu^2)\,\mathbf{L}(\nu^2) = \mathbf{1}. \tag{2.141}$$

It is quite easy to verify that the eigenvectors (2.137) provide a complete solution to this problem: using the standard orthogonality properties of the solutions of (2.13) etc., we find that the matrix $\mathbf{R}(\nu^2)$ may be expressed in the form

$$\mathbf{R}_{ll'}(\nu^2) = \frac{1}{NM}\sum_{\mathbf{q}} \frac{\mathbf{U}_{\mathbf{q}}^*\mathbf{U}_{\mathbf{q}}}{\nu^2 - \nu_{\mathbf{q}}^2} e^{i\mathbf{q}\cdot(\mathbf{l}-\mathbf{l}')}. \tag{2.142}$$

Here again, for simplicity, we suppose that the sum over all wave-vectors **q** is extended to include the various modes of phonon polarization. This formula also uses a dyadic convention to take account of the cartesian tensor character of the dynamical matrix, but this again is a relatively trivial generalization of the sort of formula we should have written down for, say, a one-dimensional chain.

In principle, we could evaluate (2.142) for all values of $(\mathbf{l}-\mathbf{l}')$. In practice, this is unnecessary, because we assume the perturbation $\delta\mathbf{L}$ to be highly localized in the neighbourhood of the imperfect site. Consider, for example, an isotopic impurity, where there is only a change in mass in (2.138). Substituting from (2.142) into (2.140) we get an equation involving only $\mathbf{u}_0$, the vector displacement on this site. This contains only the sum for $\mathbf{R}_{00}(\nu^2)$, i.e.

$$\left\{\mathbf{1} + \frac{\delta M}{M}\frac{\nu^2}{N}\sum_{\mathbf{q}}\frac{\mathbf{U}_{\mathbf{q}}^*\mathbf{U}_{\mathbf{q}}}{\nu^2 - \nu_{\mathbf{q}}^2}\right\}\cdot\mathbf{u}_0 = 0. \tag{2.143}$$

The vanishing of the 3×3 determinant of the cartesian tensor { }

imposes a condition on ν^2, thus giving the vibration frequency of a normal mode of the imperfect lattice. More complicated cases, with changes in the force constants at several sites, will obviously yield similar equations, involving determinants of higher order.

Unfortunately, these equations are difficult to solve: even the elementary case of isotopic substitution is arithmetically elaborate. But to understand what happens, let us ignore geometrical features such as the vector character of $\mathbf{u}_0$ and treat (2.143) as a scalar equation. The normal mode frequencies are roots of an equation of the form

$$1+\left(\frac{\delta M}{M}\right)\frac{1}{N}\sum_{\mathbf{q}}\frac{\nu^2}{\nu^2-\nu_{\mathbf{q}}^2}=0. \qquad (2.144)$$

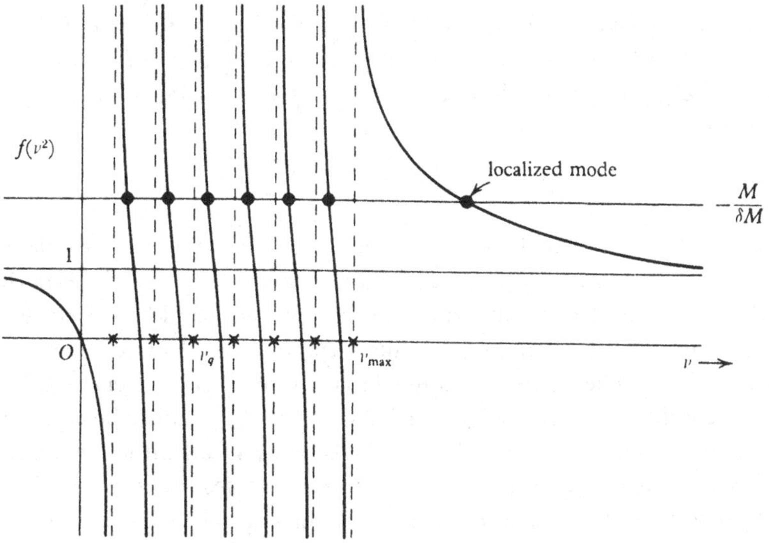

Fig. 34. Graphical solution of (2.143) for modes of lattice with isotopic impurity.

These can be discovered graphically (Fig. 34). We look for points where the function

$$f(\nu^2)\equiv\frac{1}{N}\sum_{\mathbf{q}}\frac{\nu^2}{\nu^2-\nu_{\mathbf{q}}^2} \qquad (2.145)$$

intersects the horizontal line at $(-M/\delta M)$.

Suppose, for example, that δM is negative. Each root of (2.144) must lie above a pole $\nu_{\mathbf{q}}^2$ of $f(\nu^2)$; the normal modes of the perturbed system are interleaved in frequency between those of the perfect crystal. Since the values of $\nu_{\mathbf{q}}$ form a dense *band*, whose spacing tends to zero like $1/N$, most of the new solutions are indistinguishable from the old. But looking again at Fig. 34, we notice that the highest root is not constrained, but can move away from the top of the band, ν_{max}, by a finite amount. For ν greater than ν_{max}, the sum (2.145) may be approximately represented by an integral over the spectral density (2.65), i.e.

$$\bar{f}(\nu^2) = \int^{\nu_{max}} \frac{\nu^2 \mathscr{D}(\nu')}{\nu^2 - \nu'^2} d\nu'. \tag{2.146}$$

A special mode, associated with the imperfection, occurs for a frequency where this function equals $(-M/\delta M)$.

This special mode must be *localized*. This follows from a general theorem; *the wave-number of an oscillatory motion of a lattice at a frequency outside the spectrum of normal modes is imaginary.* Take the case of the linear chain model of § 2.2. If we put $q = i\kappa$ into (2.16), we solve (2.15), with a frequency given by the analogue of (2.18), i.e.

$$\nu_\kappa = \sqrt{\left(\frac{\alpha}{M}\right)} \, 2 \sinh\left(\frac{\kappa a}{2}\right), \tag{2.147}$$

which can be satisfied for any value of ν_κ. The analytical solution is not, of course, physically permitted as a normal mode of the perfect lattice because the displacement amplitude, u_l, would grow or decay exponentially along the chain, like $\exp(\pm \kappa l)$, and hence would not satisfy the cyclic boundary conditions. In an imperfect crystal, however, conditions in the neighbourhood of the defect may be contrived so that the amplitude of vibration dies away in all directions, and is therefore negligible at the distant boundaries of the specimen. Physically speaking the impurity atom in Fig. 34, being lighter than the surrounding atoms in the crystal, would have a higher natural frequency for independent 'Einstein' oscillations. This is observed as a normal mode, above the band of extended phonon modes, in which only a few neighbouring atoms move in sympathy with the impurity (Fig. 35 (*a*)).

Various other examples of localized vibrations at the site of an imperfection can be understood similarly. If, for example, we stiffen the spring constants about some atom, we shall observe the same effect. On the other hand, a heavier atom than usual has a lower Einstein frequency, so that we ought to see a localized mode detaching itself from *below* a normal band—for example, out of the optical modes of a diatomic crystal.

Strangely enough, the theory of *surface modes* follows the same lines. We can, of course, take over from classical elasticity theory the standard discussion of the *Rayleigh waves* and *Love waves* at the free surface of a continuous medium, perhaps imposing a Debye limit on the magnitude of the wave-vector in the plane of the surface. But these waves correspond to modes of vibration localized at the surface and decaying exponentially into the interior of the crystal. Although the corresponding problem in a discrete lattice looks very complicated,

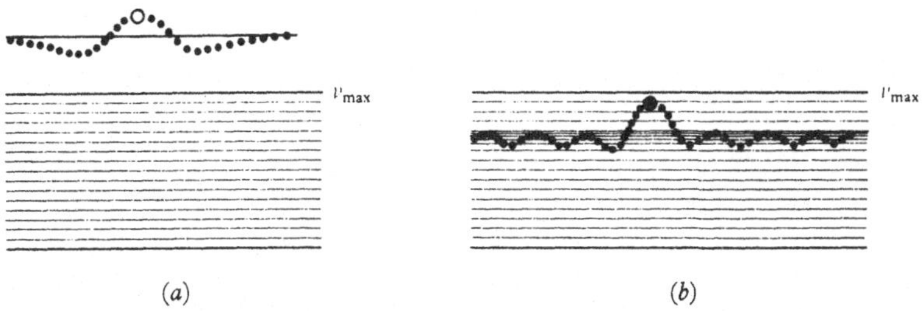

Fig. 35. (*a*) Localized mode of light impurity above phonon band. (*b*) Resonance mode of heavy impurity within phonon band.

much can be learnt by treating the system as a linear array of lattice planes, extending in one dimension away from a gross imperfection—e.g. a spring constant of zero stiffness—at the surface plane. Such modes should thus occur in a diatomic crystal, in the gap between the optical and acoustic branches of the phonon spectrum. Surface modes may make a detectable contribution to the specific heat of small crystalline particles, and are important in some electronic devices.

The characteristic frequency of a localized impurity mode is directly observable by optical absorption in the infra-red (see § 8.4). It is interesting to note, moreover, that a somewhat broadened absorption line may also be produced by an impurity with a natural vibration frequency *inside* a phonon band—e.g. with δM positive in the simple isotope case of (2.143). This is an example of a *resonance*; at the peak frequency, the amplitude of vibration is much larger at the impurity site than in the rest of the crystal, but does not die away exponentially with distance and is therefore not strictly localized (Fig. 35 (*b*)). Not surprisingly, the impurity also scatters incident phonons very strongly at this frequency.

The theory of resonances is, in fact, merely an analytical continuation of the theory of localized states. The resonance frequency is the solution of (2.144) with the sum $f(\nu^2)$ replaced by the principal part of the integral (2.146) for $\bar{f}(\nu^2)$, whilst the *width* of the resonance is given by the imaginary part of the same integral, evaluated now for $\nu < \nu_{max}$. The proof of such relations is usually given in the context of solutions of the Schrödinger equation for a quantum mechanical particle in a spherically symmetrical potential, but the argument applies generally to waves in continuous media and regular lattices.

CHAPTER 3

ELECTRON STATES

There are nine and sixty ways of constructing tribal lays,
And-every-single-one-of-them-is-right. KIPLING

3.1 Free electrons

Consider solid Na. Each atom contains 11 electrons. But 10 of these are in states which are tightly bound to the nucleus to form an ion of net positive charge $+|e|$. In the free atom the final electron moves in an orbital around this ion. When the atoms are brought together into a solid, the orbitals overlap and interact. It is argued that the overlap is so extensive that the quantization scheme in which each electron is localized on its own atom must break down. It must be replaced by a scheme in which the wave-function of each electron is a solution of the Schrödinger equation for motion in the potential of all the ions. Thus, we distinguish at the outset between the *core electrons*, which are treated as almost completely localized, and the *valence electrons* or *conduction electrons*, which are assumed to go into *Bloch states*—states extended throughout the whole crystal.

In this chapter we shall use the *one-electron model*, where we ignore any interactions between the valence electrons due to their Coulomb repulsion. These effects, of *correlation* and *exchange*, are discussed in Chapter 5. Such effects in the core, or between core and valence electrons, are usually treated as part of the potential $\mathscr{V}(\mathbf{r})$ seen by each valence electron as it moves through the crystal. Thus, $\mathscr{V}(\mathbf{r})$ is supposed to be calculated as a Hartree or Hartree–Fock potential of the ion.

An immense simplification can be made if it is assumed that $\mathscr{V}(\mathbf{r})$ is a constant. The electrons then fall automatically into the plane-wave states

$$|\mathbf{k}\rangle = \psi_{\mathbf{k}} = e^{i\mathbf{k}\cdot\mathbf{r}}, \tag{3.1}$$

with energy

$$\mathscr{E}(\mathbf{k}) = \frac{\hbar^2 k^2}{2m}. \tag{3.2}$$

As we saw in § 1.5, these states satisfy the Bloch theorem. The surfaces of constant energy are spheres in $\mathbf{k}$-space. The allowed values of $\mathbf{k}$ are distributed with density $V/8\pi^3$ in this space. But for each value of $\mathbf{k}$

there are actually two electron states, of opposite spin. If there are ZN electrons per unit volume in real space (i.e. Z electrons per atom, 1 atom per unit cell, N unit cells per unit volume), then to satisfy the Pauli principle we need to fill the states up to a wave-number k_F where

$$\tfrac{4}{3}\pi k_F^3 \frac{2}{8\pi^3} = ZN,$$

i.e.

$$k_F = (3\pi^2 ZN)^{\frac{1}{3}}. \tag{3.3}$$

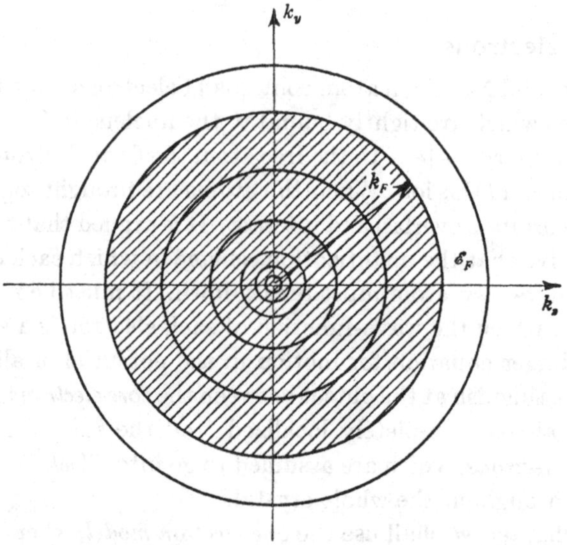

Fig. 36. The Fermi surface for free electrons.

We say that the *Fermi surface* is a sphere, of radius k_F. The *Fermi energy* or *Fermi level* is the energy corresponding to the top of this distribution,

$$\mathscr{E}_F = \frac{\hbar^2 k_F^2}{2m}. \tag{3.4}$$

We notice that the radius of the Fermi sphere is comparable with the radius of the Debye sphere. Comparing (3.3) with (2.53) we have

$$k_F = \left(\frac{Z}{2}\right)^{\frac{1}{3}} q_D. \tag{3.5}$$

This means that the wavelength of an electron at the Fermi level is comparable with the interatomic spacing (cf. (2.55)):

$$\lambda_F = \frac{2\pi}{k_F} \sim \left(\frac{2}{Z}\right)^{\frac{1}{3}} 2{\cdot}6 r_s, \tag{3.6}$$

where r_s is the radius of the Wigner–Seitz sphere.

3.2 Diffraction of valence electrons

The free-electron model is vitiated by the result we have just proved: the electrons at the Fermi level are in states of wavelength comparable with the lattice spacing. There must be strong diffraction effects, even when $\mathscr{V}(\mathbf{r})$ does not differ very markedly from a constant potential. Let us treat this potential as a perturbation on free-electron states. Let us write

$$\mathscr{E}(\mathbf{k}) = \mathscr{E}^0_{\mathbf{k}} + \langle \mathbf{k}|\mathscr{V}|\mathbf{k}\rangle + \sum_{\mathbf{k}'}{}' \frac{|\langle \mathbf{k}|\mathscr{V}|\mathbf{k}'\rangle|^2}{\mathscr{E}^0_{\mathbf{k}} - \mathscr{E}^0_{\mathbf{k}'}}, \tag{3.7}$$

where

$$\mathscr{E}^0_{\mathbf{k}} \equiv \hbar^2 k^2/2m, \tag{3.8}$$

and where the Dirac notation is used for the matrix elements of the potential between the unperturbed states (3.1).

But $\mathscr{V}(\mathbf{r})$ must have the periodicity of the lattice. As shown in § 2.6, its matrix elements vanish unless $(\mathbf{k}-\mathbf{k}')$ is a reciprocal lattice vector. From (2.78) we have at once

$$\mathscr{E}(\mathbf{k}) = \mathscr{E}^0_{\mathbf{k}} + \mathscr{V}_0 + \sum_{\mathbf{g}\neq 0} \frac{|\mathscr{V}_{\mathbf{g}}|^2}{\mathscr{E}^0_{\mathbf{k}} - \mathscr{E}^0_{\mathbf{k}-\mathbf{g}}}, \tag{3.9}$$

where $\mathscr{V}_{\mathbf{g}}$ is the Fourier component of $\mathscr{V}$ for the reciprocal lattice vector $\mathbf{g}$ (as in (1.15)).

The question is—how good is such an expansion? There are two conditions to be satisfied: (i) the Fourier components $\mathscr{V}_{\mathbf{g}}$ should tend to zero rapidly as $\mathbf{g}$ increases, (ii) there must be no degeneracy of the form

$$\mathscr{E}^0_{\mathbf{k}} = \mathscr{E}^0_{\mathbf{k}-\mathbf{g}} \tag{3.10}$$

between the unperturbed states that are mixed by the perturbation. For the moment, set aside condition (i) for later study.

The degeneracy condition (3.10) is equivalent to writing

$$|\mathbf{k}| = |\mathbf{k}-\mathbf{g}|; \tag{3.11}$$

geometrically, this means that $\mathbf{k}$ lies on the perpendicular bisector of the reciprocal lattice vector $\mathbf{g}$. That is to say—*the simple perturbation expansion* (3.9) *is not valid when* $\mathbf{k}$ *lies on (or near) a zone boundary.* This formula can only be used when the sphere of unperturbed states does not reach a boundary of the zone—in practice, even for monovalent metals, it is seldom valid.

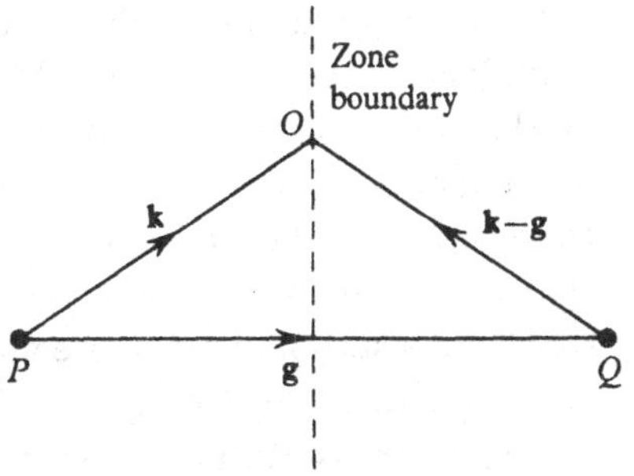

Fig. 37

Notice that Fig. 37 is simply the Ewald construction (Fig. 29 (*b*)) for coherent Bragg diffraction of an electron of wave-vector **k** into the direction $\mathbf{k}' = \mathbf{k} - \mathbf{g}$. The geometrical condition, in each case, is that *OPQ* should be an isosceles triangle with a reciprocal lattice vector **g** as the base *PQ*. We thus confirm algebraically that the valence electrons within a crystal are affected by diffraction from the lattice just as if they had been incident from the outside.

To find a solution in the neighbourhood of the zone boundary we must consider explicitly the perturbation equations. The periodicity of the lattice implies that the potential can only mix waves differing by the reciprocal lattice vector **g**, so that we are assuming that the wave-function can be expanded in the series

$$\psi_{\mathbf{k}} = \sum_{\mathbf{g}} \alpha_{\mathbf{k}-\mathbf{g}}\, e^{i(\mathbf{k}-\mathbf{g})\cdot\mathbf{r}}. \tag{3.12}$$

If we substitute this into the Schrödinger equation

$$\left\{-\frac{\hbar^2}{2m}\nabla^2 + \mathscr{V}(\mathbf{r})\right\}\psi_{\mathbf{k}} = \mathscr{E}(\mathbf{k})\,\psi_{\mathbf{k}}, \tag{3.13}$$

and multiply through by one of the terms in the expansion, we get linear equations for the coefficients α:

$$\{\mathscr{E}^0_{\mathbf{k}-\mathbf{g}} - \mathscr{E}(\mathbf{k})\}\,\alpha_{\mathbf{k}-\mathbf{g}} + \sum_{\mathbf{g}'} \mathscr{V}_{\mathbf{g}'-\mathbf{g}}\,\alpha_{\mathbf{k}-\mathbf{g}'} = 0 \tag{3.14}$$

for all values of **g** (including $\mathbf{g} = 0$ and $\mathbf{g}' = 0$). As is well known, the simple perturbation formula (3.9) is obtained from this by assuming that all $\alpha_{\mathbf{k}-\mathbf{g}}$ are small, except $\alpha_{\mathbf{k}}$. In fact, we then take

$$\alpha_{\mathbf{k}-\mathbf{g}} \approx \frac{\mathscr{V}_{-\mathbf{g}}}{\mathscr{E}^0_{\mathbf{k}} - \mathscr{E}^0_{\mathbf{k}-\mathbf{g}}}. \tag{3.15}$$

It is this assumption that breaks down when **k** lies near a zone boundary—say the boundary bisecting the vector **G**. The coefficient of the wave $\exp\{i(\mathbf{k}-\mathbf{G})\cdot\mathbf{r}\}$ becomes as important as that of the original unperturbed state. In effect there has been Bragg reflection of the electron by the lattice, just as if it were an external electron beam. The condition (3.11) is equivalent to (2.79) and (2.80) of § 2.6. To construct a stationary state we must treat both incident and diffracted beams on an equal footing.

As an approximation, let us ignore all coefficients except those two in (3.14). After shifting the origin of energy by $\mathscr{V}_0$, we get

$$\left.\begin{aligned} \{\mathscr{E}^0_{\mathbf{k}} - \mathscr{E}(\mathbf{k})\}\,\alpha_{\mathbf{k}} + \mathscr{V}_{\mathbf{G}}\,\alpha_{\mathbf{k}-\mathbf{G}} &= 0,\\ \mathscr{V}_{-\mathbf{G}}\,\alpha_{\mathbf{k}} + \{\mathscr{E}^0_{\mathbf{k}-\mathbf{G}} - \mathscr{E}(\mathbf{k})\}\,\alpha_{\mathbf{k}-\mathbf{G}} &= 0.\end{aligned}\right\} \tag{3.16}$$

The determinant of these equations is a quadratic in $\mathscr{E}(\mathbf{k})$, with solutions

$$\mathscr{E}^{\pm}(\mathbf{k}) = \tfrac{1}{2}(\mathscr{E}^0_{\mathbf{k}} + \mathscr{E}^0_{\mathbf{k}-\mathbf{G}}) \pm \tfrac{1}{2}\sqrt{\{(\mathscr{E}^0_{\mathbf{k}} - \mathscr{E}^0_{\mathbf{k}-\mathbf{G}})^2 + 4\,|\mathscr{V}_{\mathbf{G}}|^2\}} \qquad (3.17)$$

(noting that $\mathscr{V}_{-\mathbf{G}} = \mathscr{V}^*_{\mathbf{G}}$). Thus, the states $e^{i\mathbf{k}\cdot\mathbf{r}}$ and $e^{i(\mathbf{k}-\mathbf{G})\cdot\mathbf{r}}$ are combined into two other states, ψ^+ and ψ^-, with energy $\mathscr{E}^+$ and $\mathscr{E}^-$.

For simplicity, consider what happens in one dimension. Near $k = 0$ the difference in unperturbed energies is so large that we can ignore $4\,|\mathscr{V}_{\mathbf{G}}|^2$. We find

$$\mathscr{E}(\mathbf{k}) \sim \mathscr{E}^0_{\mathbf{k}}, \qquad (3.18)$$

so that ψ^- follows the free-electron parabola. At $k = \frac{1}{2}G$, which is the zone boundary, the two unperturbed waves have the same energy: thus

$$\mathscr{E}^-(\tfrac{1}{2}G) = \mathscr{E}^0_{\frac{1}{2}G} - |\mathscr{V}_G|. \qquad (3.19)$$

The state ψ^- is depressed in energy by the amount $|\mathscr{V}_G|$ below the free-electron parabola. We see this at the point A in Fig. 38.

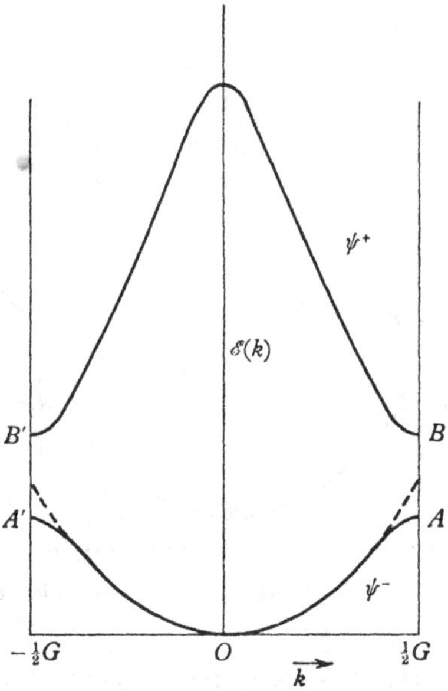

Fig. 38. Electron energy in one dimension: reduced zone scheme.

At the zone boundary the state ψ^+ has energy

$$\mathscr{E}^+(\tfrac{1}{2}G) = \mathscr{E}^0_{\frac{1}{2}G} + |\mathscr{V}_G|, \qquad (3.20)$$

so that it is raised in energy above the free-electron parabola. If we keep k within the zone, we find the energy of this state tending to

$\mathscr{E}^0_{k-G}$ as k decreases. Thus, the energy in the zone falls into two *bands* separated by the gap AB, of width $2|\mathscr{V}_G|$.

It is instructive to compare Fig. 38 with Fig. 14 of § 1.5. The rather artificial free-electron reduced zone scheme has been separated into bands by the periodic lattice potential. We could represent the result in another way, by drawing the branch ψ^+ in the second zone—in an *extended zone scheme* (Fig. 39). This shows more clearly how the energies are derived by perturbation of the free-electron parabola—but with discontinuities introduced at the zone boundaries. It is the same phenomenon as we observed in Fig. 19(*c*) of § 2.2, where the introduction of two different masses splits the vibrational spectrum of the lattice.

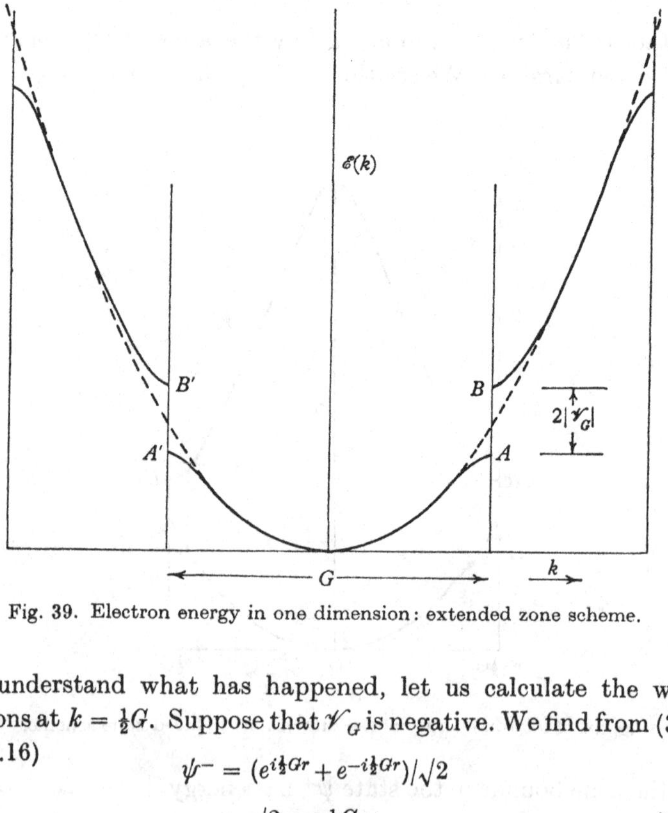

Fig. 39. Electron energy in one dimension: extended zone scheme.

To understand what has happened, let us calculate the wave-functions at $k = \frac{1}{2}G$. Suppose that $\mathscr{V}_G$ is negative. We find from (3.12) and (3.16)

$$\psi^- = (e^{i\frac{1}{2}Gr} + e^{-i\frac{1}{2}Gr})/\sqrt{2}$$

$$= \sqrt{2}\cos\tfrac{1}{2}Gr \qquad (3.21)$$

and

$$\psi^+ = \sqrt{2}i\sin\tfrac{1}{2}Gr. \qquad (3.22)$$

Thus, $|\psi^-|^2$ is a function which is large near $r = 0$, and also at each site of our linear lattice. It corresponds to a state in which the electrons are concentrated on the atoms. But this is precisely what we should expect. The *negative* value of $\mathscr{V}_G$ corresponds to a periodic potential which is negative in the neighbourhood of each atom—a potential that *attracts* electrons. Thus, ψ^- is lowered in energy by being geared to the potential wells of the atoms. Similarly ψ^+ is raised in energy, because it makes the electron density large in regions of positive potential. We say that ψ^- is *s*-like and ψ^+ is *p*-like, by analogy with atomic orbitals.

We note, in passing, that the expansion (3.12) allows $\psi_{\mathbf{k}}$ to satisfy the Bloch theorem (1.58). We can write

$$\begin{aligned}\psi_{\mathbf{k}} &= e^{i\mathbf{k}\cdot\mathbf{r}} \sum_{\mathbf{g}} \alpha_{\mathbf{k}-\mathbf{g}}\, e^{-i\mathbf{g}\cdot\mathbf{r}} \\ &= e^{i\mathbf{k}\cdot\mathbf{r}}\, u_{\mathbf{k}}(\mathbf{r}),\end{aligned} \tag{3.23}$$

where $u_{\mathbf{k}}(\mathbf{r})$ is a periodic function in the lattice as in (1.60).

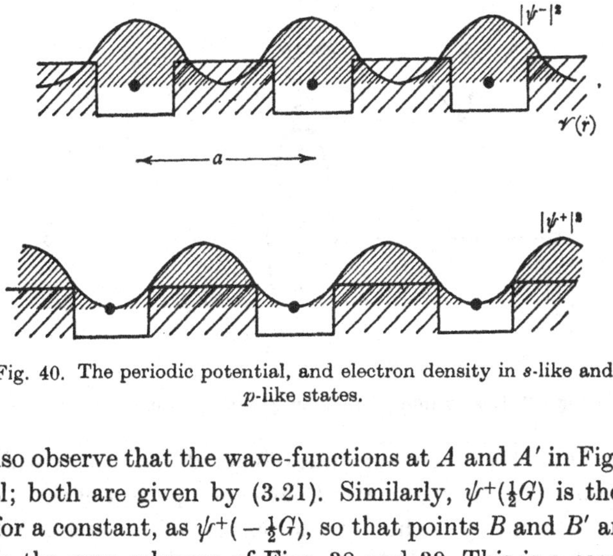

Fig. 40. The periodic potential, and electron density in *s*-like and *p*-like states.

We also observe that the wave-functions at A and A' in Fig. 39 are identical; both are given by (3.21). Similarly, $\psi^+(\frac{1}{2}G)$ is the same, except for a constant, as $\psi^+(-\frac{1}{2}G)$, so that points B and B' are equivalent in the zone schemes of Figs. 38 and 39. This is a case of the principles discussed in § 1.5, where it was shown that the wave-vector of a state is arbitrary up to a reciprocal lattice vector.

3.3 The nearly-free-electron model

The principles of the one-dimensional case can be generalized to three dimensions. If we consider the behaviour of $\mathscr{E}(\mathbf{k})$ as we go out in a particular direction in **k**-space, we find that it follows a free-electron parabola, with jump discontinuities when it crosses a Brillouin zone

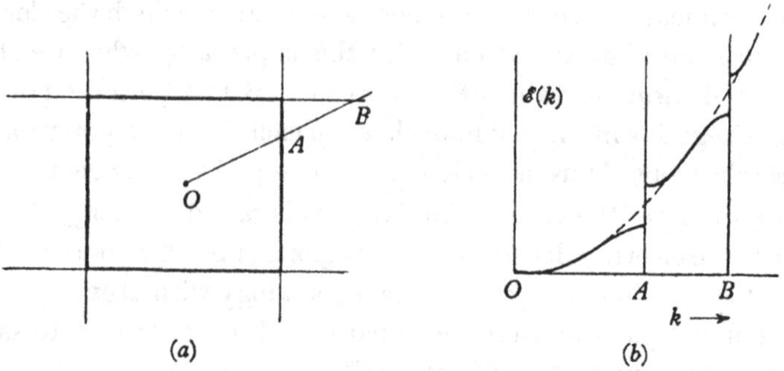

Fig. 41. Along the line OAB in k-space, there are discontinuities in $\mathscr{E}(\mathbf{k})$ at the zone boundaries.

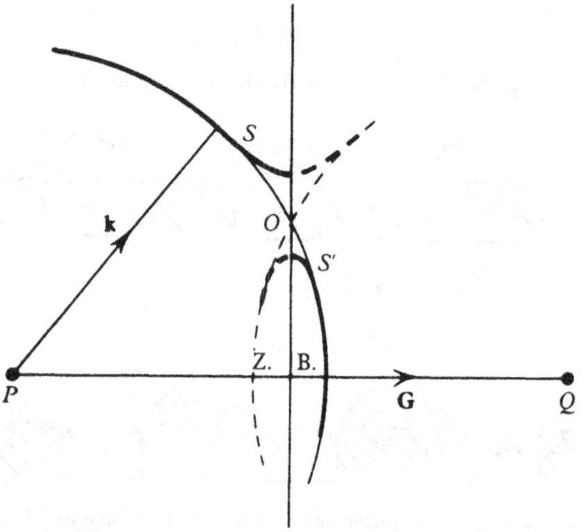

Fig. 42. Discontinuity of energy contour at zone boundary.

boundary (Fig. 41). The magnitude of each discontinuity will depend on the strength of the particular Fourier component of the potential. If the zone boundary is the normal bisector of the reciprocal lattice vector **G**, then the magnitude of the discontinuity will be of the order of $2|\mathscr{V}_{\mathbf{G}}|$. This discontinuity also appears whenever a constant energy contour crosses a zone boundary. Thus (Fig. 42) the free electron sphere that would have passed through O is split into a section of hyperboloid at S, inside the zone, generated by one of the roots of (3.17) and a section of the other branch, S', beyond the boundary.

This means that the function $\mathscr{E}(\mathbf{k})$ is continuous inside the first Brillouin zone—but that there is a jump in energy as we cross the zone boundary at any point. This makes possible one of the most striking consequences of the theory: *for some values of energy there may be no electron states*. The energy spectrum may show *gaps*. The electronic energy levels fall into *bands*. This is a fundamental property, giving rise to many of the properties of metals and semiconductors. It is a consequence of Bragg diffraction. The electron wave functions are 'pulled' into periodicity with the lattice, and split apart in energy.

For the moment we are concerned with the problem of finding $\mathscr{E}(\mathbf{k})$. The calculation of the previous section suggests a method:—use the equations (3.14) for the coefficients in the expansion (3.12): separate out all terms corresponding to the coefficients of waves that can be mixed by the zone boundaries near $\mathbf{k}$, and solve the secular determinant for the energy $\mathscr{E}(\mathbf{k})$: correct this by a perturbation expansion like (3.9) for the effects of other Fourier components of the potential.

This is called the *nearly-free-electron* (N.F.E.) method. For its validity, the series of Fourier components $\mathscr{V}_{\mathbf{g}}$ should converge rapidly. Now this does not, on the face of it, seem likely. $\mathscr{V}(\mathbf{r})$ is the potential of an array of ions. We know that the field near the core of an ion is very strong—that $\mathscr{V}(\mathbf{r})$ has a long sharp spike downward at each lattice site. This means that it has Fourier components of very short wavelength, so that $\mathscr{V}_{\mathbf{g}}$ can be large for values of $\mathbf{g}$ that are many times the dimensions of the first zone. Or we can say that $\psi_{\mathbf{k}}(\mathbf{r})$ must behave like an atomic orbital inside the core of the ion, with several oscillations corresponding to a high kinetic energy. This again means that we need many terms, down to very short wavelengths, in the representation (3.12).

These arguments led to the neglect of this method as a practical scheme for band-structure calculations. But it now appears that it can be made formally valid by the introduction of the idea of the pseudo-potential. This will be developed in § 3.6.

The N.F.E. model is very useful as a means of illustrating the behaviour of the function $\mathscr{E}(\mathbf{k})$ in $\mathbf{k}$-space. Consider the case of a rectangular lattice—in 2 dimensions—whose reciprocal lattice is also rectangular, as in Fig. 43. It is easy enough to construct the first Brillouin zone—the rectangle $RR'R''R'''$. $\mathscr{E}(\mathbf{k})$ must be continuous inside this zone, and there may be a discontinuity on any of the lines delineating this cell.

But consider the loci of discontinuities of $\mathscr{E}(\mathbf{k})$ outside the central zone. These can be the perpendicular bisectors of any vector of the reciprocal lattice. Thus, there may be a discontinuity on any of the lines in this figure—the bisector of $\mathbf{g}_3$, for example, is a diagonal line

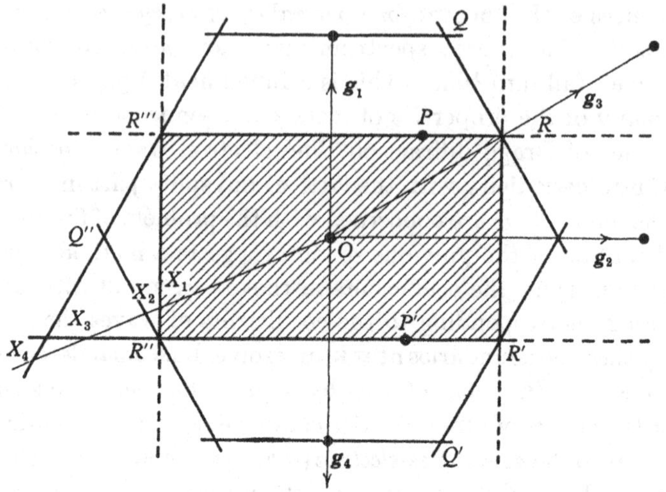

Fig. 43. Discontinuities of $\mathscr{E}(\mathbf{k})$ would occur at X_1, X_2, X_3, X_4 in the rectangular reciprocal lattice.

passing through the corner of the zone. Our first figure (Fig. 41) for the discontinuities is wrong—it does not show all that might occur, even in the case of the square lattice. Thus, there could be jumps and gaps in $\mathscr{E}(\mathbf{k})$ at the points, X_1, X_2, X_3, X_4, etc., in Fig. 43.

This seems to suggest that $\mathscr{E}(\mathbf{k})$ is a much more complicated function than is in fact the case. Suppose, for example, that we try to draw some energy contours, derived from the N.F.E. model. These would look as in Fig. 44 (confining our attention to the region within those sets of planes). We can construct these by drawing the free-electron circles and then putting in discontinuities as indicated in Fig. 42. Or, rather, we should solve the secular determinant for the energy of the waves that are mixed at that point. Thus at P there is mixing with the wave at P', because these are points that are of the same energy in the free-electron scheme, and which differ by a reciprocal lattice vector. Similarly, to find a solution at Q, we need to consider the (3×3) determinant for the mixing of waves whose wave-vectors would be OQ, OQ' and OQ'' in the free-electron model.

Now this means that at P and P' we will be solving exactly the same equation. The energy levels will be the same at both points. The lower of these we assign to the interior of the first zone; the upper level belongs to the energy contour outside the zone boundary in each case. Again, at Q, Q' and Q'' the equation is the same, and will have three energy roots. We can assign the lowest of these, at each of these points,

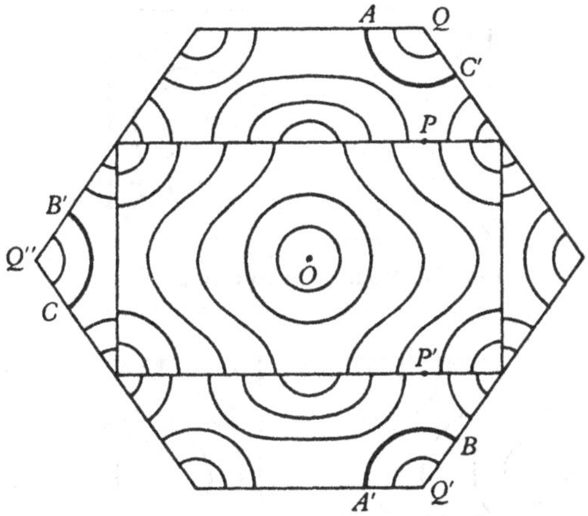

Fig. 44. Energy contours in a rectangular zone.

to the region inside the boundaries in Fig. 44. Thus, these three points are all of the same energy. We can look in the neighbourhood of these points, and show, similarly, that A and A', B and B', C and C', are pairs of points that are equivalent in this sense; A and A' are alternative labels, reducible to each other by reciprocal-lattice translations, for the same wave-function.

See now what happens when we translate the bits and pieces outside the first zone by reciprocal-lattice vectors. They fit together perfectly to make a single zone—in the Brillouin *extended zone scheme* these parts were all said to belong to the *second zone*. Moreover, the equivalent points on the boundary match up to make the contours continuous. *The energy in a reduced zone is a continuous function of* **k**.

Our scheme for the representation of the energy surfaces would now be as in Fig. 46 with a zone for each band. We might think of these as contours of different surfaces; we could pile one upon another, and thus represent $\mathscr{E}(\mathbf{k})$ as a multivalued function of **k** in the reduced zone, instead of having $\mathscr{E}(\mathbf{k})$ a single-valued function in the extended scheme.

We can go further. It can be shown (our N.F.E. model could provide a proof) that $\mathscr{E}(\mathbf{k})$ is continuous, with continuous derivatives in each band. The internal boundaries drawn in Fig. 45 do not show up on the energy surfaces; the contour lines run smoothly across them. Now let us recall that the basic zone itself is the unit cell of the reciprocal lattice, and may be repeated, by reciprocal-lattice translations, to produce a pattern covering all space. Moreover, points such as P and P', which were equivalent and therefore represented the same state

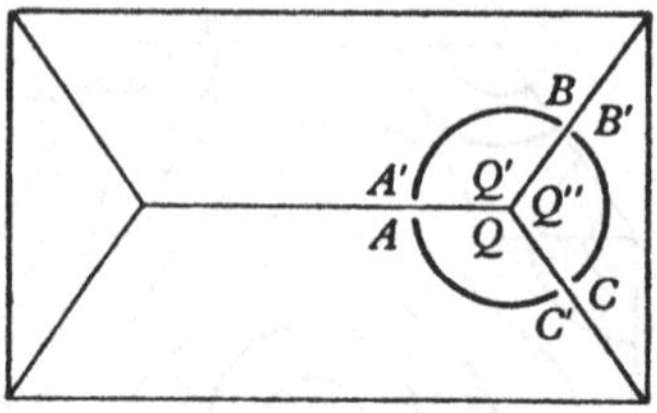

Fig. 45. Reduction to a single zone.

Fig. 46. Reduced zone scheme for Fig. 44.

and the same energies, now coincide, so that $\mathscr{E}(\mathbf{k})$ is continuous across these boundaries also. It can be shown that the energy contours join smoothly from one cell to the next. *Thus the energy in a given band is a continuous function of* **k** *with the period of the reciprocal lattice.* We call this the *repeated zone scheme.* It is just the same principle that we observed in § 2.2 in relation to surfaces of constant lattice frequency.

An interesting application of this principle is Harrison's method of constructing the Fermi surface of a metal of valence Z in the repeated zone scheme. Suppose that the perturbing potential in the N.F.E. scheme is very small. The energy surfaces must then be spheres. We calculate the radius of the sphere that contains exactly $\frac{1}{2}Z$ times the volume of a zone, and draw it with centre at the origin. We draw the same sphere about each point of the reciprocal lattice, so that we have a pattern with the periodicity of the repeated zone scheme (Fig. 48).

Now, out of this pattern we can choose various parts that can be fitted together continuously to make surfaces that repeat in each zone. Each of these separate figures then constitutes a branch of the Fermi surface, or the part of the Fermi surface in the 1st, 2nd, 3rd zone, etc. (Fig. 49). As drawn for spheres, these surfaces seem to have cusps where the parts join, but the mixing of states associated with Fourier components of the potential will round off the corners and give smooth geometrical objects. As we shall see in Chapter 9, the resulting surfaces

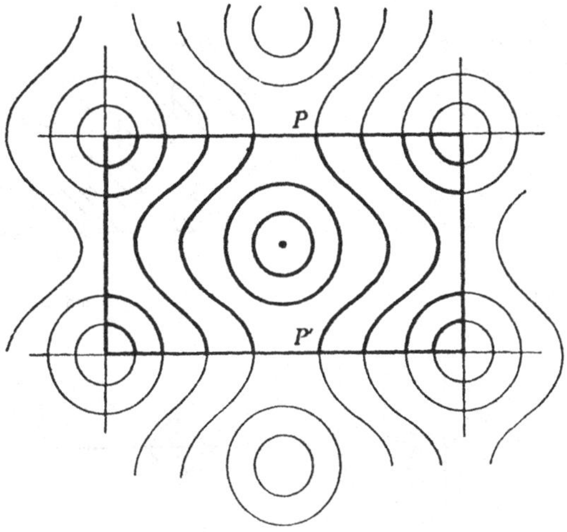

Fig. 47. Repeated first zone of Fig. 46.

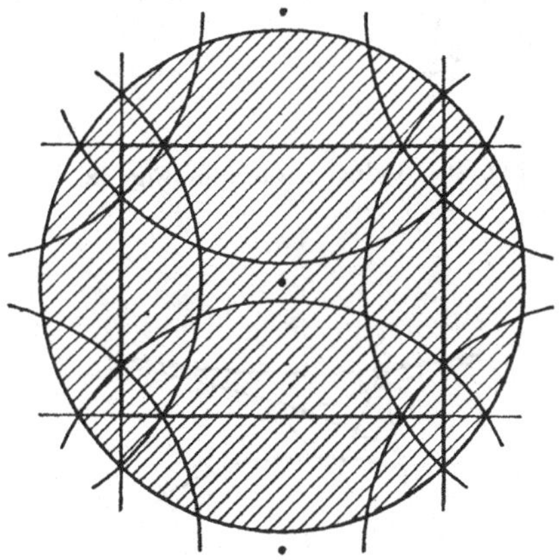

Fig. 48. Harrison's construction for free-electron Fermi surface.

are often quite like those that can be observed experimentally, and certainly give a valuable guide to the investigation of the Fermi surface in a complicated metal.

These algebraic formulae and geometrical constructions also arise in the *dynamical theory of diffraction* (§ 2.6). Very near the diffraction condition (3.11), a fast electron or X-ray beam travelling through the lattice can no longer be represented as a simple plane wave with a single wave-vector **k**. Various diffracted waves must be included in the wave-function, as in (3.12), thus changing the velocity of the

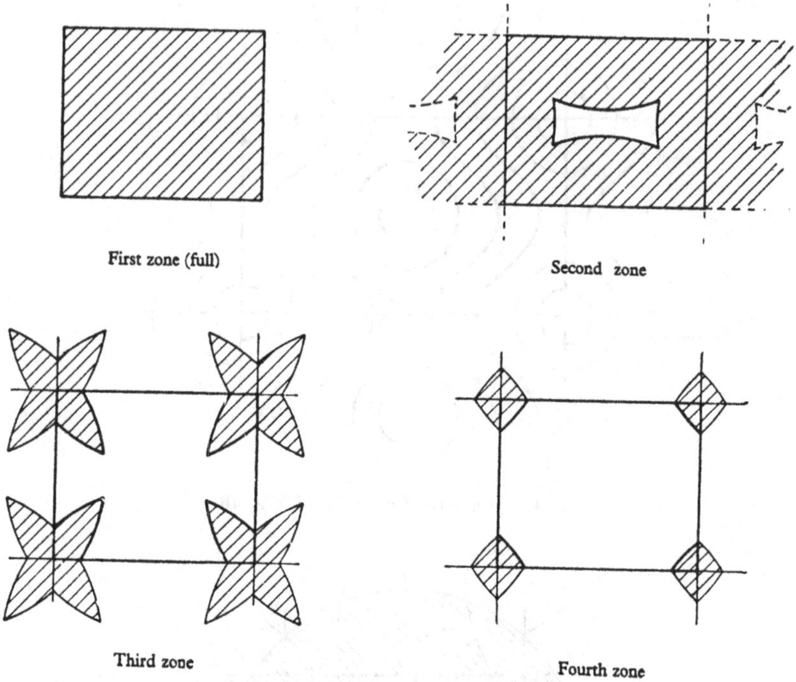

Fig. 49. Reduced zones for Fig. 48.

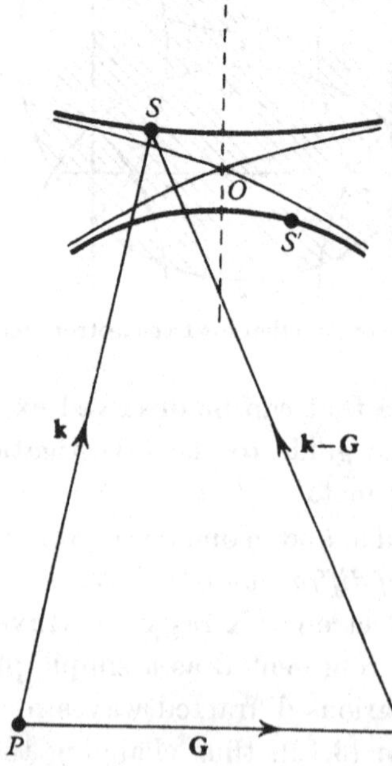

Fig. 50. Dispersion surface for dynamical diffraction of X-rays or fast electrons.

excitation. Near a simple zone boundary, for example, at least two waves must be combined, yielding a quadratic equation, like (3.17), with two roots for the electron or photon energy as a function of **k**.

These roots generate the *dispersion surface* in Fig. 50. The original 'free photon' spheres of radius ν/c centred on the reciprocal lattice points P and Q, which would have intersected at the point O on the zone boundary (cf. Figs. 29 and 37), are now split into two hyperboloids. A point such as S or S' on this diagram therefore represents an electromagnetic disturbance of frequency ν, propagating through the crystal as a combination of two plane waves with the wave-vectors SP and SQ or $S'P$ and $S'Q$. To match incident X-rays at this frequency, beams from both branches of the dispersion surface are excited, leading to the interference effects mentioned in § 2.6. Apart from a change of scale—the X-ray or fast electron wave-vector is normally much larger than a typical vector of the reciprocal lattice—Fig. 50 is really just the same as Fig. 42.

3.4 The tight-binding method

The N.F.E. method looks at the wave-functions outside the atomic cores, where they look very like plane waves. Within the cores they look like atomic orbitals. This suggests an entirely different scheme for the construction of electron wave-functions: we try to combine atomic orbitals, each localized on a particular atom, to represent a state running throughout the crystal.

Suppose that $\phi_a(\mathbf{r}-\boldsymbol{l})$ is an atomic orbital for a free atom with centre at $\boldsymbol{l}$. Then we may construct a function that satisfies the Bloch condition (1.58)

$$\phi_{\mathbf{k}}(\mathbf{r}) = \sum_{\boldsymbol{l}} e^{i\mathbf{k}\cdot\boldsymbol{l}}\phi_a(\mathbf{r}-\boldsymbol{l}). \tag{3.24}$$

This function looks like a series of strongly localized atomic orbitals, multiplied by a wavy phase factor $\exp(i\mathbf{k}\cdot\boldsymbol{l})$. Within each atom the local orbital predominates, and should be a good solution of the local Schrödinger equation.

As a first approximation, let us calculate the expectation value of the energy for this wave-function

$$\mathscr{E}(\mathbf{k}) = \frac{\int \phi_{\mathbf{k}}^* \left\{-\frac{\hbar^2}{2m}\nabla^2 + \mathscr{V}(\mathbf{r})\right\} \phi_{\mathbf{k}}\, d\mathbf{r}}{\int \phi_{\mathbf{k}}^* \phi_{\mathbf{k}}\, d\mathbf{r}}. \tag{3.25}$$

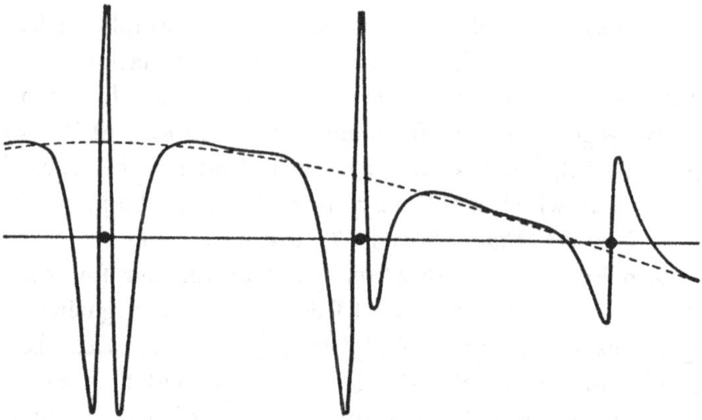

Fig. 51. Wave-function in a solid.

If there is not much overlap of orbitals in neighbouring cells, we may perhaps take the normalization factor in the denominator to be unity. The numerator can be written

$$\mathscr{E}(\mathbf{k}) \approx \sum_{ll'} e^{i\mathbf{k}\cdot(\mathbf{l}-\mathbf{l}')} \int \phi_a^*(\mathbf{r}-\mathbf{l}') \left\{ -\frac{\hbar^2}{2m}\nabla^2 + \mathscr{V}(\mathbf{r}) \right\} \phi_a(\mathbf{r}-\mathbf{l})\, d\mathbf{r}$$

$$= \sum_{\mathbf{h}} e^{i\mathbf{k}\cdot\mathbf{h}} \mathscr{E}_{\mathbf{h}}, \tag{3.26}$$

where
$$\mathscr{E}_{\mathbf{h}} = \frac{1}{v_c} \int \phi_a^*(\mathbf{r}+\mathbf{h}) \left\{ -\frac{\hbar^2}{2m}\nabla^2 + \mathscr{V}(\mathbf{r}) \right\} \phi_a(\mathbf{r})\, d\mathbf{r}. \tag{3.27}$$

We have used the periodicity of the lattice potential to make the various integrals depend only on the relative position, $\mathbf{h} = \mathbf{l} - \mathbf{l}'$, of the sites upon which the orbitals are centred.

Now suppose that $\phi_a(\mathbf{r})$ satisfies the equation

$$\left\{ -\frac{\hbar^2}{2m}\nabla^2 + v_a(\mathbf{r}) \right\} \phi_a(\mathbf{r}) = \mathscr{E}_a \phi_a(\mathbf{r}), \tag{3.28}$$

where now $v_a(\mathbf{r})$ is the potential of an isolated atom at $\mathbf{r} = 0$.

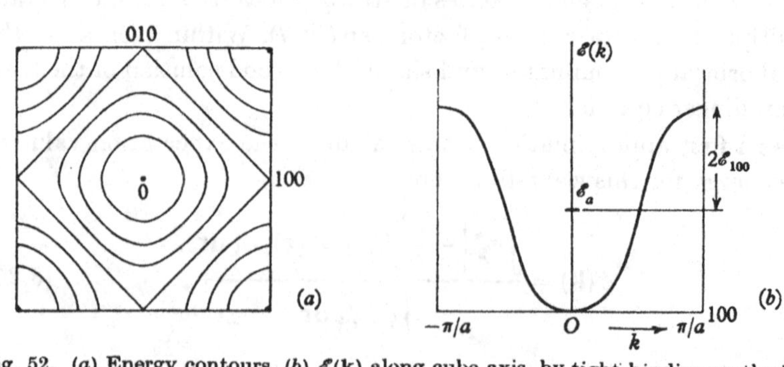

Fig. 52. (a) Energy contours, (b) $\mathscr{E}(\mathbf{k})$ along cube axis, by tight-binding method.

If we ignore various overlap integrals we get

$$\left.\begin{aligned} \mathscr{E}_0 &\approx \mathscr{E}_a, \\ \mathscr{E}_{\mathbf{h}} &\approx \frac{1}{v_c}\int \phi_a^*(\mathbf{r}+\mathbf{h})\{\mathscr{V}(\mathbf{r}) - v_a(\mathbf{r})\}\,\phi_a(\mathbf{r})\,d\mathbf{r}. \end{aligned}\right\} \tag{3.29}$$

Evidently, there must be some overlap between $\phi_a(\mathbf{r}+\mathbf{h})$ and $\phi_a(\mathbf{r})$ for $\mathscr{E}_{\mathbf{h}}$ not to vanish—and this will be enhanced by the large potential in the neighbourhood of the atom on $\mathbf{h}$. But we should expect $\mathscr{E}_{\mathbf{h}}$ to be very small except for near neighbours.

Consider a simple cubic lattice, with $\mathbf{h} = (a, 0, 0)$, etc. Then

$$\mathscr{E}(\mathbf{k}) \approx \mathscr{E}_a + 2\mathscr{E}_{100}(\cos ak_x + \cos ak_y + \cos ak_z), \tag{3.30}$$

if $\phi_a(\mathbf{r})$ has cubic symmetry. The zone in this case is, of course, a cube, and the energy surfaces are as shown in Fig. 52. It is easy to see from (3.29) that $\mathscr{E}_{100}$ is essentially negative. We get a band, of width $12\,|\mathscr{E}_{100}|$, with lowest energy $\mathscr{E}_a + 6\mathscr{E}_{100}$.

Along a cube direction we see that $\mathscr{E}(\mathbf{k})$ has a minimum at $k = 0$, and a maximum at the zone boundary, where $\mathbf{k} = (\pi/a, 0, 0)$. This is superficially similar to the curves for the N.F.E. model (Fig. 36), but the curvature at the bottom of the band depends on $\mathscr{E}_{100}$. The whole structure of the band depends upon integrals, such as (3.29), which are very sensitive to the details of the orbitals outside their atomic cores, and to the lattice spacing. It is obvious that the band will tend to become very narrow if the atomic orbitals are closely localized or if the lattice spacing becomes large.

The *tight-binding method* demonstrates an important principle. Suppose we take N atoms, and keep them far apart. On each atom there will be different atomic levels, ϕ_a, ϕ_b, ϕ_c; in the whole assembly these will be N-fold degenerate states for a single electron.

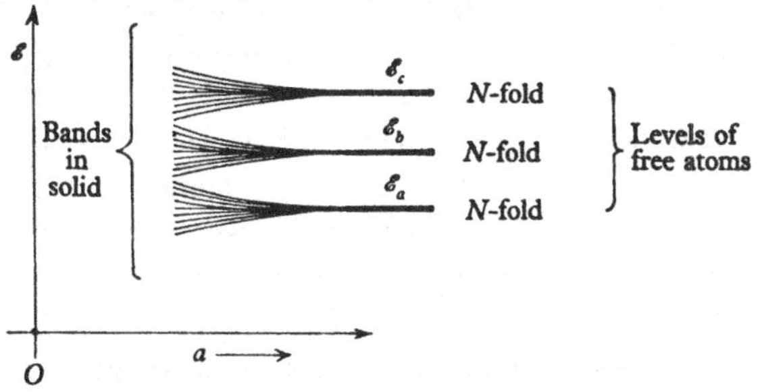

Fig. 53. Atomic levels spreading into bands as the atoms come together.

But when we bring the atoms together, the orbitals overlap—and we find that the N-fold degeneracy is split into a band of states. There will be a complete band, of N states, arising from each atomic level—so that we can refer to the $3s$-band, $4p$-band, etc., arising from the corresponding atomic states. This classification is particularly apt in the case of transition metals, where the d-states of the atom are rather compact within each core, so that they combine to form a relatively narrow and well-defined d-band.

It is clear from Fig. 53, however, that the different bands may broaden so much that they intersect. The simple scheme that we have discussed here must then be modified. We must suppose that our wave-function is a *linear combination of atomic orbitals* (L.C.A.O.)

$$\psi_{\mathbf{k}} = \sum_{lj} e^{i\mathbf{k}\cdot\mathbf{l}} \beta_j \phi_a^{(j)}(\mathbf{r}-\mathbf{l}), \tag{3.31}$$

where $\phi_a^{(j)}(\mathbf{r}-\mathbf{l})$ is one of a set of different atomic orbitals on the atom at $\mathbf{l}$.

We must then use this as a trial wave-function, for an estimate of the energy, to be minimized by suitable choice of the coefficients β_j. It is evident that we shall get sets of linear equations with coefficients involving expressions like $\mathscr{E}_{\mathbf{h}}$ in (3.26). These, in turn, involve integrals such as (3.29) but with the two atomic orbitals referring, perhaps, to different atomic levels on the two different sites.

The L.C.A.O. method is widely used in quantum chemistry to construct *molecular* wave-functions, but it is not really very satisfactory for the *quantitative* calculation of Bloch functions in solids. Not only do we run into computing difficulties with three-centre integrals and non-orthogonal basis functions; there are fundamental objections to representing the valence electron states in a semiconductor or metal by an expression such as (3.31).

When we make a solid, each atom overlaps the potential of its neighbours. Suppose, in a simple-minded way (Figs. 54 and 55), that the separate atomic potentials may be superposed, creating a relatively low and constant potential in the interstitial regions, between the main potential wells centred on the atomic nuclei. It is clear that the individual atomic orbitals used in (3.31) no longer exist, for they would now lie above the energy of the barriers between the atomic spheres. It is not a good method in principle to try to represent a Bloch function in a potential $\mathscr{V}(\mathbf{r})$ by the eigenstates of a quite different potential $v_a(\mathbf{r})$ satisfying irrelevant boundary conditions ($\phi_a(\mathbf{r}) \to 0$ rapidly as $r \to \infty$). All the nodes are displaced, and cannot easily be shifted.

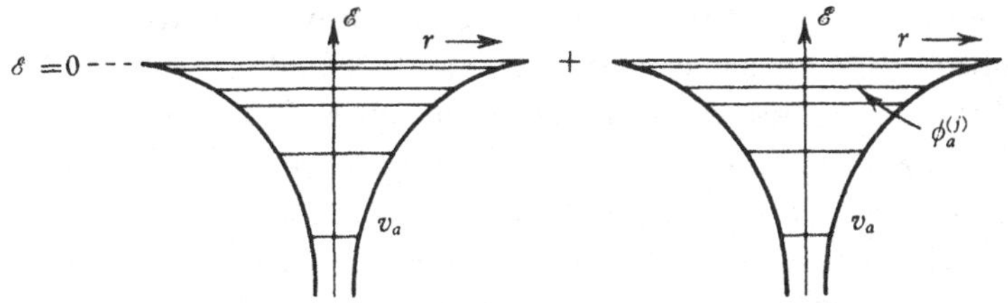

Fig. 54. Bound atomic orbitals of free atoms.

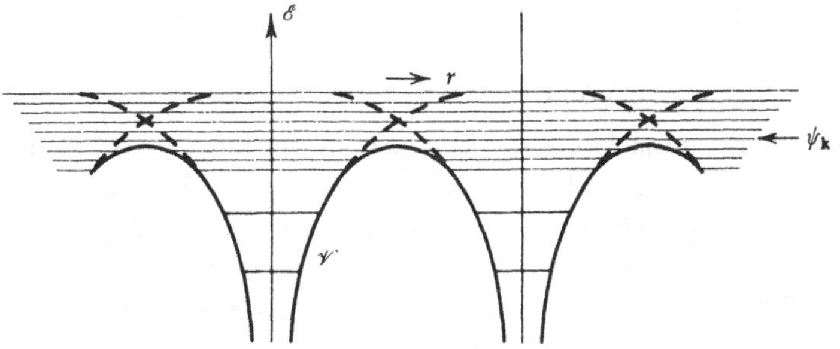

Fig. 55. Bloch states of crystal.

In any case, this set of functions is *incomplete*, since it lacks all the scattered-wave eigenstates of the Schrödinger equation in the continuum, above the energy zero of $v_a(\mathbf{r})$. Although (3.31) may be matched reasonably well to $\psi_{\mathbf{k}}$ inside the atomic core, it cannot pretend to represent a Bloch state in the interstitial region, where it must behave like a simple combination of free-electron plane waves (see § 3.6).

The method does, however, provide one important formal principle. In (3.26) we see that the energy in **k**-space can be expanded as a Fourier series

$$\mathscr{E}(\mathbf{k}) = \sum_{\mathbf{l}} e^{i\mathbf{k}\cdot\mathbf{l}}\mathscr{E}_{\mathbf{l}}. \tag{3.32}$$

One can show that an expansion of this form is always possible—that the L.C.A.O. method, if carried out exactly and completely, would give the energy in this form—although the coefficients $\mathscr{E}_{\mathbf{l}}$ would no longer be simply integrals like (3.27) or (3.29), and might not converge very rapidly for large values of $\mathbf{l}$.

Now remember that the direct lattice vectors, $\mathbf{l}$, are themselves the reciprocal lattice vectors of the reciprocal lattice $\mathbf{g}$ (§ 1.3 (iv)). This is the dual of such formulae as

$$\mathscr{V}(\mathbf{r}) = \sum_{\mathbf{g}} e^{i\mathbf{r}\cdot\mathbf{g}}\mathscr{V}_{\mathbf{g}} \tag{3.33}$$

which hold if $\mathscr{V}(\mathbf{r})$ has the periodicity $\mathscr{V}(\mathbf{r}+\mathbf{l}) = \mathscr{V}(\mathbf{r})$. Thus, *the energy function in each band must be periodic in the reciprocal lattice;*

$$\mathscr{E}(\mathbf{k}+\mathbf{g}) = \mathscr{E}(\mathbf{k}). \tag{3.34}$$

We have already noted this property in the previous section. The representation (3.32) is sometimes convenient; it makes $\mathscr{E}(\mathbf{k})$ satisfy this condition automatically. If the coefficients $\mathscr{E}_l$ are treated as adjustable parameters, one can often fit an observed Fermi surface with this formula. As an interpolation formula it has somewhat less 'physical' justification than the N.F.E. scheme where the Fourier components $\mathscr{V}_{\mathbf{g}}$ may be taken as parameters, but is easier to compute with because it expresses the energy directly as a sum of simple analytical functions.

3.5 Cellular methods

One of the objections to the L.C.A.O. method can be avoided by using functions which have been made to satisfy the Bloch condition (1.58); i.e. the wave-function changes by $\exp(i\mathbf{k}\cdot\boldsymbol{l})$ as it moves through the lattice translation $\boldsymbol{l}$ (e.g. from one face to another of a unit cell). Alternatively, one may set up the differential equation for the function $u_{\mathbf{k}}(\mathbf{r})$ in the formula

$$\psi_{\mathbf{k}}(\mathbf{r}) = e^{i\mathbf{k}\cdot\mathbf{r}}u_{\mathbf{k}}(\mathbf{r}), \tag{3.35}$$

knowing (from (1.61)) that it must be a periodic function in the direct lattice.

In one dimension this looks a simple scheme. For example, one can choose a value of $\mathscr{E}$, integrate the equation, see whether the condition is satisfied, change $\mathscr{E}$, and so on. But in three dimensions it is clumsy. The condition (3.34) has to hold over the whole face of the unit cell—by implication, at an infinite set of points. So one constructs solutions for various spherical harmonics in the cell, and one looks for a linear combination that matches at a finite set of points on the cell boundary. This can be improved by defining an average matching error over the surfaces of the cell, and minimizing this by variation of the coefficients. The procedure is mathematically tedious, but obvious enough in principle; it has been able to give quite good band structures for some solids, but it has not contributed very much in the way of new theoretical concepts.

The cellular method looks simple and direct. The unit cell is a well-defined geometrical concept, and the Bloch condition can easily be set down. But the whole idea is not quite right. One should not set up artificial cell boundaries, and create matching discontinuities in the very region where the potential is flattest and the wave-functions most nearly like those of free electrons. The matching conditions have to be defined arbitrarily, and it has sometimes turned out that apparently plausible procedures lead to results that are quite wrong, even for the elementary case of an empty lattice.

There is a predecessor of the general cellular method that is sometimes of value. This is the *Wigner–Seitz method*. We note that the unit cell of a F.C.C. or B.C.C. lattice—the Wigner–Seitz cell—is a regular polyhedron which approximates to a sphere. Suppose it were a sphere, of the same volume as the cell, i.e. of radius r_s (cf. (2.54)).

Consider the state for $\mathbf{k} = 0$. This must satisfy

$$\psi_{\mathbf{k}}(\mathbf{r}+\mathbf{l}) = \psi_{\mathbf{k}}(\mathbf{r}); \tag{3.36}$$

it must be periodic from cell to cell. This means that it should have a horizontal tangent at the boundary of the cell; we may say that it must satisfy the condition

$$\left.\frac{\partial\psi_{\mathbf{k}}}{\partial r}\right]_{r=r_s} = 0 \tag{3.37}$$

over the surface of the Wigner–Seitz sphere. Thus, if the potential is spherically symmetrical inside the sphere, we have only a radial equation to solve with this boundary condition, so that the energy $\mathscr{E}(0)$ can be defined uniquely. (Fig. 56).

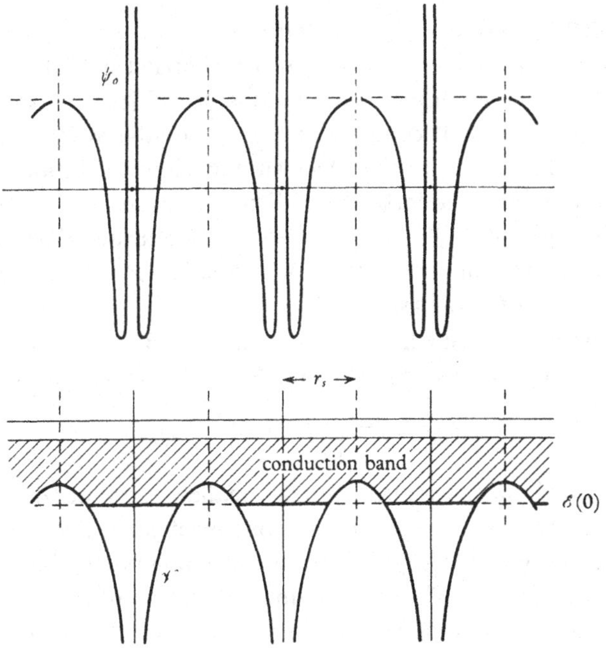

Fig. 56. The Wigner–Seitz method.

This is a neat way of locating the bottom of the conduction-band in a monovalent metal, and may be used with fair confidence in the calculation of metallic cohesion (§ 4.3). One can go a little further and write down the equation

$$-\frac{\hbar^2}{2m}\{\nabla^2+2i\mathbf{k}\cdot\nabla\}u_{\mathbf{k}}+\mathscr{V}(\mathbf{r})u_{\mathbf{k}}=\left\{\mathscr{E}(\mathbf{k})-\frac{\hbar^2k^2}{2m}\right\}u_{\mathbf{k}} \tag{3.38}$$

which must be satisfied by the periodic function $u_{\mathbf{k}}(\mathbf{r})$ in (3.35), and one can then insist that $u_{\mathbf{k}}$ should have a horizontal tangent at $r = r_s$. This gives something different from a free-electron formula for the function $\mathscr{E}(\mathbf{k})$ as measured from $\mathscr{E}(0)$, and hence something of the shape of the band.

But in this method we have entirely ignored the structure of the solid. The answer depends only on the atomic volume, not on the actual geometry of the crystal lattice. We should get just the same result for a liquid, where the environment of each atom is disordered, as for a solid metal of the same density—those beautiful diffraction effects, which we have discussed at length in previous sections, simply do not enter.

3.6 Orthogonalized plane waves

We now turn to some methods of band structure calculation that really work. For example, the fundamental objections to the tight-binding method (§ 3.4) could be met by (*a*) deliberately adding plane wave terms like (3.12) to the Bloch sum of atomic orbitals (3.31), and (*b*) excluding from the latter sum all the fictitious orbitals that have really disappeared in the overlap of atomic potentials (Figs. 54 and 55). This is what Herring actually did when he introduced the concept of an *orthogonalized plane wave*.

Consider a set of Bloch functions constructed out of *core* states, i.e. orbitals already occupied in the ion. From (3.24) we may write

$$b_{t\mathbf{k}}=\sum_{\mathbf{l}}e^{i\mathbf{k}\cdot\mathbf{l}}b_t(\mathbf{r}-\mathbf{l}). \tag{3.39}$$

A core orbital such as $b_t(\mathbf{r})$ is strongly localized, so that this is no more than a combination of degenerate one-electron states, each corresponding to the localization of the electron on a particular atom. This is actually a solution of the Schrödinger equation of the whole crystal corresponding to one of the core levels, $\mathscr{E}_t$. It could form a narrow band, which is fully occupied in the crystal.

The higher states, being solutions of the same Schrödinger equation, *must be orthogonal to* $b_{t\mathbf{k}}$:

$$\langle \psi_{\mathbf{k}}, b_{t\mathbf{k}} \rangle = 0. \tag{3.40}$$

Also, we know that $\psi_{\mathbf{k}}$ must look like a free-electron wave in the regions between the atoms. Let us try

$$\chi_{\mathbf{k}} = e^{i\mathbf{k}\cdot\mathbf{r}} - \sum_t \beta_t b_{t\mathbf{k}} \tag{3.41}$$

as a possible type of wave-function for one of our higher states. We can make this satisfy the orthogonalization condition (3.40) by a suitable choice of the coefficients β_t. That is, we make $\chi_{\mathbf{k}}$ into an *orthogonalized plane wave* (O.P.W)

$$\chi_{\mathbf{k}} = e^{i\mathbf{k}\cdot\mathbf{r}} - \sum_t \langle b_{t\mathbf{k}}, e^{i\mathbf{k}\cdot\mathbf{r}} \rangle b_{t\mathbf{k}}. \tag{3.42}$$

This is a good guess at a higher state because it looks like $\exp(i\mathbf{k}\cdot\mathbf{r})$ in the interstitial region, where each $b_{t\mathbf{k}}$ is small (the core states are highly localized), and yet within the core itself it will be orthogonal to all the core states and hence a good candidate for a higher atomic orbital. Suppose, for example, that the core states are $2s$ and $2p$, each with 1 nodal surface: $\chi_{\mathbf{k}}$ will have one extra nodal surface, just like a $3s$- or $3p$-orbital.

We now use O.P.W.'s as basis states for our wave-function. We try

$$\psi_{\mathbf{k}} = \sum_{\mathbf{g}} \alpha_{\mathbf{k}-\mathbf{g}} \chi_{\mathbf{k}-\mathbf{g}} \tag{3.43}$$

as a solution of the Schrödinger equation, just as in (3.12). We use the variational principle to minimize the expectation value of the energy and determine the coefficients $\alpha_{\mathbf{k}-\mathbf{g}}$. The procedure is familiar enough by now and need not be written down explicitly.

This process of approximation converges with surprising rapidity in the initial stages. A single O.P.W. is often quite sufficient to represent the wave-function over large regions of **k**-space; 3 or 4 terms may give quite a good picture, even in the very corners of a Brillouin zone. The method has been used for several metals and semiconductors with success, and is not restricted to spherically symmetrical potentials surrounded by empty regions.

Why is this? The following argument, due to Phillips and Kleinman, is very persuasive:

Suppose that (3.43) is the exact wave-function with some definite values of $\alpha_{\mathbf{k}-\mathbf{g}}$. Write down the function

$$\phi = \sum_{\mathbf{g}} \alpha_{\mathbf{k}-\mathbf{g}} e^{i(\mathbf{k}-\mathbf{g})\cdot\mathbf{r}}, \tag{3.44}$$

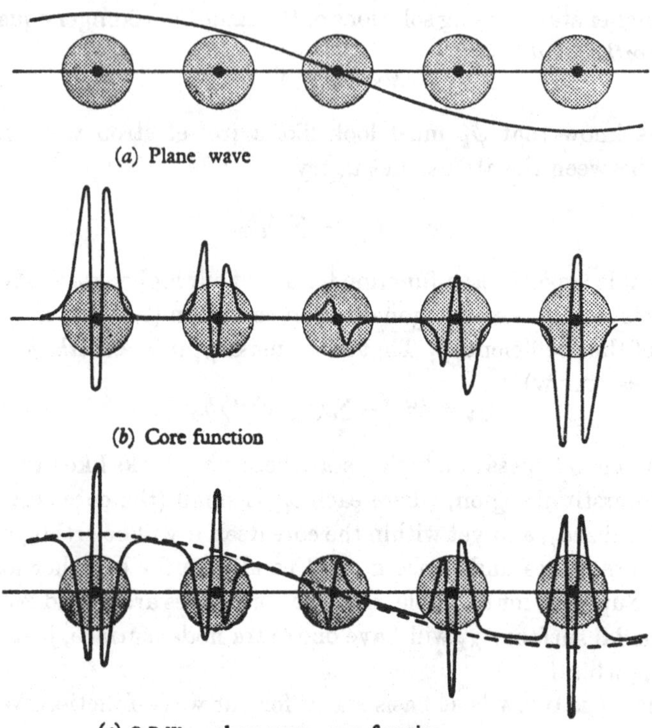

Fig. 57. Synthesis of an orthogonalized plane wave.

a combination of simple plane waves with the same coefficients. Then (3.42) and (3.43) may be written (dropping the index **k**, understood to be the same throughout)

$$\psi = \phi - \sum_t \langle b_t, \phi \rangle b_t. \tag{3.45}$$

Put this into the Schrödinger equation $\mathscr{H}\psi = \mathscr{E}\psi$,

i.e. $$\mathscr{H}\phi - \sum_t \langle b_t, \phi \rangle \mathscr{H} b_t = \mathscr{E}\phi - \mathscr{E} \sum_t \langle b_t, \phi \rangle b_t, \tag{3.46}$$

i.e. $$\mathscr{H}\phi - \sum_t \langle b_t, \phi \rangle \mathscr{E}_t b_t + \mathscr{E} \sum_t \langle b_t, \phi \rangle b_t = \mathscr{E}\phi, \tag{3.47}$$

since b_t is also an eigenstate of $\mathscr{H}$, of energy $\mathscr{E}_t$. Thus

$$\mathscr{H}\phi + \sum_t (\mathscr{E} - \mathscr{E}_t) b_t \langle b_t, \phi \rangle = \mathscr{E}\phi. \tag{3.48}$$

We may look upon this as a new Schrödinger equation

$$(\mathscr{H} + \mathscr{V}_R)\phi = \mathscr{E}\phi, \tag{3.49}$$

where we make a formal definition of an operator such that

$$\mathscr{V}_R\phi \equiv \sum_t (\mathscr{E}-\mathscr{E}_t)\, b_t \langle b_t, \phi\rangle. \tag{3.50}$$

Our 'smoothed' wave function ϕ now satisfies a new equation, of which the 'Hamiltonian' is

$$\mathscr{H}+\mathscr{V}_R = -\frac{\hbar^2}{2m}\nabla^2+\mathscr{V}+\mathscr{V}_R. \tag{3.51}$$

It is just as if we had to find the plane-wave solution for the eigenfunctions in the *pseudo-potential*,

$$\Gamma = \mathscr{V}+\mathscr{V}_R. \tag{3.52}$$

The formal machinery of § 3.2 may now be used. The coefficients $\alpha_{\mathbf{k}-\mathbf{g}}$ must satisfy equations like (3.14), except that we must use Fourier components of Γ instead of Fourier components of the true lattice potential $\mathscr{V}$.

These equations are, of course, only another form of the equations derived from varying the coefficients, $\alpha_{\mathbf{k}-\mathbf{g}}$, in the expectation value of the energy, with (3.43) as trial wave-function. But it turns out (and this can be made plausible as a general principle) that the Fourier components of Γ are small, except for the first few reciprocal-lattice vectors. The function ϕ may be called a *pseudo-wave-function*, which satisfies the equation (3.49) in which the effective potential is relatively weak. We are back again at a simple N.F.E. model.

The argument that Γ may be treated as a small localized potential is not exact. The symbol $\mathscr{V}_R$ stands for a non-localized operator

$$\mathscr{V}_R(\mathbf{r},\mathbf{r}') = \sum_t (\mathscr{E}-\mathscr{E}_t)\, b_t(\mathbf{r})\, b_t^*(\mathbf{r}'), \tag{3.53}$$

such that

$$\mathscr{V}_R\phi = \int \mathscr{V}_R(\mathbf{r},\mathbf{r}')\,\phi(\mathbf{r}')\,d\mathbf{r}'. \tag{3.54}$$

This shows us, for example, that $\mathscr{V}_R$ must be different when it operates on functions of different angular momentum; we should write

$$\mathscr{V}_R = \mathscr{V}_s+\mathscr{V}_p+\mathscr{V}_d+\ldots, \tag{3.55}$$

where $\mathscr{V}_s$ operates only on functions with s-symmetry, etc.

The point is, in essence, that $\mathscr{V}$, being the attractive potential of an atom, is negative, especially near $r = 0$, whilst $\mathscr{V}_R$, containing $\mathscr{E}-\mathscr{E}_t$, and the square of a core orbital, is positive. There is some cancellation between them, reducing the value of $\mathscr{V}_{\text{eff.}}$. This cancellation, although never complete, can be improved by suitable choice of the core functions.

Indeed, the whole process of constructing $\mathscr{V}_R$ is not unique. One can show that *the valence eigenvalues of the Hamiltonian* ($\mathscr{H}+\mathscr{V}_R$) *are the same for any operator of the form*

$$\mathscr{V}_R\phi = \sum_t \langle F_t, \phi\rangle b_t, \tag{3.56}$$

where the F_t *are completely arbitrary functions.* To see this, notice that the operator $\mathscr{V}_R$ always projects ϕ on to the manifold of the core states. But the valence eigenfunction that we are seeking is orthogonal to this manifold—so that the addition of the operator $\mathscr{V}_R$ to the Hamiltonian makes no difference to the eigenvalue problem in the space of the valence eigenfunctions.

Thus, by judicious choice of the functions F_t, one can try to get good cancellation. For example, if we put

$$F_t = -\mathscr{V}b_t, \tag{3.57}$$

then we have

$$\Gamma\phi = \mathscr{V}\phi - \sum_t \langle b_t, \mathscr{V}\phi\rangle b_t, \tag{3.58}$$

which means that we may subtract from $\mathscr{V}$ any part that can be expanded as a sum of core functions. This explains why, in many cases, Γ turns out to be quite small.

The cancellation principle, although not a rigorous theorem, thus justifies the *pseudo-potential concept: the valence electrons in metals and semiconductors behave as if they did not interact strongly with the ions of the crystal lattice.* This is the reason for the empirical success of the N.F.E. model of § 3.3, not only in providing pretty pictures, like Fig. 49, for Fermi surfaces, but also in the theory of much more complicated phenomena such as electrical resistivity and phonon spectra (see e.g. §§ 6.12, 6.13). In many ways, this has been one of the most fruitful developments in solid state theory since the work of Bloch in the early 1930's.

Nevertheless, from a theoretical point of view, this type of 'analytical' pseudo-potential, based upon the O.P.W. formalism, is not entirely satisfactory. It depends so much on energy and angular momentum: it is not a true local potential: it is not uniquely defined: it cannot deal elegantly with d-bands (see § 3.10). The main point that it establishes—that the N.F.E. model is much better than at first sight appears—follows almost as well from the 'muffin-tin' methods to be discussed in the next section, without the need to introduce pseudo-plane waves (Ψ.P.W.'s) like the function ϕ in (3.44).

3.7 Augmented plane waves

The objection to the L.C.A.O. and cellular methods is that the potential in a metal is fairly flat between the ion cores. Suppose, indeed, that it is a *muffin-tin potential*—spherically symmetrical within some radius R_s about each site and constant in the interstitial regions. We assume that R_s is rather smaller than the Wigner–Seitz radius r_s so that the spheres do not intersect. We can solve the Schrödinger equation exactly, within each sphere, in spherical harmonics, and we can also

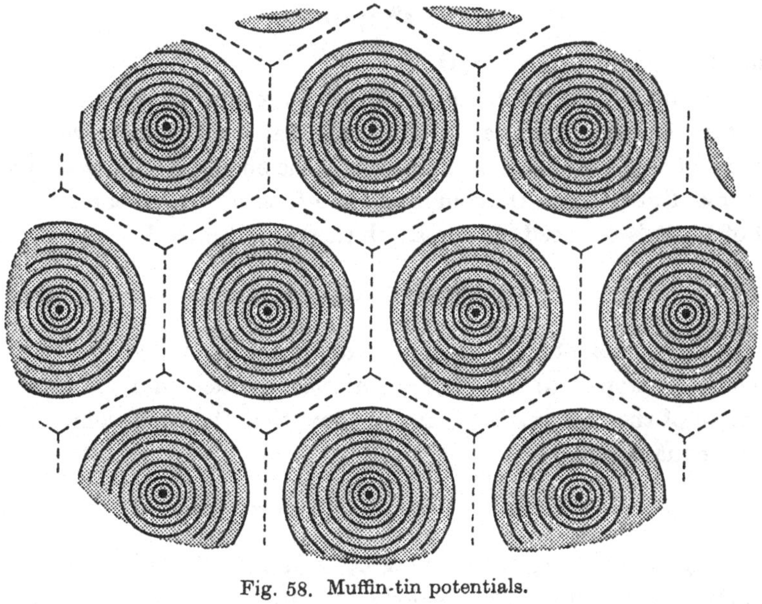

Fig. 58. Muffin-tin potentials.

find exact solutions—plane waves—in the volume between the spheres. By matching these solutions on the surface of each sphere, we can construct a solution of the Schrödinger equation throughout the crystal.

The actual procedure is as follows: suppose that $\mathscr{E}$ is the energy of our state. Within each sphere the solution of the Schrödinger equation is of the form

$$\phi(\mathbf{r}) = \sum_{lm} C_{lm} \mathscr{R}_l(r, \mathscr{E}) Y_{lm}(\theta, \phi), \tag{3.59}$$

where $Y_{lm}(\theta, \phi)$ is a spherical harmonic for the direction (θ, ϕ) of the vector $\mathbf{r}$ and $\mathscr{R}_l$ is the solution of the radial equation (in atomic units)

$$-\frac{1}{r^2}\frac{\partial}{\partial r}\left(r^2 \frac{\partial \mathscr{R}_l}{\partial r}\right) + \left\{\frac{l(l+1)}{r^2} + \mathscr{V}(r)\right\}\mathscr{R}_l = \mathscr{E}\mathscr{R}_l. \tag{3.60}$$

Let us now choose the coefficients C_{lm} so that $\phi(\mathbf{r})$ exactly matches, at the surface $r = R_s$, to a single plane wave $\exp(i\mathbf{k}\cdot\mathbf{r})$. It is easy to prove, from the standard expansion formula for a plane wave in spherical harmonics, that this can be done if

$$C_{lm} = (2l+1)\, i^l \{j_l(kR_s)/\mathscr{R}_l(R_s, \mathscr{E})\}\, Y^*_{lm}(\theta', \phi') \tag{3.61}$$

(where θ', ϕ' define the direction of $\mathbf{k}$ and j_l is a spherical Bessel function). Putting (3.61) in (3.59), we can define an *augmented plane wave* (A.P.W.), $\phi_{\mathbf{k}}(\mathbf{r})$, which is equal to $\exp(i\mathbf{k}\cdot\mathbf{r})$ in the interstitial space, and which satisfies

$$\phi_{\mathbf{k}}(\mathbf{r}) = e^{i\mathbf{k}\cdot\boldsymbol{l}}\phi_{\mathbf{k}}(\mathbf{r}-\boldsymbol{l}) \tag{3.62}$$

inside the sphere centred on $\boldsymbol{l}$.

Unfortunately, this function is still not a solution of our Schrödinger equation for the potential of the whole lattice. Indeed, we have assumed no special relationship between the energy $\mathscr{E}$ and the wave-vector $\mathbf{k}$. But we know that our final wave-function can be constructed out of A.P.W.'s—indeed, by the Bloch theorem, it must be of the form

$$\psi_{\mathbf{k}}(\mathbf{r}) = \sum_{\mathbf{g}} \alpha_{\mathbf{k}-\mathbf{g}}\, \phi_{\mathbf{k}-\mathbf{g}}(\mathbf{r}), \tag{3.63}$$

where the coefficients $\alpha_{\mathbf{k}-\mathbf{g}}$ have to be determined.

The next step is to use (3.63) as a trial function in a variational estimate of the energy. It is easy to show that the coefficients must satisfy equations like

$$\{(\mathbf{k}-\mathbf{g})^2 - \mathscr{E}\}\,\alpha_{\mathbf{k}-\mathbf{g}} + \sum_{g'} \Gamma_{\mathbf{g}\mathbf{g}'}\,\alpha_{\mathbf{k}-\mathbf{g}'} = 0, \tag{3.64}$$

where $\Gamma_{\mathbf{g}\mathbf{g}'}$ is an integral involving $\phi_{\mathbf{k}-\mathbf{g}}$, $\phi_{\mathbf{k}-\mathbf{g}'}$, and the Hamiltonian operator associated with the periodic potential.

It would take us too far into the details of the method to show the derivation of a formula for these integrals. The result is as follows

$$\begin{aligned}\Gamma_{\mathbf{g}\mathbf{g}'} = \frac{4\pi R_s^2}{v_c}\Bigg\{&-[(\mathbf{k}-\mathbf{g}')\cdot(\mathbf{k}-\mathbf{g}) - \mathscr{E}]\,\frac{j_1(|\mathbf{g}-\mathbf{g}'|\,R_s)}{|\mathbf{g}-\mathbf{g}'|}\\ &+\sum_{l=0}^{\infty}(2l+1)\,P_l(\cos\theta_{\mathbf{g}\mathbf{g}'})\,j_l(|\mathbf{k}-\mathbf{g}|\,R_s)\,j_l(|\mathbf{k}-\mathbf{g}'|\,R_s)\\ &\qquad\times\frac{\mathscr{R}_l'(R_s, \mathscr{E})}{\mathscr{R}_l(R_s, \mathscr{E})}\Bigg\}\end{aligned} \tag{3.65}$$

(where $\theta_{\mathbf{g}\mathbf{g}'}$ is the angle between $(\mathbf{k}-\mathbf{g})$ and $(\mathbf{k}-\mathbf{g}')$, and where $\mathscr{R}_l'$ is the derivative of the radial function $\mathscr{R}_l$). The essence of the procedure is to find a contribution like the first main term in (3.65) from the regions outside the spheres; the rest comes from the gradient operator acting on the discontinuity of slope in each A.P.W. at the surface of each sphere.

The band-structure problem is now solved by seeking a solution of the linear equations (3.64)—a solution that can only exist if the secular determinant vanishes. Now suppose we choose a value of $\mathbf{k}$. All the coefficients will depend on $\mathscr{E}$—either explicitly as in (3.64), or implicitly through (3.65). One can compute the determinant for a given value of $\mathscr{E}$, which can then be varied until a root is discovered. This value of $\mathscr{E}$ is our estimate of $\mathscr{E}(\mathbf{k})$.

The A.P.W. method is a sound practical procedure for the calculation of the band structure of metals. It requires a good deal of computing but it works. For example, the result is not very sensitive to the choice of R_s, the radius of the atomic potential and remains valid in the presence of d-bands (see § 3.10). The point of interest here, however, is to notice how, in the end, we are back at formulae that are similar in structure to those of the N.F.E. method. Equation (3.64) is exactly the same as equation (3.14), except that we have replaced the Fourier components of the potential of an atom by a more complicated expression,

$$\mathscr{V}_{\mathbf{g}-\mathbf{g}'} \rightarrow \Gamma_{\mathbf{g}\mathbf{g}'}.$$

In other words, $\Gamma_{\mathbf{g}\mathbf{g}'}$ is the Fourier component of an effective atomic potential. If it should happen not to depend very much on $\mathscr{E}$ and $\mathbf{k}$, but were a function of $|\mathbf{g}-\mathbf{g}'|$, then we could construct our Fermi surfaces, etc., by the N.F.E. method directly.

Unfortunately, the A.P.W. pseudo-potential components (3.65) are not usually as small as we should like. This follows from the empty lattice test, where we solve for $\mathscr{E}(\mathbf{k})$ in a 'lattice' with $\mathscr{V}(\mathbf{r}) \equiv 0$. The radial derivatives of the functions R_l in (3.65) do not automatically vanish at $r = R_s$, so that the coefficients $\Gamma_{\mathbf{g}\mathbf{g}'}$ do not go to zero even when the Fourier Components $\mathscr{V}_{\mathbf{g}-\mathbf{g}'}$ of the true potential are zero. A great deal of computation would then have to be done merely to arrive at the trivial result $\mathscr{E}(\mathbf{k}) = k^2$. The practical successes of the A.P.W. method cannot, therefore, be immediately justified by an appeal to the pseudo-potential concept of § 3.6. This point will be discussed further in § 3.8.

3.8 The Green function method

A technique that looks very different in principle, and yet works out quite similarly, is to start from the standard integral equation for the wave-function in any potential $\mathscr{V}(\mathbf{r})$:

$$\psi(\mathbf{r}) = -\frac{1}{4\pi}\int \frac{\exp(i\kappa|\mathbf{r}-\mathbf{r}'|)}{|\mathbf{r}-\mathbf{r}'|}\mathscr{V}(\mathbf{r}')\,\psi(\mathbf{r}')\,d\mathbf{r}', \tag{3.66}$$

where
$$\kappa^2 = \mathscr{E} \tag{3.67}$$

so that κ is real for $\mathscr{E} > 0$ and imaginary for $\mathscr{E} < 0$.

This says that, self-consistently, the wave-function is scattered, by the potential, into itself; the wavelets coming from all different points $\mathbf{r}'$, with strength $\mathscr{V}(\mathbf{r}')\psi(\mathbf{r}')$, combine to give $\psi(\mathbf{r})$ again. Now suppose that we have a muffin-tin potential in which

$$\mathscr{V}(\mathbf{r}) = \sum_{l} v_a(\mathbf{r}-\mathbf{l}). \tag{3.68}$$

Substituting in (3.66) we have

$$\psi(\mathbf{r}) = -\frac{1}{4\pi}\sum_{l}\int \frac{\exp(i\kappa|\mathbf{r}-\mathbf{r}'|)}{|\mathbf{r}-\mathbf{r}'|}v_a(\mathbf{r}'-\mathbf{l})\psi(\mathbf{r}')\,d\mathbf{r}', \tag{3.69}$$

where the integral is over only the cell around $\mathbf{l}$, since $v_a(\mathbf{r}-\mathbf{l})$ vanishes elsewhere.

We are looking for a Bloch function. We can use this property to move the origin of $\psi(\mathbf{r}')$ to $(\mathbf{r}'-\mathbf{l})$, which we relabel as $\mathbf{r}''$. Thus

$$\begin{aligned}\psi_{\mathbf{k}}(\mathbf{r}) &= -\frac{1}{4\pi}\sum_{l}\int \frac{\exp(i\kappa|\mathbf{r}-\mathbf{r}'|)}{|\mathbf{r}-\mathbf{r}'|}v_a(\mathbf{r}'-\mathbf{l})\,e^{i\mathbf{k}\cdot\mathbf{l}}\psi_{\mathbf{k}}(\mathbf{r}'-\mathbf{l})\,d\mathbf{r}' \\ &= -\frac{1}{4\pi}\int \{G(\kappa,\mathbf{k};\mathbf{r}-\mathbf{r}'')\}v_a(\mathbf{r}'')\psi_{\mathbf{k}}(\mathbf{r}'')\,d\mathbf{r}''.\end{aligned} \tag{3.70}$$

In effect we are now saying that the wave-function at $\mathbf{r}$ is the sum of contributions from wavelets scattered from all other cells at $\mathbf{l}$, etc., with appropriate phase factors. Since all cells are identical, we have been able to introduce the *structural Green function* or *Greenian*

$$G(\kappa,\mathbf{k};\mathbf{r}-\mathbf{r}'') \equiv \sum_{l}\frac{\exp(i\kappa|\mathbf{r}-\mathbf{r}''+\mathbf{l}|)}{|\mathbf{r}-\mathbf{r}''+\mathbf{l}|}e^{i\mathbf{k}\cdot\mathbf{l}}, \tag{3.71}$$

which gives the effect at $\mathbf{r}$ due to waves scattered from $\mathbf{r}''$, and from all equivalent points in all other cells (Fig. 59*b*). Thus, we need only integrate through a single cell to get the whole wave-function.

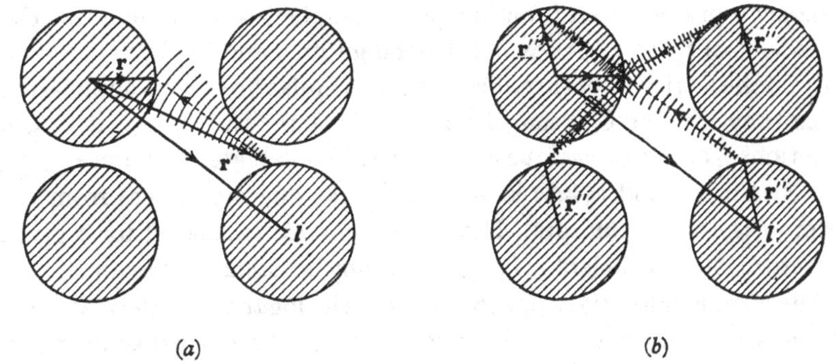

Fig. 59. The ordinary Green function, (a), for propagating from $\mathbf{r}'$ to $\mathbf{r}$ is replaced by a structural Green function, (b), which combines waves from all equivalent points $\mathbf{r}''$ in the lattice.

The next step in the argument is to construct a functional which gives the integral equation (3.70) upon variation. This is easily proved to be

$$\Lambda = N\int \psi_{\mathbf{k}}^*(\mathbf{r})\, v_a(\mathbf{r})\, \psi_{\mathbf{k}}(\mathbf{r})\, d\mathbf{r}$$
$$+\frac{1}{4\pi}\iint \psi_{\mathbf{k}}^*(\mathbf{r})\, v_a(\mathbf{r})\, G(\kappa,\mathbf{k}\,;\mathbf{r}-\mathbf{r}'')v_a(\mathbf{r}'')\psi_{\mathbf{k}}(\mathbf{r}'')\, d\mathbf{r}\, d\mathbf{r}''. \qquad (3.72)$$

The integrations need only cover a single cell—and because the potential vanishes outside each atomic sphere, it can be confined in each case to $r, r'' \leqslant R_s$.

We now assume a trial wave-function of the same form as (3.59) inside each cell

$$\psi_{\mathbf{k}}(\mathbf{r}) = \sum_{lm} C_{lm}\, \mathscr{R}_l(r)\, Y_{lm}(\theta,\phi). \qquad (3.73)$$

It is evident that (3.72) will become, eventually, a quadratic function

$$\Lambda = \sum_{lm;\,l'm'} \Lambda_{lm;\,l'm'}\, C_{lm}\, C^*_{l'm'} \qquad (3.74)$$

in the coefficients C_{lm}. The variational condition gives a set of linear equations whose determinant $|\Lambda_{lm;\,l'm'}|$ must vanish. This determinant is a function of the wave-vector $\mathbf{k}$ and of the energy $\mathscr{E} = \kappa^2$. The roots of the determinant give the required relationships for the branches of $\mathscr{E}(\mathbf{k})$.

The actual form of this determinant is of some interest. Its terms are

$$\Lambda_{lm;\,l'm'} = A_{lm;\,l'm'} + \kappa\delta_{ll'}\,\delta_{mm'}\frac{n_l' - n_l L_l}{j_l' - j_l L_l}. \qquad (3.75)$$

The coefficients $A_{lm;\,l'm'}$ are structural in origin. They come from the expansion of the Green function (3.71) in spherical harmonics for the directions of **r** and **r**″. They are dependent on κ and **k**—but otherwise only on the lattice structure. The actual formulae are very complicated, since they involve Clebsch–Gordan coefficients and spherical Bessel functions j_l and n_l of various arguments. To evaluate the lattice sums, the Ewald method (§ 2.3) is also useful (as one may guess from the form of the denominators). Nevertheless, once tabulated for a given structure, the coefficients $A_{lm;l'm'}$ need not be calculated again.

The atomic potential appears through the logarithmic derivative L_l of the radial function $\mathscr{R}_l$ at the surface of the sphere—just as in (3.66). This arises because the trial function (3.73) satisfies the Schrödinger equation inside the atomic sphere, so that we may write

$$v_a(\mathbf{r})\,\psi_{\mathbf{k}}(\mathbf{r}) = (\nabla^2+\mathscr{E})\,\psi_{\mathbf{k}}(\mathbf{r}) \tag{3.76}$$

in the integrands of (3.70) and (3.72). Use of Green's theorem then yields a surface integral in which there are derivatives of the Green function (these appear as n_l' and j_l' in (3.75)) and of the radial function. In this method the assumption of a spherically symmetrical muffin-tin potential for each atom is essential.

It is clear from this discussion that the *Green function method* (also known as the method of Korringa and of Kohn and Rostoker, or KKR method) is very similar to the A.P.W. method. In the A.P.W. method we expand as far as we like in spherical harmonics, and then have to solve a secular determinant for contributions from different reciprocal lattice vectors. In the KKR method we sum over lattice vectors (actually, over reciprocal lattice vectors, by a Fourier transformation of $G(\kappa, \mathbf{k}; \mathbf{r}-\mathbf{r}'')$), but then have a secular determinant in the contributions from different spherical harmonics. Both methods require a high-speed computer and they both work quite well. The advantage of the KKR method is that structural and atomic properties are segregated in the terms of the determinant, so that they can be evaluated separately and combined quickly. The advantage of A.P.W.'s is that they give a better picture of the wave-functions outside the atomic sphere.

3.9 Model pseudo-potentials

We saw in § 3.6 that the effect of each atomic potential on the electrons seemed no more than that of a weak pseudo-potential. How does this important principle enter in the band structure techniques based upon a muffin-tin lattice?

The atomic potentials do not enter explicitly in the A.P.W. and KKR formulae (3.65) and (3.75), which depend only on the gradient of $\mathscr{R}_l$ at the surface of the atomic sphere. This is a quantity that also occurs in the partial wave theory of scattering by a spherical potential (see § 5.5). We match each radial solution $\mathscr{R}_l(r, \mathscr{E})$ to a free electron wave of the same energy $\mathscr{E} = \kappa^2$ and angular momentum l, thus defining a *phase shift*, $\eta_l(\mathscr{E})$, related to the logarithmic derivative of $\mathscr{R}_l$ by the formula

$$L_l \equiv \frac{\mathscr{R}_l'(R_s, \mathscr{E})}{\mathscr{R}_l(R_s, \mathscr{E})} = \frac{j_l'(\kappa r) - \tan \eta_l(\mathscr{E}) . n_l'(\kappa r)}{j_l(\kappa r) - \tan \eta_l(\mathscr{E}) . n_l(\kappa r)}\bigg|_{r=R_s}. \tag{3.77}$$

The spherical Bessel function j_l and n_l (with derivatives j_l' and n_l') are, of course, free-space solutions of the radial Schrödinger equation (3.60), outside the muffin-tin well.

Knowledge of the phase shifts of all the partial waves at all energies would thus give us all the information we need about the atomic potential; in (3.75), for example, the diagonal elements of the secular determinant contain simply $\kappa \cot \eta_l(\mathscr{E})$. This is natural enough; the phase shifts also give the scattering cross-section of each atom, and hence the diffraction pattern of the lattice for electrons of this energy. We might expect each atom to behave as if it had a *t-matrix* or *scattering amplitude*, for scattering through the angle θ, of the standard partial wave form

$$t_a(\theta) = -\left(\frac{4\pi N}{\kappa}\right) \sum_l (2l+1) \sin \eta_l \exp(i\eta_l) P_l(\cos\theta). \tag{3.78}$$

A deep potential may well give rise to a large phase shift—but in (3.77) and (3.78) we may obviously ignore all integral multiples of π in each η_l. It is only the remainder that counts—and this remainder can leave the scattering very weak. In the partial wave theory, the cross-section, $|t_a(\theta)|^2$ does not increase enormously as the potential well is deepened; it fluctuates, as bound states are brought in, and is seldom larger than a few electron wavelengths. In other words, *the Born approximation, in which the cross-section for scattering from* $\mathbf{k}$ *to* $\mathbf{k}'$ *is proportional to* $|\langle \mathbf{k}' | v_a | \mathbf{k} \rangle|^2$, *is not valid when* $v_a(\mathbf{r})$ *is a very deep potential*. We must then replace the matrix elements of the potential by the partial wave formula (3.78).

That is why the simple N.F.E. program of § 3.2 does not converge. We are using perturbation theory in an attempt to describe the scattering by each atom of the crystal. But the high Fourier coefficients needed to describe the rapid oscillations of the wave-function inside the atom are largely irrelevant to the general diffraction pattern of the Bloch states outside the core. Each node of $\mathscr{R}_l$ adds π to η_l, but makes little difference in (3.65), (3.75), or (3.78). The O.P.W. pseudo-potential (3.50) is thus an analytical device for removing these inner nodes by the subtraction of orthogonalized core states. The residual weak pseudo-potential Γ is just an operator whose matrix elements give rise to essentially the same wave-function outside the atomic core as the original atomic potential v_a.

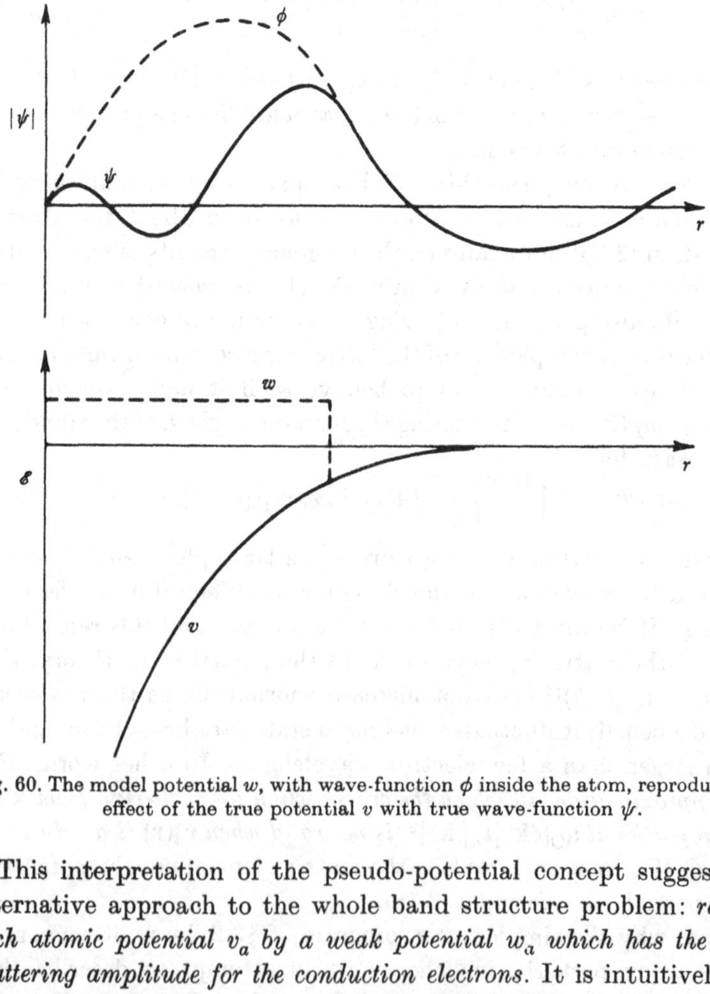

Fig. 60. The model potential w, with wave-function ϕ inside the atom, reproduces the effect of the true potential v with true wave-function ψ.

This interpretation of the pseudo-potential concept suggests an alternative approach to the whole band structure problem: *replace each atomic potential v_a by a weak potential w_a which has the same scattering amplitude for the conduction electrons.* It is intuitively obvious that $\mathscr{E}(\mathbf{k})$ in our original crystal will be identical with the band structure in this hypothetical material, where the N.F.E. formalism should now converge. All we need to do is to replace $\mathscr{V}_{\mathbf{g}'-\mathbf{g}}$ in (3.14), or

$\Gamma_{\mathbf{gg'}}$ in (3.64) by the corresponding Fourier component, of our *model potential* w_a.

This programme encounters the difficulties of non-locality, energy dependence and arbitrariness, already discovered with analytical pseudo-potentials of the form (3.56). For a given energy $\mathscr{E}$ and angular momentum l there are infinitely many localized potentials, $w_l(\mathbf{r}, \mathscr{E})$ in which the solution of the radial Schrödinger equation (3.60) exactly reproduces the true radial function $\mathscr{R}_l$ outside the atomic core. But the same model potential will not usually work for a different value of l or at a different energy, so that we must introduce operators that separate out the various angular momentum components in a wave-function just as in (3.55). In practice, the functional form of each $w_l(\mathbf{r}, \mathscr{E})$ in a *model pseudo-potential* is chosen more for computational convenience than for deep analytical reasons.

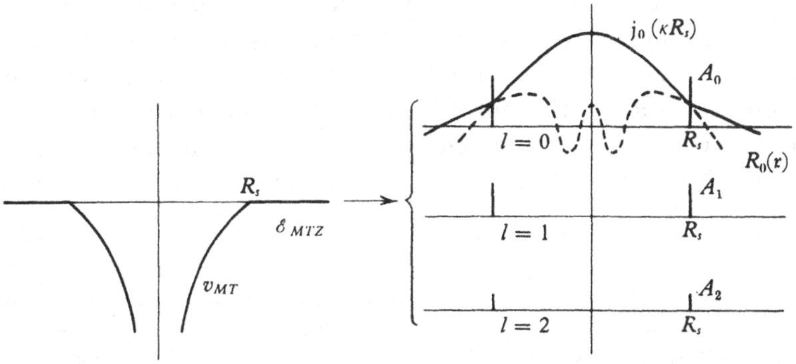

Fig. 61. Model pseudo-potential for KKR matrix elements.

In a muffin-tin lattice, an obvious choice is a delta-function singularity at the surface of the atomic sphere, of sufficient strength, for each value of l, to reproduce the scattering phase shift $\eta_l(\mathscr{E})$ (Fig. 61). Using standard analytical properties of plane waves and spherical Bessel functions, we find 'pseudo-potential' matrix elements of the form

$$\Gamma^{\mathrm{KKR}}_{\mathbf{gg'}} = -\frac{4\pi N}{\kappa}\sum_l (2l+1)\tan\eta'_l \frac{j_l(|\mathbf{k}-\mathbf{g}|\,R_s)\,j_l(|\mathbf{k}-\mathbf{g}'|\,R_s)}{\{j_l(\kappa R_s)\}^2} P_l(\cos\theta_{\mathbf{gg'}}), \tag{3.79}$$

where $$\cot\eta'_l \equiv \cot\eta_l - n_l(\kappa R_s)/j_l(\kappa R_s). \tag{3.80}$$

This formula closely resembles (3.78), the scattering amplitude of a single isolated atom, to which it tends, in the limit of small phase shifts, when $|\mathbf{k}-\mathbf{g}| \sim |\mathbf{k}-\mathbf{g}'| \sim \kappa$ near a zone boundary.

Strangely enough, a set of N.F.E. equations with these matrix elements may be derived from the secular determinant of the KKR equations, by a succession of matrix transformations from the angular momentum representation (3.64) to a reciprocal lattice representation. This formula also has algebraic connections with the A.P.W. matrix elements (3.65), which cannot themselves be derived from a set of simple local potentials.

In practice, the calculation of a muffin-tin potential distribution for a given crystal is a fairly complicated process, where screening, correlation and exchange effects play an important part (see Chapter 5). It is very convenient to replace the deep potential in the core of each atom, by a model pseudo-potential w at the very beginning, before, so to speak, bringing the atoms together to make a lattice. If this pseudo-potential is allowed to behave outside the core like the coulomb function $-Ze^2/r$ seen by the valence electrons, we may even calculate the free atom spectroscopic levels, and adjust the parameters of w to give the observed values. This procedure, which bypasses many theoretical computations of self-consistent atomic potentials etc., is the basis of the *Heine–Abarenkov*, and *Shaw*, *pseudo-potentials* sketched in Fig. 62. In w^{HA} the interior of the 'core', of arbitrary radius is made

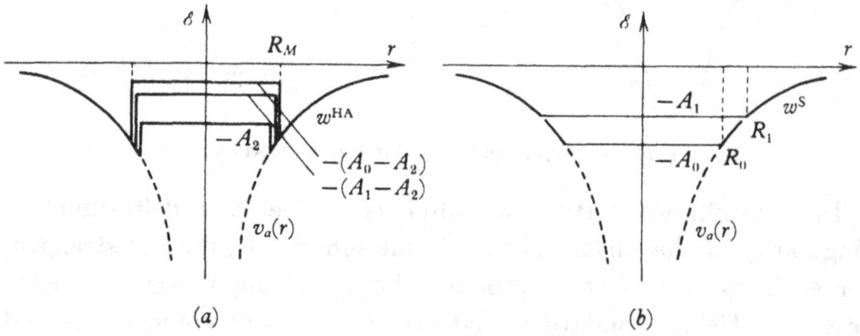

Fig. 62. Model pseudo-potentials: (*a*) Heine–Abarenkov model; (*b*) Shaw model.

flat, but raised or lowered to give the correct electron eigenstates for each angular momentum: in w^{S}, a radius R_l is chosen for each value of l so that there is no sharp discontinuity in the potential at the 'edge of the core'. Difficulties arise, however, in the choice of these parameters at the energy $\mathscr{E}$ of the Bloch state, which does not necessarily lie within the set of spectroscopic energy levels.

3.10 Resonance bands

In early applications of the tight-binding method (§ 3.4), it was noticed that the d-orbitals are so closely concentrated within the core of an atom that they produce only narrow bands by overlap with neighbours. In the *transition elements* these inner d-states are not all filled, but lie very close to the s or p 'valence' states, which themselves combine to form an ordinary conduction band. The narrow 'd-band', with a density of states capable of holding up to 10 electrons per atom, lies within the 's-p-band', and *hybridizes* with it where they cross. (Fig. 63). This situation is, of course, very important because of the associated magnetic phenomena.

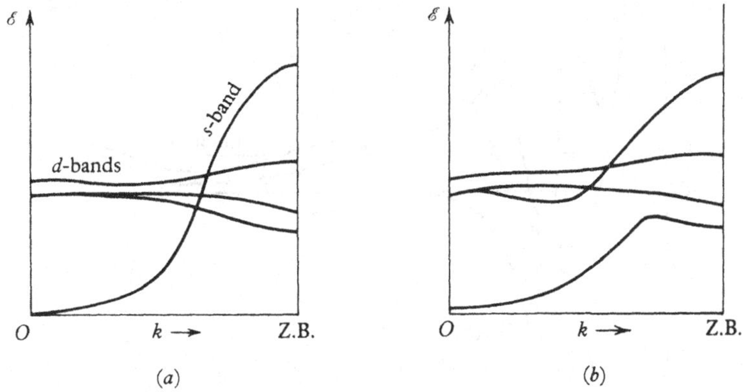

Fig. 63. (*a*) d-bands crossing s-bands; (*b*) s–d hybridization.

It is not difficult to set up an interpolation scheme, or *model Hamiltonian* with adjustable parameters, in which the d-band is represented by the matrix of an L.C.A.O. expression such as (3.31) with suitable pseudo-potential components. The extent of hybridization is then defined by further arbitrary matrix elements linking these two sub-matrices.

But a proper quantitative analysis must allow for the point emphasized in § 3.4 (see Fig. 54): as we make the crystal, the atomic orbitals themselves are destroyed by the overlap of potentials. This destruction is not, however, always complete; *virtual bound states* may still be observed at about the energy of the original d-orbitals. In the theory of lattice dynamics (§ 2.12) we referred to a *resonance*—an excitation that is strongly concentrated in the neighbourhood of an impurity, but that has finite amplitude throughout the crystal.

Similarly, the electron wave-function at a resonance energy may have large amplitude within the muffin-tin well, but is not strictly localized therein. Consideration of the radial Schrödinger equation (3.60) shows that this can easily occur when $l \neq 0$; the 'centrifugal potential' $l(l+1)/r^2$ provides a barrier through which the electron may tunnel from its virtual state (Fig. 64).

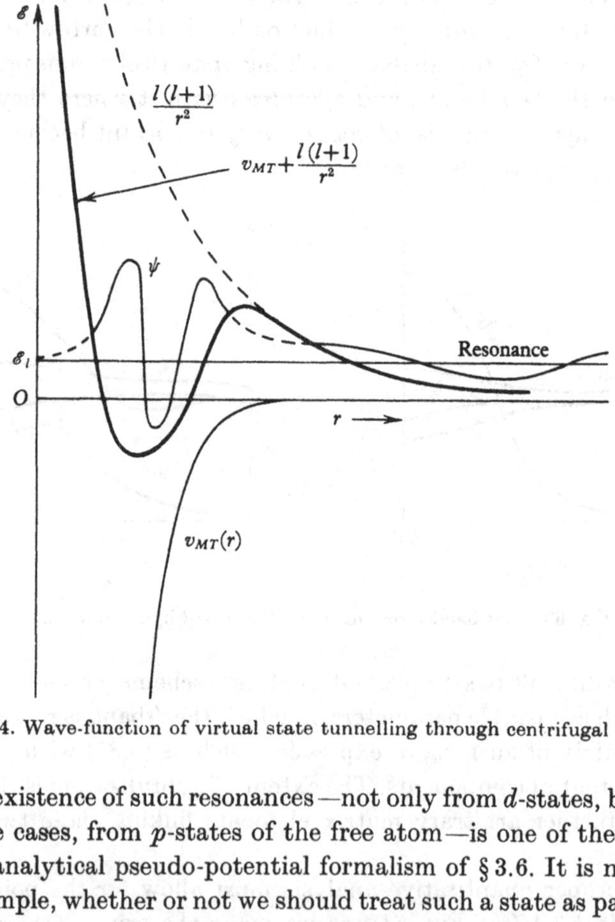

Fig. 64. Wave-function of virtual state tunnelling through centrifugal barrier.

The existence of such resonances—not only from d-states, but even, in some cases, from p-states of the free atom—is one of the hazards of the analytical pseudo-potential formalism of § 3.6. It is not clear, for example, whether or not we should treat such a state as part of the 'core' for orthogonalization. To understand the situation, it is essential to turn, once more, to general scattering theory, where resonances are familiar phenomena. In the neighbourhood of a resonance energy $\mathscr{E}_l$, the phase shift of the lth partial wave passes through $\pi/2$, according to the formula

$$\tan \eta_l(\mathscr{E}) \approx \frac{W_l}{\mathscr{E}-\mathscr{E}_l}. \tag{3.81}$$

Putting this formula into, say, an A.P.W. or KKR program, we may crank the handles and arrive at a band structure. The result is to produce complex structure in $\mathscr{E}(\mathbf{k})$, centred on $\mathscr{E}_l$, just as in the tight-binding picture (Fig. 63). But the position and width of this *resonance band* within the conduction band are now determined by the value of $\mathscr{E}_l$, measured relative to the muffin-tin zero, and by the width, W_l of the resonance, neither of which can be deduced accurately from the L.C.A.O. model.

If we now substitute from (3.81) into the KKR pseudo-potential (3.79), we find a singularity in $\Gamma_{\mathbf{gg}'}$ in the neighbourhood of $\mathscr{E}_l$, quite spoiling the convergence of the N.F.E. equations. This is not an artefact of the method. Strong energy dependence of pseudo-potential matrix elements, even at a considerable distance in energy from the resonance level, is an essential feature of the physics of electronic band structure in many solids, and is not always allowed for fully in the computations.

3.11 Crystal symmetry and spin-orbit interaction

In the calculation of electron-band structure, the symmetry of the crystal structure is of great importance. To discuss this properly requires group theory, which is too big a subject for us to embark upon here. The basic theorem is, in general terms, that *the energy function in the Brillouin zone has the full point group of the crystal.* Any operation, such as rotating the crystal around an axis, that leaves it invariant, also transforms the function $\mathscr{E}(\mathbf{k})$ into itself.

This can be used to classify and simplify the energy levels and wave-functions corresponding to various points in the Brillouin zone. To see how this can happen consider a square lattice with a square zone. Consider some general point like $\mathbf{k}$. Suppose we are attempting an A.P.W. or O.P.W. expansion of the wave-function in the form

$$\psi_{\mathbf{k}} = \sum_{\mathbf{g}} \alpha_{\mathbf{k}-\mathbf{g}} \phi_{\mathbf{k}-\mathbf{g}} \tag{3.82}$$

as in (3.43) or (3.63). We might decide that only reciprocal vectors of the set illustrated in Fig. 65(*a*) need be included in the sum. But then all the vectors $(\mathbf{k}-\mathbf{g})$ will be different in magnitude and direction, so that all the coefficients $\alpha_{\mathbf{k}-\mathbf{g}}$ will be different, without any obvious relations between them. We cannot avoid having to solve a determinantal equation of the maximum degree of complexity.

But suppose we were looking for a solution at some point like Δ, on a symmetry axis of the zone. It is obvious that there will be many coefficients in (3.82) the same. For example, the pairs ($\mathbf{g}_5$ and $\mathbf{g}_8$), ($\mathbf{g}_2$ and $\mathbf{g}_4$), ($\mathbf{g}_6$ and $\mathbf{g}_7$) are symmetrically placed with respect to Δ, and so will have the same coefficient in the expansion. Instead of 8 separate coefficients, we can assume that there are only 5. Thus, the determinant will be of lower degree, and can thus be solved more readily.

The same will hold at a point such as Σ, on the diagonal. At X and M, there will be further simplifications, because these are half-way along reciprocal lattice vectors, so that other equivalences between coefficients can be used. In effect, the higher the symmetry of the point in the zone, the lower the degree of the equation to be solved. *For a given amount of computing, we can get more accurate values of* $\mathscr{E}(\mathbf{k})$ *at points of high symmetry, than we can at an arbitrary point in the zone.*

Group theory tells us rather more than this. It will generate for us the functions that have the same coefficient—or, in more complicated

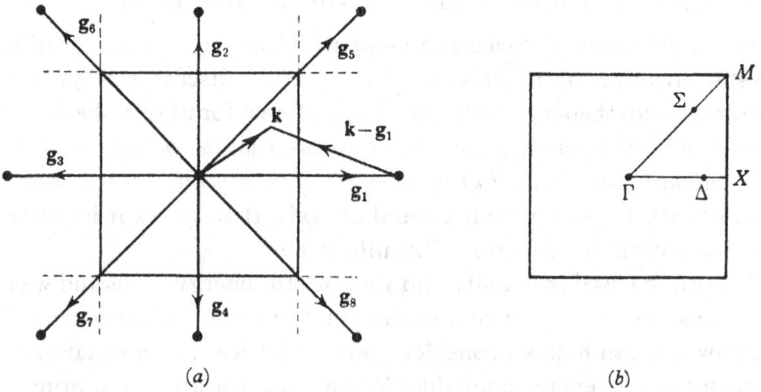

Fig. 65. (*a*) General point **k** in reciprocal space. (*b*) Symmetry points and lines in a square zone.

cases, the same coefficient multiplied by a number such as $1/\sqrt{3}$. It also tells us at what points we may expect degeneracy between the levels. For example, at the point M there are, say, four O.P.W.'s or A.P.W.'s that are degenerate in energy (corresponding to $\mathbf{k}$, $(\mathbf{k}-\mathbf{g}_5)$, $(\mathbf{k}-\mathbf{g}_1)$ and $(\mathbf{k}-\mathbf{g}_2)$ in Fig. 65). These will be split by the perturbation associated with the lattice potential—but one can prove that a two-fold degeneracy will always remain. When we draw energy surfaces, this information is of great help.

There is one case where group theory is of great importance. In heavy atoms we know that there is a strong *spin-orbit interaction*—an operator of the form $\lambda \mathbf{L}\cdot\mathbf{S}$, where $\mathbf{S}$ is the spin operator for the electrons and $\mathbf{L}$ is the operator for the orbital angular momentum. In the free atom this has the effect of removing the degeneracy between some states of the same space wave-function but opposite electron spin. For example, an atomic P-state is split into two states, $P_{\frac{3}{2}}$ and $P_{\frac{1}{2}}$, corresponding to the total angular momenta $j = \frac{3}{2}$ and $j = \frac{1}{2}$, respectively.

The assumption we have made in constructing energy bands is that the spin is irrelevant; we just use a wave-function in space, and for each $\psi_{\mathbf{k}}$ assume that we can put two electrons, of opposite spins, into the two states with this wave-function. The spin-orbit interaction may resolve the degeneracy between these states. Notice, however, that the fundamental quantum principle of *time reversal symmetry* always leaves *Kramers degeneracy* between any Bloch state $\psi_{\mathbf{k}}(\mathbf{r})$ and its complex conjugate $\psi_{\mathbf{k}}^{*}(\mathbf{r})$, which describes a state in which both the wave-vector and the spin of the electron have been reversed. For 'spinless' particles, this implies $\mathscr{E}(\mathbf{k}) = \mathscr{E}(-\mathbf{k})$, regardless of point group symmetries of the lattice.

At a general point in the zone, where the $\psi_{\mathbf{k}}$ themselves are not degenerate, the spin-orbit interaction does not separate the states of opposite spin if the lattice potential has a centre of inversion. In effect, reversing the spin in the state $\psi_{\mathbf{k}}(\mathbf{r})$ is equivalent to looking at $\psi_{\mathbf{k}}(-\mathbf{r})$, which has the same energy.

But there are special points in the zone where the spin-orbit effect may be quite important. The most interesting case is at the centre of the zone, a point of cubic symmetry. In the tight-binding model we could use p-orbitals to construct a 'p-band'. But there are three degenerate p-states of an atom, so that we have three such bands—corresponding, say, to the p_x, p_y, and p_z orbitals of the atom. These three bands are degenerate at $\mathbf{k} = 0$; they are simply transformed into each other by the cubic symmetry at that point. Each band state is doubly spin-degenerate, so that there is six-fold degeneracy altogether.

When we apply the spin-orbit interaction we find that this is split into a four-fold and a two-fold level. In effect, we should now use the atomic states $p_{\frac{3}{2}}$, which is four-fold degenerate, and $p_{\frac{1}{2}}$ which is two-fold degenerate, to construct our bands. It is interesting to see what happens as we go away from $\mathbf{k} = 0$ in some arbitrary direction. If there is inversion symmetry, then each level is still doubly degenerate, as in Fig. 66 (*b*). We can say that the $j = \frac{3}{2}$ band splits again into the band with $m_j = \pm\frac{3}{2}$ (measured along the direction of propagation, say), and a band with $m_j = \pm\frac{1}{2}$. Without inversion symmetry, these would again split as in Fig. 66 (*c*); through the spin-orbit interaction the electron spin becomes sensitive to the amount of 'screw' in the crystal potential along the direction of propagation.

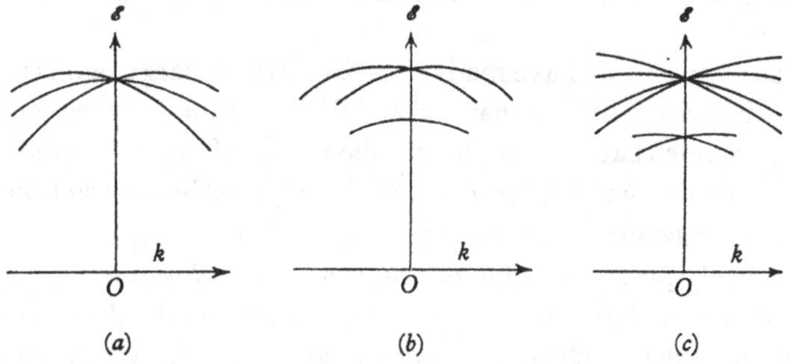

Fig. 66. Effect of spin-orbit interaction on *p*-type levels near centre of zone: (*a*) 6 levels degenerate at $\mathbf{k} = 0$; (*b*) in lattice with inversion symmetry, spin-orbit interaction leaves all states doubly degenerate; (*c*) in absence of inversion symmetry, degeneracy is completely resolved.

These effects are important at the top of the valence band in a semiconductor such as Ge or InSb. In the former case, Fig. 66 (*b*) holds: in the latter there is no centre of inversion, and the degeneracy at the centre of the zone is resolved as we go in an arbitrary direction. The actual sign of the splitting, and its magnitude, can be guessed from the sign and magnitude of the corresponding effects in the levels of the free atom.

To make a quantitative calculation, it is convenient to treat the spin-orbit interaction $\lambda\mathbf{L}\cdot\mathbf{S}$ as a perturbation. Near the centre of the zone, this acts mainly on the periodic part, $u_{\mathbf{k}}$ of the Bloch function, which satisfies the equation (3.38). In fact we may exploit the similarity of this equation to an ordinary Schrödinger equation by treating the extra term

$$-\frac{i\hbar^2}{m}\mathbf{k}\cdot\nabla = \frac{\hbar}{m}\mathbf{k}\cdot\boldsymbol{p} \tag{3.83}$$

as an additional perturbation on the state u_0, for small values of $\mathbf{k}$, and thus calculate the spin-orbit splitting in detail without having to do a complete calculation of the band structure. This is known as the ***k·p** method*.

CHAPTER 4

STATIC PROPERTIES OF SOLIDS

Stand still you ever-moving spheres.... MARLOWE

4.1 Types of solid: band picture

We have gone into the theory of electronic structure at some length, because this determines the type of solid and its macroscopic properties.

Consider a completely full band, with an energy gap above it. This is the ground state, and from its symmetry it must be such that no

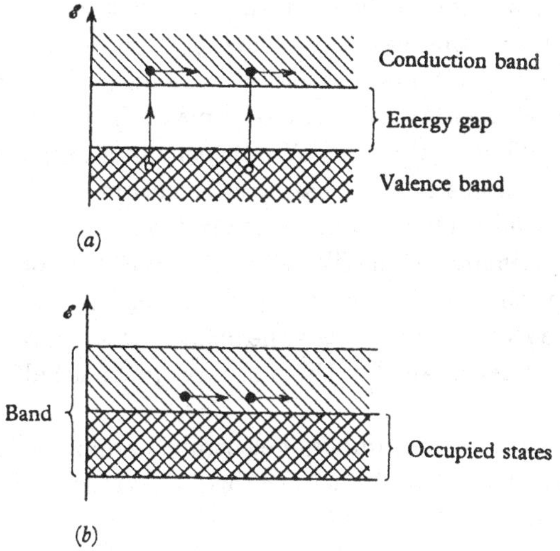

Fig. 67. (*a*) Excitation of carriers in semiconductor. (*b*) Current-carrying electrons in a metal.

electric current is flowing. To make a current one applies an electric field, which excites some electrons into states representing a net current. But if there is an energy gap, a finite excitation energy is required to carry the electrons up over the gap into the next band. This cannot be supplied by a small constant electric field, so that the solid would be an *insulator*.

But suppose the energy gap $\mathscr{E}_{gap}$ is small. At the temperature T there will be a small, but not zero, density of electrons excited by thermal fluctuations into the upper band—a density characterized by a Boltzmann factor like $\exp(-\mathscr{E}_{\text{gap}}/\mathbf{k}T)$. These electrons can easily carry a current, so that the material would have observable electrical conductivity, which would increase rapidly at higher temperatures. We say that we have a *semiconductor*.

If a band is not full, as in a monovalent metal, then although the true ground state of the system will be symmetrical as between $\psi_{\mathbf{k}}$ and $\psi_{-\mathbf{k}}$, thus again carrying no current, there are current-carrying states in the band an infinitesimal distance in energy above the top of the occupied levels. The conductivity will be high and will not depend very much on the temperature, except in so far as this governs the mechanism by which the electrons are scattered. We then have a typical *metal*.

To decide whether a solid of given structure is likely to be a metal, semiconductor, or effectively an insulator we recall (§ 1.6) that a Brillouin zone contains exactly as many 'allowed **k**-vectors', i.e. distinct electron wave-functions, as there are unit cells in the crystal. But each allowed spatial wave-function may contain 2 electrons, of opposite spins. Thus, if there are N unit cells in a unit volume, then there are $2N$ electron states per band, or 2 electron states per unit cell per band. By counting the number of electrons per unit cell, we can make certain estimates of the likely properties of the solid.

(i) A solid with *one* free electron per unit cell is always a metal. Typical are the *monovalent metals*—the *alkali metals* Li, Na, K, Rb, Cs and the *noble metals* Cu, Ag, Au. These all have just one half-filled band, or half-filled zone.

(ii) A solid with an *odd* number of electrons per unit cell is always a metal. Thus, in Al, Ga, In, Tl, we have three electrons per atom, which can only fill one and a half bands.

But note that As, Sb, Bi, with 5 electrons per atom, crystallize into a structure with 2 atoms per unit cell, so that this rule does not apply. In fact, these are *semi-metals*; the 10 electrons almost exactly fill 5 bands, but the 5th band is not quite full and there is a little overlap into the 6th band, so that there are always a few electrons available to carry a current, regardless of the temperature.

(iii) A solid with an *even* number of electrons per unit cell is not necessarily an insulator. This is because the bands may *overlap* in energy. It may be energetically advantageous to leave the lower band unfilled, and let some of the electrons go into the upper band.

In one dimension this does not happen: the bands are all distinct

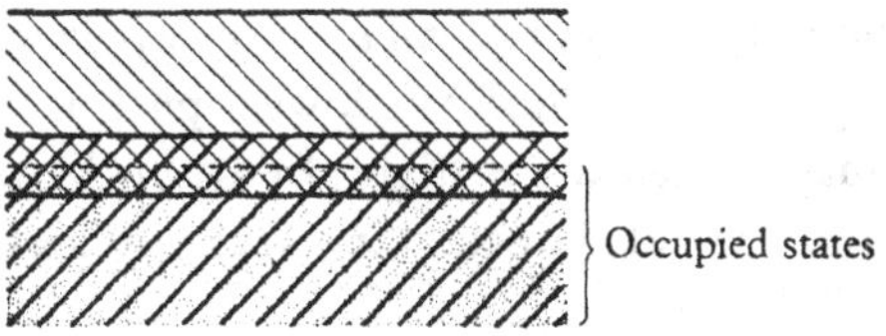

Fig. 68. Overlapping bands.

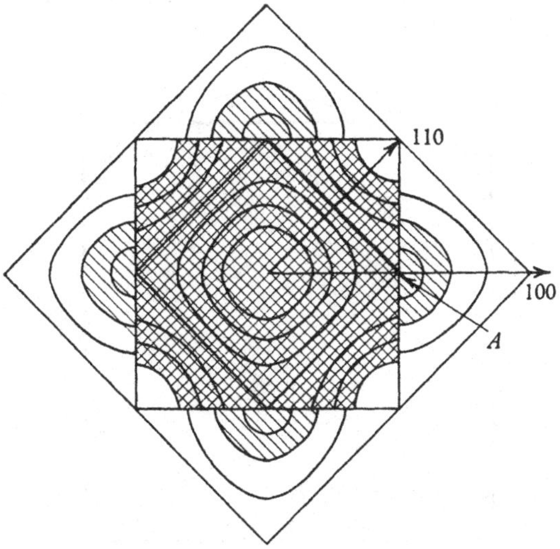

(a)

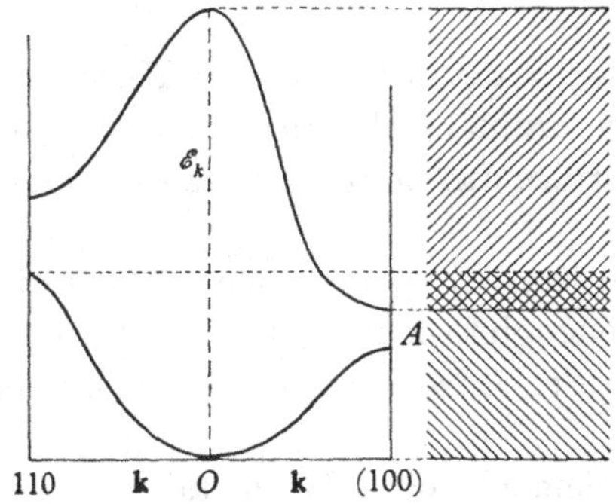

(b)

Fig. 69. Overlapping bands for divalent metal: (a) extended zone scheme; (b) energy as function of **k** in reduced zone.

(see Fig. 38). But in a real metal we may have the situation sketched in Fig. 69. Although there may be an energy gap as we cross the zone boundary at any particular place, the lowest part of the 2nd band (at A, for example) may lie below the highest part of the 1st band, so that electrons 'spill over' the edge of the zone at A and leave the corners of the zone unoccupied.

This is easily seen in the N.F.E. model in the extended zone scheme (cf. § 3.3). A sphere containing $2N$ states would intersect the zone boundaries in various places. If the corresponding Fourier components of potential (or pseudo-potential) are small, the effect is just to disconnect the portions of the sphere as they cross each zone

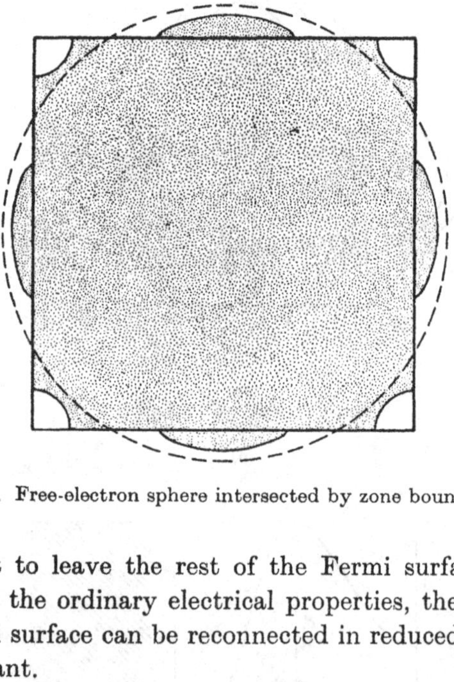

Fig. 70. Free-electron sphere intersected by zone boundaries.

boundary—but to leave the rest of the Fermi surface practically unaltered. For the ordinary electrical properties, the fact that the pieces of Fermi surface can be reconnected in reduced and repeated zones is irrelevant.

(iv) It turns out that all *divalent* elements are metals: the energy gaps are not large enough to contain the electrons within a single zone. But some of them (Sr, Ba) are poor conductors, where the overlap is probably small.

(v) *Tetravalent* elements are either metals or semiconductors. The interesting case is Sn, which is metallic in one phase and semiconducting in another. The change of crystal structure changes the shape of the zone, and hence the possibilities of large enough energy gaps to hold all the electrons. There is an interesting progression in Group IV of the Periodic Table, from C, which, as diamond, forms a semiconductor with such a large energy gap that in practice it is an insulator, to Si and Ge which are semiconductors, to Sn, which is both metal and semiconductor, to Pb which is a metal. The energy band structure of all these elements can be represented by N.F.E. models; the semiconductors are 'metals' with large energy gaps and no overlap.

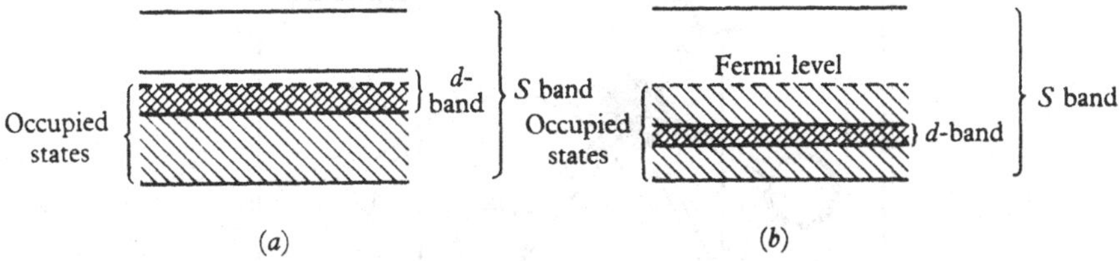

Fig. 71. (*a*) Transition metal. (*b*) Noble metal.

(vi) The *transition elements*, such as the iron group, Cr, Mn, Fe, Co, Ni, and other groups higher in the Periodic Table, are characterized by their incomplete inner *d*-shells. In atomic Fe, for example, only 6 of the 10 states in the 3*d*-shell are filled, despite the 2 other electrons in the outer 4*s*-state. When the atoms are brought together to make a solid or liquid, these outer valence electrons go into a broad '*s*-band', not very different from the N.F.E. conduction band in an ordinary metal. In some ways (see § 10.5) the *d*-electrons still behave as if localized in their atomic cores: for the 4*f*-electrons in the *rare earth* group this is actually a very good approximation. But the *d*-orbitals must actually overlap from atom to atom to form a narrow '*d*-band', capable of holding up to 10 electrons per atom. As we saw in § 3.10 this band crosses and hybridizes with the *s*-band; strictly speaking, this complicated band structure is associated with a resonance at the former atomic *d*-level, and cannot be analysed into separate *s*- and *d*-bands.

Nevertheless, it is natural to say, for example, that neither of the bands is full, so that the material is metallic, with conduction mainly by the '*s*-electrons' (Fig. 71 (*a*)). In the *noble metals* the *d*-states are in fact full, but they generate a resonance *d*-band within the ordinary N.F.E. band of the valence electrons, a few volts below the Fermi level (Fig. 31 (*b*)). This has significant effects in what would otherwise be simple monovalent metals (see § 9.4).

4.2 Types of solid: bond picture

This is about as far as we can go easily with a band model. But let us consider the semiconductors Si and Ge again. These have the '*diamond structure*' which is like F.C.C. (see § 1.3) except that there

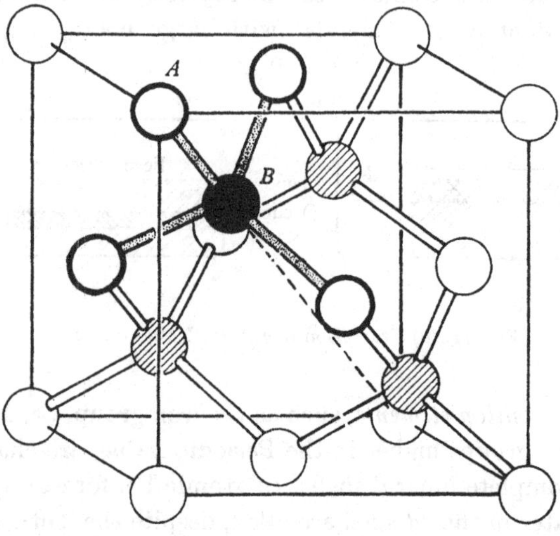

Fig. 72. Tetrahedral bonds in the diamond structure.

are two atoms per unit cell. Take an F.C.C. lattice, cornered at A, and put another atom at B a point $\frac{1}{4}$ of the way along the main diagonal of the cube. Make this new atom the corner of the second F.C.C. lattice, interpenetrating the first.

This structure has the same Brillouin zone as the F.C.C. lattice. We now have 8 electrons per unit cell, so that we need to fill 4 bands. If we look at these from the point of view of a tight-binding model, we find that they come from the 4 orbitals of the free atom—the $3s$- and the three $3p$-orbitals in Si; the $4s$- and $4p$-orbitals in Ge. These cross over and combine when the atoms are brought together, and make 4 bands which cover much the same range of energy (we noticed in § 3.11 that three of these would be degenerate at the centre of the zone, if it were not for spin-orbit interaction). The next band is separated from these *valence bands* by an energy gap of the order of 1 eV.

But this combination of s- and p-orbitals is well known in the theory of chemical bonds. It is known that these wave-functions may be combined to give four orbital wave-functions, each directed towards the vertex of a tetrahedron. It is known that the tetrahedral symmetry of a molecule such as CH_4 is associated with the formation of *covalent bonds*, by the sharing of electron pairs in these *s*-*p*-hybrid orbitals. Looking again at Fig. 72, we see that the atom at *B* has just 4 nearest neighbours, arranged at the vertices of a tetrahedron. Each of these is linked similarly to its neighbours, and so on: in other words, *the whole crystal is like one enormous covalently bonded molecule.*

Band-structure calculations confirm this. One can get quite a good match to the experimental band structure by a few O.P.W.'s with suitable pseudo-potential coefficients. The resulting wave-functions tend to concentrate the electron density along the 'bonds'. Schematically, we represent the electronic structure as in Fig. 73. Each Ge or Si ion shares 8 electrons with its neighbours. These electrons are not actually localized in the bonds. Like free electrons in a metal their wave-functions extend throughout the crystal. But there is always a highly concentrated electron cloud, equivalent to a pair of electrons with opposite spins, in the regions between neighbouring ions.

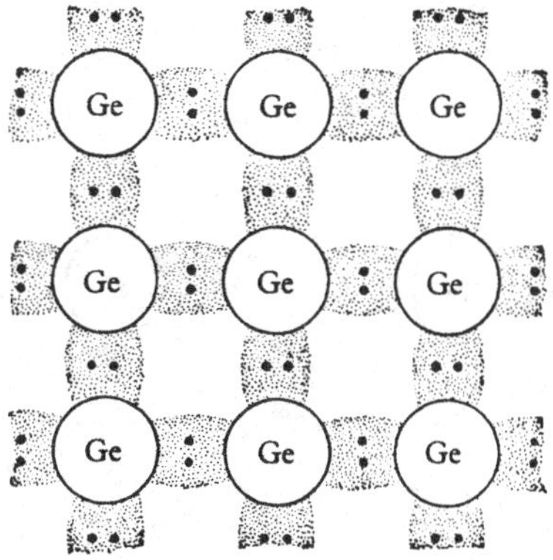

Fig. 73. Electronic structure of Ge (schematic).

Notice, however, that this is to some extent a self-fulfilling prophecy. In a lattice of low co-ordination number, where the muffin-tin approximation (§ 3.7) would not be appropriate, the potential energy of an electron would not be uniform throughout the interstitial regions. In any attempt at a self-consistent calculation (cf. § 5.2) electron charge would automatically accumulate along the line between neighbouring atomic spheres, where the interstitial potential would be lowest. The general chemical argument seems essential if we are to explain the occurrence of such crystal structures in the first place.

It is interesting now to modify this picture, by changing from Ge to InSb. We recall that In and Sb are in the same row of the Periodic Table as Ge, so they have the same basic core of closed shells. But In has only 3 outer electrons, and Sb has 5. This also forms a diamond-

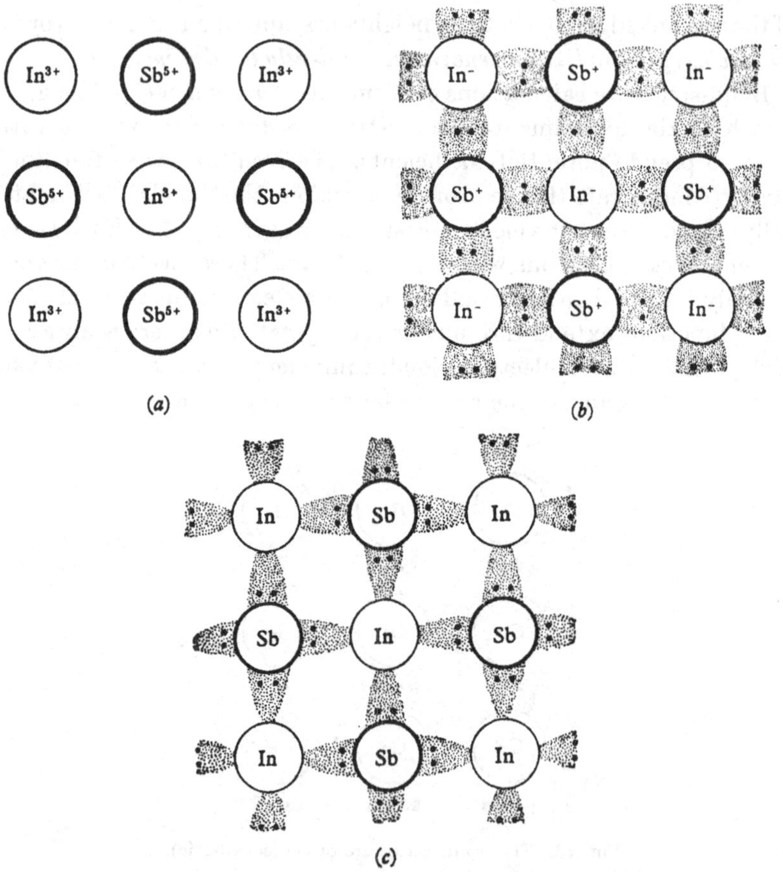

Fig. 74. Indium antimonide (schematic). (*a*) As bare ions. (*b*) With symmetrical bonds. (*c*) Bond electrons polarized by residual charge on ions.

like lattice, except now that the In occupies all the *A* sites and the Sb all the *B* sites. Schematically, we have an arrangement of ions as in Fig. 74 (*a*).

In many ways this is so like the Ge structure that we can put the 8 electrons per atom into the same 'bond–band' orbitals, as in Fig. 74 (*b*). But this leaves some residual charges, on the Sb, not quite neutralized by the average of 4 electrons around it, and puts a little extra negative charge on the In. However, these effects can easily be accommodated by a slight shrinking of the charge clouds towards the Sb, and away from the In, as in Fig. 74(*c*). This compound is a semiconductor, like Ge in its basic band structure, although, as it turns out, with a smaller energy gap. It is a typical *III–V intermetaltic compound*.

Now turn to ZnS, which is a *II–VI compound*—Zn has 2 electrons and S has 6. It forms a crystal exactly similar to InSb—indeed, this is called the *zinc-blende* structure—and is also a semiconductor. But now the electrostatic attraction of the electrons to the S^{6+} ions is enhanced by *electron affinity*: the 8 electrons that this ion can see tend to pull in still further, and try to form a complete closed shell leaving the Zn^{2+} ion practically free of electrons.

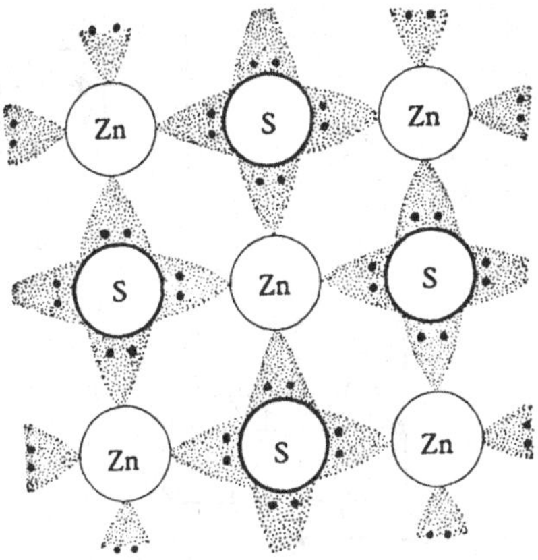

Fig. 75. Zinc sulphide (schematic).

This process goes the whole way in CuCl, say. The Cu atom completely loses its electron, which goes to make a Cl^- closed shell. We now have practically independent ions, Cu^+ and Cl^-, without any free electrons left over. This will crystallize in an alternating lattice as in Fig. 76. But the cohesion now is not associated with bonds. It comes essentially from the electrostatic attraction of the ions, as one might calculate by the method of § 2.3. The tetrahedral arrangement of nearest neighbours is now irrelevant—the main thing is to pack as many positive ions as possible round a negative ion, and vice versa. *Ionic solids* tend to crystallize in structures such as those of NaCl and CsCl.

It is a nice point whether it makes much sense to talk of a 'band structure' for the valence electrons in such a crystal. The wave-functions of these electrons are closely localized on their ions and do not show 'Bloch function' properties such as free carrier 'hole' mobility (cf. § 6.6); the 'energy gap' would just be something like the second ionization potential of the metal atom. Electrons excited above this gap would, of course, behave much more like mobile current carriers in an ordinary conduction band.

In practice, there is no clear-cut distinction between ionic and covalent solids. In a III–V or II–VI compound there is always some degree of ionic character. Even an ionic crystal can show properties akin to those of semiconductors.

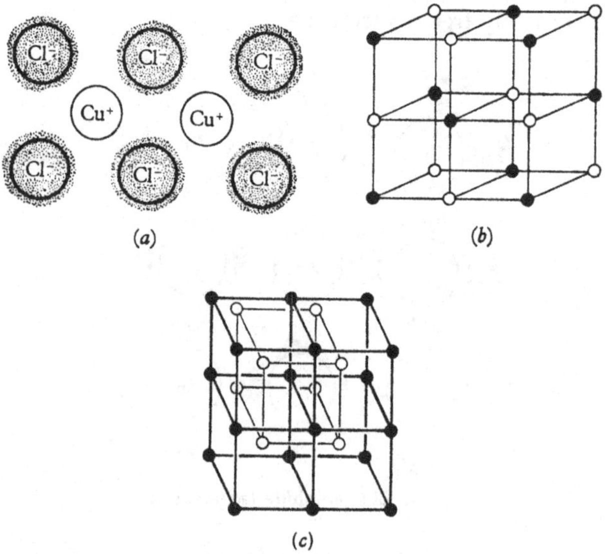

Fig. 76. (a) Copper chloride (schematic). (b) NaCl structure. (c) CsCl structure.

This does not exhaust the classification of solids. There are *molecular crystals*, in which the building blocks are stable, neutral molecules, such as CH_4, which cohere by their mutual van der Waals attraction. There is a special class called *hydrogen-bonded structures*, in which the forces between molecular groups are taken through a shared proton. These solids are usually dominated by the chemical properties of the molecules. Except for the rare gas solids, which are useful as a paradigm of a simple crystal, they tend to have complicated structures, and properties that fall outside the scope of this book.

4.3 Cohesion

One of the major goals of the theory of solids is to show why solids are as they are—that is, to demonstrate that a whole lot of, say, Na atoms put into a box will condense to form a crystal of sodium metal. This goal is too ambitious; but one can study some aspects of the energetics of solids. In particular, one can attempt to calculate the cohesive energy of the solid, measured from the level of an assembly of separate atoms, and compare this with experiment, or with the energy of other putative solids of the same atoms, with different lattice constant, or different crystal lattice.

For example, we confirm the weakness of the van der Waals forces by noting the small cohesive energy of molecular crystals—of the order of 0·1 eV./molecule. Hydrogen-bonded crystals (typically H_2O) have somewhat higher cohesion—say, 0·5 eV./molecule. Both these types of solid have low melting points. Metals are much more typically 'solid', and have cohesive energies between 1 and 5 eV./atom, whilst ionic and covalent solids are very strongly bound, having energies of the order of 10 eV./atom.

The principles by which one sets about calculations of cohesive energy will be apparent from previous chapters. For example, in the case of ionic crystals, we calculate the Madelung constant (2.27) for the structure, and hence determine the contribution of the electrostatic forces, in the form

$$\mathscr{E} = -\tfrac{1}{2}N\alpha\frac{e^2}{a}. \tag{4.1}$$

This energy obviously tends to make the solid collapse, to decrease the lattice constant a. But this force is opposed by the repulsion between the ion cores—a repulsion that is often represented by a potential of the form

$$\phi(R) = B\exp(-R/\rho) \tag{4.2}$$

for ions distant R apart.

It is a trivial calculation to put these together and estimate the total energy of the solid as a function of a. This can be compared with experimental values of the total cohesive energy, and of the compressibility of the solid. Unfortunately the parameters B and ρ for a given pair of ions cannot be calculated with any great accuracy from first principles; the most we can do is to derive a set of empirical values more or less consistent with the observations on a variety of different crystals where the ions are paired in all possible combinations. This phenomenological scheme may then be interpreted as yielding characteristic *ionic radii* for a number of elements.

The cohesive energy of covalent solids is a much more complicated problem. It can, in principle, be found by calculating the band structure and finding the contribution from the energy of each electron state. In practice, it is easier to confirm that the total energy of the crystal is quite close to the sum of the energies of the covalent bonds and that the lattice spectrum can be derived from the forces required to bend and/or stretch the bonds in a phonon mode. The agreement, in the case of diamond, is excellent. If we understand the energetics of the covalent bond then we are a long way towards understanding the cohesive energy of semiconductors.

Since the cohesion of a covalent solid depends almost entirely on the chemical saturation of the bonds, it is not necessary that the structure should have crystalline order. Each atom can often be given the correct number of neighbours, in very nearly the correct geometrical configuration, in a network without long range order. Ordinary silica *glass*, for example, is an assembly in which each Si atom lies at the centre of a tetrahedron of oxygens, whilst each oxygen atom lies nearly on the line between two Si atoms (Fig. 77). Although this network is obviously stable chemically and dynamically, it is not clear how one should describe its cohesion from the band point of view.

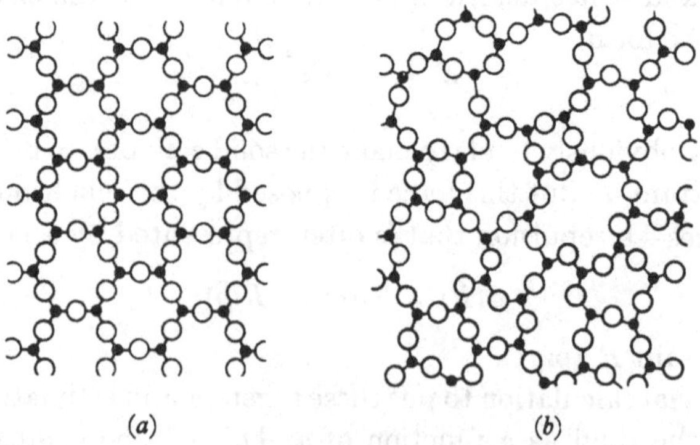

Fig. 77. Covalently bonded networks: (*a*) Crystalline; (*b*) Glassy. This is the two-dimensional analogue of SiO_2.

Metallic cohesion is a much more subtle phenomenon. In a general way, we may say that by freeing each electron from localization on a particular atom we are increasing the uncertainty in its position coordinates; hence, by the Heisenberg principle, the spread of momentum is reduced. This lowers the average kinetic energy of the electrons, and thus binds the system together. But calculations of total cohesive energy are amongst the most difficult to get right, and to verify in principle, in the whole theory of metals. There are many different ways of dividing up the cake, so as to take account of complex many-body effects such as correlation and exchange in the electron gas (see § 5.8).

In the simplest analysis, we consider two main contributions, working in opposite directions. As the atoms approach and overlap, each electron is able to move in a larger region, where it has, on the average, a lower potential energy. On the other hand, as we compress the electron gas the Fermi energy is pushed up, thus increasing the average kinetic energy of the electrons. At the metallic density these forces are in equilibrium.

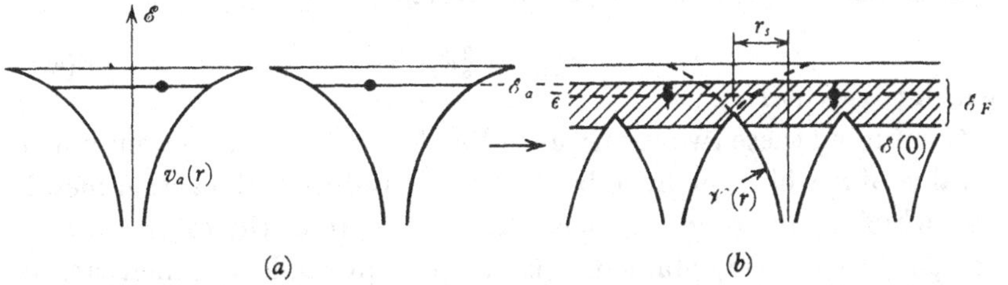

Fig. 78. Wigner–Seitz model for cohesion: (a) monatomic gas; (b) metal.

To illustrate this argument, suppose we start with free atoms of an alkali metal such as Na. Each atom contains a valence electron in the lowest bound state, $\mathscr{E}_a$, of an atomic potential $v_a(r)$, which would behave like the coulomb potential $-e^2/r$ outside the ion core. The crystal potential $\mathscr{V}(\mathbf{r})$ seen by a conduction electron in the solid metal would obviously be much more complicated than a mere superposition of such functions, for it would now have to include contributions from the 'other' conduction electrons, etc. As a crude approximation, however, let us assume that the chosen electron, by its electrostatic repulsion, temporarily excludes all the other valence electrons from the whole of the Wigner–Seitz cell (§ 3.5) in which it happens to be; but all the other cells of the crystal then contain electrons which exactly neutralize them electrostatically. In other words, the effects of atomic overlap and of electron correlation etc. are simulated by taking $\mathscr{V}(\mathbf{r})$ to be equal to $v_a(r)$ for $r \leqslant r_s$ in each cell (Fig. 78).

At the actual boundary of the atomic sphere, this is obviously a poor approximation. Nevertheless, the net effect is a substantial lowering of the average potential energy of the electron in the crystal. More precisely, an application of the Wigner–Seitz cellular approximation shows that $\mathscr{E}(0)$, the bottom of the conduction band, must lie a little below the energy of this barrier—i.e. at some level below $-e^2/r_s$. We are thus gaining potential energy as r_s decreases.

The theory of the repulsive term is trivial for a free-electron gas. From the formulae of § 3.1 we may easily prove that the *density of states in energy*—the number of electron states per unity energy range —is given by

$$\mathscr{N}(\mathscr{E}_F) = \frac{3}{2}\frac{n}{\mathscr{E}_F} \tag{4.3}$$

for a density n of electrons with Fermi energy $\mathscr{E}_F$. It follows that

$$\mathscr{N}(\mathscr{E}) \propto \mathscr{E}^{\frac{1}{2}}, \tag{4.4}$$

whence the average energy per electron is

$$\overline{\mathscr{E}}_{\text{kin.}} = \tfrac{3}{5}\mathscr{E}_F, \tag{4.5}$$

which varies as $1/r_s^2$.

Combining these two terms, as in Fig. 79, we observe a minimum at a value of r_s which ought to be the atomic radius in the actual metal, and which ought to give a reasonable estimate of the total cohesive energy. The results obtained from the present crude argument are at least of the right magnitude, showing that this general explanation is sound in principle.

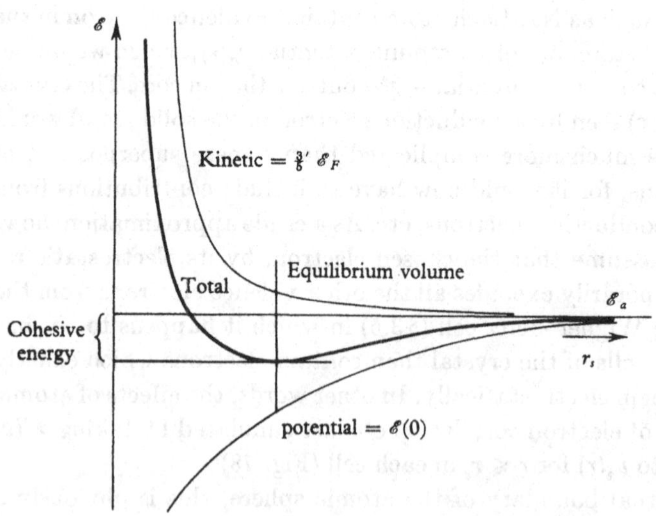

Fig. 79. Contributions to metallic cohesion, as a function of interatomic distance.

It is interesting to note, however, that the cohesive energy is very close to $\frac{2}{5}\mathscr{E}_F$ in all alkali metals. From this empirical fact, we deduce from (4.5) the approximate coincidence of the Fermi level in the metal with the energy of the atomic level from which the band supposedly derived. This is just what would be expected for a half-filled band in the L.C.A.O. theory, as in (3.30). An alternative explanation of metallic cohesion would then refer to the ability of the electrons to go into Bloch states in pairs, with opposite spins, without paying a penalty in electrostatic energy as they would in separate atomic orbitals.

The character of the cohesive forces influences the elastic and plastic properties of solids. Covalent solids are stiff, and brittle, because the directional character of the bonds inhibits shearing motion, and prevents the movement of one atom past another as in the transit of a dislocation through the lattice. Ionic crystals are much more plastic, if they are perfectly pure (though ordinary crystals may be brittle because of the imperfections grown into them). The electrostatic forces are non-directional, so they allow the ions to move about, except in so far as they are hindered by their sizes. Metals are the most plastic solids, allowing the free motion of dislocations, because the Fermi energy of the electron gas tends to keep the ions well apart, and the cohesive energy is a function mainly of density of packing. Local deviations from strict lattice regularity are easily accommodated.

4.4 Rigid band model and density of states

Why do many metals occur in different phases, with different crystal structures, and change from one phase to another as the temperature varies? The energy associated with the thermal excitations of a crystal is small—of the order of 0·1 eV./atom. The differences in energy of different structures must be small. Indeed, our crude arguments above, based on the Wigner–Seitz and tight-binding formulae, cannot distinguish between phases of different crystal structure but of the same density. It is interesting to note that metals do not change much in density when they melt; the above arguments are so insensitive to structure that they would just hold as well for a liquid metal, which may be described quite simply as a randomly packed assembly of spherical ions, held together by a glue of free electrons.

There is, however, a phenomenon which illustrates a principle concerning the structural energy of solids. Consider the following table:

Structure	B.C.C.	'γ-brass'	Hexagonal close-packed
Alloys	AgZn Cu_3Al Cu_5Sn	Ag_5Zn_8 Cu_9Al_4 $Cu_{31}Sn_8$	$AgZn_3$ — Cu_3Sn
Electrons / Atom	$\frac{3}{2} = 1{\cdot}5$	$\frac{21}{13} = 1{\cdot}615$	$\frac{7}{4} = 1{\cdot}75$

Each column refers to a series of alloys, each with the same basic crystal structure and each forming a stable phase of more or less definite composition. What have they in common? In the bottom row we show the ratio of electrons to atoms, which is constant for each column. Thus, in Cu_9Al_4 there are 9 electrons from the Cu atoms, and 3 electrons from each of the 4 Al atoms, making 21 electrons for 13 atoms.

This phenomenon exemplifies the *Hume–Rothery rules* which are often discussed on the basis of the *rigid band* theory of alloys. It is argued that the valence electrons of the constituent elements are all poured into a single N.F.E. band, similar to that of an ordinary pure metal (§ 3.3). In $Cu_{31}Sn_8$, for example, this hypothetical metal would have 1·615 electrons per atom.

We now suppose that the Brillouin zone structure of the alloy depends on the basic crystal lattice, and not much on the actual ions that occupy the lattice sites. Now it turns out that for the 'γ-brass' structure this zone is of such a shape and size that it just touches a free-electron sphere drawn to contain 1·615 conduction electrons per atom. Thus, in these solids the Fermi surface might be expected to touch the zone boundaries over substantial areas, but would probably not have broken through into the next zone. Since states just inside the zone boundary are lower in energy than free electrons with the same wave-vector, there is a gain in cohesive energy if the metal takes this structure. This explanation of the crystal structure of many pure metals and complex alloy phases is very neat, and has often been regarded as one of the triumphs of the electron theory of metals. Unfortunately, there are objections to the use of the N.F.E. model for metals such as Cu, Ag and Au, where the Fermi surface already touches the zone boundary in the pure metal (see § 9.4). The theoretical status of the rigid band model itself is also in some doubt (see § 5.3), despite the strong evidence for 'zone boundary effects' in particular cases.

In effect, we are saying that the structure is such that the Fermi

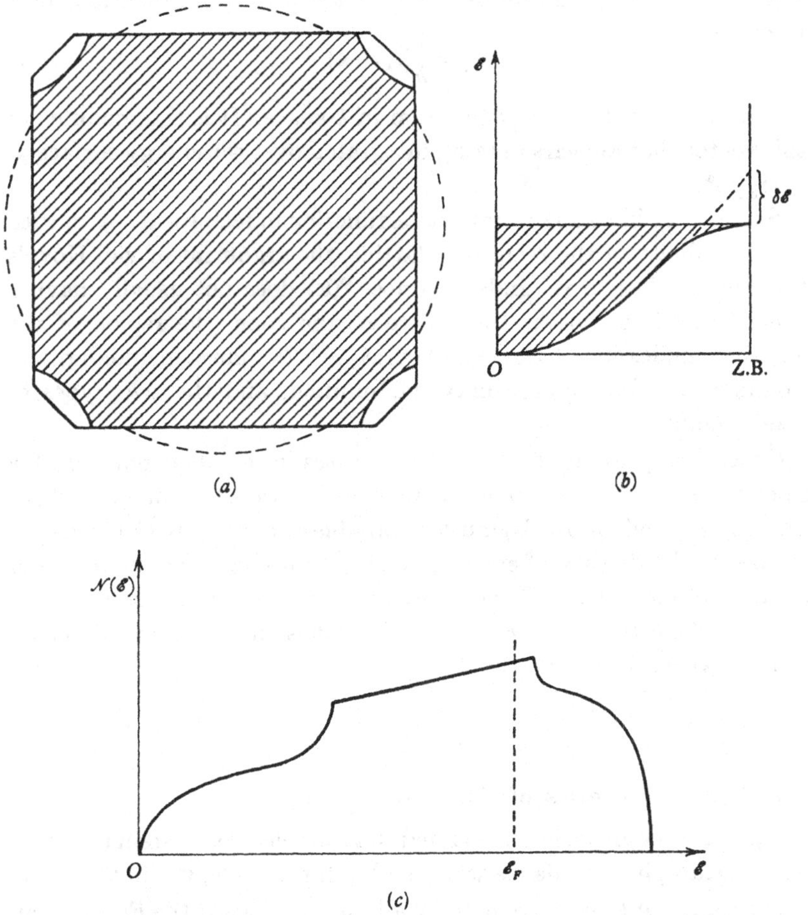

Fig. 80. (*a*) Fermi surface touches zone boundary. (*b*) Energy reduced below free electron value. (*c*) Fermi level comes near maximum of density of states.

level comes near the top of a range of energies where the *density of states* in energy is high. This function has already been given in (4.3), for a free-electron gas. To calculate $\mathscr{N}(\mathscr{E})$ for an actual metal we have to find the volume enclosed between successive energy surfaces in the Brillouin zone. The problem is precisely the same in form as the calculation of the lattice spectrum in § 2.5.

Following the argument of (2.66)–(2.69), we can write

$$\mathscr{N}(\mathscr{E}) = \frac{1}{4\pi^3 N\hbar}\int \frac{dS}{|\mathbf{v_k}|}, \tag{4.6}$$

where for the moment we define a vector with the dimensions of a velocity,

$$\mathbf{v}_{\mathbf{k}} = \frac{1}{\hbar}\nabla_{\mathbf{k}}\mathscr{E}(\mathbf{k}), \tag{4.7}$$

the gradient of the energy function in k-space. It is the modulus of this vector that appears in (4.6), integrated over the energy surface of energy $\mathscr{E}$.

Since $\mathscr{E}(\mathbf{k})$, like $\nu_{\mathbf{q}}$, is a continuous function in reciprocal space, with the periodicity of the reciprocal lattice, the argument of van Hove's theorem applies to it. Indeed, since $\mathscr{E}(\mathbf{k})$ does not usually have a special multiple root at the centre of the zone (the region of the acoustic modes in $\nu_{\mathbf{q}}$) the singularities in $\mathscr{E}(\mathbf{k})$ must include critical points of all four types—maxima, minima, and the two types of saddle-point.

Generally speaking, the density of states in metals is not very far from the free-electron value, except for the transition metals, where a narrow d-band (or bands) must be capable of holding 10 electrons per atom—so the density of states in energy must be much higher than in the ordinary s-band. There is evidence for this, and also evidence that the d-electrons contribute to the cohesion of the metal, as we should expect from this model.

4.5 Fermi statistics of electrons

Up to this point we have assumed that our electron system is at zero temperature; in accordance with the Pauli principle, we fill the levels in succession until we have used up all the electrons at the Fermi level, of energy $\mathscr{E}_F$. But we know from statistical mechanics that electrons obey Fermi–Dirac statistics; the probability of occupation of a level of energy $\mathscr{E}$ is

$$f^0(\mathscr{E}) = \frac{1}{e^{(\mathscr{E}-\zeta)/kT}+1}, \tag{4.8}$$

where ζ is the *Fermi potential*, i.e. the chemical potential, or free energy per electron, which must be constant throughout the material if the electron gas is in thermodynamic equilibrium.

Suppose we have $\mathscr{N}(\mathscr{E})\,d\mathscr{E}$ states, in the range of energy $d\mathscr{E}$. Then if n is the total density of electrons (per unit volume of crystal), we must have

$$n = \int_0^\infty f^0(\mathscr{E})\mathscr{N}(\mathscr{E})\,d\mathscr{E}. \tag{4.9}$$

These two equations fix the values of ζ, and hence the whole distribution.

At absolute zero, this is trivial: it is easy to see that $f^0(\mathscr{E})$ will be equal to unity if $\mathscr{E} < \zeta$, and that it will vanish for $\mathscr{E} > \zeta$. We then have

$$n = \int_0^{\zeta} \mathscr{N}(\mathscr{E})\,d\mathscr{E} = \int_0^{\mathscr{E}_F} \mathscr{N}(\mathscr{E})\,d\mathscr{E}, \tag{4.10}$$

which is a mathematical formulation of the definition of the Fermi level

$$\zeta = \mathscr{E}_F \quad \text{at} \quad T = 0. \tag{4.11}$$

To find ζ as a function of T requires, in principle, a knowledge of $\mathscr{N}(\mathscr{E})$. But in metals the distribution is highly *degenerate*, i.e. $\boldsymbol{k}T \ll \mathscr{E}_F$. We can then use the following mathematical device:—

Suppose we have to evaluate integrals of the form

$$I = \int_0^{\infty} g(\mathscr{E}) f^0(\mathscr{E})\,d\mathscr{E}, \tag{4.12}$$

where $g(\mathscr{E})$ is some function of energy. Integrate by parts. We have

$$I = [G(\mathscr{E}) f^0(\mathscr{E})]_0^{\infty} - \int_0^{\infty} G(\mathscr{E}) \frac{\partial f^0}{\partial \mathscr{E}}\,d\mathscr{E}, \tag{4.13}$$

where

$$G(\mathscr{E}) \equiv \int_0^{\mathscr{E}} g(\mathscr{E})\,d\mathscr{E}. \tag{4.14}$$

The first term vanishes in both limits (we may suppose $g(0) = 0$, if we take our origin of energy sufficiently low).

The remaining integral is more manageable; because $(\partial f^0/\partial \mathscr{E})$ has a symmetrical peak at $\mathscr{E} = \zeta$ it behaves like a delta-function of width $\boldsymbol{k}T$ there. Thus, we expand $G(\mathscr{E})$ in a Taylor series about this point

$$G(\mathscr{E}) = G(\zeta) + (\mathscr{E} - \zeta) G'(\zeta) + \tfrac{1}{2}(\mathscr{E} - \zeta)^2 G''(\zeta) + \ldots, \tag{4.15}$$

and substitute in (4.13). We get

$$I = G(\zeta) \int_0^{\infty} \left(-\frac{\partial f^0}{\partial \mathscr{E}}\right) d\mathscr{E} + G'(\zeta) \int_0^{\infty} (\mathscr{E} - \zeta) \left(-\frac{\partial f^0}{\partial \mathscr{E}}\right) d\mathscr{E} + \ldots. \tag{4.16}$$

But

$$\int_0^{\infty} \left(-\frac{\partial f^0}{\partial \mathscr{E}}\right) d\mathscr{E} = f^0(0) - f^0(\infty) = 1, \tag{4.17}$$

so the first approximation is

$$I \approx G(\zeta) = \int_0^{\zeta} g(\mathscr{E})\,d\mathscr{E}, \tag{4.18}$$

which is what we should have had at $T = 0$.

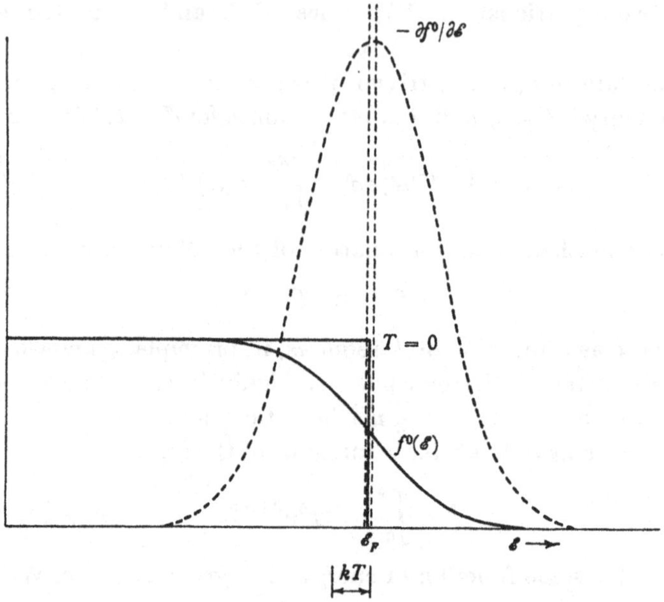

Fig. 81. Fermi–Dirac function, and its derivative, at $T = 0$, and at a finite temperature.

The general term in (4.16) contains the integral

$$\begin{aligned} F_n &= \frac{1}{n!}\int_0^\infty (\mathscr{E}-\zeta)^n\left(-\frac{\partial f^0}{\partial \mathscr{E}}\right) d\mathscr{E} \\ &= \frac{(kT)^n}{n!}\int_0^\infty \left(\frac{\mathscr{E}-\zeta}{kT}\right)^n \frac{\exp\{(\mathscr{E}-\zeta)/kT\}}{[\exp\{(\mathscr{E}-\zeta)/kT\}+1]^2}\, d\left(\frac{\mathscr{E}-\zeta}{kT}\right) \\ &= \frac{(kT)^n}{n!}\int_{-\infty}^\infty \frac{z^n\,dz}{(e^z+1)(1+e^{-z})} \\ &= \left\{\begin{matrix} 2c_n(kT)^n & \text{for} & n\,\text{even}, \\ 0 & \text{for} & n\,\text{odd}, \end{matrix}\right\} \end{aligned} \tag{4.19}$$

where the coefficients c_n are readily evaluated as summable series. In practice, we seldom go beyond the second term,

$$2c_2 = \tfrac{1}{6}\pi^2. \tag{4.20}$$

Thus

$$\int_0^\infty g(\mathscr{E})f^0(\mathscr{E})\,d\mathscr{E} \approx \int_0^\zeta g(\mathscr{E})\,d\mathscr{E} + \frac{\pi^2}{6}(kT)^2\left[\frac{\partial g(\mathscr{E})}{\partial \mathscr{E}}\right]_{\mathscr{E}=\zeta} + \dots \tag{4.21}$$

For example, to calculate the variation of ζ with temperature we use (4.9), and (4.10):

$$\int_0^{\mathscr{E}_F} \mathscr{N}(\mathscr{E})\,d\mathscr{E} = \int_0^{\infty} \mathscr{N}(\mathscr{E}) f^0(\mathscr{E})\,d\mathscr{E}$$

$$= \int_0^{\zeta} \mathscr{N}(\mathscr{E})\,d\mathscr{E} + \frac{\pi^2}{6}(kT)^2 \left[\frac{\partial \mathscr{N}(\mathscr{E})}{\partial \mathscr{E}}\right]_{\mathscr{E}=\zeta} + \ldots, \quad (4.22)$$

which has the approximate solution (check by differentiating with respect to ζ)

$$\zeta \approx \mathscr{E}_F - \frac{\pi^2}{6}(kT)^2 \left[\frac{\partial}{\partial \mathscr{E}} \ln \mathscr{N}(\mathscr{E})\right]_{\mathscr{E}=\mathscr{E}_F}. \quad (4.23)$$

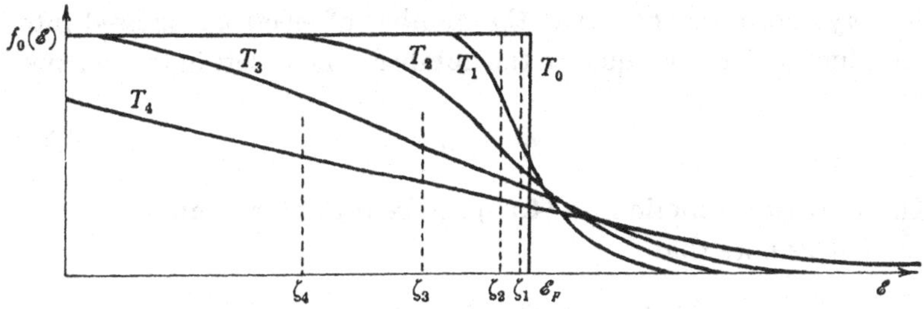

Fig. 82. At high temperatures, Fermi–Dirac function becomes classical Boltzmann distribution.

We notice that $\zeta < \mathscr{E}_F$, except at $T = 0$—but the correction is small when $kT \ll \mathscr{E}_F$, as in a metal at ordinary temperatures. The value of ζ has to decrease as the temperature rises, because the Fermi–Dirac function must tend to a classical distribution

$$f^0(\mathscr{E}) \propto e^{-\mathscr{E}/kT} \quad (4.24)$$

above the degeneracy temperature of the electron gas, i.e. when $kT > \mathscr{E}_F$. It is obvious, from the form of (4.8), that this holds only for energies above ζ; if it is to hold for all energies in the band, the value of ζ must lie below $\mathscr{E} = 0$.

4.6 Statistics of carriers in a semiconductor

In a semiconductor the density of states function has a large gap. In effect, the range of integration of energy is divided into two—up to $\mathscr{E}_v$, the top of the valence band, and then onwards from $\mathscr{E}_c$, the bottom of the conduction band. Thus, the condition equivalent to (4.9) and (4.10) is

$$\int_0^{\mathscr{E}_v} \mathscr{N}_v(\mathscr{E})\, d\mathscr{E} = \int_0^{\mathscr{E}_v} f^0(\mathscr{E})\, \mathscr{N}_v(\mathscr{E})\, d\mathscr{E} + \int_{\mathscr{E}_c}^{\infty} f^0(\mathscr{E})\, \mathscr{N}_c(\mathscr{E})\, d\mathscr{E}, \quad (4.25)$$

where $\mathscr{N}_c(\mathscr{E})$ and $\mathscr{N}_v(\mathscr{E})$ are densities of states in conduction and valence bands, respectively.

We can write this in the form

$$\int_0^{\mathscr{E}_v} \{1 - f^0(\mathscr{E})\}\, \mathscr{N}_v(\mathscr{E})\, d\mathscr{E} = \int_{\mathscr{E}_c}^{\infty} f^0(\mathscr{E})\, \mathscr{N}_c(\mathscr{E})\, d\mathscr{E}, \quad (4.26)$$

which says no more than that the number of electrons excited into the conduction band is equal to the number of holes left in the valence band;

$$n_h = n_e. \quad (4.27)$$

The statistical functions in (4.26) can be rewritten in an interesting form. Let us note that

$$\begin{aligned} 1 - f^0(\mathscr{E} - \zeta) &= 1 - \frac{1}{\exp\{(\mathscr{E} - \zeta)/kT\} + 1} \\ &= \frac{1}{\exp\{-(\mathscr{E} - \zeta)/kT\} + 1} \\ &= f^0(\mathscr{E}_h + \zeta_h), \end{aligned} \quad (4.28)$$

where

$$\mathscr{E}_h + \zeta_h = -(\mathscr{E} - \zeta). \quad (4.29)$$

In other words, the probability of finding a hole at the energy level $\mathscr{E}$, distant $|\mathscr{E} - \zeta|$ *below* the Fermi level of the electrons, is the same as the probability of finding an electron at an energy $|\mathscr{E} - \zeta|$ *above* the Fermi level. The holes, like the electrons, obey Fermi–Dirac statistics —but as if their energy were measured *downwards*. It is convenient to measure all electron energies from $\mathscr{E}_c$; we use the variable $\mathscr{E}_e$ for $\mathscr{E} - \mathscr{E}_c$. Hole energies are then measured from the top of the valence band: $\mathscr{E}_h = -(\mathscr{E} - \mathscr{E}_v)$. The Fermi level for electrons is at some point $\mathscr{E}_e = -\zeta_e$; the Fermi level for holes corresponds to $\mathscr{E}_h = -\zeta_h$. Our formula (4.26) then reads

$$\int_0^{\infty} f^0(\mathscr{E}_h + \zeta_h)\, \mathscr{N}_v(\mathscr{E}_h)\, d\mathscr{E}_h = \int_0^{\infty} f^0(\mathscr{E}_e + \zeta_e)\, \mathscr{N}_c(\mathscr{E}_e)\, d\mathscr{E}_e. \quad (4.30)$$

The further conditions on these functions is that ζ_e and ζ_h should in fact correspond to the same over-all Fermi potential ζ. Since they are measured from energy origins distant $\mathscr{E}_{\text{gap}}$ apart, we have†

$$\zeta_e + \zeta_h = \mathscr{E}_{\text{gap}}. \tag{4.31}$$

As we shall show in a moment, ζ tends to lie in the gap, so that both ζ_e and ζ_h are positive, and rather larger than kT. The Fermi functions for electrons and holes are then practically indistinguishable from classical distributions in the range where $\mathscr{E}_e$ and $\mathscr{E}_h$ are positive

$$f^0(\mathscr{E}_e + \zeta_e) \approx \exp\{-(\mathscr{E}_e + \zeta_e)/kT\}. \tag{4.32}$$

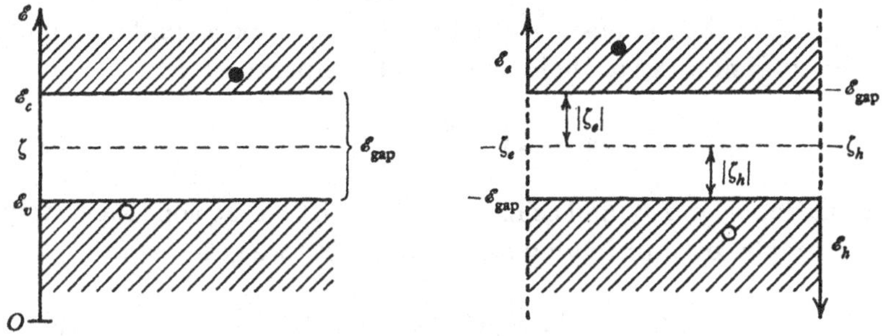

Fig. 83. (a) Absolute energy scheme for semiconductor.
(b) 'Electron' and 'hole' energy schemes.

This somewhat simplifies the evaluation of the integrals in (4.30). According to the van Hove theorem (§ 2.5), the form of the density of states near the bottom of the conduction band must be

$$\mathscr{N}_c(\mathscr{E}_e) = \frac{1}{2\pi^2}\left(\frac{2m_e}{\hbar^2}\right)^{\frac{3}{2}} \mathscr{E}_e^{\frac{1}{2}}. \tag{4.33}$$

This is actually the result for a gas of free electrons of *effective mass* m_e, but in general this parameter is

$$m_e = (m_1 m_2 m_3)^{\frac{1}{3}}, \tag{4.34}$$

where m_1, m_2, m_3 are the coefficients in an expansion of the energy about the minimum of $\mathscr{E}(\mathbf{k})$ in $\mathbf{k}$-space, referred to local principal axes:

$$\mathscr{E}_e = \frac{\hbar^2 k_1^2}{2m_1} + \frac{\hbar^2 k_2^2}{2m_2} + \frac{\hbar^2 k_3^2}{2m_3}. \tag{4.35}$$

The proof of this is exactly as in (2.70)–(2.72).

† It seems most convenient here to define ζ_e and ζ_h so that they are both usually positive quantities.

Putting (4.32) and (4.33) into (4.30) we have

$$n_e = \int_0^\infty \exp\{-(\mathscr{E}_e+\zeta_e)/kT\}\frac{1}{2\pi^2}\left(\frac{2m_e}{\hbar^2}\right)^{\frac{3}{2}}\mathscr{E}_e^{\frac{1}{2}}\,d\mathscr{E}_e$$

$$= \frac{1}{2\pi^2}\left(\frac{2m_e}{\hbar^2}\right)^{\frac{3}{2}} e^{-\zeta_e/kT}(kT)^{\frac{3}{2}}\int_0^\infty e^{-z}z^{\frac{1}{2}}\,dz$$

$$= 2\left(\frac{m_e kT}{2\pi\hbar^2}\right)^{\frac{3}{2}} e^{-\zeta_e/kT}. \tag{4.36}$$

In simple models, where, for example, spin-orbit splitting (§ 3.11) is neglected, the top of the valence band will also have a density of states of the same functional form as (4.33), characterized by a similar parameter, m_h, of the dimensions of mass. There will be a similar formula for the total number of 'holes excited at temperature T';

$$n_h = 2\left(\frac{m_h kT}{2\pi\hbar^2}\right)^{\frac{3}{2}} e^{-\zeta_h/kT}, \tag{4.37}$$

Equating n_e and n_h, and using (4.31), we can find each one separately. For example, by multiplying (4.36) and (4.37) we get

$$n_e n_h = 4\left(\frac{kT}{2\pi\hbar^2}\right)^3 (m_e m_h)^{\frac{3}{2}} e^{-\mathscr{E}_{\text{gap}}/kT}, \tag{4.38}$$

which is independent of the position of the Fermi level and is therefore quite generally true, provided the Fermi level stays in the band gap. Taking the square root of this

$$n_e = n_h = 2\left(\frac{kT}{2\pi\hbar^2}\right)^{\frac{3}{2}} (m_e m_h)^{\frac{3}{4}} e^{-\mathscr{E}_{\text{gap}}/2kT}, \tag{4.39}$$

showing that the carrier density does indeed vary with a Boltzmann factor, as if it required the energy $\frac{1}{2}\mathscr{E}_{\text{gap}}$ to activate an electron into the conduction band.

From (4.36), (4.37), and (4.39), one can easily calculate ζ_e and ζ_h separately; it is obvious by symmetry that ζ will come in the middle of the gap if $m_e = m_h$. So long as $\mathscr{E}_g \gg kT$, these formulae will hold for an *intrinsic semiconductor* where the carriers are all created by thermal excitation.

In this work we are confining the discussion mainly to pure, regular solids, but it should be mentioned that most semiconductors are not quite pure—indeed, that we almost always use them in a state where they have been *doped* by the addition of impurities. The standard case is that of a *donor* impurity—an atom of As introduced substitutionally in a crystal of Ge. From the argument of § 4.2, it is obvious that this can be done without harm to the bond structure of the lattice if the 5th electron of the As is set free into the conduction band of the crystal. We can thus make an *n-type* specimen; we have added a few electrons without creating any holes in the valence band.

If there are N_d such *donor atoms* per unit volume, then we must have

$$n_e - n_h = N_d \tag{4.40}$$

as our condition from which to calculate n_e, n_h, ζ_e, ζ_h, as functions of T. In particular, if the material is heavily doped by adding a large number of donors, there will be an overwhelming preponderance of electrons. The Fermi level must then lie in the energy range occupied by these *majority carriers*—i.e. near the bottom of the conduction band. With sufficient doping this gas of carriers may reach metallic densities, where the distribution becomes degenerate even at ordinary temperatures.

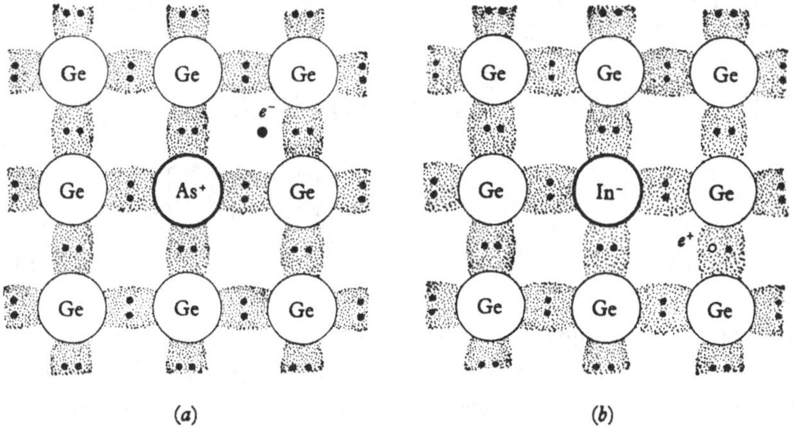

Fig. 84. (*a*) Donor in Ge. (*b*) Acceptor in Ge.

There is a dual of this type of impurity. If we add In, with only 3 electrons per atom, in place of Ge, we cannot quite fill the bonds; a hole is left in the valence band. A density of N_a *acceptor* impurities in the crystal creates a *p-type* specimen, in which

$$n_e - n_h = -N_a. \tag{4.41}$$

In general, we may have both donors and acceptors, so that

$$n_e - n_h = N_d - N_a \tag{4.42}$$

in a *compensated* sample. However, these impurities also introduce localized levels into the energy gap (see § 6.4), so that the statistical theory at low temperatures, when these levels may be occupied by a significant proportion of the excess carriers, becomes more complicated.

4.7 Electronic specific heat

The electrons in a metal must contribute to the specific heat. To calculate this contribution let us calculate the mean energy of the electron distribution using the formula (4.21), since we assume that the system is highly degenerate:

$$\bar{\mathscr{E}} = \int \mathscr{E} f^0(\mathscr{E}) \mathscr{N}(\mathscr{E})\, d\mathscr{E}$$

$$= \int_0^\zeta \mathscr{E}\mathscr{N}(\mathscr{E})\, d\mathscr{E} + \frac{\pi^2}{6}(kT)^2 \left[\frac{\partial}{\partial \mathscr{E}}\{\mathscr{E}\mathscr{N}(\mathscr{E})\}\right]_{\mathscr{E}=\zeta} + \dots \tag{4.43}$$

We differentiate with respect to temperature, acknowledging (cf. (4.22)) that ζ is a function of T:

$$C_{\text{el.}} = \frac{\partial \bar{\mathscr{E}}}{\partial T} = \zeta \mathscr{N}(\zeta)\frac{d\zeta}{dT} + \frac{\pi^2}{3} k^2 T \left[\mathscr{N}(\mathscr{E}) + \zeta \frac{\partial \mathscr{N}(\mathscr{E})}{\partial \mathscr{E}}\right]_{\mathscr{E}=\zeta} + O(T^2)$$

$$= \frac{\pi^2}{3} k^2 T \mathscr{N}(\mathscr{E}_F) + \zeta \mathscr{N}(\zeta) \left[\frac{d\zeta}{dT} + \frac{\pi^2}{3} \frac{k^2 T}{\mathscr{N}(\mathscr{E})} \frac{\partial \mathscr{N}(\mathscr{E})}{\partial \mathscr{E}}\right]_{\mathscr{E}=\zeta}$$

$$= nk \frac{\pi^2}{3} kT \frac{\mathscr{N}(\mathscr{E}_F)}{n} \tag{4.44}$$

from (4.22). The higher terms in T^2, etc., are negligible.

This is a very important result. Compare (4.44) with the specific heat of a classical gas of particles, say $\frac{3}{2}nk$. The quantum result is much smaller. For free electrons the density of states (4.3) is $\frac{3}{2}n/\mathscr{E}_F$, so that

$$C_{\text{el.}}/C_{\text{classical}} = \frac{\pi^2}{3}\frac{kT}{\mathscr{E}_F}. \tag{4.45}$$

In ordinary metals, at ordinary temperatures, this factor is about 1/100. It shows why the total specific heat of a metal is given quite accurately by the lattice term of § 2.4, and why the Dulong and Petit Law holds at high temperatures.

We also note that $C_{\text{el.}}$ is linear in T. At very low temperatures this linear term, usually written

$$C_{\text{el.}} = \gamma T, \tag{4.46}$$

can be distinguished against the lattice term, which goes to zero, as T^3, more rapidly. Measurement of γ provides direct information about $\mathcal{N}(\mathcal{E}_F)$, the density of states at the Fermi level. For example, high values of γ are observed in the transition metals, as discussed in § 4.4.

We can understand the linear law as follows. Look at the Fermi distribution (Fig. 85); the effect of temperature is to excite a few electrons to higher levels. But this effect is only noticeable in a range of energy of the order of $\boldsymbol{k}T$ around $\mathcal{E}_F$. We can say that about $\boldsymbol{k}T\mathcal{N}(\mathcal{E}_F)$ electrons each receive an energy of about $\boldsymbol{k}T$. Thus, the whole solid gains the energy

$$\delta\overline{\mathcal{E}} \sim \boldsymbol{k}^2T^2\mathcal{N}(\mathcal{E}_F). \tag{4.47}$$

This corresponds to

$$C_{\text{el.}} = \frac{\partial\delta\overline{\mathcal{E}}}{\partial T} \sim 2\boldsymbol{k}^2T\mathcal{N}(\mathcal{E}_F). \tag{4.48}$$

Apart from a numerical factor this is essentially our result (4.44).

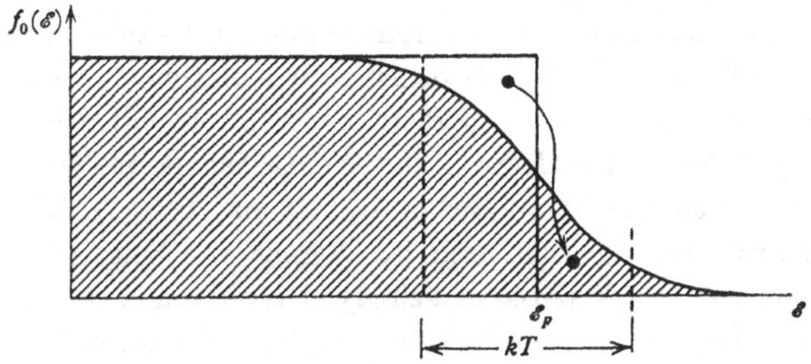

Fig. 85. Thermal excitation of electrons in metal.

What we are saying, in effect, is that the electrons deep down in the distribution are scarcely conscious of the temperature. Their situation is determined by the Pauli principle which insists that they fill up all the levels, but does not permit them to poach on each other's state. It is no more surprising than that the electrons deep down in bound states of the ion cores need not be counted as making a separate contribution to the specific heat of the solid—at least until the temperature becomes so enormous that these electrons can be thermally excited.

CHAPTER 5

ELECTRON-ELECTRON INTERACTION

'The whole thing is a low put-up job on our noble credulity' said Sam. NORMAN LINDSAY, *The Magic Pudding*

5.1 Perturbation formulation

The theory of electronic structure, as presented in Chapter 3, is a *one-electron model*; each electron is treated as an independent particle, moving in a well-defined potential, and the interactions between conduction electrons are ignored. But we know that these interactions are strong, and are of long range, being the *Coulomb force* between the charges and the so-called *exchange force* associated with the anti-symmetry of the wave-functions.

Naïvely we assume that these interactions can be taken care of by a Hartree or Hartree–Fock self-consistent calculation that adjusts the atomic potentials for the charge distribution of the valence electrons (as well, of course, as the electrons in the closed shells of the ion cores). But this is not easy to do properly—and we may have to fall back on some assumption, like that used by Wigner and Seitz (§ 4.3), that the electron sees the potential of a singly charged ion in the cell where it happens to be, but that neighbouring cells are electrically neutral.

In recent years, therefore, a lot of effort has been expended on the *many-body problem* of a gas of electrons interacting via their Coulomb potential, and the basic effects of the interaction are now well understood. Much of the theory is expressed in complicated formal language; the main results are surprisingly simple, and can be derived by elementary arguments.

We consider a free-electron gas, subject to a time-dependent perturbation. Suppose that the potential seen by an electron at $\mathbf{r}$, at time t, is given by

$$\delta\mathscr{U}(\mathbf{r},t) = \mathscr{U}\, e^{i\mathbf{q}\cdot\mathbf{r}}\, e^{i\omega t}\, e^{\alpha t}. \tag{5.1}$$

We have imposed an oscillation, of frequency ω, wave-vector $\mathbf{q}$, growing slowly with time-constant α.

Acting on the state $|\mathbf{k}\rangle = \exp i\{\mathbf{k}\cdot\mathbf{r} + \mathscr{E}(\mathbf{k})t/\hbar\}$ this mixes in other states, so that the wave-function becomes

$$\psi_{\mathbf{k}}(\mathbf{r},t) = |\mathbf{k}\rangle + b_{\mathbf{k}+\mathbf{q}}(t)|\mathbf{k}+\mathbf{q}\rangle, \tag{5.2}$$

where the coefficients may be calculated, in first order, by perturbation theory;

$$b_{\mathbf{k}+\mathbf{q}}(t) = \frac{\langle \mathbf{k}+\mathbf{q}|\,\delta\mathscr{U}\,|\mathbf{k}\rangle}{\mathscr{E}(\mathbf{k}) - \mathscr{E}(\mathbf{k}+\mathbf{q}) + \hbar\omega - i\hbar\alpha}$$

$$= \frac{\mathscr{U} e^{i\omega t}\, e^{\alpha t}}{\mathscr{E}(\mathbf{k}) - \mathscr{E}(\mathbf{k}+\mathbf{q}) + \hbar\omega - i\hbar\alpha}. \tag{5.3}$$

This follows directly by consideration of the time-dependent Schrödinger equation for $\psi_{\mathbf{k}}$, where the unperturbed Hamiltonian—just the kinetic energy operator, with eigenvalues $\mathscr{E}(\mathbf{k})$—is augmented by (5.1), which is supposed to be small.

Now consider the change in charge density due to this change in the electron wave-functions. We assume that we are dealing with *jellium*, a uniform positively charged medium against which the electrons move.

$$\delta\rho(\mathbf{r},t) = e\sum_{\mathbf{k}}\{|\psi_{\mathbf{k}}(\mathbf{r},t)|^2 - 1\}$$

$$= e\sum_{\mathbf{k}}[\{e^{-i\mathbf{k}\cdot\mathbf{r}} + b^*_{\mathbf{k}+\mathbf{q}}(t)\, e^{-i(\mathbf{k}+\mathbf{q})\cdot\mathbf{r}}\}\{e^{i\mathbf{k}\cdot\mathbf{r}} + b_{\mathbf{k}+\mathbf{q}}(t)\, e^{i(\mathbf{k}+\mathbf{q})\cdot\mathbf{r}}\} - 1]$$

$$\approx e\sum_{\mathbf{k}}\{b_{\mathbf{k}+\mathbf{q}}(t)\, e^{i\mathbf{q}\cdot\mathbf{r}} + b^*_{\mathbf{k}+\mathbf{q}}(t)\, e^{-i\mathbf{q}\cdot\mathbf{r}}\}, \tag{5.4}$$

if we drop the term in $|b|^2$. The summation here is over all occupied electron states.

Since $\delta\rho$ is, by its definition, real, we find that our wave (5.1) creates two types of charge disturbance: one that travels with the wave and one that is 180° out of phase with it. If we assume that in fact we had a real perturbation, i.e. if we add to (5.1) its complex conjugate

$$\delta\mathscr{U}^*(\mathbf{r},t) = \mathscr{U}\, e^{-i\mathbf{q}\cdot\mathbf{r}}\, e^{-i\omega t}\, e^{\alpha t}, \tag{5.5}$$

then we find that the variation in charge density follows the total perturbation, without introducing any extra Fourier components, i.e.

$$\delta\rho = e\sum_{\mathbf{k}}\left\{\frac{\mathscr{U}}{\mathscr{E}(\mathbf{k}) - \mathscr{E}(\mathbf{k}+\mathbf{q}) + \hbar\omega - i\hbar\alpha} + \frac{\mathscr{U}}{\mathscr{E}(\mathbf{k}) - \mathscr{E}(\mathbf{k}-\mathbf{q}) - \hbar\omega + i\hbar\alpha}\right\} e^{i\mathbf{q}\cdot\mathbf{r}}\, e^{i\omega t}\, e^{\alpha t} + \text{complex conjugate}. \tag{5.6}$$

To make this slightly more general, let us introduce $f_0(\mathbf{k})$ as the probability that $|\mathbf{k}\rangle$ is occupied in the unperturbed metal, for example, this might be the Fermi–Dirac function (4.8). By writing $\mathbf{k}$ for $\mathbf{k}-\mathbf{q}$ as labels in the second term of (5.6), we can rewrite the sum in the form

$$\delta\rho = e\mathscr{U}\sum_{\mathbf{k}}\left\{\frac{f^0(\mathbf{k})-f^0(\mathbf{k}+\mathbf{q})}{\mathscr{E}(\mathbf{k})-\mathscr{E}(\mathbf{k}+\mathbf{q})+\hbar\omega-i\hbar\alpha}\right\}e^{i\mathbf{q}\cdot\mathbf{r}}\,e^{i\omega t}\,e^{\alpha t}$$
$$+\text{complex conjugate}, \quad (5.7)$$

where the sum now is over all states $|\mathbf{k}\rangle$, 'occupied' or 'unoccupied'.

This charge distribution gives rise to a potential-energy field acting on the electrons, via the Coulomb interaction. If we call this $\delta\Phi(\mathbf{r},t)$ then we must have, by Poisson's equation,

$$\nabla^2(\delta\Phi) = -4\pi e\delta\rho. \quad (5.8)$$

We may assume that $\delta\Phi$ has the same space and time variation as $\delta\rho$, i.e. we write

$$\delta\Phi(\mathbf{r},t) = \Phi\, e^{i\mathbf{q}\cdot\mathbf{r}}\,e^{i\omega t}\,e^{\alpha t}+\text{complex conjugate}. \quad (5.9)$$

Combining (5.7), (5.8), and (5.9) we have

$$-q^2\Phi = -4\pi e^2\mathscr{U}\sum_{\mathbf{k}}\frac{f^0(\mathbf{k})-f^0(\mathbf{k}+\mathbf{q})}{\mathscr{E}(\mathbf{k})-\mathscr{E}(\mathbf{k}+\mathbf{q})+\hbar\omega-i\hbar\alpha}, \quad (5.10)$$

i.e.

$$\Phi = \left\{\frac{4\pi e^2}{q^2}\sum_{\mathbf{k}}\frac{f^0(\mathbf{k})-f^0(\mathbf{k}+\mathbf{q})}{\mathscr{E}(\mathbf{k})-\mathscr{E}(\mathbf{k}+\mathbf{q})+\hbar\omega-i\hbar\alpha}\right\}\mathscr{U}. \quad (5.11)$$

This is the potential (energy) associated with the charge redistribution created by the original potential $\delta\mathscr{U}$. But this new potential, $\delta\Phi$, should really have been counted as itself a perturbation on the electron distribution. To make our calculation self-consistent, our assumed perturbation $\delta\mathscr{U}$ should already have contained $\delta\Phi$. In other words

$$\delta\mathscr{U}(\mathbf{r},t) = \delta\mathscr{V}(\mathbf{r},t)+\delta\Phi(\mathbf{r},t), \quad (5.12)$$

where $\delta\mathscr{V}(\mathbf{r},t)$ is the actual external potential that we thought we had applied.

If we assume that $\delta\mathscr{V}$ is of the form

$$\delta\mathscr{V} = \mathscr{V}e^{i\mathbf{q}\cdot\mathbf{r}}\,e^{i\omega t}\,e^{\alpha t}+\text{complex conjugate}, \quad (5.13)$$

then we have, from (5.12) and (5.11),

$$\mathscr{U} = \mathscr{V}+\left\{\frac{4\pi e^2}{q^2}\sum_{\mathbf{k}}\frac{f^0(\mathbf{k})-f^0(\mathbf{k}+\mathbf{q})}{\mathscr{E}(\mathbf{k})-\mathscr{E}(\mathbf{k}+\mathbf{q})+\hbar\omega-i\hbar\alpha}\right\}\mathscr{U} \quad (5.14)$$

or

$$\mathscr{U} = \frac{\mathscr{V}}{\epsilon(\mathbf{q},\omega)}, \quad (5.15)$$

where
$$\epsilon(\mathbf{q},\omega) \equiv 1+\frac{4\pi e^2}{q^2}\sum_{\mathbf{k}}\frac{f^0(\mathbf{k})-f^0(\mathbf{k}+\mathbf{q})}{\mathscr{E}(\mathbf{k}+\mathbf{q})-\mathscr{E}(\mathbf{k})-\hbar\omega+i\hbar\alpha} \tag{5.16}$$

(the sign of the denominator in the sum has been changed for convenience).

In other words, the effective potential $\mathscr{U}$ acting on the electrons is not the applied potential $\mathscr{V}$, but is divided by a *dielectric constant* $\epsilon(\mathbf{q},\omega)$, which is a function of the wavelength and frequency of the applied perturbation. We have deduced this result for a single Fourier component, but because we have, at various points such as (5.2) and (5.4), linearized the equations, we can sum the effects of different Fourier components. Thus, if we express $\delta\mathscr{V}$ by a Fourier integral

$$\delta\mathscr{V}(\mathbf{r},t) = \iint \mathscr{V}(\mathbf{q},\omega)\, e^{i\mathbf{q}\cdot\mathbf{r}} e^{i\omega t}\, d\mathbf{q}\, d\omega, \tag{5.17}$$

we can calculate the effective potential seen by the electrons from

$$\delta\mathscr{U}(\mathbf{r},t) = \iint \frac{\mathscr{V}(\mathbf{q},\omega)}{\epsilon(\mathbf{q},\omega)}\, e^{i\mathbf{q}\cdot\mathbf{r}} e^{i\omega t}\, d\mathbf{q}\, d\omega. \tag{5.18}$$

The formula (5.16) is known as *Lindhard's expression*. In the next sections of this chapter we shall show that it contains the seeds of a number of fruitful descriptions of physical phenomena.

5.2 Static screening

Let us consider the effect of a static perturbation where $\omega = 0$. Consider first the form of $\epsilon(\mathbf{q},0)$ near $\mathbf{q} = 0$. We can approximate in (5.16) by writing

$$\mathscr{E}(\mathbf{k}+\mathbf{q})-\mathscr{E}(\mathbf{k}) \approx \mathbf{q}\cdot\nabla_{\mathbf{k}}\mathscr{E}(\mathbf{k}) \tag{5.19}$$

and, remembering that $f^0(\mathbf{k})$ is a function only of $\mathscr{E}(\mathbf{k})$,

$$f^0(\mathbf{k})-f^0(\mathbf{k}+\mathbf{q}) \approx -\mathbf{q}\cdot\frac{\partial f^0}{\partial \mathscr{E}}\nabla_{\mathbf{k}}\mathscr{E}(\mathbf{k}). \tag{5.20}$$

The summation in (5.16) can be written as an integral (remember that in principle it is now over all states, occupied and empty):

$$\begin{aligned}\epsilon(\mathbf{q},0) &\to 1+\frac{4\pi e^2}{q^2}\int\frac{\{\mathbf{q}\cdot\nabla_{\mathbf{k}}\mathscr{E}(\mathbf{k})\}}{\{\mathbf{q}\cdot\nabla_{\mathbf{k}}\mathscr{E}(\mathbf{k})\}}\left(-\frac{\partial f^0}{\partial\mathscr{E}}\right)d\mathbf{k}\\ &= 1+\frac{4\pi e^2}{q^2}\int\left(-\frac{\partial f^0}{\partial\mathscr{E}}\right)\mathscr{N}(\mathscr{E})\,d\mathscr{E}\\ &= 1+\frac{\lambda^2}{q^2},\end{aligned} \tag{5.21}$$

where $$\lambda^2 = 4\pi e^2 \mathcal{N}(\mathscr{E}_F) \tag{5.22}$$

since $(-\partial f^0/\partial \mathscr{E})$ is a function only of energy, and, by (4.13)–(4.18), is practically a delta-function at the Fermi level.

This result shows that $\epsilon \to \infty$ as $q \to 0$. Looking at (5.15), this means that $\mathscr{U} \to 0$ as $q \to 0$, for fixed $\mathscr{V}$. In other words, *an external field of long wavelength is screened out, almost entirely, by the flow of electrons.*

The same result may be obtained by a more elementary argument, in the *Thomas–Fermi approximation.* Suppose there is a perturbing potential $\delta\mathscr{U}$ at some point $\mathbf{r}$. Then the Fermi distribution of the

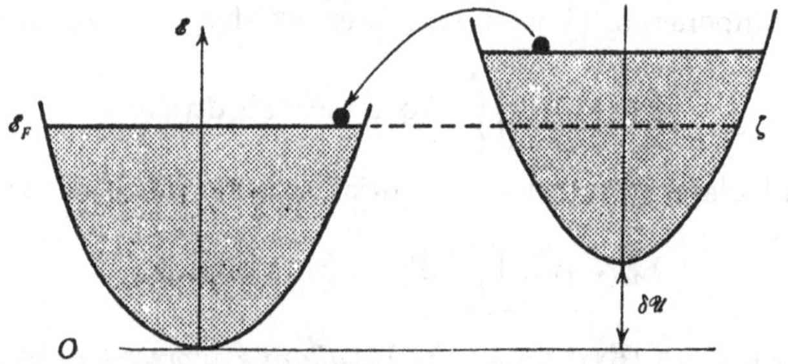

Fig. 86. Effect of perturbation on Fermi distribution.

electrons will be lifted by this amount, relative to the distribution at a point where $\delta\mathscr{U}$ is zero. But this means that the Fermi level, ζ, will change unless electrons flow away from this region. Indeed, this must happen, because ζ, being a chemical potential, is constant throughout the whole volume.

If $\delta\mathscr{U}$ is small we lose a layer of this thickness off the top of the distribution, i.e. the local electron density changes by the amount

$$\delta n(\mathbf{r}) = -\mathcal{N}(\mathscr{E}_F)\,\delta\mathscr{U}(\mathbf{r}). \tag{5.23}$$

But changes in electron density correspond to changes in local charge density, which give rise, in their turn, to a potential $\delta\Phi(\mathbf{r})$. This must satisfy Poisson's equation

$$\begin{aligned} \nabla^2(\delta\Phi) &= -4\pi e^2 \delta n(\mathbf{r}) \\ &= 4\pi e^2 \mathcal{N}(\mathscr{E}_F)\,\delta\mathscr{U}(\mathbf{r}) \\ &= \lambda^2 \delta\mathscr{U}. \end{aligned} \tag{5.24}$$

This is the equivalent, before a Fourier transformation, of (5.11), with $\mathcal{N}(\mathscr{E}_F)$ in place of the summation over $\mathbf{k}$: we get back to (5.21).

Note that for a *classical gas*—for example, the carrier in a semiconductor—we can argue that (5.23) should be replaced by

$$\delta n(\mathbf{r}) \approx n_0 \exp\{-\delta\mathscr{U}(\mathbf{r})/kT\} - n_0$$

$$\approx -\frac{n_0}{kT}\delta\mathscr{U}, \tag{5.25}$$

where n_0 is the local average density of carriers. This follows from (4.32), say. We get the same type of formula as (5·24), but now with

$$\lambda^2 = \frac{4\pi e^2 n_0}{kT}. \tag{5.26}$$

The density of states for a free-electron gas can be written

$$\mathscr{N}(\mathscr{E}_F) = \frac{3}{2}\frac{n}{kT_F}, \tag{5.27}$$

where T_F is the *Fermi temperature* of the gas (i.e. $\mathscr{E}_F/k$). Thus, the quantum theory formula (5.22) for the *screening parameter* λ is equivalent to the classical *Debye–Hückel formula* (5.26), upon the assumption that the electrons have a very high effective temperature, $\sim T_F$. The same result (5.26) may be obtained directly by taking f^0 to be a Boltzmann distribution in the derivation of (5.21).

5.3 Screened impurities and neutral pseudo-atoms

A typical case of a perturbing field in a solid is the electrostatic field around a charged impurity. If we substitute a Zn ion for a Cu ion in the metal lattice, we put a charge $+2|e|$ where previously there was only a single charge. We treat this impurity as if it were a net charge $+|e|$ in a neutral background, and calculate the effect upon the free-electron gas.

The simplest way of calculating this is to assume, in (5.24), that $\delta\mathscr{U}$ and $\delta\Phi$ are the same, except at the very centre of the impurity. The equation

$$\nabla^2(\delta\mathscr{U}) = \lambda^2(\delta\mathscr{U}) \tag{5.28}$$

has a solution of the form

$$\delta\mathscr{U} = \frac{e^2}{r}\exp(-\lambda r), \tag{5.29}$$

where we have made the potential behave like that of the point charge near $r = 0$. This is a typical result; we have a *screened Coulomb potential*, which falls off exponentially with distance with a *screening radius* $1/\lambda$. In accordance with our general screening principle, the impurity cannot be 'seen' by the electrons beyond a few multiples of this radius. In a metal this radius is of the order of an interatomic spacing; in a semiconductor it is much larger.

Let us obtain this result from our dielectric formulation. The 'external' field $\delta\mathscr{V}(\mathbf{r})$ is a bare Coulomb potential,

$$\delta\mathscr{V}(\mathbf{r}) = \frac{e^2}{r}. \tag{5.30}$$

The Fourier transform (5.17) of this function in a three-dimensional box of unit volume is well known;

$$\mathscr{V}(\mathbf{q}) = \frac{4\pi e^2}{q^2}. \tag{5.31}$$

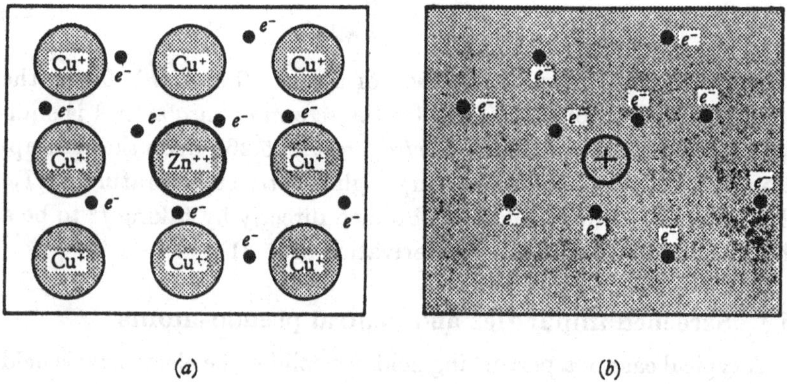

Fig. 87. (a) Divalent impurity in a monovalent metal is replaced by (b) point charge in a continuum.

We note that this is singular at $q = 0$, corresponding to the infinite 'range' of the Coulomb potential.

From (5.15) and (5.21), we have for the screened potential

$$\mathscr{U}(\mathbf{q}) = \frac{4\pi e^2}{q^2(1+\lambda^2/q^2)}$$

$$= \frac{4\pi e^2}{\lambda^2+q^2}. \tag{5.32}$$

It is well known that this is the Fourier transform, (5.18), of the potential $\delta\mathscr{U}(\mathbf{r})$ which we have already derived in (5.29). The singularity in $\epsilon(\mathbf{q}, 0)$ as $q \to 0$ cancels the singularity in $\mathscr{V}(\mathbf{q})$, removing the long-range part of the bare Coulomb potential (5.30).

This formalism may also be applied to the effects of electron–electron interaction on the potential distribution with a pure metal. In principle, we may compute matrix-elements of a self-consistently screened potential $u_s(\mathbf{r})$, such as we should need in an N.F.E. band structure calculation (§ 5.3), by dividing the corresponding Fourier components of a bare ion potential, $\mathscr{V}_b(r)$, by the dielectric function, as in (5.18).

For actual deep core potentials this procedure would not be valid. But suppose we follow the pseudo-potential procedure, and represent the field of each bare ion by a model pseudo-potential $w_b(r)$ defined as in § 3.9. Provided that ion cores do not overlap, we may write the total 'potential' before screening as a superposition of ionic pseudo-potentials, i.e.

$$\mathscr{V}_b(\mathbf{r}) = \sum_l w_b(\mathbf{r}-\mathbf{R}_l). \tag{5.33}$$

By (2.86) and (5.18), the corresponding self-consistent potential has Fourier components

$$\mathscr{U}_s(\mathbf{K}) = \frac{1}{N}\sum_l e^{i\mathbf{K}\cdot\mathbf{R}_l}\frac{w_b(\mathbf{K})}{\epsilon(K,0)}, \tag{5.34}$$

where now $w_b(\mathbf{K})$ is a Fourier transform or matrix element of the pseudo-potential w_b. In a regular lattice, where the structure factor (2.92) merely picks out the reciprocal lattice vectors, this formula gives us screened pseudo-potential matrix elements for the coefficients Γ in an N.F.E. formalism such as (3.64).

By an appropriate choice of structure factor, we may also use (5.34) as an estimate of the matrix element for the net scattering of the conduction electrons by deviations from crystalline order—for example, by phonons (§ 6.12) or even in liquid metals. In such cases, it is instructive to rewrite this formula as the Fourier transform of a real space function, analogous to (5.33): the linear screening approximation allows us to interchange the operations of screening and superposition to yield

$$\mathscr{U}_s(\mathbf{r}) \doteq \sum_l w_s(\mathbf{r}-\mathbf{R}_l), \tag{5.35}$$

where each atomic centre $\mathbf{R}_l$ now seems to carry a screened ionic pseudo-potential, $w_s(r)$, whose Fourier transform is $w_b(\mathbf{K})/\epsilon(K,0)$, (Fig. 88).

As we have seen, the main consequence of screening is to transform a long range Coulomb potential into a screened Coulomb function (5.29). From (5.7) or (5.23), we could easily show that each ion now carries with it a cloud of electron charge, whose density falls off exponentially with range λ. Seen from a distance, this cloud must contain just enough negative charge to neutralize the valence charge $+Z|e|$. In other words, we have created a *neutral pseudo-atom* which can be treated as a single entity in many metallic systems. In a thermal vibration,

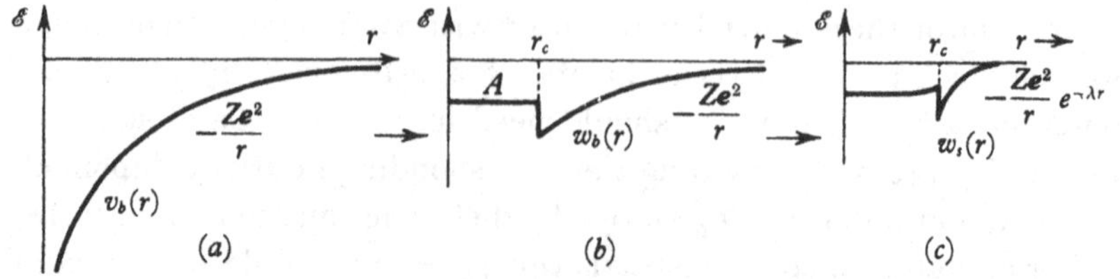

Fig. 88. (*a*) Bare ion potential; (*b*) Bare ion pseudo-potential; (*c*) Screened pseudo-potential.

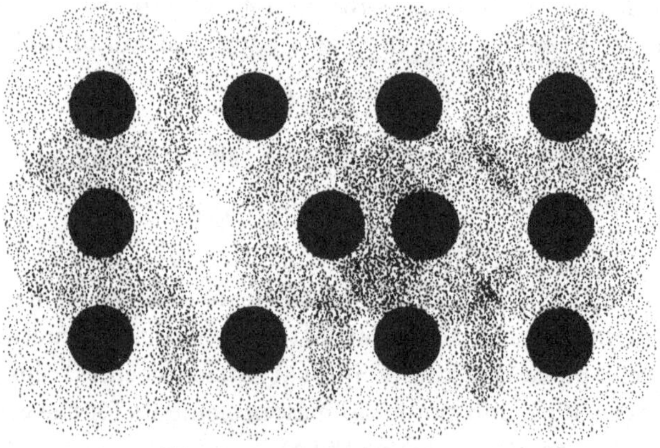

Fig. 89. Neutral pseudo-atoms.

for example (Fig. 89), the displacement of an ion from its lattice site produces an electric field that polarizes the electron gas and hence causes further charge transfer: within the linear response approximation, all such effects are accounted for self-consistently by merely moving the pseudo-atom as a whole.

Notice, however, that the screening clouds of adjacent ions must always overlap and add up to something like the assumed free electron density in the interstitial regions. In a true 'atomic' assembly, such as solid argon, the electrons are localized in atomic orbitals which tend to repel each other strongly when forced into contact. A screening cloud in a metal is no more than a local accumulation of *average* charge density contributed by the innumerable *non-localized* states of the gas of conduction electrons.

The pseudo-atom method is only an approximate procedure for removing the divergences in a bare ionic potential, deep in the core and at long range. Nevertheless, for simple metals it is intuitively sound and gives excellent 'zeroth order' estimates of many, observable quantities (see §§ 6.12, 6.13).

5.4 The singularity in the screening: Kohn effect

The formula (5.21) for the dielectric constant is only an approximation. If we wish to study screening at short distances we need to evaluate the sum (5.16) for large values of $\mathbf{q}$. This depends on the detailed structure of the energy surfaces $\mathscr{E}(\mathbf{k})$. For our free-electron model at absolute zero it is not very difficult to evaluate the sum by a straightforward integration over $\mathbf{k}$-space. The result is as follows:

$$\epsilon(\mathbf{q}, 0) = 1 + \frac{4\pi e^2}{q^2}\frac{n}{\frac{2}{3}\mathscr{E}_F}\left\{\frac{1}{2} + \frac{4k_F^2 - q^2}{8k_F q}\ln\left|\frac{2k_F + q}{2k_F - q}\right|\right\}, \tag{5.36}$$

where k_F is the radius of the Fermi sphere.

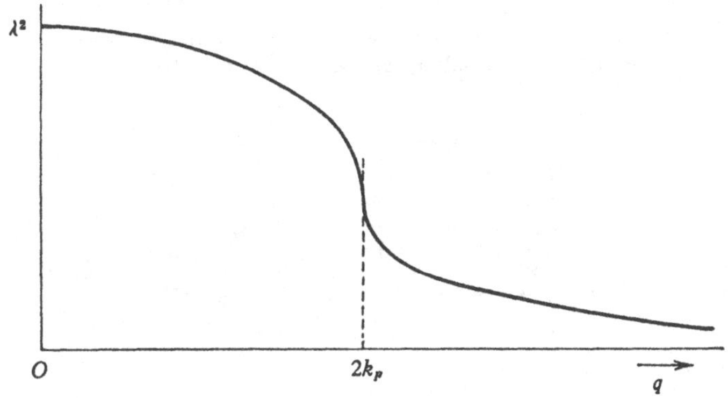

Fig. 90. Variation of screening parameter with wave-number.

If we express this formally as in (5.21), then we see that λ becomes a function of q, tending to (5.22) when $q \to 0$. The *effective screening length*, $1/\lambda$ tends to increase as q increases. It is becoming more and more difficult to make the electron distribution screen out potentials of short wavelength.

This formula is interesting for what happens at $q = 2k_F$, where the logarithmic term is singular. This is a real effect. It comes from the form of the sum (5.16), which contains $f^0(\mathbf{k}) - f^0(\mathbf{k}+\mathbf{q})$. Thus, we need only consider values of $\mathbf{k}$ where $|\mathbf{k}\rangle$ is occupied and $|\mathbf{k}+\mathbf{q}\rangle$ empty, or vice versa. For small values of q, these lie in two regions covering the surface of the sphere (Fig. 91 (*a*)). As q is increased, these regions expand steadily, so that the sum increases. But a value of q is then reached when the whole sphere contributes to the sum. This occurs when

$$q = q_c = 2k_F. \tag{5.37}$$

Beyond this point there are no new terms in the sum over **k**, so that the functional form of the sum changes, even though each term remains a continuous function of **q**. However, the contribution from the last few points on the sphere, before we reach $2k_F$, is small, so that the singularity is not serious. The dielectric function itself remains continuous, but $\partial\epsilon/\partial q$ has a logarithmic infinity at $q = q_c$. This demonstration may be generalized to the case of a Fermi surface that is not a sphere. Any

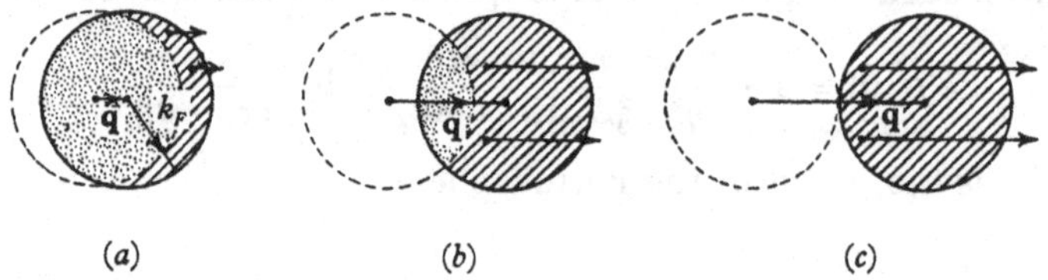

Fig. 91. Regions of the Fermi sphere contributing to the dielectric constant: (a) small q; (b) $q < 2k_F$; (c) $q > 2k_F$.

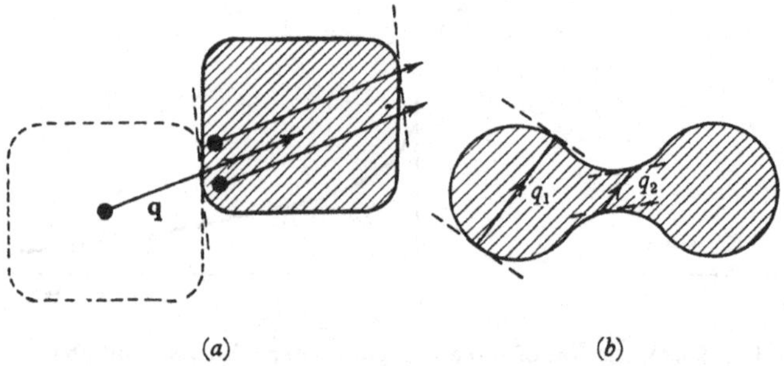

Fig. 92. (a) Kohn anomaly occurs when **q** joins points on the Fermi surface where tangents are parallel. (b) There may be several different values of q_c for a given direction of **q**.

sum of the general form of (5.16) will have a similar singularity for values of **q** that just span the surface. If **q** is fixed in direction, we expect q_c to come at the extreme caliper dimension of the surface in that direction. If we can measure q_c as a function of direction, we may map out the Fermi surface although (Fig. 92(*b*)) there may be several values of q_c for a given direction, corresponding to minima as well as maxima in the 'diameter'.

This is the origin of the *Kohn effect* in the lattice spectrum. A phonon of wave-vector q sets up a potential with components like (5.16) due to the motion of the ions. The electrons move to screen this field. The ions now interact with one another via this screened field, which will be inversely proportional to $\epsilon(\mathbf{q})$. Thus, the forces between the ions will be modified and the lattice frequency of this mode of vibration will depend on $\epsilon(\mathbf{q})$ (see § 6.11). The singularity in ϵ will thus be reflected in the phonon frequency.

The same effect is likely to be observed in any other system of waves propagated in the solid, e.g. spin waves (§ 10.10), if they interact with the conduction electrons. In principle, this offers a general method for the investigation of the Fermi surface—although in practice the detection of the effect may not be feasible.

5.5 The Friedel sum rule

The singularity in $\epsilon(\mathbf{q}, o)$ has another interesting consequence. Suppose that we put (5.36) into our formula for the screened potential of a point charge. Following (5.18) and (5.31) we should have

$$\mathscr{U}(\mathbf{r}) = 4\pi e^2 \int \left\{ q^2 + \frac{4\pi e^2 n}{\frac{2}{3}\mathscr{E}_F} \left[\frac{1}{2} + \frac{4k_F^2 - q^2}{8k_F q} \ln \left| \frac{2k_F + q}{2k_F - q} \right| \right] \right\}^{-1} e^{-i\mathbf{q}\cdot\mathbf{r}} d\mathbf{q} \tag{5.38}$$

for the potential at distance r. Without actually doing this calculation, we can readily believe that the singularity at $q = 2k_F$ will show up as a special contribution to $\mathscr{U}(\mathbf{r})$; instead of this being a smooth exponential function it will contain oscillations of wave-number $2k_F$.

Instead of deriving this effect from (5.38), let us follow a rather different line of argument due to Friedel. Consider a spherically symmetric potential $\mathscr{U}(\mathbf{r})$. We know that there is a solution of the Schrödinger equation in the presence of such a potential, of the form

$$\psi_{k,l} \sim \frac{1}{r} \sin\left(kr - \tfrac{1}{2}l\pi + \eta_l\right) P_l(\cos\theta) \tag{5.39}$$

at large distances. Here $P_l(\cos\theta)$ is an ordinary spherical harmonic of order l. We are not considering a scattering problem now, but a stationary state of a 'free' electron, in a space that is empty except for the potential $\mathscr{U}(\mathbf{r})$ at the centre.

The existence of solutions like (5.39), shifted in phase by the amount η_l as compared with the corresponding solution when $\mathscr{U}(\mathbf{r}) = 0$, is at the heart of the *partial wave method* of solving the problem of scattering from such a potential (cf. § 3.9). As is well known, functions like (5.39) are matched to a plane wave, $\exp(i\mathbf{k}\cdot\mathbf{r})$, representing the initial state, and the scattered part can be calculated. The *differential scattering cross-section is given by*

$$\sigma(\theta) = \frac{1}{k^2} \left| \sum_{l=0}^{\infty} (2l+1)\, e^{i\eta_l} \sin\eta_l\, P_l(\cos\theta) \right|^2. \tag{5.40}$$

But to return to our stationary states (5.39). Suppose that we use these as the basic eigenstates of the system and fill them up, in order of increasing energy, just as we do for the plane wave states. It is convenient in the counting of states to use a different type of boundary condition—the obvious rule is $\psi_k = 0$ at $r = R$, as if we had a large sphere, of radius R, with the atom in the centre. Thus the 'allowed' values of k must satisfy

$$kR - \tfrac{1}{2}l\pi + \eta_l = \text{integer} \times \pi. \tag{5.41}$$

If our sphere had been empty, so that all the phase shifts η_l were zero, then the 'allowed' values of k would be the set

$$k_n = \frac{(n + \frac{1}{2}l)\,\pi}{R}. \tag{5.42}$$

For each value of n, there are, of course, many different values of l, and for each l there are $(2l+1)$ different wave-functions. But one can show—indeed, it follows from a general theorem on the distribution

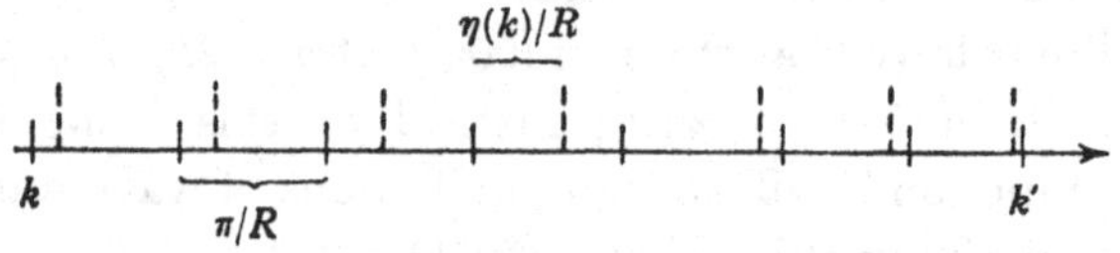

Fig. 93

of eigenvalues—that the total number of states with energy less than $\mathscr{E}_F$, i.e. of wave-vector less than k_F, is the same in this scheme as it is in the more familiar case of the cubical box used in § 1.6. But with the atom at the centre the η_l are not zero, so the 'allowed' values of k do not quite come at k_n; they will be shifted by an amount η_l/R. But η_l varies with k. It is easy to see from Fig. 93 that the new set will contain

$$\{\eta_l(k) - \eta_l(k')\}/\pi$$

new allowed values between two points k and k' on the k-axis.

Now suppose (as is reasonable) that η_l tends to zero at $k = 0$. The total number of new electrons required to fill up the levels to the same Fermi wave-vector $k = k_F$ must be

$$Z = \frac{2}{\pi}\sum_l (2l+1)\,\eta_l(k_F) \tag{5.43}$$

allowing for the $(2l+1)$ states of given l, and the 2 spin states of the electron. We have put this number equal to Z, the valence difference between the impurity and the solvent metal, because we need precisely this number of extra electrons in the neighbourhood of the impurity to neutralize its charge.

This is the *Friedel sum rule*. It depends upon two general principles: the electric charge of an impurity must be neutralized by an excess of electrons within a finite distance; the Fermi wave-vector k_F at large distances from the impurity must be the same as in a pure and perfect crystal.

This formula provides a most useful method for calculating the scattering cross-section of an impurity. We choose some function, such as a screened Coulomb potential (5.29), with an adjustable parameter λ. We calculate the phase shifts, and find what value of λ will make them satisfy (5.43). With these values of η_l, we may find the scattering cross-section from the partial wave formula (5.40). This procedure gives quite good results, which are not very sensitive to the shape of the function $\mathscr{U}(\mathbf{r})$. It is certainly a better method, in principle, than the use of the matrix elements of (5.38), say, in Born approximation. The conduction electrons are too slow for this to be valid.

But now consider the electron density associated with phase-shifted waves like (5.39). At a large distance r from the atom we shall find the extra charge

$$\begin{aligned}
\delta\rho &= e\sum_l (2l+1)\int_0^{k_F} \{|\psi_{k,l}(\eta_l)|^2 - |\psi_{k,l}(0)|^2\}\frac{2R}{\pi}\,dk \\
&= \frac{e}{2\pi^2}\sum_l (2l+1)\int_0^{k_F} \{\sin^2(kr-\tfrac{1}{2}l\pi+\eta_l) - \sin^2(kr-\tfrac{1}{2}l\pi)\}\frac{1}{r^2}\,dk \\
&= \frac{e}{2\pi^2}\sum_l (2l+1)\int_0^{k_F} \sin(2kr-l\pi+\eta_l)\sin\eta_l\frac{1}{r^3}\,d(kr) \\
&\sim \frac{e}{2\pi^2}\sum_l (2l+1)(-1)^l\sin\eta_l\frac{\cos(2k_F r+\eta_l)}{r^3}
\end{aligned} \tag{5.44}$$

using the normalization factor $(2\pi R)^{-\frac{1}{2}}$ for the wave-functions (5.39) in the sphere of radius R, and dropping terms going to zero more rapidly at large values of r.

This is the oscillating charge density associated with the singularity in (5.38). It tends to zero only as $1/r^3$, so that it is not a negligible effect. It is interesting that the charge in the neighbourhood of the impurity is not simply heaped up; there are regions where $\delta\rho$ is negative, where the electrons are actually driven away. This is significant in phenomena associated with the interaction between impurities and with the Knight shift in alloys (see § 10.2).

What is happening is that the localized charge is not that of one or two electrons in a bound state round the impurity. It is due to the attraction to the impurity of the fast electrons of the Fermi gas, which tend to linger a little in the neighbourhood, and crowd closer there. The sharp cut-off in electron wavelength then gives rise to interference effects, which cast haloes of charge around the scattering centre.

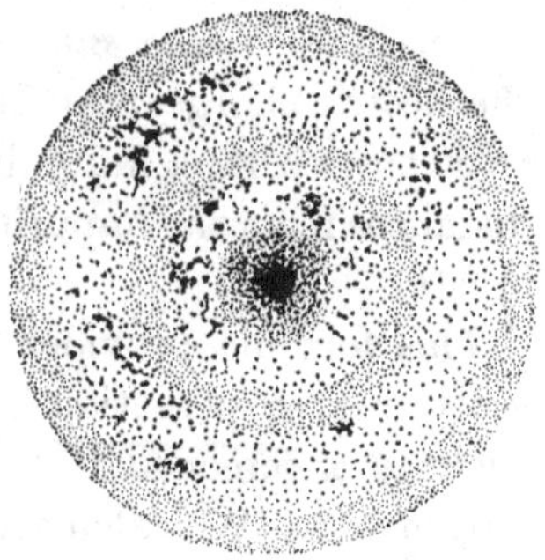

Fig. 94. Haloes of charge round an impurity.

In the case of a *transition metal* dissolved in a 'normal' metal, we must take account of the incomplete d shell. As we saw in § 3.10, the d-orbitals of the free atom appear as a *resonance*, where, as in (3.81), the d-phase-shift passes rapidly through $\pi/2$ over a range of energy of width W. From the Friedel sum rule (5.43), it follows that a charge of up to 10 electrons rapidly builds up on the impurity as we go through the resonance. But the total charge on the impurity must stabilize itself so as to neutralize its nuclear charge; we must have approximately the same number of 'd-electrons' in the ion as we had in the free atom. This means that the Fermi level of the solvent metal must lie within the resonance region.

The electrons in a d-shell of a free atom are always magnetically polarized: in accordance with Hund's rule, their spins line up to form the maximum moment consistent with the Pauli principle. In Mn, for example, all $5d$-electrons may have the same spin. This tendency may persist when the atom is dissolved in another metal, even though the spin polarization can no longer be ascribed to a stationary atomic level.

Suppose, for example, that the impurity happens to have an excess of 'up' spin electrons. For another electron of the same spin, encountering the atom, this appears as a resonance lying below the Fermi level, into which it is attracted (Fig. 95). The 'up' spin density is thus enhanced and the polarized state is stabilized. For a 'down' spin electron, on the other hand, the *d-d exchange interaction* pushes up the energy of the resonance, thus depleting the down spin density within the impurity. The net effect must be to compensate the total charge of the ion; but in a suitable solvent we may also observe a dynamical many-electron magnetic moment, slowly fluctuating in direction but comparable in magnitude with the moment of the free atom. This process explains many of the special electrical and magnetic properties of dilute alloys of transition metals (see § 10.6).

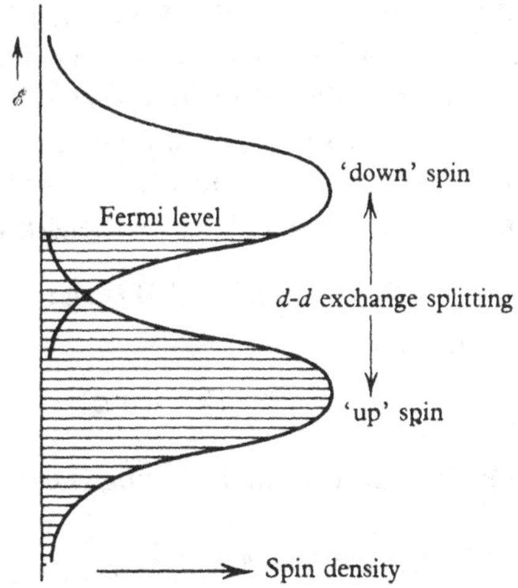

Fig. 95. Resonances for an electron at a magnetic impurity polarized in the 'up' direction.

5.6 Dielectric constant of a semiconductor

The macroscopic dielectric constant of a metal tends to infinity as $q \to 0$, because the metal has a high conductivity. This result follows from (5.16) and (5.21). In a semiconductor or insulator, however, the conductivity is small; apart from thermal excitation of carriers it may be taken to be zero. The static dielectric constant $\epsilon(\mathbf{q}, 0)$ should remain finite at long wavelengths.

This is due to the energy gap between filled and empty states. The denominator in a sum like (5.16) does not vanish at $\mathbf{q} = 0$ but remains finite.

To calculate $\epsilon(\mathbf{q}, 0)$ for an actual semiconductor requires a detailed model of the band structure. We cannot use the simple theory of § 5.1 because the unperturbed electron states $|\mathbf{k}\rangle$ are no longer simple plane waves. It is easy enough to show, however, that we need only introduce the square of the matrix element of the perturbing wave $\exp(i\mathbf{q}\cdot\mathbf{r})$ into the sum (5.16) to get the correct formula:

$$\epsilon(\mathbf{q}, \omega) = 1 + \frac{4\pi e^2}{q^2} \sum_{\mathbf{k}, \mathbf{g}} \frac{|\langle\mathbf{k}|\, e^{i\mathbf{q}\cdot\mathbf{r}}\, |\mathbf{k}+\mathbf{q}+\mathbf{g}\rangle|^2 \{f^0(\mathbf{k}) - f^0(\mathbf{k}+\mathbf{q}+\mathbf{g})\}}{\mathscr{E}(\mathbf{k}+\mathbf{q}+\mathbf{g}) - \mathscr{E}(\mathbf{k}) - \hbar\omega + i\hbar\alpha}. \tag{5.45}$$

We need to include terms for which the reciprocal lattice vector $\mathbf{g} \neq 0$, because the matrix element of $\exp(i\mathbf{q}\cdot\mathbf{r})$ between Bloch functions does not vanish when they differ in wave-vector by $\mathbf{q}+\mathbf{g}$. The situation is essentially the same as in the diffraction theory of § 2.7.

We shall not attempt to evaluate (5.45) exactly, but we can find a rough approximation as follows. Let us use a reduced zone scheme. The only empty states available when $|\mathbf{k}\rangle$ is a full state are thus in the next band; if $\mathbf{q}$ is small $|\mathbf{k}+\mathbf{q}+\mathbf{g}\rangle$ now refers to a state that lies nearly vertically 'above' $|\mathbf{k}\rangle$. Suppose that the energy denominator is the same for all values of $\mathbf{k}$: we take it to be the gap width

$$\mathscr{E}(\mathbf{k}+\mathbf{q}+\mathbf{g})-\mathscr{E}(\mathbf{k}) \approx \mathscr{E}_{\text{gap}}. \tag{5.46}$$

Now we want to calculate

$$\sum_{\mathbf{g}} |\langle \mathbf{k}|\, e^{i\mathbf{q}\cdot\mathbf{r}}\, |\mathbf{k}+\mathbf{q}+\mathbf{g}\rangle|^2 \tag{5.47}$$

for each value $\mathbf{k}$. We can use the following theorem:—

$$\sum_{n} (\mathscr{E}_n-\mathscr{E}_s)\, |\langle n|\, e^{i\mathbf{q}\cdot\mathbf{r}}\, |s\rangle|^2 = \frac{\hbar^2 q^2}{2m}, \tag{5.48}$$

for the matrix element of $\exp(i\mathbf{q}\cdot\mathbf{r})$ between the eigenstates $|n\rangle$ and $|s\rangle$ of a Hamiltonian, $\mathscr{H}$, where the kinetic energy term is $-(\hbar^2/2m)\nabla^2$. The proof of this theorem (which is a generalization of the well-known *sum rule for oscillator strengths*) follows from an expansion of the expectation value of the double commutator

$$[[\mathscr{H}, e^{i\mathbf{q}\cdot\mathbf{r}}], e^{-i\mathbf{q}\cdot\mathbf{r}}] \tag{5.49}$$

in the state $|s\rangle$.

If we take $|s\rangle$ to be, in fact, our state $|\mathbf{k}\rangle$, and if we assume, as in (5.46), that

$$\mathscr{E}_n-\mathscr{E}_s = \mathscr{E}(\mathbf{k}+\mathbf{q}+\mathbf{g})-\mathscr{E}(\mathbf{k}) \approx \mathscr{E}_{\text{gap}} \tag{5.50}$$

for the only important states $|n\rangle$ in the sum, then we have

$$\sum_{\mathbf{g}} |\langle \mathbf{k}|\, e^{i\mathbf{q}\cdot\mathbf{r}}|\mathbf{k}+\mathbf{q}+\mathbf{g}\rangle|^2 \approx \frac{1}{\mathscr{E}_{\text{gap}}}\frac{\hbar^2 q^2}{2m}. \tag{5.51}$$

Putting (5.46) and (5.51) into (5.45), and adding contributions from each of the n electrons in the full band, we have (including also terms in which $|\mathbf{k}\rangle$ is empty and $|\mathbf{k}+\mathbf{q}+\mathbf{g}\rangle$ is full)

$$\begin{aligned}\epsilon(\mathbf{q},0) &\approx 1+\frac{4\pi e^2}{q^2}\frac{n}{\mathscr{E}_{\text{gap}}}\frac{\hbar^2 q^2}{2m}\frac{2}{\mathscr{E}_{\text{gap}}}\\ &= 1+\frac{4\pi n e^2\hbar^2}{m(\mathscr{E}_{\text{gap}})^2}\\ &= 1+\left(\frac{\hbar\omega_p}{\mathscr{E}_{\text{gap}}}\right)^2,\end{aligned} \tag{5.52}$$

where we define the *plasma frequency*,

$$\omega_p = \left(\frac{4\pi ne^2}{m}\right)^{\frac{1}{2}} \tag{5.53}$$

—a parameter whose physical significance will shortly be revealed.

This formula (5.52) indicates that the static dielectric constant tends to a constant at $q = 0$. This should be the observed macroscopic dielectric constant of the solid. The smaller the energy gap, the more polarizable is the medium, i.e. the larger the value of ϵ. Of course we cannot take the assumed value of $\mathscr{E}_{\text{gap}}$ to be the same as the optical gap between the bottom of the conduction band and the top of the valence band; it is more like the average distance in energy between states lying vertically above each other in the reduced zone.

The electrostatic field around an impurity in a semiconductor is thus not screened as it is in a metal: it is simply reduced to $e^2/\epsilon r$ where ϵ is the static dielectric constant. This allows bound electron states to form around the impurity—the *impurity levels*, which will be discussed again in § 6.4. Of course, if there is a residual density n_0 of carriers in the medium, we may find classical screening of the impurity at large distances, as in the Debye–Hückel formulae (5.26), (5.29).

5.7 Plasma oscillations

Now we consider time-dependent effects, in which the frequency ω of the 'external' perturbing field is not zero. There are several important phenomena arising here. The most obvious is where $\hbar\omega$, in (5.16) or (5.45), happens to coincide with the energy difference between a pair of states mixed by the perturbation, i.e. where, for some value of $\mathbf{k}$ in the occupied levels,

$$\mathscr{E}(\mathbf{k}+\mathbf{q}+\mathbf{g}) - \mathscr{E}(\mathbf{k}) = \hbar\omega. \tag{5.54}$$

We then have a singularity in $\epsilon(\mathbf{q}, \omega)$—or the imaginary part $i\hbar\alpha$ becomes dominant. This corresponds simply to *optical absorption* (or emission) by the electrons, with the transition of an electron from state $|\mathbf{k}\rangle$ to state $|\mathbf{k}+\mathbf{q}+\mathbf{g}\rangle$. This we call a *single-particle excitation* of the system. We shall come back to these effects in § 8.5.

But suppose that $\hbar\omega$ is large—much larger than any of the energy differences in the denominator of (5.16). We can rearrange the sum in the form

$$\epsilon(\mathbf{q}, \omega) = 1 + \frac{4\pi e^2}{q^2} \sum_{\mathbf{k}} \frac{2f^0(\mathbf{k})\{\mathscr{E}(\mathbf{k}) - \mathscr{E}(\mathbf{k}+\mathbf{q})\}}{(\hbar\omega)^2 - \{\mathscr{E}(\mathbf{k}) - \mathscr{E}(\mathbf{k}+\mathbf{q})\}^2}. \tag{5.55}$$

Then we can take the denominator to be just $(\hbar\omega)^2$, and expand the numerator in powers of $\mathbf{q}$. The first terms vanish: we arrive at

$$\begin{aligned}\epsilon(\mathbf{q},\omega) &= 1+\frac{4\pi e^2}{q^2}\sum_{\mathbf{k}}\frac{f^0(\mathbf{k})}{(\hbar\omega)^2}\left(-q^2\frac{\partial^2\mathscr{E}}{\partial k^2}\right)\\ &= 1-\frac{4\pi ne^2\hbar^2}{\hbar^2\omega^2 m}\\ &= 1-\frac{\omega_p^2}{\omega^2},\end{aligned} \tag{5.56}$$

where ω_p is the plasma frequency defined by (5.53).

This is derived for free electrons. But the same result can be obtained from (5.45), which can be rearranged to read

$$\epsilon(\mathbf{q},\omega) = 1+\frac{4\pi e^2}{q^2}\sum_{\mathbf{k},\mathbf{g}}\frac{2f^0(\mathbf{k})\,|\langle\mathbf{k}|\,e^{i\mathbf{q}\cdot\mathbf{r}}\,|\mathbf{k}+\mathbf{q}+\mathbf{g}\rangle|^2\{\mathscr{E}(\mathbf{k})-\mathscr{E}(\mathbf{k}+\mathbf{q}+\mathbf{g})\}}{(\hbar\omega)^2-\{\mathscr{E}(\mathbf{k})-\mathscr{E}(\mathbf{k}+\mathbf{q}+\mathbf{g})\}^2}. \tag{5.57}$$

If the denominator can be treated as a constant, i.e. if

$$\hbar\omega \gg \mathscr{E}(\mathbf{k})-\mathscr{E}(\mathbf{k}+\mathbf{q}+\mathbf{g}) \tag{5.58}$$

for all occupied $\mathbf{k}$, and all $\mathbf{g}$ for which the matrix element is not negligible, then we can apply the sum rule (5.48) to the sum over $\mathbf{g}$, and get a term $(4\pi e^2/m\omega^2)$ for each of the n electrons in the occupied states.

The formula (5.56) shows that $\epsilon \to 0$ when $\omega \to \omega_p$. But when $\epsilon = 0$, we have, from (5.15),

$$\mathscr{U} = \frac{\mathscr{V}}{\epsilon} \to \infty. \tag{5.59}$$

In other words, an infinitesimal 'external field' $\mathscr{V}$ gives rise to a large effective field $\mathscr{U}$: the system is self-exciting. The frequency ω_p is a natural mode of oscillation of the electron gas—called a *plasma mode*.

This result can be derived by an elementary argument. Suppose that electrons of volume density n are moved bodily through a distance $\mathbf{x}$, relative to the fixed background of positive charge of the lattice. This sets up a polarization,

$$\mathbf{P} = ne\mathbf{x}, \tag{5.60}$$

which gives rise, in turn, to an electric field,

$$\mathbf{E} = -4\pi\mathbf{P}. \tag{5.61}$$

The equation of motion of each charge is thus

$$m\ddot{\mathbf{x}} = e\mathbf{E} = -4\pi ne^2\mathbf{x}, \tag{5.62}$$

which defines a simple harmonic oscillation of frequency ω_p.

The plasma frequency corresponds to quite a large energy, 10–20 eV., so these modes are not excited at thermal energies. But if one fires electrons through thin films of metal one observes energy losses corresponding to the excitation of one or more quanta in these modes. It is interesting to note that, even in a semiconductor, ω_p depends only on the total electron density in the valence band. This is because the condition (5.58) will hold if the plasma energy is rather higher than the energy difference between valence and conduction bands—usually something less than 5 eV. On the other hand, electrons in the cores of the ions will not contribute; they are too tightly bound. The energy differences between these levels and the states in the conduction band are much larger than $\hbar\omega_p$.

We can expand $\epsilon(\mathbf{q}, \omega)$ to higher powers of q, and find the value of ω, as a function of $\mathbf{q}$, that makes ϵ vanish. The result is, for free electrons,

$$\omega(q) = \omega_p\left[1 + \frac{3}{10}\frac{q^2 v_F^2}{\omega_p^2} + \ldots\right], \tag{5.63}$$

where v_F is the electron velocity at the Fermi surface. Thus, the plasma oscillations can occur at various wavelengths—one can refer to quanta in these modes as *plasmons*. But the dispersion law (5.63) shows that the frequency does not depend strongly on q; plasmons tend to be localized oscillations that move only slowly through the medium.

Strictly speaking $\epsilon(\mathbf{q}, \omega)$ is complex: (5.55) and (5.57) represent its real part (assuming α to be small). The imaginary part is given by

$$\epsilon_2(\mathbf{q}, \omega) = \frac{-4\pi^2 e^2}{q^2} \sum_{\mathbf{k}} \{f^0(\mathbf{k}) - f^0(\mathbf{k}+\mathbf{q})\}\,\delta\{\mathscr{E}(\mathbf{k}+\mathbf{q}) - \mathscr{E}(\mathbf{k}) - \hbar\omega\}, \tag{5.64}$$

with a delta-function coming from the pole near the real axis at

$$\hbar\omega = \mathscr{E}(\mathbf{k}+\mathbf{q}) - \mathscr{E}(\mathbf{k}) \tag{5.65}$$

in the integration (5.16). The interesting point about this formula is that it vanishes when

$$\omega \geqslant q v_F, \tag{5.66}$$

i.e. when ω is too large to excite a single particle out of the Fermi distribution with only the change $\mathbf{q}$ of wave-vector. This shows that the plasma modes are not dissipated rapidly above this frequency.

Another interesting phenomenon associated with plasma oscillations can be deduced from (5.56). If this represents the actual dielectric constant of the metal, as seen by a high frequency electric field of long wavelength, we can write down a *complex refractive index*, N, of the medium

$$N^2 = \epsilon(\mathbf{q}, \omega) \approx 1 - \frac{\omega_p^2}{\omega^2}. \tag{5.67}$$

If $\omega < \omega_p$, then ϵ is negative, and N pure imaginary. This gives rise to *total reflection* of light. But if $\omega > \omega_p$, ϵ is positive, and N is real; the metal becomes *transparent* if viewed normal to the surface. This theory is changed a bit when scattering mechanisms are introduced for the electrons (see § 8.6), but the phenomenon of *ultra-violet transparency* is observed in the alkali metals.

5.8 Quasi-particles and cohesive energy

The above analysis, based upon the dielectric constant is, of course, only an approximation. The essential steps were taken in § 5.1, where it was assumed that each Fourier component of the potential could be treated independently. This is known technically as the *Random Phase Approximation* and has to be assumed in almost all more detailed analyses of the problem.

We shall not discuss these more complicated theories, except to make some general remarks. It turns out that the strong long-range interaction between the electrons, associated with the Coulomb potential, can be reduced to a screened interaction like (5.29). Not only do the electrons form charge clouds about impurities, and screen out the field at large distances; each electron also carries around with it, so to speak, its own charge cloud—actually a 'positive' cloud, corresponding to the effective exclusion of 'other' electrons from its neighbourhood.

Conversely, a *positron* moving about in the crystal would attract to itself a 'negative' charge cloud of electrons. The rate of *positron annihilation* in a metal is thus somewhat higher, due to the enhanced electron probability density at the positron centre, than it would be for an unperturbed uniform electron gas. But many-body effects do not seriously perturb the simple interpretation of the angular correlation between the directions of the two γ-ray quanta emitted in the annihilation process. Since the positron is believed to come to rest in the lattice before annihilation, any deviation from emission in opposite directions must be due to momentum supplied by the annihilating electron (Fig. 96). The distribution of the angle θ therefore provides a measure of the distribution of momentum in the gas of conduction electrons, which may prove a useful check on theories of electronic structure in metals and alloys.

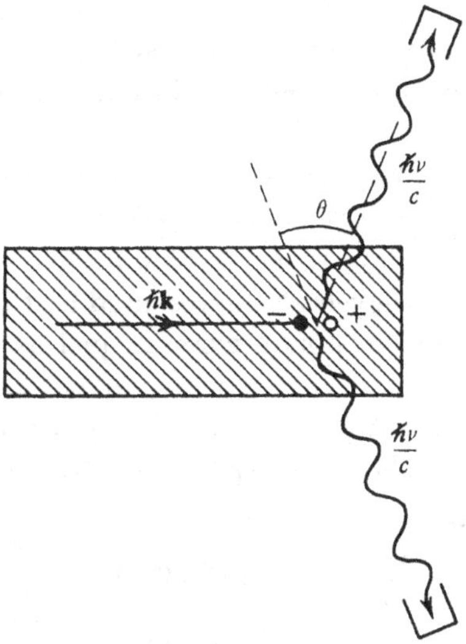

Fig. 96. Momentum conservation in positron annihilation.

Strictly speaking, an electron with its halo of positive charge requires a very complicated wave-function for its proper description. However, it does behave very like a particle, in that it carries the electron charge e. The effect of the 'other' electrons is to modify the relationship between energy and wave-vector for this quantum state; we may have, for example,

$$\mathscr{E}(\mathbf{k}) = \frac{\hbar^2 k^2}{2m^*}, \tag{5.68}$$

where m^* is not quite equal to the ordinary free-electron mass m. It turns out that these *quasi-particles* are quite stable when they are near the Fermi level, but tend to break down and be dissipated when far away from the Fermi surface. This is an important point in the formal justification of transport theory, and of the notion of a Fermi surface.

The quasi-particle picture is really a representation of the spectrum of *elementary excitations* of the electron gas. Although we cannot calculate this spectrum accurately for a real metal, we can make quite good estimates of the properties of an ideal electron gas, of the corresponding density, in a positive jellium (§ 5.1). The ground state energy of this model, for example, tells us something about the cohesive energy of the metal (§ 4.3). The crude assumption of Wigner and Seitz, that the electron drives all other conduction electrons from the cell it is visiting, may be replaced by a direct calculation of the electrostatic interaction between the electron and its screening cloud, thus making a much better estimate of the effects of *electron correlation* in metallic bonding.

Similarly, we should expect two quasi-particles to interact quite weakly, more or less as if by a screened Coulomb interaction. Our system therefore conforms to the *Landau model* of a *Fermi liquid*, in which the lower energy levels are represented by the excitation of one or more quasi-particles into higher states, just as if we had a gas of *independent* fermions at a very low temperature. But as we excite more and more particles above the Fermi level, we must allow for their mutual interactions. These are represented by an additional energy term of the form

$$E_2 = \tfrac{1}{2}\sum_{\mathbf{k},\boldsymbol{\sigma};\,\mathbf{k}',\boldsymbol{\sigma}'} f(\mathbf{k},\boldsymbol{\sigma};\mathbf{k}',\boldsymbol{\sigma}')\,\delta n(\mathbf{k},\boldsymbol{\sigma})\,\delta n(\mathbf{k}',\boldsymbol{\sigma}') \tag{5.69}$$

where $\delta n(\mathbf{k},\boldsymbol{\sigma})$ is the change in the occupation number of the quasi-particle mode of wavevector $\mathbf{k}$ and spin $\boldsymbol{\sigma}$ as we go from the ground state to some higher energy level of the system.

The parameters of the Landau model determine the spectrum of the gas of conduction electrons, and hence such properties as the electron specific heat (§4.7). An important feature of the interaction term $f(\mathbf{k},\boldsymbol{\sigma};\mathbf{k}',\boldsymbol{\sigma}')$ is that it really does depend on the mutual spins of the two electrons. At ordinary metallic densities, this *exchange* force favour parallel alignment—essentially because the Pauli principle keeps two electrons with the same spin away from the same point in space, thereby enhancing the correlation effect and slightly reducing their mutual electrostatic potential energy. This effect is not large, but obviously plays an important part in the theory of the magnetic properties of metals (see §§10.2–10.5).

5.9 The Mott transition

Another qualitative effect due essentially to the force between electrons is not often observed, but is of importance in principle. Suppose we bring together slowly, a large number of hydrogen atoms (or Na atoms—or any array of atoms with an odd number of electrons per unit cell). According to the principles enunciated in §4.1, the electron wave-functions on each atom will overlap, bands will be formed, and the solid should immediately become a conductor. This should happen, it would seem, however far apart the atoms might be.

This is a paradoxical result, difficult to believe. Indeed, it is probably not so. The formal equivalence of, say, a set of localized orbitals $\phi_a(\mathbf{r}-\mathbf{l})$ to a set of Bloch functions $\phi_{\mathbf{k}}(\mathbf{r})$, suggested in § 3.4, is only valid if we neglect the Pauli principle and the electron–electron interaction. If we have *two* electrons per atom, then the determinantal wave-function of all the states $\phi_{\mathbf{k}}(\mathbf{r})$ in a full band is the same as the determinant of all the localized atomic orbitals† $\phi_a(\mathbf{r}-\mathbf{l})$. But if we have only one electron per atom, the determinant of the states in the half-filled band is not the same as the determinant of the localized orbitals. It will tend to deposit two electrons on some atoms, and leave some atoms nearly empty—a distribution of charge that may contribute large positive terms to the total Coulomb energy of the system. Thus, the set of localized states may be a better trial wave-function for the system than the set of delocalized band states. The former type of wave-function describes an insulator, the second type a conductor.

To distinguish between these we may use the following argument. Suppose we take an electron out of one of the array of localized states. It will leave behind it an empty ion, i.e. a positive charge. The electron will be attracted to this ion, and may even form a bound state there, so that it will not really be free to move through the crystal and carry a current. But suppose there are already a number of such electrons present. They will tend to screen the charge of the ion, whose potential may be taken to be of the form

$$-\frac{e^2}{r}e^{-\lambda r}, \tag{5.70}$$

as in (5.29). If the density of ionized electrons is n, then, according to (5.22)

$$\lambda^2 \approx \frac{4me^2n^{\frac{1}{3}}}{\hbar^2} \tag{5.71}$$

(using (3.3), (3.4) and (4.3)).

An ordinary Coulomb field can bind an electron. The lowest state—the ground state of a hydrogen atom—has radius

$$a_H = \frac{\hbar^2}{me^2}. \tag{5.72}$$

If the screening radius $1/\lambda$ is less than a_H, then even this state cannot be bound. Thus, the condition that the ionized atom with potential

† Strictly speaking, we should refer to Wannier functions here (see § 6.2). But the argument is the same as if we assume that the tight-binding approximation is valid when the atoms are far apart.

(5.70) should be capable of recapturing the electron that has been removed from it is

$$\lambda < a_H^{-1},$$

i.e.

$$n^{-\frac{1}{3}} > 4a_H. \tag{5.73}$$

If the average spacing of our atoms is greater than 4 atomic units, then the metallic state is not self-sustaining and the system should become an insulator. Moreover, this is a co-operative effect; every extra free electron helps loosen the remainder. Even if we have not correctly calculated the number 4 in (5.73), it seems likely that the transition will be quite sharp in principle; bringing our hydrogen or Na atoms together, we should see them suddenly transform from an insulator to a metal at a definite atomic spacing.

The *Mott transition* has been observed in certain transition metal oxides, which become conducting, very suddenly, above a certain temperature. This theory can also be applied to impurity centres in a semiconductor, with modification of a_H by multiplying by the bulk dielectric constant (as in § 5.5) of the semiconductor (see § 6.4).

The above argument is, indeed, closely related to *Wigner's hypothesis* that an electron gas at very low density will tend to 'crystallize'. Let r_0 be the average radius of the sphere occupied by an electron, i.e.

$$\tfrac{4}{3}\pi r_0^3 = 1/n. \tag{5.74}$$

The potential energy of each electron, because of the sphere of background positively charged jellium around it, will be of the order of

$$-e^2/2r_0. \tag{5.75}$$

But each electron will have a zero-point energy, of the order of

$$\hbar^2/mr_0^2 \tag{5.76}$$

because of its confinement to the region of this sphere. Thus, the potential energy will win when

$$e^2/2r_0 \geqslant \hbar^2/mr_0^2,$$

i.e.

$$r_0 \geqslant 2a_H. \tag{5.77}$$

The numerical factor, in this as in (5.73), need not be taken seriously, and it is not at all clear under what conditions the transition might actually be observed. Nevertheless, it is important to realize these limitations of the simple band theory of Chapter 4, and of the simple theory of electron correlation in the present chapter.

CHAPTER 6

DYNAMICS OF ELECTRONS

Freedom is the recognition of necessity. ENGELS

6.1 General principles

So far we have discussed the static properties of solids. But many of the most important physical phenomena, such as electrical and thermal conductivity, are dynamical effects—the results of external forces and fields. We need a theory of such phenomena, especially as applied to conduction electrons.

The elementary principles are very simple. We first ask to calculate the *velocity* of an electron in the state $|\mathbf{k}\rangle$. This has energy $\mathscr{E}(\mathbf{k})$; in elementary wave mechanics we associate this with a time-varying factor, as if it had frequency

$$\nu_{\mathbf{k}} = \frac{1}{\hbar}\mathscr{E}(\mathbf{k}). \tag{6.1}$$

But in wave theory, if we suppose we have a dispersive medium, the *group velocity* of a wave packet near this frequency would be

$$\mathbf{v}_{\mathbf{k}} = \nabla_{\mathbf{k}}\nu_{\mathbf{k}} = \frac{1}{\hbar}\frac{\partial\mathscr{E}(\mathbf{k})}{\partial\mathbf{k}}. \tag{6.2}$$

The velocity of an electron 'in the state $|\mathbf{k}\rangle$' is just the gradient of $\mathscr{E}(\mathbf{k})$ in $\mathbf{k}$*-space.*

For free electrons this works perfectly:

$$\mathscr{E}(\mathbf{k}) = \frac{\hbar^2k^2}{2m}, \tag{6.3}$$

i.e.

$$\mathbf{v}_{\mathbf{k}} = \frac{\hbar\mathbf{k}}{m} = \frac{\mathbf{p}}{m}, \tag{6.4}$$

where $\mathbf{p}$ is the momentum. It is well known, from elementary wave mechanics, that this is the velocity of an electron considered as a wave packet moving freely in space.

Then again, the quantity $\hbar\mathbf{k}$ behaves very like a *momentum* in diffraction phenomena (§ 2.8). It is natural to suppose that the dynamical law of acceleration is of the form

$$\hbar\dot{\mathbf{k}} = \mathbf{F}, \tag{6.5}$$

where **F** is the force acting on the electron. We have a Newtonian law; *the rate of change of crystal momentum is equal to the force.*

These two principles, (6.2) and (6.5) are indeed true, but they need to be justified. We need to show how it is that the presence of the crystal lattice profoundly modifies the propagation properties of the electron, and we need to get some idea of the approximations involved in postulating these simple laws. The proof is rather complicated, and takes several steps of argument.

6.2 Wannier functions

Remember our definition (§ 3.4) of a Bloch function in terms of atomic orbitals ϕ_n:

$$\psi_{\mathbf{k},n} = \frac{1}{\sqrt{N}} \sum_{l} e^{i\mathbf{k}\cdot\boldsymbol{l}} \phi_n(\mathbf{r}-\boldsymbol{l}). \tag{6.6}$$

This is quite a neat formula because it shows a wave, $\exp(i\mathbf{k}\cdot\boldsymbol{l})$, superposed on sets of localized atomic orbitals, giving the phase and amplitude at each lattice site.

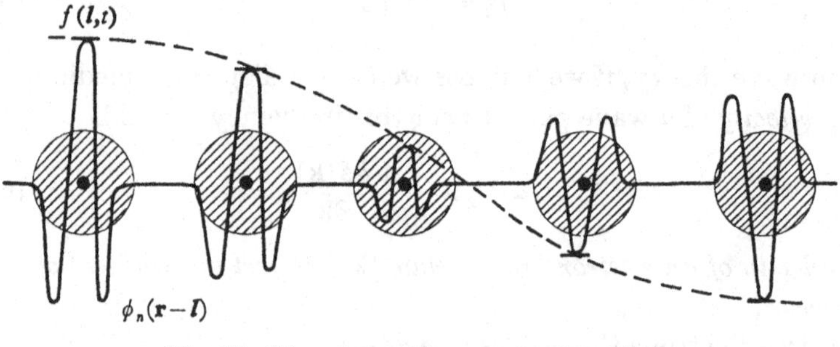

Fig. 97. Wave imposed upon atomic orbitals.

To represent dynamical effects we might try to construct a time-dependent wave-function on the same model

$$\psi(\mathbf{r},t) \sim \sum_{l} f(\boldsymbol{l},t)\,\phi_n(\mathbf{r}-\boldsymbol{l}), \tag{6.7}$$

where $f(\boldsymbol{l},t)$ would be an envelope function showing the strength of the wave in the neighbourhood of $\boldsymbol{l}$ impressed on the local atomic orbital $\phi_n(\mathbf{r}-\boldsymbol{l})$.

But the tight-binding formulation as such will not quite do, because of difficulties of orthogonalization. Indeed, the function defined in (6.6) is not a solution of the Schrödinger equation of the crystal potential. We know that we need to make a linear combination of atomic orbitals, such as

$$\psi_{\mathbf{k}} = \frac{1}{\sqrt{N}} \sum_{l, m} e^{i\mathbf{k}\cdot\mathbf{l}} c_m \phi_m(\mathbf{r}-\mathbf{l}), \tag{6.8}$$

with coefficients to be determined to satisfy the Schrödinger equation. Thus, we could not use the representation (6.7) without having already solved the L.C.A.O. problem—a most tedious preliminary.

However, it is possible to *define* functions that will have the properties that we wanted for the localized atomic orbitals. *Let us suppose that a function* $a_n(\mathbf{r})$ *exists such that the true Bloch functions in the n-th energy band are given by*

$$\psi_{\mathbf{k}, n} = \frac{1}{\sqrt{N}} \sum_{l} e^{i\mathbf{k}\cdot\mathbf{l}} a_n(\mathbf{r}-\mathbf{l}). \tag{6.9}$$

We call this new function a_n a *Wannier function*. It is a function of position, but is independent of $\mathbf{k}$ within a Brillouin zone. There will be a different Wannier function for each electron energy band, just as each atomic orbital in the tight-binding formulation gives rise to a different band.

The formula (6.9) can easily be inverted. If we have solved the Schrödinger equation by some means, such as the O.P.W. method, for the Bloch functions $\psi_{\mathbf{k},n}$, then we can multiply (6.9) by a suitable wave factor and sum over all values of $\mathbf{k}$ in a zone;

$$\begin{aligned}\sum_{\mathbf{k}} e^{-i\mathbf{k}\cdot\mathbf{l}'} \psi_{\mathbf{k},n}(\mathbf{r}) &= \frac{1}{\sqrt{N}} \sum_{\mathbf{k},l} e^{i\mathbf{k}\cdot(\mathbf{l}-\mathbf{l}')} a_n(\mathbf{r}-\mathbf{l}) \\ &= \sqrt{N} \sum_{l} \delta_{ll'} a_n(\mathbf{r}-\mathbf{l}) \\ &= \sqrt{N} . a_n(\mathbf{r}-\mathbf{l}'). \end{aligned} \tag{6.10}$$

In this we have used a result which we have not actually proved but which can easily be deduced by analogy with (2.92):

$$\sum_{\mathbf{k}} e^{i\mathbf{k}\cdot(\mathbf{l}-\mathbf{l}')} = 0, \tag{6.11}$$

for a summation over allowed values in a zone, unless $\mathbf{l}-\mathbf{l}' = 0$.

We may write (6.10) in the form

$$a_n(\mathbf{r}-\mathbf{l}) = \frac{1}{\sqrt{N}} \sum_{\mathbf{k}} e^{-i\mathbf{k}\cdot\mathbf{l}} \psi_{\mathbf{k},n}(\mathbf{r}). \tag{6.12}$$

It is easy to prove, from the orthogonality of $\psi_{\mathbf{k},n}$ and $\psi_{\mathbf{k},n'}$, where n and n' belong to different *bands*, that $a_n(\mathbf{r}-\mathbf{l})$ is orthogonal to

$a_{n'}(\mathbf{r}-\boldsymbol{l})$. But also Wannier functions centred on different *sites* are always orthogonal, even when they belong to the same energy band. Thus

$$\begin{aligned}\int a_n^*(\mathbf{r}-\boldsymbol{l})\,a_n(\mathbf{r}-\boldsymbol{l}')\,d\mathbf{r} &= \frac{1}{N}\sum_{\mathbf{k},\mathbf{k}'} e^{i\mathbf{k}\cdot\boldsymbol{l}}\,e^{-i\mathbf{k}'\cdot\boldsymbol{l}'}\int \psi_{\mathbf{k},n}^*\,\psi_{\mathbf{k}',n}\,d\mathbf{r}\\ &= \frac{1}{N}\sum_{\mathbf{k}} e^{i\mathbf{k}\cdot(\boldsymbol{l}-\boldsymbol{l}')}\\ &= \delta_{\boldsymbol{l}\boldsymbol{l}'} \qquad (6.13)\end{aligned}$$

(we have used the known orthogonality of Bloch functions of different wave-vector).

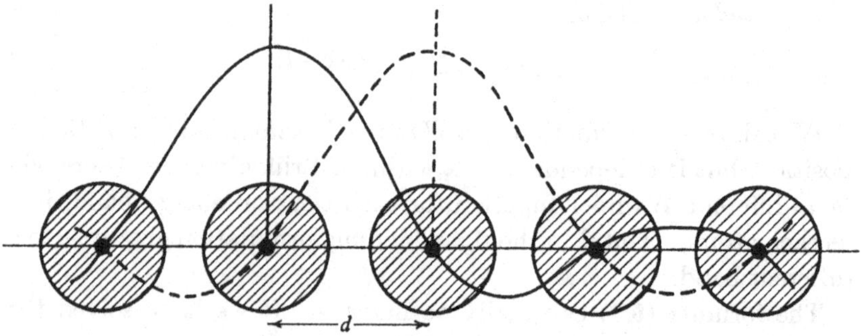

Fig. 98. Wannier functions on adjacent sites.

The Wannier functions are thus a unitary transformation of the Bloch functions, and are formally equivalent as a representation of the electron states. As we have indicated in § 5.8, they might perhaps actually constitute a more appropriate representation in an insulator, where the electrons are well localized, and are confined to their own atoms by correlation effects. In a metal or semiconductor, however, the Bloch states, with the electrons travelling rapidly through the whole crystal, are more appropriate as descriptions of the particles, quasi-particles, single-electron excitations, carriers, or whatever they are.

The actual form of a Wannier function may be seen if we assume the formula (1.60)

$$\psi_{\mathbf{k}}(\mathbf{r}) = N^{-\frac{1}{2}}u(\mathbf{r})\,e^{i\mathbf{k}\cdot\mathbf{r}} \qquad (6.14)$$

and take the periodic function $u(\mathbf{r})$ to be approximately the same for all Bloch states in a band. It is easy to prove, for a simple cubic lattice of side d, that the Wannier function at the origin is

$$a(\mathbf{r}) = u(\mathbf{r})\,\frac{\sin(\pi x/d)\,.\,\sin(\pi y/d)\,.\,\sin(\pi z/d)}{(\pi x/d)\,.\,(\pi y/d)\,.\,(\pi z/d)}. \qquad (6.15)$$

This function looks like $u(\mathbf{r})$ in the centre of the cell, but it spreads out a long way with gradually decreasing oscillations. These oscillations are needed to ensure orthogonality, but they have no direct physical significance. Wannier functions are convenient for certain mathematical purposes, but do not help us in thinking directly about physical phenomena.

6.3 Equations of motion in the Wannier representation

We want a solution of the time-dependent Schrödinger equation for an electron moving in the potential of the lattice, plus the potential of an external field. We want to satisfy

$$(\mathscr{H}^0 + \mathscr{U})\,\psi(\mathbf{r},t) = i\hbar\frac{\partial\psi(\mathbf{r},t)}{\partial t}, \tag{6.16}$$

where $\mathscr{H}^0$ is the Hamiltonian of the electron in a perfect lattice and $\mathscr{U}$ is the perturbing potential.

Let us try to construct a solution like (6.7), but with Wannier functions as localized functions: we assume

$$\psi(\mathbf{r},t) = \sum_{n,\mathbf{l}} f_n(\mathbf{l},t)\,a_n(\mathbf{r}-\mathbf{l}), \tag{6.17}$$

where we have a separate envelope function for each energy band—that is, for each type of local Wannier function.

Substitute (6.17) in the Schrödinger equation, multiply by $a^*_{n'}(\mathbf{r}-\mathbf{l}')$, and integrate over the whole crystal. We get

$$\sum_{n,\mathbf{l}} \int a^*_{n'}(\mathbf{r}-\mathbf{l}')\,(\mathscr{H}^0+\mathscr{U})\,a_n(\mathbf{r}-\mathbf{l})f_n(\mathbf{l},t) = i\hbar\frac{\partial f_{n'}(\mathbf{l}',t)}{\partial t} \tag{6.18}$$

(we have used the orthogonality relation (6.13) on the right-hand side).

But the Wannier functions were themselves deduced from solutions of the equations for the unperturbed crystal, i.e.

$$\mathscr{H}^0\psi_{\mathbf{k},n} = \mathscr{E}_n(\mathbf{k})\,\psi_{\mathbf{k},n} \tag{6.19}$$

in the nth band. Applying this to (6.9) and (6.12), we have

$$\begin{aligned}
\mathscr{H}^0 a_n(\mathbf{r}-\mathbf{l}) &= \frac{1}{\sqrt{N}}\sum_{\mathbf{k}} e^{-i\mathbf{k}\cdot\mathbf{l}}\mathscr{H}^0\psi_{\mathbf{k},n} \\
&= \frac{1}{\sqrt{N}}\sum_{\mathbf{k}} e^{-i\mathbf{k}\cdot\mathbf{l}}\mathscr{E}_n(\mathbf{k})\,\psi_{\mathbf{k},n} \\
&= \frac{1}{N}\sum_{\mathbf{k}} e^{-i\mathbf{k}\cdot\mathbf{l}}\mathscr{E}_n(\mathbf{k})\sum_{\mathbf{l}'} e^{i\mathbf{k}\cdot\mathbf{l}'}a_n(\mathbf{r}-\mathbf{l}') \\
&= \sum_{\mathbf{l}'} \mathscr{E}_{n,\mathbf{l}-\mathbf{l}'}\,a_n(\mathbf{r}-\mathbf{l}'),
\end{aligned} \tag{6.20}$$

where we *define* the Fourier transform of the energy, just as in (3.32), i.e.

$$\mathscr{E}_{n,\boldsymbol{l}} \equiv \frac{1}{N}\sum_{\mathbf{k}} \mathscr{E}_n(\mathbf{k})\, e^{-i\mathbf{k}\cdot\boldsymbol{l}}. \tag{6.21}$$

It is quite interesting to look at (6.20) and (6.21), by analogy with (3.27). The coefficient $\mathscr{E}_{n,\boldsymbol{l}}$ is the overlap integral of the lattice Hamiltonian between Wannier functions in the nth energy band, from sites distant $\boldsymbol{l}$ apart. The orthogonality of the Wannier functions makes these results exact, whereas they were only approximate for atomic orbitals.

Putting (6.20) into our equation of motion (6.18), we get

$$\sum_{n,\boldsymbol{l}}\{\delta_{nn'}\mathscr{E}_{n,\boldsymbol{l}-\boldsymbol{l}'} + \mathscr{U}_{nn'}(\boldsymbol{l},\boldsymbol{l}')\} f_n(\boldsymbol{l},t) = i\hbar\frac{\partial f_{n'}(\boldsymbol{l}',t)}{\partial t}, \tag{6.22}$$

where
$$\mathscr{U}_{nn'}(\boldsymbol{l},\boldsymbol{l}') \equiv \int a^*_{n'}(\mathbf{r}-\boldsymbol{l}')\,\mathscr{U}(\mathbf{r})\,a_n(\mathbf{r}-\boldsymbol{l})\,d\mathbf{r} \tag{6.23}$$

is a matrix element of the perturbing potential between Wannier functions.

These equations are exact. We can take one further step before we approximate. Consider the following formal argument

$$\begin{aligned}
\mathscr{E}_n(-i\nabla)f(\mathbf{r}) &= \sum_{\boldsymbol{l}} \mathscr{E}_{n,\boldsymbol{l}}\, e^{i\boldsymbol{l}\cdot(-i\nabla)} f(\mathbf{r}) \\
&= \sum_{\boldsymbol{l}} \mathscr{E}_{n,\boldsymbol{l}}[1+\boldsymbol{l}\cdot\nabla+\tfrac{1}{2}(\boldsymbol{l}\cdot\nabla)^2\ldots]f(\mathbf{r}) \\
&= \sum_{\boldsymbol{l}} \mathscr{E}_{n,\boldsymbol{l}}\left[f(\mathbf{r})+\boldsymbol{l}\cdot\nabla f(\mathbf{r})+\frac{1}{2}\left\{l_x^2\frac{\partial^2 f(\mathbf{r})}{\partial x^2}+\ldots\right\}\ldots\right] \\
&= \sum_{\boldsymbol{l}} \mathscr{E}_{n,\boldsymbol{l}} f(\mathbf{r}+\boldsymbol{l}).
\end{aligned} \tag{6.24}$$

In making this deduction we are assuming that $\mathscr{E}_n(\mathbf{k})$ in a given band is a continuous analytical function of $\mathbf{k}$, not near any singularity or branch point. We express the corresponding function of the operator $(-i\nabla)$ by means of the inverse of (6.21), i.e.

$$\mathscr{E}_n(\mathbf{k}) = \sum_{\boldsymbol{l}} \mathscr{E}_{n,\boldsymbol{l}}\, e^{i\mathbf{k}\cdot\boldsymbol{l}}, \tag{6.25}$$

and we expand the exponential as a power series in the operator ∇. This series turns out to be the Taylor expansion of $f(\mathbf{r}+\boldsymbol{l})$. The result (6.24) will be valid if $f(\mathbf{r})$ is continuous, etc.

Putting (6.24) into (6.22) we get

$$\left[\mathscr{E}_{n'}(-i\nabla)f_{n'}(\mathbf{r},t) - i\hbar\frac{\partial f_{n'}(\mathbf{r},t)}{\partial t}\right]_{\mathbf{r}=\boldsymbol{l}'} + \sum_{n,\boldsymbol{l}} \mathscr{U}_{nn'}(\boldsymbol{l},\boldsymbol{l}')\, f_n(\boldsymbol{l},t) = 0. \tag{6.26}$$

This equation—still exact within a band—governs the envelope function $f_n(\mathbf{r}, t)$. It is interesting, and useful, because we now treat this envelope function as if it were a continuous function of $\mathbf{r}$, like a continuous wave travelling through the crystal, and not merely as if it were defined only at the lattice sites. The rather clumsy difference equations (6.22) are replaced by a differential equation.

Moreover, this differential equation is not unlike a time-dependent Schrödinger equation. For example, for free electrons we have

$$\mathscr{E}(\mathbf{k}) = \frac{\hbar^2 k^2}{2m}, \tag{6.27}$$

whereupon (6.26) becomes

$$\left[-\frac{\hbar^2}{2m}\nabla^2 - i\hbar\frac{\partial}{\partial t}\right] f_{n'}(\mathbf{r}, t) + \sum_{n,\,\boldsymbol{l}} \mathscr{U}_{nn'}(\boldsymbol{l}, \mathbf{r}) f_n(\boldsymbol{l}, t) = 0, \tag{6.28}$$

which is like the ordinary time-dependent Schrödinger equation in which $f_{n'}(\mathbf{r}, t)$ is the ordinary wave-function of an electron.

6.4 The equivalent Hamiltonian: impurity levels

To make best use of (6.26), it is usual to make several approximations. The first is to suppose that we may represent the moving electron quite satisfactorily by using Wannier functions out of only one band. This means that we ignore all matrix elements $\mathscr{U}_{nn'}(\boldsymbol{l}, \boldsymbol{l}')$ where $n \neq n'$. In other words, *we assume that the perturbation is not strong enough, or sharp enough to induce interband transitions.* This is not always true. For example, a high-frequency electric field might induce an optical transition to a higher band. A very strong electric field can give rise to the *Zener effect* (see § 6.8).

We shall also assume that $\mathscr{U}_{nn}(\boldsymbol{l}, \boldsymbol{l}')$ can be ignored when $\boldsymbol{l} \neq \boldsymbol{l}'$. This, again, assumes that *the perturbation varies slowly with position in the lattice*, so that (6.23) vanishes by virtue of the orthogonality of the Wannier functions on different sites. This approximation is liable to lead to serious errors if $\mathscr{U}(\mathbf{r})$ is a rapidly varying function, such as the field in the neighbourhood of an impurity in a metal or semiconductor.

We can now write

$$\mathscr{U}_{nn}(\boldsymbol{l}, \boldsymbol{l}) = [\mathscr{U}(\mathbf{r})]_{\mathbf{r}=\boldsymbol{l}}, \tag{6.29}$$

the value of the perturbing potential at the $\boldsymbol{l}$th lattice site. Our equation of motion becomes equivalent to

$$\left\{\mathscr{E}_n(-i\nabla) - i\hbar\frac{\partial}{\partial t}\right\} f_n(\mathbf{r}, t) + \mathscr{U}(\mathbf{r}) f_n(\mathbf{r}, t) = 0 \tag{6.30}$$

evaluated for $\mathbf{r}$ at each lattice site. It is natural to interpolate between lattice sites, so to speak, and treat $f_n(\mathbf{r}, t)$ as an ordinary continuous function of position, playing the role of the wave-function of the electron in the local potential $\mathscr{U}(\mathbf{r})$.

The difference between (6.30), and our starting equation (6.16) is that we do not need to include, explicitly, the Hamiltonian $\mathscr{H}^0$ for the electron moving in the field of the pure perfect crystal lattice. This operator is replaced by an *equivalent Hamiltonian*, the operator $\mathscr{E}_n(-i\nabla)$. If we know $\mathscr{E}(\mathbf{k})$ in a single band, we can discuss all the dynamical effects of slow perturbing fields by ignoring the lattice potential as such; we treat the electron as if it were free, but with a modified kinetic energy operator of the form $\mathscr{E}_n(-i\nabla)$.

In the case of ordinary free electrons (i.e. in an empty lattice) we have already verified in (6.28) that (6.30) reduces to the ordinary Schrödinger equation. If, as in a semiconductor, we found that the energy at the bottom of a band were still parabolic, but of the form

$$\mathscr{E}(\mathbf{k}) = \frac{\hbar^2 k^2}{2m^*}, \tag{6.31}$$

where m^* is not the ordinary mass of an electron, then our equivalent Hamiltonian would be the same as that of a free electron, except that m^* would everywhere replace m.

This is the heart of the idea of *impurity levels* in a semiconductor. We have already seen (§§ 4.6, 5.5) that an impurity such as As in Ge gives rise to an effective field

$$\mathscr{U}(\mathbf{r}) = \frac{-e^2}{\epsilon r}, \tag{6.32}$$

where ϵ is the static dielectric constant of the medium. If this field can be treated as slowly varying, and if (6.31) describes adequately the energy of electrons near the bottom of the conduction band, then our equivalent Schrödinger equation is the same as for a hydrogen atom, except that m^* replaces the mass of the electron, and $e/\sqrt{\epsilon}$ replaces its charge.

We can thus take over the well-known theory of the atomic levels of hydrogen. The bottom of the conduction band counts as the zero of kinetic energy; there will be bound states at

$$W_n = -\frac{m^* e^4}{2\epsilon^2 \hbar^2} \frac{1}{n^2} \tag{6.33}$$

below this level. But this is a small energy. If, for example, $m^* = m/10$, and $\epsilon = 10$, then the *ionization energy of the impurity level* will be 10^{-3} rydbergs. Thus, it lies only just below the bottom of the conduction band, within the energy gap.

We also note that the 'Bohr radius' of the equivalent wave-function (i.e. of the envelope function $f(\mathbf{r}, t)$) is enlarged by a factor $(m\epsilon/m^*)$, which would be about 100. This large size is what justifies the use of the approximate Hamiltonian (6.30) to describe a perturbation, (6.32), that really varies rapidly near $r = 0$.

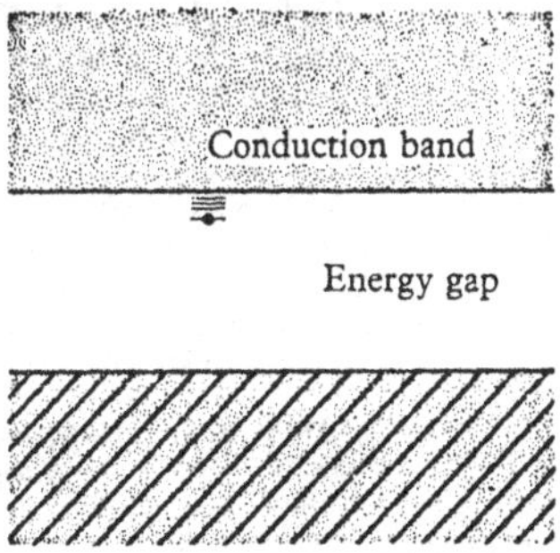

Fig. 99. Impurity levels.

Levels like (6.33) are certainly observed, although the details of the theory, and the observed effects, are more complicated than we have here suggested. For example, the assumption that the impurities do not affect one another is only justifiable at very low relative concentrations. With no more than one atom of As or Sb in 10^5 atoms of Ge or Si, electron states on nearby impurities must begin to overlap. Although the 'atoms' are not arranged in a regular lattice, this assembly is similar to the tight-binding model of § 3.4: the separate 'atom orbitals' should combine and broaden into a narrow band (Fig. 100). At higher concentrations the overlap integrals (cf. (3.27)) exceed the ionization energy of the impurity levels, so that this band eventually merges with the bottom of conduction band, from which, in fact, the impurity states had originally been drawn. The theory of this *impurity band* is interesting, both as an extreme case of electron states in a disordered system and as a system where a Mott transition (§ 5.9) from insulating to metallic conduction may be engineered by adjustment of the concentration of various impurities.

But the equivalent Hamiltonian formalism breaks down in the analogous case of a *negative ion vacancy* or *F-centre* in an ionic crystal. It is easy to argue that the absence of a negative ion in an electrostatically neutral lattice is equivalent to a localized positive charge, attracting an electron by the coloumb field (6.32). But the effective mass, m^*, of a conduction electron is not small enough, nor is ϵ big enough, to justify the use of hydrogenic wave functions. The electron spends most of its time in the immediate vicinity of the vacancy, where it is best treated as an ordinary bare particle moving in the superposed

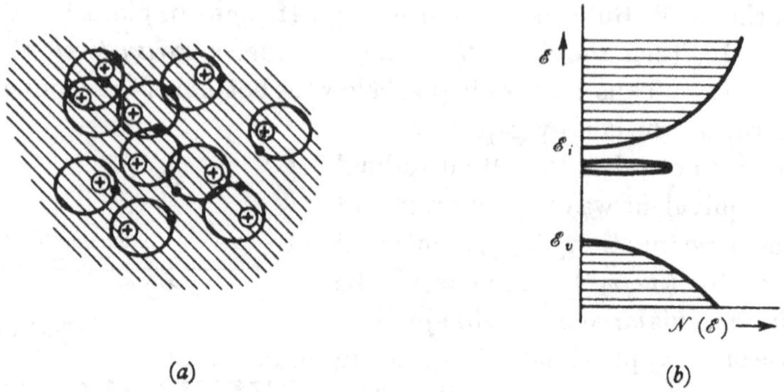

Fig. 100. (a) Overlapping impurity states give rise to (b) impurity band in the energy gap.

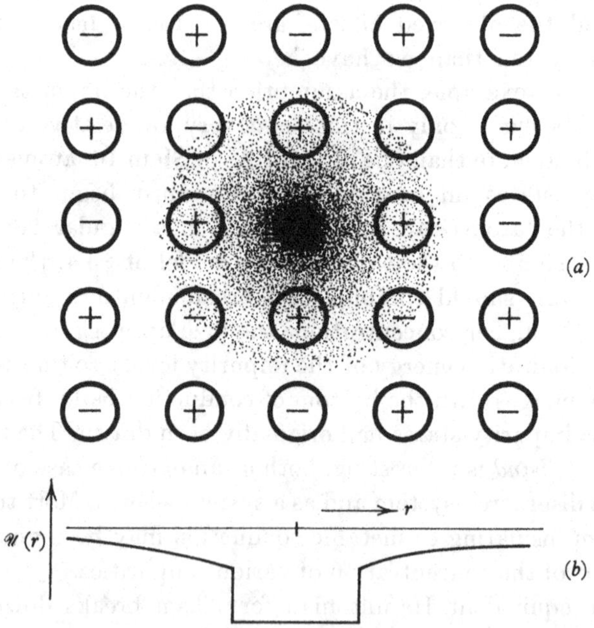

Fig. 101. (a) Electron charge cloud in F-centre. (b) Point-ion model potential well.

electrostatic fields of the various neighbouring ions—a potential that looks like a flat-bottomed square well without any central singularity (Fig. 101). This *point-ion model*, despite its simplicity, provides a good quantitative basis for the theory of electron states in colour centres (see § 8.5), of which the F-centre is the simplest example.

6.5 Quasi-classical dynamics

In any problem involving electrons moving in a lattice we could set about solving the equivalent Schrödinger equation (6.30). But if $\mathscr{E}(\mathbf{k})$ is a complicated function of $\mathbf{k}$ this could lead to rather difficult mathematics.

We can avoid most of these difficulties by an application of the correspondence principle. It is well known that wave-packet solutions of the Schrödinger equation follow the same trajectories as classical particles obeying the equations of motion derived from the corresponding classical Hamiltonian. According to the Schrödinger formulation of quantum mechanics, the Hamiltonian *operator* in the wave equation is obtained by putting $-i\hbar\nabla$ in place of the classical momentum variable, $\mathbf{p}$, in the classical Hamiltonian *function*. Reversing these steps, we may replace the operator $-i\nabla$, in our equivalent Hamiltonian operator $\mathscr{E}(-i\nabla)$, by the momentum variable $\mathbf{p}/\hbar$, and call the result the equivalent classical Hamiltonian function. Thus,

$$\mathscr{E}(-i\nabla)+\mathscr{U}(\mathbf{r}) \rightarrow \mathscr{E}(\mathbf{p}/\hbar)+\mathscr{U}(\mathbf{r}) = \mathscr{H}(\mathbf{r},\mathbf{p}). \tag{6.34}$$

Our electron wave packets will behave like classical point particles subject to the equations of motion generated by this Hamiltonian function.

Hamilton's equations read, symbolically,

$$\dot{r} = \frac{\partial \mathscr{H}}{\partial p}, \quad \dot{p} = -\frac{\partial \mathscr{H}}{\partial r}. \tag{6.35}$$

The first of these becomes an equation for the velocity,

$$\begin{aligned}\mathbf{v} = \frac{\partial \mathscr{H}}{\partial \mathbf{p}} &= \frac{\partial}{\partial \mathbf{p}}\{\mathscr{E}(\mathbf{p}/\hbar)+\mathscr{U}(\mathbf{r})\} \\ &= \frac{\partial \mathscr{E}(\mathbf{p}/\hbar)}{\partial \mathbf{p}} \\ &= \frac{1}{\hbar}\frac{\partial \mathscr{E}(\mathbf{k})}{\partial \mathbf{k}}\end{aligned} \tag{6.36}$$

in terms of our more usual variable $\mathbf{k}$. This is (6.2), derived at last. We see that

$$\hbar\mathbf{k} = \mathbf{p} \tag{6.37}$$

does play the role of the classical momentum; even though it is not a uniquely defined variable, it labels the 'state' in which the wave packet happens to be mainly concentrated.

The second Hamiltonian equation can be written

$$\hbar\dot{\mathbf{k}} = -\frac{\partial \mathscr{H}}{\partial \mathbf{r}} = -\nabla \mathscr{U}(\mathbf{r}). \tag{6.38}$$

But if $\mathscr{U}(\mathbf{r})$ is, say, the potential energy of an electron in a fixed electrostatic field, the right-hand side is simply the classical force acting on the electron. For example, in the electric field $\mathbf{E}$, we have

$$\hbar\dot{\mathbf{k}} = e\mathbf{E}. \tag{6.39}$$

This proves our second dynamical law (6.5). The variable $\hbar\mathbf{k}$ plays the role of momentum in that the effect of a constant force is to increase $\hbar\mathbf{k}$ at a constant rate. Here again, the fact that $\mathbf{k}$ is not unique does not matter, since it is only its derivative that comes into these equations.

For the effect of a magnetic field, $\mathbf{H}$, we assume that the *Lorentz force* equation is valid, i.e.

$$\hbar\dot{\mathbf{k}} = e\left(\mathbf{E} + \frac{1}{c}\mathbf{v}\wedge\mathbf{H}\right), \tag{6.40}$$

where $\mathbf{v}$ is the velocity of the wave packet, as calculated by (6.36) and c is the velocity of light. This formula follows by analogy from (6.39), but its proper derivation is far more difficult. It is believed to hold quite well up to very high magnetic fields (see § 9.8).

6.6 The mass tensor: electrons and holes

Let us combine the kinematic and dynamical equations (6.36) and (6.39), to calculate the *acceleration* of an electron,

$$\begin{aligned}\dot{\mathbf{v}} &= \frac{\partial}{\partial t}\left\{\frac{1}{\hbar}\frac{\partial \mathscr{E}(\mathbf{k})}{\partial \mathbf{k}}\right\} \\ &= \frac{1}{\hbar}\frac{\partial}{\partial \mathbf{k}}\left\{\frac{1}{\hbar}\frac{\partial \mathscr{E}(\mathbf{k})}{\partial \mathbf{k}}\right\}\cdot\frac{\hbar\,\partial \mathbf{k}}{\partial t} \\ &= \frac{1}{\hbar^2}\frac{\partial^2 \mathscr{E}(\mathbf{k})}{\partial \mathbf{k}\,\partial \mathbf{k}}\cdot e\mathbf{E}. \end{aligned} \tag{6.41}$$

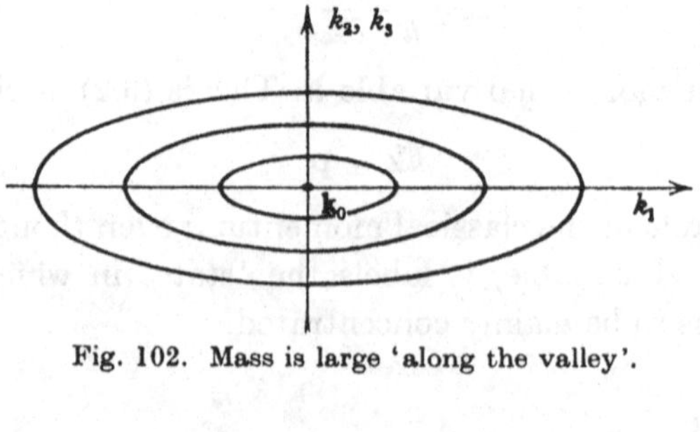

Fig. 102. Mass is large 'along the valley'.

If we compare this with the conventional Newtonian equation,

$$m\dot{\mathbf{v}} = e\mathbf{E}, \tag{6.42}$$

we see that the equivalent of the mass of an electron is now the inverse of a tensor;

$$\frac{1}{m} \rightarrow \frac{1}{\hbar^2}\frac{\partial^2 \mathscr{E}(\mathbf{k})}{\partial \mathbf{k}\,\partial \mathbf{k}}. \tag{6.43}$$

This tensor is the nearest equivalent to a dynamical mass for the electron. For example, in a semiconductor the energy surface may be of the form (4.35), i.e.

$$\mathscr{E}(\mathbf{k}) = \frac{\hbar^2 k_1^2}{2m_1} + \frac{\hbar^2 k_2^2}{2m_2} + \frac{\hbar^2 k_3^2}{2m_3} \tag{6.44}$$

referred to some local principal axes. The effect of an electric field in accelerating an electron will then be very different in different directions. If the energy surface is elongated into an ellipsoid in the k_1 direction, i.e. if $m_1 \gg m_2 \sim m_3$, the mass will appear to be much greater when measured along the 'valley' than in the transverse direction. This is important when one comes to calculate the contribution of various states to the average mobility of the electrons. The simple hydrogenic formula (6.33) for donor impurity levels must also be modified to take account of the anisotropy of the equivalent Hamiltonian for each valley and the multiplicity of valleys of the same energy at the bottom of the conduction band in many semiconductors.

The *inverse mass tensor* is not necessarily positive definite. Where (as in the noble metals, Cu, Ag, Au) the Fermi surface makes contact with the zone boundary, we can represent it approximately in the form (6.44), but with m_3 *negative* to give hyperbolic energy surfaces (which can be made to run right through the zone boundary in the repeated zone scheme of § 3.3). The dynamics of electrons at such a point in **k**-space would look very odd from a classical point of view.

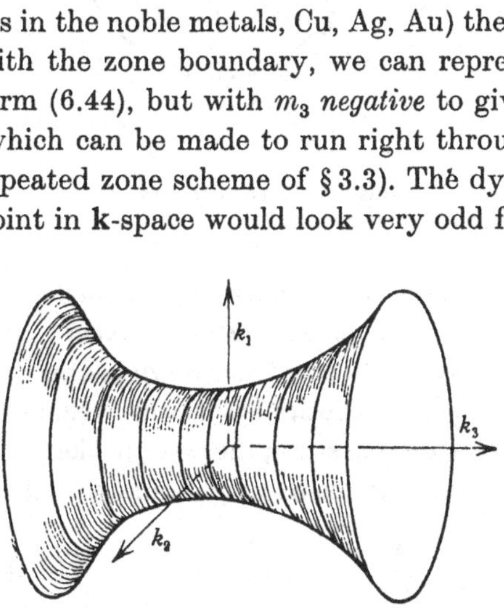

Fig. 103. Hyperboloidal energy surface ('neck').

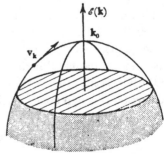

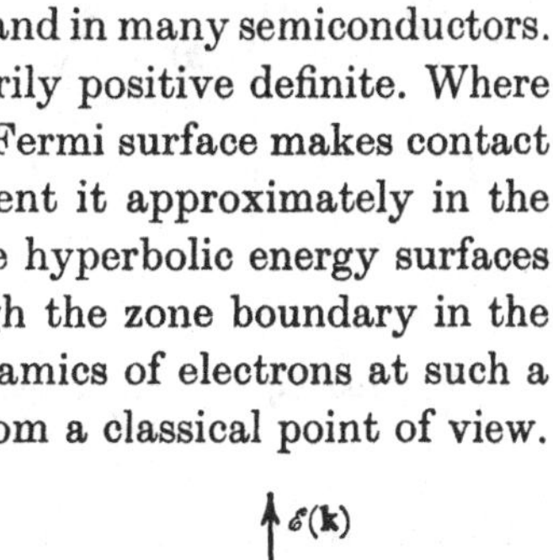

Fig. 104. A 'pocket of holes' at the top of a nearly-filled band.

They are profoundly altered by the strong Bragg reflection from the zone boundary. In effect, they tend to be propagated parallel to the planes of the lattice, as if by a series of wave-guides.

A more usual case is that of a *nearly full band* as in a semiconductor or semimetal. Since $\mathscr{E}(\mathbf{k})$ is a continuous periodic function in the repeated zone, the states will be occupied nearly up to a well-defined maximum of $\mathscr{E}(\mathbf{k})$. If this maximum occurs at $\mathbf{k}_0$, say, then $\mathbf{v}_{\mathbf{k}}$, being the gradient of $\mathscr{E}(\mathbf{k})$, will be zero at $\mathbf{k} = \mathbf{k}_0$, and the principal components of $\partial^2\mathscr{E}/\partial\mathbf{k}\,\partial\mathbf{k}$ will be *negative* in all directions near $\mathbf{k}_0$.

The properties of electrons with *negative effective mass* are so unlike those of ordinary particles that it is much easier to invert the argument and discuss the properties of a few *holes* in the full band. We have already noted, in § 4.6, that the number of such holes can be calculated rather simply by using Fermi–Dirac statistics with the energy measured *downwards*; we shall now show that if we treat holes as *positively* charged particles, the same energy convention will describe their dynamical properties in a classically comprehensible form.

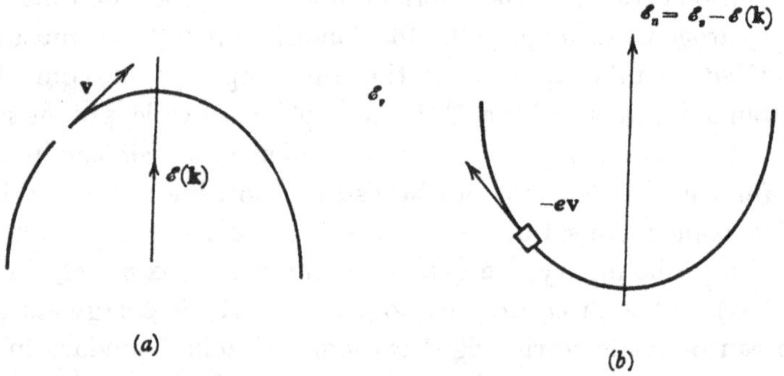

Fig. 105. (*a*) A gap in the electron distribution. (*b*) The corresponding 'hole', with inverted energy scale.

We first remark that a 'hole in the state **k**' must be described by a determinant containing all wave-functions in the band except the state $\psi_{\mathbf{k}}$. One can go further, and actually construct wave packets of such determinants. But the main point to notice is that the 'velocity' of the hole is the same as that of the missing state; the wave packet will be carried through space at the same rate as its neighbours on either side of the hole;

$$\mathbf{v}_{\mathbf{k}} = \frac{1}{\hbar}\frac{\partial\mathscr{E}(\mathbf{k})}{\partial\mathbf{k}}. \qquad (6.45)$$

Now consider the electric current associated with a hole. The current of a full band is zero. The current of an electron in state $\mathbf{k}$ is $-|e|\,\mathbf{v}_\mathbf{k}$. Therefore the current of a full band with one electron missing must be

$$\mathbf{J}_h = 0-(-|e|\,\mathbf{v}_\mathbf{k}) = |e|\,\mathbf{v}_\mathbf{k}. \tag{6.46}$$

A hole thus behaves like a *positively charged particle*.

Again, the dynamics of a hole must be the same as that of the states on either side of it. The velocity will be changed at the rate defined by (6.41). If we write this in terms of current, we have

$$\begin{aligned} |e|\,\dot{\mathbf{v}} &= |e|\frac{1}{\hbar^2}\frac{\partial^2\mathscr{E}(\mathbf{k})}{\partial\mathbf{k}\,\partial\mathbf{k}}\cdot(-|e|)\,\mathbf{E} \\ &= |e|\left\{-\frac{1}{\hbar^2}\frac{\partial^2\mathscr{E}(\mathbf{k})}{\partial\mathbf{k}\,\partial\mathbf{k}}\right\}\cdot|e|\,\mathbf{E}, \end{aligned} \tag{6.47}$$

which is the acceleration equation for a particle of positive charge $|e|$, but of mass m^*, where

$$\frac{1}{m^*} \sim -\frac{1}{\hbar^2}\frac{\partial^2\mathscr{E}(\mathbf{k})}{\partial\mathbf{k}\,\partial\mathbf{k}}. \tag{6.48}$$

But because we are near the top of the band, all the components of curvature are negative, so that m^* is now a positive quantity. Thus, the dynamical properties of holes may not be isotropic, but they will be essentially those of classical particles of positive mass.

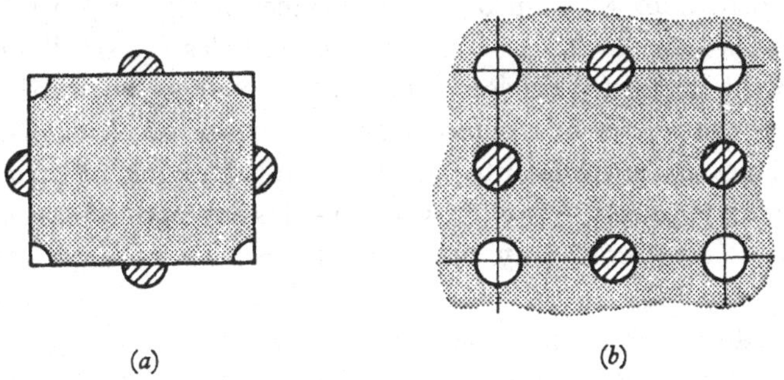

Fig. 106. Band with small overlap: (*a*) extended zone scheme; (*b*) repeated zone scheme.

Conduction by holes at the top of the valence band is, of course, a characteristic property of semiconductors. In Si and Ge, where the energy surfaces at the centre of the zone are split by spin-orbit interaction (cf. Fig. 66), we may actually observe both 'light' and 'heavy' holes, corresponding to the two branches of $\mathscr{E}(\mathbf{k})$, of different curvatures, that touch at the maximum.

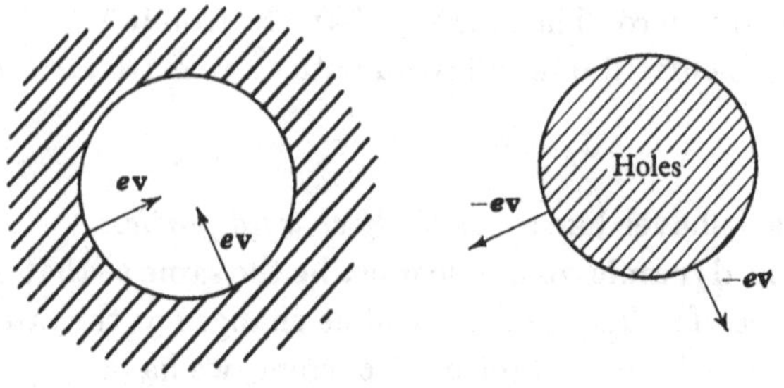

Fig. 107. (a) An empty sphere in the electron distribution becomes (b) a 'sphere of holes'.

It is evident that the velocity of a hole increases as we go away from the position of the maximum, i.e. as $\mathscr{E}(\mathbf{k})$ decreases. It is again convenient to think of the kinetic energy of a hole as $\mathscr{E}_h = \mathscr{E}_v - \mathscr{E}(\mathbf{k})$, increasing outward from an origin at the top of the valence band. This is the convention introduced in § 4.6.

In the case of a semimetal such as Bi, with an even number of electrons per unit cell, we may encounter a similar situation, if there is a small overlap into a higher band, and a small number of electrons is spilled over. An equal number of holes will then be left in the nearly-full band—and these will be contained by bits of Fermi surface that can be combined as in § 3.3, into closed surfaces in the repeated zone scheme. It is easier to think of such a *pocket* of holes as a small number of positively charged particles, with velocity vectors outwards from the centre than it is to discuss a Fermi surface of electrons with velocities drawn inwards into the unoccupied region of space. It should be emphasized, however, that the classification of states into 'holes' and 'electrons' may break down—as, for example, on the 'necks' of Fig. 103.

It is doubtful whether a hole can move like a free particle in an ionic crystal. The small overlap between valence orbitals on neighbouring ions implies a very narrow valence band (§ 3.4) with a correspondingly high effective mass. This seems to permit *self-trapping* of the hole—for example, in KCl, by localization between two Cl^- ions which are pulled closer together into a stable *V centre*, from which the hole may be released only by thermal excitation. Such a *small polaron* (§ 8.3) is so sensitive to the graininess of the lattice that it is mobile only by 'hopping' from cell to cell. For this reason, the concept of a 'valence band' has little meaning in an ionic crystal (§ 4.2).

6.7 Excitons

The existence of two types of carrier, of opposite charge, but otherwise with essentially similar properties, is at the heart of the applied physics of semiconductors. We have already shown, in § 4.6, how one

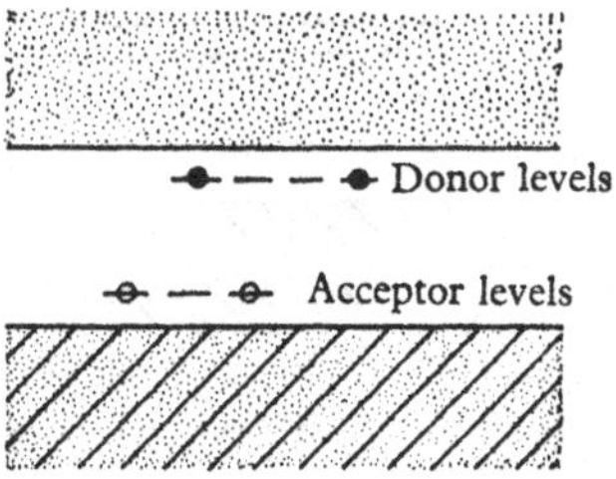

Fig. 108. Impurity levels in a semiconductor.

can, by doping with acceptor impurities, increase the number of holes to any desired amount, making a *p*-type sample.

In the case of donor impurities we saw in § 6.4 that impurity levels are formed, just below the bottom of the conduction band, by the attraction of the charged impurity for extra electrons. There is an exactly equivalent theory for acceptor impurities, which are negatively charged (because they have to take up a fourth valence electron to maintain the bond structure) and which therefore attract holes. The effect is to create bound states for holes at a small distance below the zero of their kinetic energy—that is, at a small distance above the top of the valence band. The 'ionization energy' of such an *acceptor level* can be calculated approximately just as in (6.33)—although here again the assumption that $\mathscr{E}(\mathbf{k})$ is a simple quadratic function, like (6.31) inverted, may be too naïve.

In a *compensated sample*, where there are substantial numbers of both donor and acceptor impurities, some of the electrons can 'fall' from the donor levels and annihilate the holes on the acceptors. As pointed out in (4.42), the net number of carriers, and their sign, will be the difference $N_d - N_a$. It is worth remarking, however, that the 'empty' impurities are not electrically neutral objects; the donors will be positively charged and the acceptors negatively charged, so that they will attract or repel the remaining carriers.

Suppose that we have an intrinsic semiconductor, or one that is so carefully compensated that the numbers of electrons and holes are given by the thermal excitation formula (4.39), and are nearly equal. These are oppositely charged particles—so they attract one another. One can set up hydrogen-like wave-functions in which electron and

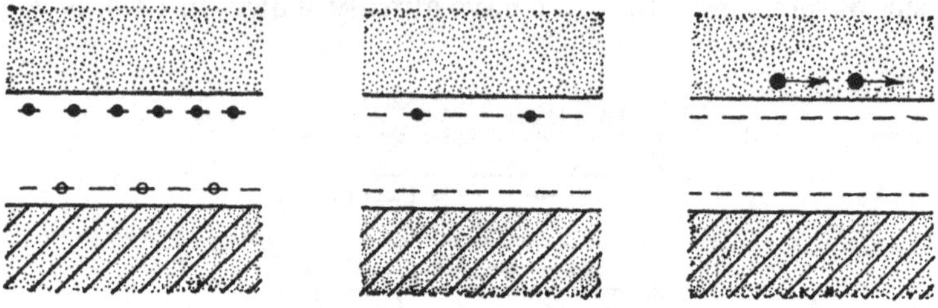

(a) (b) (c)

Fig. 109. (a) Donor and acceptor levels, occupied by electrons and holes. (b) Electrons and holes combine. (c) Excess electrons thermally excited into conduction band.

hole circle about their joint centre of mass; the reduction in energy of the pair will be of the same form as (6.33), except that m^* should now be a reduced effective mass.

This means that we have an energy level for the electron a bit below the bottom of the conduction band associated with a hole level just above the valence band—indeed, a whole series of such levels, corresponding to different quantum numbers of the hydrogen-like state—indeed, a whole band of *exciton states*, corresponding to the addition of kinetic energy of motion of the joint centre of mass of the electron-hole pair (Figs. 110, 111 (a)).

An alternative way of thinking of an exciton is in terms of excited atoms. One of the electrons on an atom or ion is raised to an excited state, $\phi_e(\mathbf{r})$, say. We then make a Bloch function, of the form

$$\psi_{\mathbf{k}} = \sum_{\mathbf{l}} e^{i\mathbf{k}\cdot\mathbf{l}}\phi_e(\mathbf{r}-\mathbf{l}) \tag{6.49}$$

to allow the excitation to travel through the crystal (Fig. 111 (b)). At first sight, this seems to be no more than a state in the conduction band. But the tight-binding theory (§ 3.4) from which we should derive this argument, is a single-electron theory; it takes no account of electron–electron interaction. The exciton state is more complicated than (6.49); the travelling electron carries with it a local polarization of the other electrons in the valence band—in other words, it carries its own hole around with it. A further small amount of energy is therefore required to separate this electron-hole pair into independent carriers.

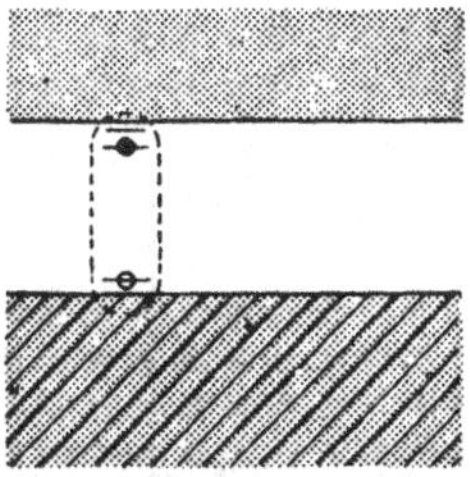

Fig. 110. Exciton states.

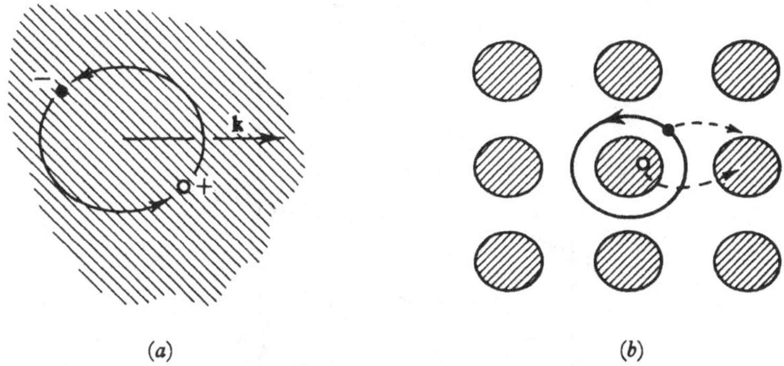

Fig. 111. (*a*) Wannier exciton. (*b*) Frenkel exciton.

Although these two descriptions of an exciton are essentially equivalent, they have different ranges of practical use. In a semiconductor, with light carriers and a high dielectric constant, the expanded hydrogenic orbits of the *Wanner exciton* are the better approximation; in a molecular crystal, such as solid argon, or anthracene, it is more accurate to start with a *Frenkel exciton* model, whose wave-function is mostly within a single unit cell. An exciton in an ionic crystal such as NaCl is of intermediate size, being described quite nicely by the transfer of an electron from a Cl^- ion to a neighbouring Na^+ site.

Exciton lines may be detected in the optical spectra of semiconductors and insulators (§ 8.5), but since the excitation is electrically neutral it does not contribute directly to electrical conduction. Although excitons are mobile, and should in principle be capable of transporting energy through the crystal, the experimental demonstration of this phenomenon is not easy.

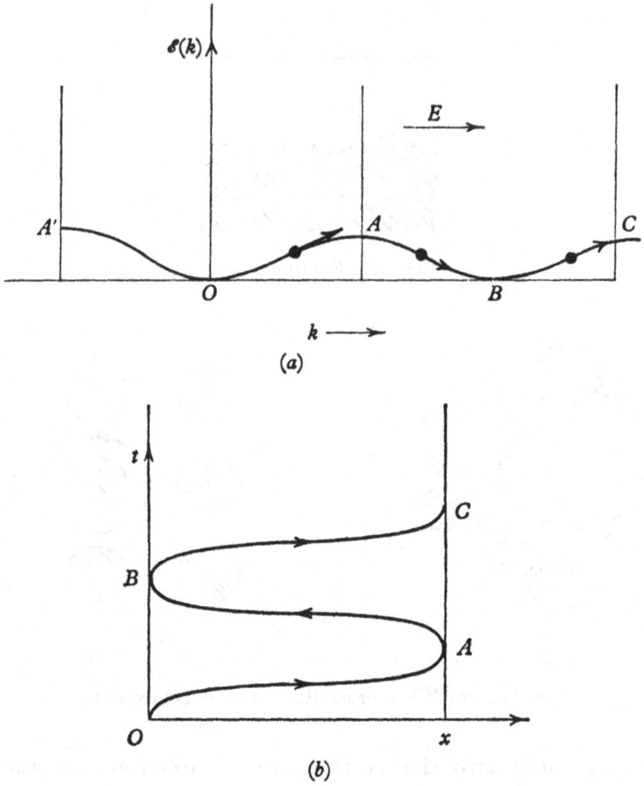

Fig. 112. Electron moving in electric field, in one dimension.
(a) Trajectory in k-space. (b) Path in real space.

6.8 Zener breakdown: tunnelling

The theory of the equivalent Hamiltonian as set forth in § 6.4, depends upon the assumption that interband transitions can be ignored. This is only true if the electric field is not too strong; we must be able to drop matrix elements like

$$\mathscr{U}_{nn'}(\mathbf{l}, \mathbf{l}) = \int a^*_{n'}(\mathbf{r}-\mathbf{l})\, \mathscr{U}(\mathbf{r})\, a_n(\mathbf{r}-\mathbf{l})\, d\mathbf{r}, \tag{6.50}$$

on the grounds of the orthogonality of the Wannier functions in different bands.

But if $\mathscr{U}(\mathbf{r})$ varies rapidly with $\mathbf{r}$, the vanishing of the matrix element (6.50) cannot be assured. From a purely mathematical point of view, it is evident that the probability of an interband transition will depend on the magnitude of $\nabla\mathscr{U}$.

We could, presumably, calculate this probability from the formulation of § 6.4. But it is easier to look upon this phenomenon in a more 'physical' way. Let us first consider what happens to an electron subject to a weak electric field. If it is free, it is continuously accelerated, its velocity increases without limit, and it travels on until it is scattered, or until it passes out of the specimen.

But suppose the electron is in a single simple band—we can think of a one-dimensional model with the electric field along the x-axis. The

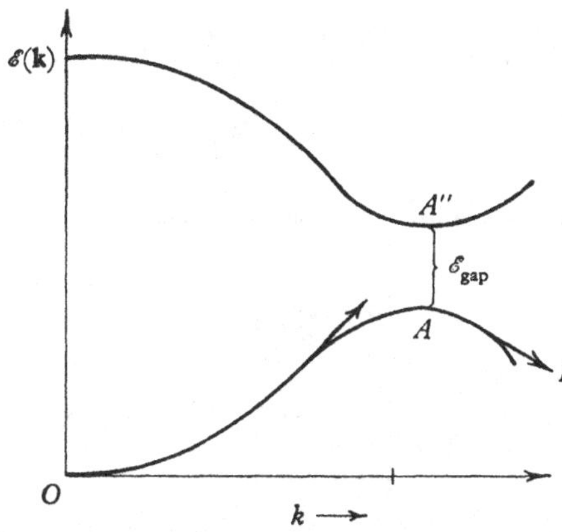

Fig. 113. Instead of going on to B, the electron may 'jump' the gap to A''.

representative point moves steadily on, from O to A. This is equivalent to A'—so we could have it suddenly jumping back there—or, if we use the repeated zone scheme of § 3.3, we can see it moving up and down the periodic curve $OABC$, etc.

In real space, the electron is accelerated from O, but as it approaches A it slows down. Then it reverses direction, until it reaches B, is accelerated again, and so on. At A we have a Bragg reflection; the electron cannot go into the region beyond A, because its energy would be in the energy gap. It is confined to a limited range of the x-axis.

But now suppose there is another band, whose lowest point, A'', is at an energy $\mathscr{E}_{gap}$ above A. There is always the possibility that the electron may gain so much energy from the electric field that it makes a transition to A from A''.

To see that this is possible, suppose the electric field is constant. We can plot, along the x-axis, the energy eEx that the electron has gained from the field. Suppose it has reached the point A, at the top of the lower band. Then normally it will be reflected back. But if it can travel a further distance

$$d = \frac{\mathscr{E}_{\text{gap}}}{eE} \tag{6.51}$$

to the point A'', it will have gained enough energy from the field to surmount the gap (Fig. 114).

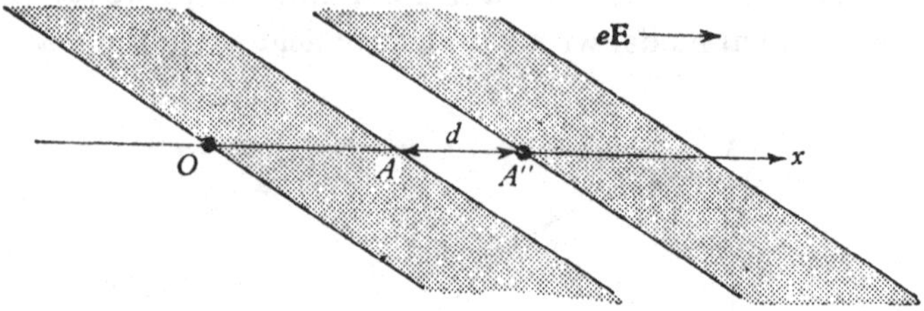

Fig. 114. Bands tilted by strong electric field.

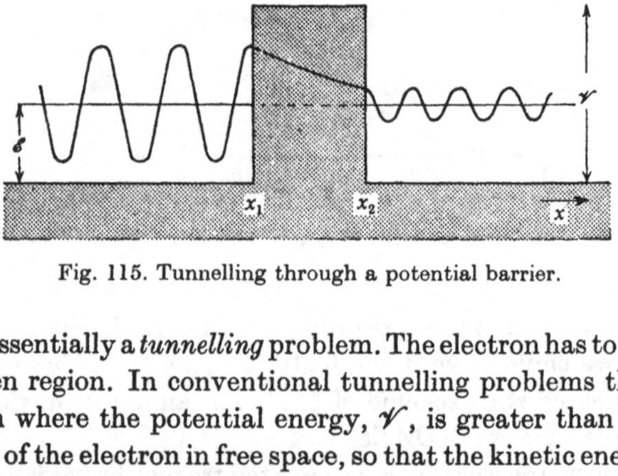

Fig. 115. Tunnelling through a potential barrier.

This is essentially a *tunnelling* problem. The electron has to penetrate a forbidden region. In conventional tunnelling problems the barrier is a region where the potential energy, $\mathscr{V}$, is greater than the total energy, $\mathscr{E}$, of the electron in free space, so that the kinetic energy of the electron would be negative inside the barrier. Quantum mechanically, this means that the electron wave-function acquires a real exponential factor

$$\psi = \psi_0 e^{-\beta x}, \tag{6.52}$$

where

$$\beta^2 = \mathscr{V} - \mathscr{E}. \tag{6.53}$$

If $\mathscr{V}$ is a function of x, we may use the W.K.B. method to derive a transition probability, i.e. a ratio of transmission to reflection, of the form

$$P = \exp\left\{-2\int_{x_1}^{x_2} \beta(x)\, dx\right\}, \tag{6.54}$$

where x_1 and x_2 are the limits of the barrier region, and where the factor 2 comes from squaring the wave-function for the probability density.

To use this argument for our present problem, we need to consider the nature of the electron wave-functions in an energy gap. This is a point that we have not previously considered. The partial differential equation

$$-\frac{\hbar^2}{2m}\nabla^2\psi + \{\mathscr{V}(\mathbf{r}) - \mathscr{E}\}\psi = 0 \tag{6.55}$$

certainly has solutions for any value of $\mathscr{E}$, even when $\mathscr{V}(\mathbf{r})$ is a periodic potential such that

$$\mathscr{V}(\mathbf{r}+\mathbf{l}) = \mathscr{V}(\mathbf{r}), \tag{6.56}$$

only these solutions do not satisfy the Bloch conditions and are not, therefore, acceptable as eigenfunctions to be fitted to periodic boundary conditions.

There is no reason, however, why we should not use the Fourier analysis of the potential, as in § 3.2, and try a solution of the form

$$\psi = \sum_{\mathbf{g}} \alpha_{\mathbf{k}-\mathbf{g}}\, e^{i(\mathbf{k}-\mathbf{g})\cdot\mathbf{r}}, \tag{6.57}$$

but with the components of **k** complex numbers. We shall get the same set of linear equations as (3.14) for the coefficients—except that we must write

$$\mathscr{E}^0_{\mathbf{k}-\mathbf{g}} \equiv \frac{\hbar^2}{2m}(\mathbf{k}-\mathbf{g})^2 \tag{6.58}$$

explicitly, since this is no longer, strictly, an 'unperturbed energy', but only the result of operating with $(-\hbar^2/2m)\nabla^2$ on the terms of (6.57).

Let us look at a one-dimensional model in the region of an energy gap, i.e. where **k** is in the neighbourhood of the zone boundary corresponding to the wave-vector

$$\tfrac{1}{2}G = \frac{\pi}{a} \tag{6.59}$$

for planes spaced a apart transverse to the electric field. For free electrons the energy would be

$$\mathscr{E}^0 = \frac{\hbar^2}{2m}(\tfrac{1}{2}G)^2 \tag{6.60}$$

but, as in (3.16), with the perturbation $\mathscr{V}_G$ we must solve the determinantal equation

$$\left\{\frac{\hbar^2}{2m}k^2 - \mathscr{E}\right\}\left\{\frac{\hbar^2}{2m}(k-G)^2 - \mathscr{E}\right\} = |\mathscr{V}_G|^2 \tag{6.61}$$

As we saw in (3.19) and (3.20), if k is real this has no roots for $\mathscr{E}$ in the energy gap, i.e. for

$$\mathscr{E}^0 - |\mathscr{V}_G| < \mathscr{E} < \mathscr{E}^0 + |\mathscr{V}_G|. \tag{6.62}$$

But now we treat $\mathscr{E}$ as an arbitrary parameter and solve for k. If we write

$$k = \tfrac{1}{2}G + \kappa, \quad \mathscr{E} = \mathscr{E}^0 + \epsilon, \tag{6.63}$$

where κ and ϵ are supposed to be small, then we find an approximate solution

$$\kappa^2 \approx \frac{2m}{\hbar^2}\left[\frac{\epsilon^2 - |\mathscr{V}_G|^2}{4\mathscr{E}^0}\right]. \tag{6.64}$$

It is obvious that κ will be pure imaginary if $|\epsilon| < |\mathscr{V}_G|$—that is, by (6.62) and (6.63), *an electron with energy in the energy gap has a wave-function containing a real exponential factor.* This is a perfectly proper solution of the Schrödinger equation—but is usually physically irrelevant because it cannot be made to fit periodic or other simple boundary conditions over the width of a large crystal. The electron simply cannot propagate any distance through the solid. This general property of wave-functions in a periodic system has already been noted for the case of a lattice vibration with frequency outside the phonon spectrum (§ 2.12).

Returning to our tunnelling problem, we see that we may calculate the transmission through the forbidden region by using (6.54) with the coefficient β equal to the imaginary part of κ in (6.64). If the electric field causes the electron energy to increase linearly across the forbidden region, so that ϵ is proportional to distance, x, from the centre of the potential barrier (whose total width is d), we have

$$\beta(x) = \frac{|\mathscr{V}_G|}{2\mathscr{E}^0}(\tfrac{1}{2}G)\left(1 - \frac{4x^2}{d^2}\right)^{\frac{1}{2}},$$

which integrates to

$$P = \exp - \left\{\frac{\pi d\,|\mathscr{V}_G|\,(\tfrac{1}{2}G)}{2\mathscr{E}^0}\right\}$$

$$= \exp - \left\{\frac{\pi^2}{4}\frac{\mathscr{E}^2_{\text{gap}}}{\mathscr{E}^0 eEa}\right\}, \tag{6.65}$$

using (6.51), (6.59), and the fact that $2\,|\mathscr{V}_G|$ is the width of the energy gap.

This formula shows that *Zener breakdown* ought to occur if the electric field $\mathbf{E}$ is so strong that the electron can gain the fraction $\mathscr{E}_{\text{gap}}/\mathscr{E}^0$ of the gap energy in going one lattice spacing a. It is uncertain,

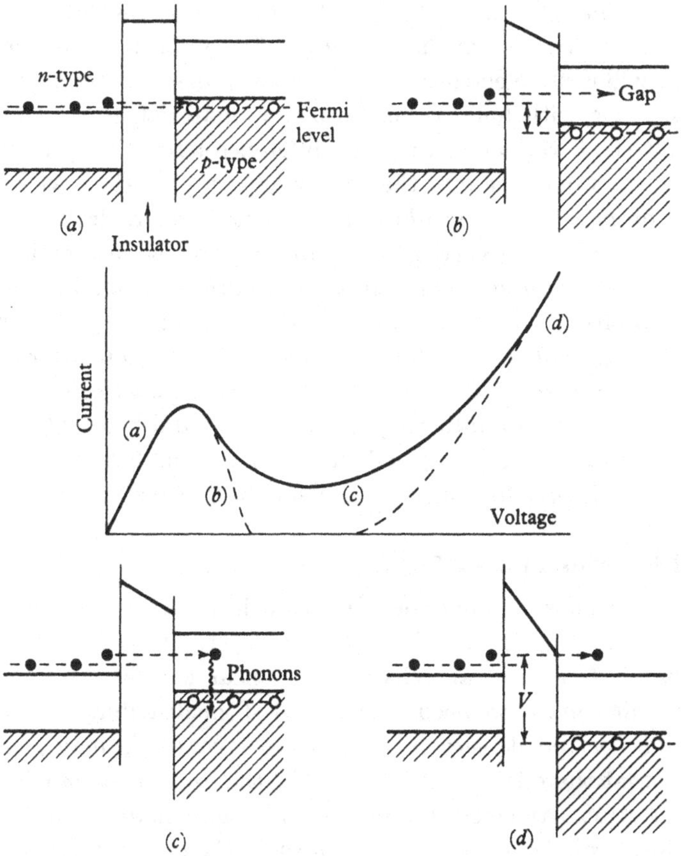

Fig. 116. Tunnel diode characteristic: (*a*) near-zero bias; (*b*) tunnelling into gap; (*c*) phonon-assisted tunnelling; (*d*) minority carrier injection.

however, whether this effect is observable in practice, being masked by other more complex phenomena such as *avalanche breakdown*, i.e. the multiplication of carriers by ionization of impurity levels in a current of energetic electrons or holes.

A very similar effect occurs in a *tunnel* or *Esaki diode*, where an external electric field is applied across a thin film of insulator separating heavily doped p-type and n-type materials. The tunnelling rate is governed primarily by the thickness and nature of the barrier, but is mediated by the density of states available in the band to which the carrier is proceeding—and this is sensitive to the exact energy to which the electron is carried by the applied electric field as it crosses the forbidden region.

At zero bias, for example (Fig. 116(*a*)), the Fermi level must be the same on each side of the barrier (§ 4.6), so that a small 'forward' voltage will cause electrons from the n-type region to tunnel to meet and combine with holes from the p-type material, thus allowing a current to flow. But a larger voltage difference across the barrier drives carriers from one side into the energy gap on the other, where they cannot propagate (Fig. 116(*b*)). The current therefore drops—but not to zero because the carriers may lose energy by interaction with phonons (*phonon-assisted tunnelling*, Fig. 116(*c*)). An analogous effect with photons is observed optically as the *Franz–Keldysh effect*. At larger voltages, of course, states become available again after tunnelling (Fig. 116(*d*)), so that the current rises as *minority carriers* are *injected*—i.e. electron into the p-type material and holes into n-type.

Junctions between superconductors behave similarly but also exhibit *coherent tunnelling* phenomena which will be discussed in § 11.10.

6.9 Electrons at a surface

The electron wave functions discussed in Chapter 3 are supposed to satisfy the cyclic boundary conditions (1.76), as if deep in the interior of a large crystal. What happens near a surface?

In the first place, we need some model of the distribution of atoms, ions and electron density in the surface layers. We may generally assume that there is a rise in potential—the *work function* (Fig. 117) for an electron leaving the Fermi level in the bulk material and carried to a considerable distance away from the surface—but it is extremely difficult to calculate the exact form of the barrier to be climbed. Even when all chemical impurities have been removed—a formidable technical task—there may be rearrangements of structure and complex shifts of electron density due to unsaturated chemical bonds, image forces, etc.

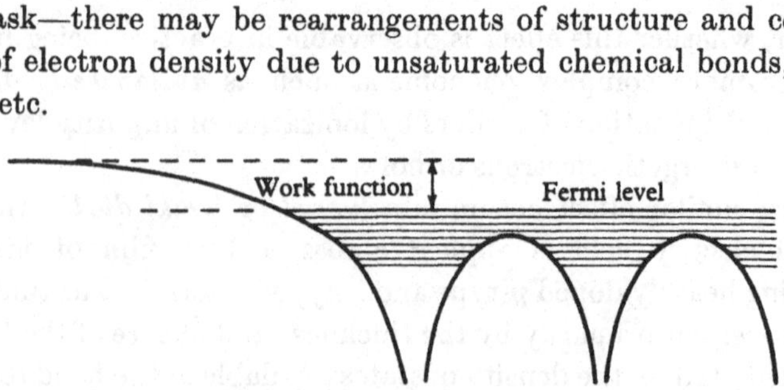

Fig. 117. The work function barrier at a surface.

It is easy to show, however, that the distribution of ordinary Bloch states will be scarcely affected by the conditions at the surface. A wave-function decaying exponentially out of the crystal can be matched to a combination of internal waves that are already nearly degenerate, with only an infinitesimal shift of energy. The exact details of the matching will depend on the precise shape of the barrier, but this has no effect on the band structure etc., of the bulk material.

But consider an electron with energy in a band gap. As shown in the previous section, the solution of the Schrödinger equation *within* the crystal is also dominated by an exponential function of distance (6.52). Suppose we take this function to be decaying into the interior, and then succeed in matching it, in amplitude and gradient, to the simple exponential function for an electron trying to escape through the surface barrier. We shall have constructed an eigenstate satisfying the general conditions of finiteness, etc., but localized at the crystal surface (Fig. 118).

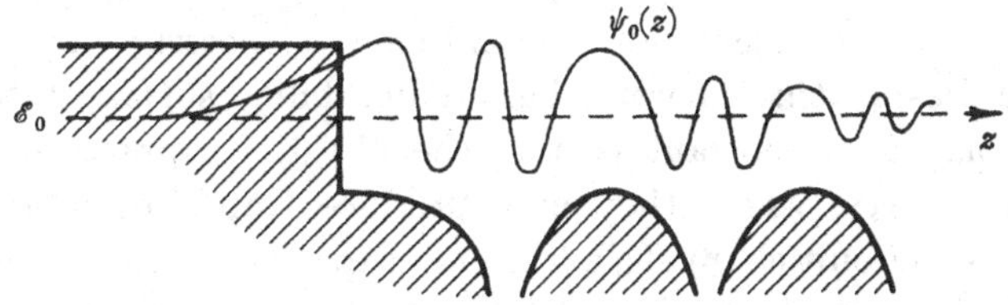

Fig. 118. Surface state.

Suppose that we have discovered a localized state $\psi_0(z)$ at the energy $\mathscr{E}_0$ in a band gap of a one-dimensional model, representing perhaps the average atomic potential or pseudo-potential, plane by plane as we go along the z-direction away from the surface. Since the crystal is still, presumably, periodic in directions parallel to the surface, Bloch's theorem (1.58) must still hold for translations in this plane. The localized state in one dimension must therefore expand into a whole band of *surface states* or *Tamm states* of the type

$$\psi_{\mathbf{k}} = \psi_0(z) \exp i(k_x x + k_y y), \tag{6.66}$$

where k_x and k_y would be components of a wave-vector measured in the plane of the surface. In the N.F.E. case, for example, we should expect this band to behave like

$$\mathscr{E}(\mathbf{k}) \sim \mathscr{E}_0 + (\hbar^2/2m)(k_x^2 + k_y^2) \tag{6.67}$$

out to some two-dimensional analogue of the zone boundary in the reciprocal space of k_x and k_y.

Electron surface states are, in fact, exactly the same in principle as the surface modes of lattice vibration discussed in § 2.12. By a suitable relabelling of the symbols, we could even interpret the algebra of (2.138)–(2.145) to describe electron surface states in a tight binding model under appropriate boundary conditions. But the exact energy at which such states can occur depends on the details of the model. Even in one dimension, it is obvious that the shape of the barrier and its position with respect to the nearest 'atom' of the lattice, will have a critical influence on the local curvature of ψ, and hence on the possibility of making a good match. In three dimensions the situation is even more complicated, since the matching must be done over the whole surface, which could never be a simple plane discontinuity of electron potential. We can discuss in principle the conditions for the appearance of such states in a given band gap, but their sensitivity to structural details, chemical impurities and space charge fields impedes quantitative calculations of the energy spectrum and other properties. Despite this theoretical unpredictability, surface states are found to play an important part in the physics of practical semiconductor devices, where they may produce traps, recombination centres, short-circuit channels, additional capacitance, and other undesirable effects. Such states are also important in surface catalysis and other chemical phenomena at interfaces.

A true surface state can only exist when there is a band gap below the work-function barrier. At higher energies the corresponding *evanescent surface waves* are only quasi-stationary in time, being analogous to resonances (cf. § 2.12). The typical surface phenomena are then the emission of electrons, either thermally or by optical activation (see § 8.5), and the converse process in which electrons are fired at the surface from the outside.

The diffraction of high energy electrons passing through thin films was discussed in §§ 2.6, 3.3. The phenomena of *Low Energy Electron Diffraction* (L.E.E.D.) are much more difficult to interpret quantitatively. For an electron that has been reflected back from the surface, the normal component of crystal momentum is no longer a good quantum number. The essential condition for a diffraction peak is the two-dimensional analogue of (2.79), i.e.

$$\mathbf{k}'_{\parallel} = \mathbf{k}_{\parallel} + \mathbf{g}_{\parallel}, \tag{6.68}$$

where the initial and final wave-vectors and the reciprocal lattice vector are all represented by their projections in the plane parallel to the surface—the (x, y)-plane of (6.66). The normal component of the reflected beam can always be adjusted to satisfy the condition $|\mathbf{k}'|^2 = |\mathbf{k}|^2$ for energy conservation, so that diffraction is allowed whatever the energy or direction of the incident beam.

But to calculate the intensity of diffraction we need to match the incoming and reflected waves to a combination of all solutions of the Schrödinger equation within the crystal, i.e. to all components of the *complex band structure* of the solid. The relative intensities of different diffracted beams will then depend on the relative proportions of the corresponding plane waves in these internal wave-functions. The strongest peaks, for example, must occur when the incident electron is trying to enter the crystal at the energy of a band gap in the ordinary Bloch states—that is, near to the full three-dimensional Bragg–von Laue diffraction condition (2.79). But energy-sensitive 'resonance' effects can also occur in this region, where the evanescent surface waves can 'channel' the electrons into different diffracted beams. Impurity scattering, absorption, and other inelastic effects are additional complications in the interpretation of L.E.E.D. observations.

6.10 Scattering of electrons by impurities

In §§ 5.2, 5.3 it was shown that the potential of a charged impurity in a metal is screened, and takes the form (5.29),

$$\mathscr{U}(\mathbf{r}) = \frac{Ze^2}{r} e^{-\lambda r}, \tag{6.69}$$

where λ is given by (5.22). It is easy to show, by elementary time-dependent perturbation theory, that this gives rise to a differential scattering cross-section

$$\sigma(\theta) = \left(\frac{2mZe^2}{\hbar^2}\right)^2 \frac{1}{(K^2+\lambda^2)^2}, \tag{6.70}$$

where $K = 2k_F \sin \tfrac{1}{2}\theta$.

A similar result should hold for carriers scattered in a semiconductor, with λ now given by (5.26), and with m^* for m, and a further factor $1/\epsilon^2$ to allow for the dielectric constant of the medium. This is no more than a direct application of the equivalent-Hamiltonian principle.

However, these formulae should not be taken too literally. They assume Born approximation for solution of the scattering problem; this is not very good for electrons at the Fermi energy being scattered by the deep potential (6.69). It would be better to use partial waves, as in (5.40), and the Friedel sum rule. Again, there is no guarantee that the electron wave-functions are really simple plane waves—and the potential is too strong to use the equivalent-Hamiltonian properly. Finally, the scattering is not all to be attributed to the charge difference between solute and solvent; there may be scattering from the difference of the core potentials—a difference, shall we say, in the effective potentials of the impurity ion and the ion for which it is substituted. Thus, there is scattering when Ag is substituted in Au, even though these have the same valency and almost exactly the same atomic volume.

Scattering by a transition metal impurity is dominated by the d-resonance (§ 5.5). From (3.81) and (5.40) it is clear that the scattering cross-section must go through a maximum at the resonance energy where the d-phase shift goes through $\pi/2$. Notice, however, that this resonance may depend upon the magnetic polarization of the impurity relative to the spin orientation of the scattered electron, as in Fig. 95.

In some cases, there is an additional small *s–d-exchange* interaction between the spin $\mathbf{s}$ of the scattered conduction electron and the total spin $\mathbf{S}$ of the d-electrons at the impurity. As will be shown in § 10.6, the Hamiltonian of this interaction ought to be of the form

$$\mathscr{H}_{s-d} = -(J/N)\,\mathbf{S}\cdot\mathbf{s}, \tag{6.71}$$

where the 'exchange' parameter, J, is usually negative. This has important physical consequences, because the spin of the incident electron may be 'flipped' when it is scattered.

In a semiconductor there may also be scattering from a neutral impurity—where, for example, an electron has settled in a donor impurity level, or a hole is resident on an acceptor level (§§ 6.4, 6.7). The calculation of the scattering is then a substantial problem, equivalent to the calculation of the scattering of slow electrons by neutral atoms of hydrogen.

6.11 Adiabatic principle

We now want to consider the interaction between conduction electrons and lattice waves. This is a dynamical problem of apparently great complexity—until we see that it can be treated by elementary perturbation theory. The justification for this is the *Born–Oppenheimer theorem*, which we now proceed to derive.

Let us write down the Hamiltonian operator of the whole system of ions and electrons. As in Chapter 2, let $\mathbf{u}_l$ represent the position of the lth ion, relative to the lth site (for simplicity, we assume one atom per unit cell); let $\mathbf{r}_i$ represent the position of the ith electron. The total Hamiltonian is

$$\mathscr{H} = -\sum_l \frac{\hbar^2}{2M}\frac{\partial^2}{\partial \mathbf{u}_l^2} - \sum_i \frac{\hbar^2}{2m}\frac{\partial^2}{\partial \mathbf{r}_i^2} + \sum_{i<j}\frac{e^2}{|\mathbf{r}_i - \mathbf{r}_j|} + \mathscr{V}(\mathbf{u},\mathbf{r}) + G(\mathbf{u}). \quad (6.72)$$

The first two terms are the kinetic energy operators for the ions and for the electrons. The next term is the electron–electron interaction. Then we have the potential energy of the electrons in the field of the ions (in their displaced positions). Finally, $G(\mathbf{u})$ symbolizes the potential energy of any direct forces between the ions.

Let us try the following as an eigenfunction for this Hamiltonian:

$$\Psi = \psi(\mathbf{u},\mathbf{r})\,\Phi(\mathbf{u}), \quad (6.73)$$

where we make ψ satisfy the Schrödinger equation for the electrons in a static lattice, frozen with the lth ion at $\mathbf{u}_l$,

$$\left\{\sum_i -\frac{\hbar^2}{2m}\frac{\partial^2}{\partial \mathbf{r}_i^2} + \sum_{ij}\frac{e^2}{|\mathbf{r}_i - \mathbf{r}_j|} + \mathscr{V}(\mathbf{u},\mathbf{r})\right\}\psi(\mathbf{u},\mathbf{r}) = \mathscr{E}_e(\mathbf{u})\,\psi(\mathbf{u},\mathbf{r}). \quad (6.74)$$

This may be—in fact, must be—a many-electron wave-function; we still assume that we can find its eigenvalues, $\mathscr{E}_e(\mathbf{u})$, for example, as a set of quasi-particle levels. But these eigenvalues will be functions of $\mathbf{u}_l$; the energy of the electron gas will depend on the positions of the ions.

Now apply the operator $\mathscr{H}$ to this wave-function:

$$\begin{aligned}\mathscr{H}\Psi &= -\sum_l \frac{\hbar^2}{2M}\frac{\partial^2}{\partial \mathbf{u}_l^2}\Psi + \mathscr{E}_e(\mathbf{u})\Psi + G(\mathbf{u})\Psi \\ &= \psi(\mathbf{u},\mathbf{r})\left\{-\sum_l \frac{\hbar^2}{2M}\frac{\partial^2}{\partial \mathbf{u}_l^2} + \mathscr{E}_e(\mathbf{u}) + G(\mathbf{u})\right\}\Phi(\mathbf{u}) \\ &\quad -\sum_l \frac{\hbar^2}{2M}\left\{2\frac{\partial \Phi}{\partial \mathbf{u}_l}\cdot\frac{\partial \psi}{\partial \mathbf{u}_l} + \Phi\frac{\partial^2 \psi}{\partial \mathbf{u}_l^2}\right\}.\end{aligned} \quad (6.75)$$

If now the second line of this expression can be ignored, we can solve our complete eigenvalue problem $\mathscr{H}\Psi = \mathscr{E}\Psi$ by making $\Phi(\mathbf{u})$ satisfy a Schrödinger-type equation

$$\left\{-\sum_l \frac{\hbar^2}{2M}\frac{\partial^2}{\partial \mathbf{u}_l^2} + \mathscr{E}_e(\mathbf{u}) + G(\mathbf{u})\right\}\Phi(\mathbf{u}) = \mathscr{E}\Phi(\mathbf{u}). \quad (6.76)$$

This is an equation for a wave-function of the ions alone. It is what we should write down if we set about trying to solve the lattice dynamical problem of § 2.1 directly by quantum mechanics—except that we must add to the direct interaction between the ions the term $\mathscr{E}_e(\mathbf{u})$, which is the total energy of the electron system as a function of the positions of the ions. This is the *adiabatic* contribution of the electrons to the lattice energy. It is the change in their energy caused by their need to follow the lattice motion. In principle, this term would depend on the precise electron configuration, for example, on the number of quasi-particles excited at the temperature of the metal—but in practice it is very insensitive to such differences and can usually be computed upon the assumption that the electrons are in their ground state.

It is still necessary to prove that the non-adiabatic terms in (6.75) can be ignored. It is easy to prove that they contribute almost nothing to the expectation value of the energy of the system in the state Ψ. The first such term vanishes: it would produce integrals like

$$\int \psi^* \frac{\partial \psi}{\partial \mathbf{u}_l} \mathbf{dr} = \frac{1}{2} \frac{\partial}{\partial \mathbf{u}_l} \int \psi^* \psi \, \mathbf{dr}$$

$$= \frac{1}{2} \frac{\partial n_e}{\partial \mathbf{u}_l}, \tag{6.77}$$

where n_e is the total number of electrons. The second term is small because, at worst, the electrons would be tightly bound to their ions

$$\psi(\mathbf{u}_l, \mathbf{r}_i) = \psi(\mathbf{r}_i - \mathbf{u}_l), \tag{6.78}$$

which would give a contribution like

$$-\int \psi^* \frac{\hbar^2}{2M} \frac{\partial^2 \psi}{\partial \mathbf{u}_l^2} \mathbf{dr} = -\int \psi^* \frac{\hbar^2}{2M} \frac{\partial^2 \psi}{\partial \mathbf{r}_i^2} \mathbf{dr}$$

$$= -\frac{m}{M} \int \psi^* \frac{\hbar^2}{2m} \frac{\partial^2 \psi}{\partial \mathbf{r}_i^2} \mathbf{dr}. \tag{6.79}$$

This is just m/M times the kinetic energy of the electrons—an energy that is negligible by comparison with ordinary thermal energies, since m/M is of the order of 10^{-4} or 10^{-5}.

This argument, which we have only sketched out very abstractly, does not consider the off-diagonal non-adiabatic terms. In particular, the first of these, containing $\partial \psi / \partial \mathbf{u}_l$, can give rise to transitions between electron states as the ions move. This is none other than the *electron–phonon interaction*, which will be discussed in § 6.13. All that we have shown in (6.77) is that it has zero diagonal elements. We would get contributions to the lattice energy from second-order perturbations involving the off-diagonal elements of this term.

The adiabatic principle is important because it allows us to separate ionic from electronic motion, leaving only a residual interaction between the electrons and the phonons. It justifies our intuitive notion that the energy of the electron gas is an important part of the total elastic energy of the solid, as was assumed in the discussion of metallic cohesion (§ 4.3). After allowing for this energy term, we can treat electrons and lattice waves as nearly independent entities.

6.12 Renormalization of velocity of sound

The adiabatic principle invites the following simple argument. Suppose we calculate the vibration spectrum of a lattice of bare ions. What is the effect of adding the electrons?

As it happens, we can derive one result without doing any work. A long-wave longitudinal vibration of a lattice of point charges gives rise to volume changes, and polarization fields, that may be calculated macroscopically, exactly as in (5.60)–(5.62). The equations of motion for the displacement of positive ions of density N per unit volume, mass M, and charge Ze are

$$M\ddot{\mathbf{x}} = -4\pi NZ^2e^2\mathbf{x}. \tag{6.80}$$

That is, for a longitudinal wave of wave-vector $\mathbf{q}$, the frequency is the plasma frequency of the ions

$$\Omega_p = \left(\frac{4\pi NZ^2e^2}{M}\right)^{\frac{1}{2}}, \tag{6.81}$$

and is nearly independent of q.

Now suppose we pour in the electron gas. As shown in Chapter 5, this modifies all the effective electrostatic fields from other sources. For example, an applied field of wave-vector $\mathbf{q}$ and frequency ω is divided by the dielectric constant $\epsilon(\mathbf{q}, \omega)$ defined in (5.16). This should apply to the forces due to polarization of the plasma of ions; we should divide the right-hand side of (6.80) by $\epsilon(\mathbf{q}, \omega)$ where $\mathbf{q}$ is the wave-vector of the vibration being considered and ω is its actual frequency. This means that the observed frequency will be

$$\nu_q^2 = \frac{\Omega_p^2}{\epsilon(\mathbf{q}, \nu_{\mathbf{q}})}. \tag{6.82}$$

In practice $\nu_{\mathbf{q}}$ is small and can be neglected in ϵ (i.e. the plasma of ions vibrates at a low frequency compared with all electronic events). With the approximation (5.21) for $\epsilon(\mathbf{q}, 0)$ we have

$$\nu_{\mathbf{q}}^2 \approx \frac{4\pi NZ^2e^2}{M} \frac{q^2}{q^2 + 4\pi e^2\mathcal{N}(\mathscr{E}_F)}, \tag{6.83}$$

i.e.
$$\nu_{\mathbf{q}} \to q\left(\frac{NZ^2}{M\mathcal{N}(\mathscr{E}_F)}\right)^{\frac{1}{2}}. \tag{6.84}$$

In other words, the velocity of longitudinal acoustic modes tends to a constant, which, for a free-electron gas of density NZ, is simply

$$s = \left(\frac{mZ}{3M}\right)^{\frac{1}{2}} v_F, \tag{6.85}$$

where v_F is, as usual, the Fermi velocity of the electrons. This is not a bad estimate of the velocity of sound in most metals, and is especially good for the alkali metals. This formula, (6.82), shows how the Kohn effect (§ 5.4) arises. The group velocity is the derivative of $\nu_{\mathbf{q}}$ with respect to q—and the derivative of $\epsilon(\mathbf{q}, 0)$ contains a logarithmic singularity at $q = 2k_F$.

For a more systematic approach to the lattice dynamics of metals, valid for all polarizations and at all wavelengths, we ought to proceed as follows: (i) Calculate the electronic band structure of the perfect crystal; (ii) calculate the band structure of the same crystal with the ions displaced in a phonon pattern of wave vector $\mathbf{q}$, as in (2.49); (iii) treat the difference in total energy of these two structures as the 'potential energy' change in this pattern of displacements, as in (2.2), and hence calculate the frequency $\omega_{\mathbf{q}}$ of the mode.

This looks a formidable procedure, because the adiabatic principle is justified only if each structure is calculated self-consistently. But suppose we are dealing with a transverse phonon mode, in which the local atomic volume stays constant. Most of the cohesive energy of the crystal (cf. § 4.3) is unchanged, and we can concentrate on small contributions from distortions of the Fermi surface, such as might be described as perturbations produced by scattering the electron from the displaced ions (cf. §§ 2.7, 6.13). For a simple N.E.F. metal, the screened pseudo-potential scheme of § 5.3 describes the effects of such displacements precisely, with complete self-consistency. We thus discover a relationship between frequency $\omega_{\mathbf{q}}$ and various Fourier components $w_s(\mathbf{q} + \mathbf{g})$ of the screened ionic pseudo-potential.

This relationship has been used with some success to interpret the complicated dispersion curves observed in the transverse lattice spectra of metals by neutron diffraction (§ 2.8). We can even go back into a real space representation, and show that the metal behaves dynamically as if there were pair-wise elastic forces between the 'neutral pseudo-atoms'—forces that may also be interpreted as the screened electrostatic interaction between the 'pseudo-charge' clouds of the bare metal ions. But this picture is misleading when applied to the long wave longitudinal modes, where volume effects predominate, and is also quite inappropriate for transition metals and covalently bonded semiconductors (§ 4.2).

6.13 The electron–phonon interaction

The simplest way of treating this is to assume that the conduction electrons in the metal are inelastically diffracted by the lattice vibrations in exactly the same way as we assume for the inelastic diffraction of X-rays, neutrons, or fast electrons. This treatment seems, at first sight, altogether too naive, but may be justified by an appeal to the adiabatic principle of § 6.11. We take over all the theory of §§ 2.7, 2.8. In (2.86) and (2.97) we reached the following formula: the matrix element for the transition from state $\mathbf{k}$ to state $\mathbf{k}'$ can be written

$$\mathscr{M}_{\mathbf{k}'\mathbf{k}} = i\mathscr{V}_a(\mathbf{K}).(\mathbf{K}\cdot\mathbf{U}_\mathbf{q}), \tag{6.86}$$

where $\mathbf{K} = \mathbf{k}'-\mathbf{k}$, where $\mathbf{q} = \mathbf{K}$ or $\mathbf{q} = \mathbf{K}-\mathbf{g}$, according as we have N-processes or U-processes, where $\mathbf{U}_\mathbf{q}$ is the vector amplitude of the lattice vibration of wave-vector $\mathbf{q}$, and where $\mathscr{V}_a(\mathbf{K})$ is the matrix element of the potential of a single atom or ion for this transition.

We can express this very simply as follows. Take the square modulus of $\mathscr{M}_{\mathbf{k}'\mathbf{k}}$ and calculate a scattering probability by conventional perturbation theory. Then each atom in the lattice will behave as if it had an average effective differential scattering cross-section

$$\bar{\sigma}(\mathbf{K}) = \sigma_a(\mathbf{K})\,N|\mathbf{K}\cdot\mathbf{U}_\mathbf{q}|^2, \tag{6.87}$$

where $\sigma_a(\mathbf{K})$ is the differential cross-section of an *isolated* atom for scattering through the vector $\mathbf{K}$.

This formula is quite transparent. It is as if each atom could scatter electrons independently—but the amount of scattering depends on its relative displacement from its neighbours in the lattice. The *average* of the factor $|\mathbf{K}\cdot\mathbf{U}_\mathbf{q}|^2$ over all $\mathbf{q}$ occurs in the theory of the Debye–Waller factor, discussed in § 2.9; as shown in (2.107) and (2.118), it is just the mean-square displacement of an ion from its lattice site, expressed as a fraction of the lattice spacing. But when we look at each process in detail—that is, when we choose a particular value of $\mathbf{K}$—we have to consider the scattering from a set of lattice displacements with a well-defined wave-vector; the correlation between displacements of neighbouring ions then becomes important, and the scattering by long-wave modes is reduced.

We can see this even more clearly by reference to (2.86) and (2.103). The factor $N|\mathbf{K}\cdot\mathbf{U}_\mathbf{q}|^2$ is an approximation for the square of the structure factor, which is proportional to $S(\mathbf{K})$, the Fourier transform of the pair correlation function for atomic positions; i.e.

$$\bar{\sigma}(\mathbf{K}) = \sigma_a(\mathbf{K})\,S(\mathbf{K}). \tag{6.88}$$

This result holds for a liquid just as well as for a solid; the scattering of a conduction electron depends on the radial distribution function of the atoms in the liquid.

When we come to discuss electrical resistivity, etc., we shall consider this structure factor further, and analyse contributions from various lattice frequencies. We shall also need to take account of the change of energy of the electron when it is scattered, as in (2.101). For the moment, we merely remark that the adiabatic principle allows us to justify (6.87), because it allows us to assume the lattice to be frozen in any configuration, and then to calculate scattering by the electrons in that slightly non-periodic arrangement. The only dynamical effect of the motion of the ions is the energy change to which we have just referred.

The problem of the electron–phonon interaction thus reduces mainly to the calculation of $\sigma_a(\mathbf{K})$. If the atom were just a shallow potential $\mathscr{V}(\mathbf{r})$, carried bodily, then we should have, from elementary wave mechanics, in the Born approximation

$$\frac{\partial\sigma_a}{\partial\Omega} = \left(\frac{m}{2\pi\hbar^2 N}\right)^2 |\mathscr{V}_a(\mathbf{K})|^2 \tag{6.89}$$

for the differential scattering per unit solid angle Ω. In this formula

$$\mathscr{V}_a(\mathbf{K}) = N\int \mathscr{V}_a(\mathbf{r})\,e^{-i\mathbf{K}\cdot\mathbf{r}}\,d\mathbf{r}; \tag{6.90}$$

it is the Fourier transform of the potential of the ion, normalized to one unit cell of the crystal.

But as we saw in § 3.9, the actual potential of an atom or ion is too deep for the Born approximation for the scattering cross-section. The partial-wave formula (3.78), or some similar exact formula for the scattering amplitude, would be much more appropriate. In particular we may adopt the pseudo-potential formalism of §§ 3.6, 3.9 to calculate the electron–phonon interaction within the same framework as the band structure calculation.

This suggests that the Fourier components of $\mathscr{V}_a$ in (6.89) should be replaced by the corresponding components of an atomic pseudo-potential. But the change in potential produced by displacing an ion must include any effects due to the shift in the charge distribution of the electron gas. As we saw in § 5.3, this can be taken care of approximately by dividing the Fourier components of the bare ion potential by the dielectric function

To cut a long story short, the correct procedure is to start our whole discussion of the electron–phonon interaction from the superposed *pseudo-atom* potentials (5.35), instead of the superposed 'atomic' potentials (2.83). In place of (6.89), we obtain

$$\frac{\partial \sigma_a}{\partial \Omega} = \left(\frac{m}{2\pi\hbar^2 N}\right)^2 |w_s(K)|^2, \tag{6.91}$$

where $w_s(K)$ is a Fourier transform of a neutral pseudo-atom 'potential'. This formula is not exact to all orders in the perturbation series for scattering, and rather uncertain at large K because linear screening is not valid in the atomic core, but it is a good self-consistent approximation for simple N.F.E. metals.

The behaviour of $w_s(K)$ is much the same for all simple metals. As we saw in § 5.3, the pseudo-potential for a *bare* ion, $w_b(r)$ includes the long range Coulomb field of the valence charge Ze In Fourier transform, therefore, (cf. (5.31)), this must behave like

$$w_b(K) \approx -\frac{4\pi Z N e^2}{K^2} + \mathscr{V}'(K) \tag{6.92}$$

where $\mathscr{V}'(\mathbf{K})$ remains finite as $\mathbf{K} \to 0$, whatever the form of $\mathscr{V}(\mathbf{r})$ inside the ion core.

The singularity in (6.92) is, of course, cancelled by the singularity in $\epsilon(\mathbf{K})$ at $\mathbf{K} = 0$, just as in (5.32). Putting together these various results, we may write

$$w_s(K) = \frac{w_b(K)}{\epsilon(K)} \approx \frac{-4\pi Z N e^2/K^2 + \mathscr{V}'(\mathbf{K})}{1 + 4\pi e^2 \mathscr{N}(\mathscr{E}_F)/K^2}$$
$$\approx \frac{-\frac{2}{3}\mathscr{E}_F + (K^2/\lambda^2)\mathscr{V}'(\mathbf{K})}{1 + (K^2/\lambda^2)}, \tag{6.93}$$

where λ, the screening parameter, comes from (5.22), or more precisely, from (5.21) and (5.36). In essence, this is the formula derived by Bardeen in 1937 for the matrix element of the electron–phonon interaction in the alkali metals. We now see that this pseudo-atom form factor (Fig. 119) plays a central role in the theory of simple

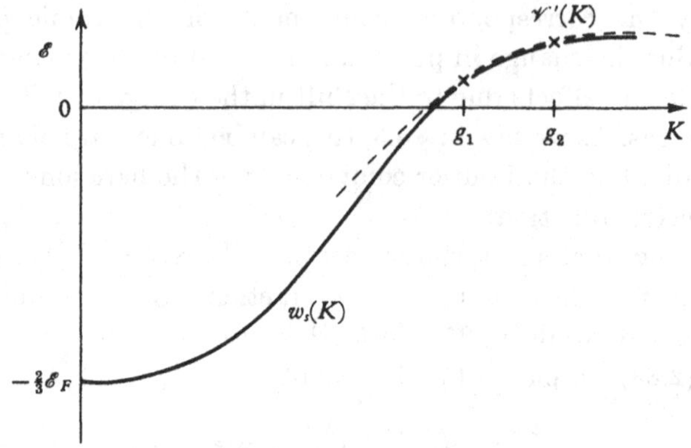

Fig. 119. Pseudo-atom form factor. Band structure depends only on values at reciprocal lattice vectors, g_1, g_2, etc.

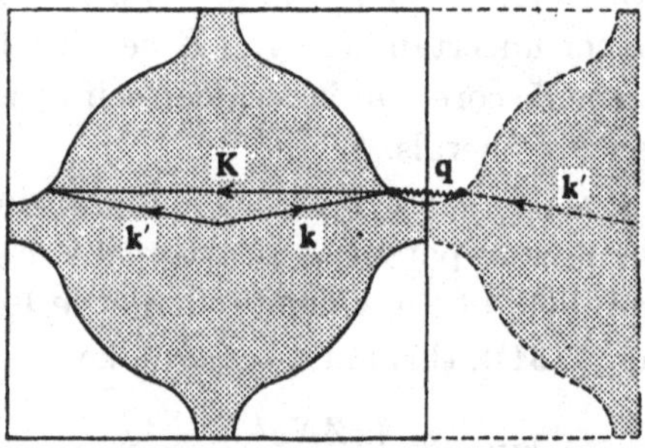

Fig. 120. In the repeated zone scheme, U-processes can be reduced to N-processes.

N.F.E. metals. Knowing $w_s(K)$ we can in principle calculate the band structure, (§ 3.9), the lattice spectrum (§ 6.12), the electrical transport properties of the crystalline and liquid phases (§ 7.5), and even the superconducting transition temperature (§ 11.1). Unfortunately, the core contribution, $\mathscr{V}'(\mathbf{K})$ in (6.93) is neither 'local' (§ 3.6) nor uniquely defined, so that the correlation between these properties is not as perfect as we should like. Life is seldom that simple!

There is one point to be careful of in this discussion. Suppose we ask for the electron–phonon interaction between two electron states, $\mathbf{k}$ and $\mathbf{k}'$, which lie nearly a reciprocal lattice vector apart. It would then certainly be wrong to use a free-electron formula such as (6.89) for the differential scattering cross-sections. The two states, seeming to differ by a large wave-vector, are actually quite close, as may be seen on a repeated zone scheme. Because they are both near a zone boundary, each is a complicated mixture of plane waves, as pointed out in §§ 3.2, 3.3. One finds (as one must for consistency) that the value of $\mathscr{V}_a(\mathbf{K})$, when computed as a matrix element between the properly mixed wave-functions, tends to zero as $\mathbf{K} \to \mathbf{g}$.

We can put this another way. A transition that looks like an Umklapp process when mapped in a single reduced zone, can be made to look like a normal process in the repeated zone scheme—and the matrix element, which would now be proportional to $\mathbf{q} \cdot \mathbf{U}_\mathbf{q}$, tends to zero as the phonon wave-vector, $\mathbf{q}$, tends to zero.

6.14 Deformation potentials

In a semiconductor, where the wave-functions are complicated, it is convenient to use a phenomenological theory, in which we treat the crystal as a continuous elastic medium. The lattice vibrations may now be described as waves of elastic strain, e.g.

$$\mathsf{W}_{ij}(\mathbf{r}) = \mathsf{W}^0_{ij} e^{i\mathbf{q}\cdot\mathbf{r}} \tag{6.94}$$

for the Cartesian components of the strain tensor at $\mathbf{r}$. We then argue that the local strain changes the energy of a carrier by the amount

$$\delta\mathscr{E}(\mathbf{r}) = \sum_{ij} \mathscr{E}_{ij} \mathsf{W}_{ij}(\mathbf{r}), \tag{6.95}$$

where $\mathscr{E}_{ij}$ is the *deformation potential* tensor.

We can easily calculate a matrix element for scattering of electrons. As in (2.84), say,

$$\begin{aligned} \mathscr{M}_{\mathbf{k}'\mathbf{k}} &= \int e^{-i\mathbf{k}'\cdot\mathbf{r}} \delta\mathscr{E}(\mathbf{r}) e^{i\mathbf{k}\cdot\mathbf{r}} \mathbf{dr} \\ &= (\sum_{ij} \mathsf{W}^0_{ij} \mathscr{E}_{ij}) \delta_{\mathbf{k}-\mathbf{k}'+\mathbf{q}}, \end{aligned} \tag{6.96}$$

in a crystal of unit volume.

The scattering conserves crystal momentum; this is natural enough, since we have smoothed away the lattice. The scattering probability goes as $|\mathbf{W}^0|^2$, which is the same sort of quantity as the structure factor (6.88) in the more precise theory. To evaluate the coefficients $\mathscr{E}_{ij}$, we can think of subjecting the whole solid to a uniform strain $\mathbf{W}^0$, and measuring the change in the width of the energy gap.

For a metal one can easily determine the dilatation component of $\mathscr{E}_{ij}$. The argument is essentially the one used in § 5.2. A varying dilatation $\Delta(\mathbf{r})$ changes the density of electrons near $\mathbf{r}$ from n_0 to $n_0(1-\Delta)$. This would change the Fermi level by

$$\delta\zeta(\mathbf{r}) = \frac{n_0\Delta(\mathbf{r})}{\mathscr{N}(\mathscr{E}_F)} = \tfrac{2}{3}\mathscr{E}_F\Delta(\mathbf{r}) \tag{6.97}$$

in a free-electron gas. For long-wave oscillations of density, we can get back to a uniform Fermi level by allowing a few electrons to flow back out of the denser regions. This small deviation from local electrical neutrality gives rise to an electric field whose potential just brings the Fermi level back to constancy over the whole crystal. This field must be $-\delta\zeta(\mathbf{r})$ which we identify with $\delta\mathscr{E}(\mathbf{r})$, i.e. the trace of the deformation potential tensor must be

$$\sum_i \mathscr{E}_{ii} = -\tfrac{2}{3}\mathscr{E}_F. \tag{6.98}$$

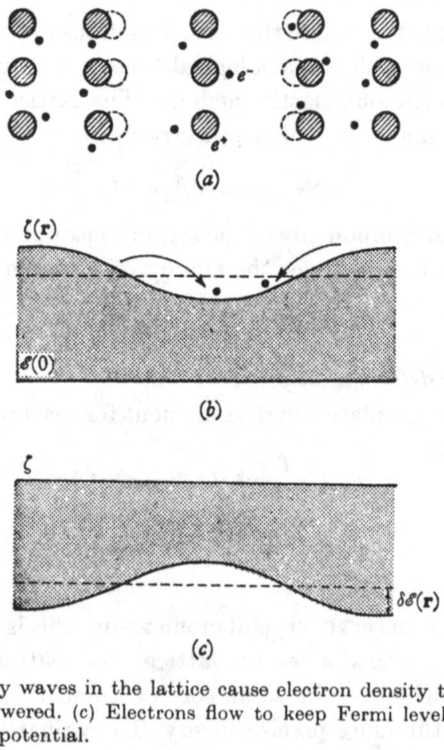

Fig. 121. (a) Density waves in the lattice cause electron density to vary. (b) Fermi level is raised or lowered. (c) Electrons flow to keep Fermi level constant, giving rise to deformation potential.

This is a rigorous derivation of the limiting value of $w_s(K)$ when $K \to 0$ in (6.93), showing that this feature of the pseudo-atom form factor is entirely self consistent and exact to all orders in perturbation theory. Any attempt to use muffin-tin potentials (§ 3.7) in a deformed crystal must also take account of this effect as a change in the muffin-tin zero.

CHAPTER 7

TRANSPORT PROPERTIES

There was no 'One, two, three, and away', but they began running when they liked and left off when they liked, so that it was not easy to know when the race was over.

LEWIS CARROLL

7.1 The Boltzmann equation

The carriers in a metal or semiconductor can be affected by external fields, and by temperature gradients. They also suffer scattering from impurities, lattice waves, etc. These effects have to be balanced against each other—we have to consider situations in which the electron is accelerated by a field, but loses its extra energy and momentum by scattering. In this chapter we shall consider the 'ordinary' transport properties, such as are observed when constant fields are applied.

Much the simplest approach to this problem, in general, is to set up the *transport equation* or *Boltzmann equation*. We study a quantity $f_{\mathbf{k}}(\mathbf{r})$, the local concentration of carrier in the state $\mathbf{k}$ in the neighbourhood of the point $\mathbf{r}$ in space. Strictly speaking, this quantity can only be defined in terms of fine-grained distributions, ensemble averages, density matrices, etc. There is a considerable literature upon this point—but it belongs more to the formal theory of quantum statistical mechanics than to the theory of solids.

Now consider how $f_{\mathbf{k}}(\mathbf{r})$ can change with time. There are three types of effect:

(i) Carriers move in and out of the region $\mathbf{r}$. Suppose that $\mathbf{v}_{\mathbf{k}}$ is the velocity of a carrier in state $\mathbf{k}$. Then, in an interval t, the carriers in this state move a distance $t\mathbf{v}_{\mathbf{k}}$. Thus, assuming Liouville's theorem on the invariance of volume occupied in phase space, the number of carriers in the neighbourhood of $\mathbf{r}$ at time t is equal to the number of them in the neighbourhood of $\mathbf{r}-t\mathbf{v}_{\mathbf{k}}$ at time 0:

$$f_{\mathbf{k}}(\mathbf{r},t) = f_{\mathbf{k}}(\mathbf{r}-t\mathbf{v}_{\mathbf{k}},0). \tag{7.1}$$

This means that the rate of change of the distribution due to *diffusion* is

$$\left.\frac{\partial f_{\mathbf{k}}}{\partial t}\right]_{\text{diff.}} = -\mathbf{v}_{\mathbf{k}}\cdot\frac{\partial f_{\mathbf{k}}}{\partial \mathbf{r}} = -\mathbf{v}_{\mathbf{k}}\cdot\nabla f_{\mathbf{k}}. \tag{7.2}$$

(ii) External fields will change the **k**-vector of each carrier, at the rate (6.40), i.e.

$$\dot{\mathbf{k}} = \frac{e}{\hbar}\left(\mathbf{E} + \frac{1}{c}\mathbf{v}_{\mathbf{k}} \wedge \mathbf{H}\right). \tag{7.3}$$

We can look upon this as the velocity of the carrier in **k**-space, so that, by analogy with (7.1),

$$f_{\mathbf{k}}(\mathbf{r}, t) = f_{\mathbf{k}-\dot{\mathbf{k}}t}(\mathbf{r}, 0); \tag{7.4}$$

the distribution changes, because of the *fields*, at the rate

$$\left.\frac{\partial f_{\mathbf{k}}}{\partial t}\right]_{\text{field}} = -\dot{\mathbf{k}}\cdot\frac{\partial f_{\mathbf{k}}}{\partial \mathbf{k}}$$
$$= -\frac{e}{\hbar}\left(\mathbf{E} + \frac{1}{c}\mathbf{v}_{\mathbf{k}} \wedge \mathbf{H}\right)\cdot\frac{\partial f_{\mathbf{k}}}{\partial \mathbf{k}} \tag{7.5}$$

(where we write $\partial/\partial\mathbf{k}$ for the gradient in **k**-space—the operator $\nabla_{\mathbf{k}}$).

(iii) The effect of scattering is more complicated. But we shall here confine ourselves mostly to elastic scattering. This gives rise to a rate of change of $f_{\mathbf{k}}$

$$\left.\frac{\partial f_{\mathbf{k}}}{\partial t}\right]_{\text{scatt.}} = \int \{f_{\mathbf{k}'}(1-f_{\mathbf{k}}) - f_{\mathbf{k}}(1-f_{\mathbf{k}'})\}\, Q(\mathbf{k}, \mathbf{k}')\, d\mathbf{k}'. \tag{7.6}$$

The process of scattering from **k** to **k**′, decreases $f_{\mathbf{k}}$. The probability of this process depends on $f_{\mathbf{k}}$, the number of carriers in the state **k**, and on $(1-f_{\mathbf{k}'})$, the number of vacancies available in the final state. There is also the inverse process, from **k**′ into **k**, which increases $f_{\mathbf{k}}$, and which is weighted with $f_{\mathbf{k}'}(1-f_{\mathbf{k}})$. We sum over all possible other states **k**′. For each value of **k** and **k**′, however, there is a basic transition probability $Q(\mathbf{k}, \mathbf{k}')$, which would measure the rate of transition if, say, **k** were known to be occupied and **k**′ known to be empty. The principle of *microscopic reversibility* tells us that the same function measures the transition rate from **k**′ to **k**, so this is a common factor in the integrand.

The Boltzmann equation says that at any point, and for any value of **k**, the net rate of change of $f_{\mathbf{k}}(\mathbf{r})$ is zero; i.e.

$$\left.\frac{\partial f_{\mathbf{k}}}{\partial t}\right]_{\text{scatt.}} + \left.\frac{\partial f_{\mathbf{k}}}{\partial t}\right]_{\text{field}} + \left.\frac{\partial f_{\mathbf{k}}}{\partial t}\right]_{\text{diff.}} = 0. \tag{7.7}$$

Notice that this is a *steady state*; it is not the *equilibrium state*, which we label $f_{\mathbf{k}}^0$ and which holds when fields and temperature gradients are absent. If the external field is itself varying with time, then these three terms must add up to the time variation of $f_{\mathbf{k}}$ in response to this force (see § 8.6).

But we assume that the steady-state distribution does not depart very far from equilibrium. We write

$$g_{\mathbf{k}} = f_{\mathbf{k}} - f_{\mathbf{k}}^0, \tag{7.8}$$

where, as in § 4.5,

$$f_{\mathbf{k}}^0 = \frac{1}{\exp\{(\mathscr{E}_{\mathbf{k}} - \zeta)/kT\} + 1} = f^0(\mathscr{E}_{\mathbf{k}}). \tag{7.9}$$

We must be a little bit careful here; how is $f_{\mathbf{k}}^0$ defined when the temperature varies from point to point? We suppose that there is a well-defined temperature $T(\mathbf{r})$ at each point, and write

$$g_{\mathbf{k}}(\mathbf{r}) = f_{\mathbf{k}}(\mathbf{r}) - f_{\mathbf{k}}^0\{T(\mathbf{r})\}. \tag{7.10}$$

If there is difficulty in discovering what $T(\mathbf{r})$ should be, we can insist that our final solution should satisfy some subsidiary condition, for example, that

$$\int g_{\mathbf{k}}(\mathbf{r})\, d\mathbf{k} = 0. \tag{7.11}$$

With (7.8) in (7.7), and using (7.2) and (7.5), the Boltzmann equation becomes

$$-\mathbf{v}_{\mathbf{k}} \cdot \frac{\partial f_{\mathbf{k}}}{\partial \mathbf{r}} - \frac{e}{\hbar}\left(\mathbf{E} + \frac{1}{c}\mathbf{v}_{\mathbf{k}} \wedge \mathbf{H}\right) \cdot \frac{\partial f_{\mathbf{k}}}{\partial \mathbf{k}} = -\left.\frac{\partial f_{\mathbf{k}}}{\partial t}\right]_{\text{scatt.}}, \tag{7.12}$$

i.e.

$$-\mathbf{v}_{\mathbf{k}} \cdot \frac{\partial f_{\mathbf{k}}^0}{\partial T} \nabla T - \frac{e}{\hbar}\left(\mathbf{E} + \frac{1}{c}\mathbf{v}_{\mathbf{k}} \wedge \mathbf{H}\right) \cdot \frac{\partial f_{\mathbf{k}}^0}{\partial \mathbf{k}}$$

$$= -\left.\frac{\partial f_{\mathbf{k}}}{\partial t}\right]_{\text{scatt}} + \mathbf{v}_{\mathbf{k}} \cdot \frac{\partial g_{\mathbf{k}}}{\partial \mathbf{r}} + \frac{e}{\hbar}\left(\mathbf{E} + \frac{1}{c}\mathbf{v}_{\mathbf{k}} \wedge \mathbf{H}\right) \cdot \frac{\partial g_{\mathbf{k}}}{\partial \mathbf{k}}. \tag{7.13}$$

With the help of (7.9), and the kinematic principle (6.2), we can write this

$$\left(-\frac{\partial f^0}{\partial \mathscr{E}}\right) \mathbf{v}_{\mathbf{k}} \cdot \left\{-\frac{\mathscr{E}(\mathbf{k}) - \zeta}{T} \nabla T + e\left(\mathbf{E} - \frac{1}{e}\nabla \zeta\right)\right\}$$

$$= -\left.\frac{\partial f_{\mathbf{k}}}{\partial t}\right]_{\text{scatt.}} + \mathbf{v}_{\mathbf{k}} \cdot \frac{\partial g_{\mathbf{k}}}{\partial \mathbf{r}} + \frac{e}{\hbar c}(\mathbf{v}_{\mathbf{k}} \wedge \mathbf{H}) \cdot \frac{\partial g_{\mathbf{k}}}{\partial \mathbf{k}}. \tag{7.14}$$

This is the linearized Boltzmann equation. We have dropped $(\mathbf{E} \cdot \partial g_{\mathbf{k}}/\partial \mathbf{k})$, which would be of order E^2, and correspond to deviations from Ohm's Law. We may also remove a term $(\mathbf{v}_{\mathbf{k}} \cdot \mathbf{v}_{\mathbf{k}} \wedge \mathbf{H})$, which vanishes identically; the magnetic field does not appear on the left-hand side.

If we were to put (7.6) into (7.14), we should see that the Boltzmann equation is a linear integro-differential equation for the 'out of balance' of the distribution function, $g_{\mathbf{k}}(\mathbf{r})$. The value of this function is determined by the strength of the electric field, and the temperature gradient, through the driving terms on the left. The rest of this chapter will be concerned with finding solutions of this equation under conditions of increasing complexity.

The critical reader should by now have revolted against this simple-minded use of quasi-classical concepts to describe the behaviour of a quantum-mechanical system. Can one really ignore interference between electron wave-packets in this cavalier fashion?

To answer such questions, we need a more advanced technique using density matrices and/or Green functions. As in § 5.1, the external field is treated as a small perturbation on the equilibrium state of our many-particle system, eliciting a *linear response*, whose magnitude measures the corresponding transport coefficient. The electrical conductivity tensor for example, may be expressed abstractly by the *Kubo formula*:

$$\sigma_{\mu\nu} = \frac{1}{kT}\int_0^\infty \langle j_\mu(t)\, j_\nu(0)\rangle\, dt. \tag{7.15}$$

In other words the conductivity depends on the *time correlation* (cf. § 2.8) between a component of the current operator $j_\nu(0)$ at time zero and the component $j_\mu(t)$ at some later time t, integrated over all time and evaluated as the average of the expectation value of the product over the equilibrium ensemble. The average value of $\mathbf{j}(t)$ itself is, of course, zero, but the decay of *fluctuations* in the current depends on precisely the same features of the system—impurity scattering, etc.—as would govern the response to an external field.

Formulae of this kind are very elegant, and provide an exact basis for various theorems involving the transport coefficients—e.g. the Onsager relations for irreversible processes (§ 7.9). But the evaluation of an expression such as (7.15) in terms of scattering cross-sections, etc., is peculiarly subtle, and has only been carried out in a few simple standard cases. Fortunately, these confirm almost all the results obtained much more directly by the Boltzmann method.

7.2 Electrical conductivity

Suppose we have only an electric field $\mathbf{E}$, in an 'infinite' medium kept at constant temperature. The equation becomes

$$\begin{aligned}\left(-\frac{\partial f^0}{\partial \mathscr{E}}\right)\mathbf{v}_\mathbf{k}\cdot e\mathbf{E} &= -\frac{\partial f_\mathbf{k}}{\partial t}\bigg]_{\text{scatt.}}\\ &= \int (f_\mathbf{k}-f_{\mathbf{k}'})\,Q(\mathbf{k},\mathbf{k}')\,d\mathbf{k}'\\ &= \int (g_\mathbf{k}-g_{\mathbf{k}'})\,Q(\mathbf{k},\mathbf{k}')\,d\mathbf{k}'\end{aligned} \tag{7.16}$$

by (7.6). This is a simple integral equation for the unknown function $g_\mathbf{k}$.

Instead of solving this equation directly, let us make the phenomenological assumption

$$-\frac{\partial f_\mathbf{k}}{\partial t}\bigg]_{\text{scatt.}} = \frac{1}{\tau}g_\mathbf{k}. \tag{7.17}$$

That is, we introduce a *relaxation time* τ. If the field were turned off, then any out of balance, $g_\mathbf{k}$, would decay to zero according to

$$-\frac{\partial g_\mathbf{k}}{\partial t} = \frac{g_\mathbf{k}}{\tau},$$

i.e.

$$g_\mathbf{k}(t) = g_\mathbf{k}(0)\,e^{-t/\tau}. \tag{7.18}$$

This is a plausible assumption—but requires justification, which we give in § 7.3.

Putting (7.17) into (7.16) gives

$$g_\mathbf{k} = \left(-\frac{\partial f^0}{\partial \mathscr{E}}\right)\tau\mathbf{v}_\mathbf{k}\cdot e\mathbf{E}. \tag{7.19}$$

To calculate the electrical conductivity, we need the corresponding current density,

$$\begin{aligned}\mathbf{J} &= 2\int e\mathbf{v}_\mathbf{k} f_\mathbf{k}\,d\mathbf{k}\\ &= 2\int e\mathbf{v}_\mathbf{k} g_\mathbf{k}\,d\mathbf{k}\quad \left(\text{since}\int e\mathbf{v}_\mathbf{k} f^0_\mathbf{k}\,d\mathbf{k} \equiv 0\right)\\ &= \frac{1}{4\pi^3}\iint e^2\tau\mathbf{v}_\mathbf{k}(\mathbf{v}_\mathbf{k}\cdot\mathbf{E})\left(-\frac{\partial f^0}{\partial \mathscr{E}}\right)\frac{dS}{\hbar v_\mathbf{k}}\,d\mathscr{E},\end{aligned} \tag{7.20}$$

using (2.66) and (4.6) to transform an integral over a volume of $\mathbf{k}$-space into integrations over surfaces of constant energy.

By § 4.5, in a metal the function $(-\partial f^0/\partial\mathscr{E})$ behaves like a delta-function at the Fermi level; we are left with an integral over the Fermi

surface. Thus
$$\mathbf{J} = \frac{1}{4\pi^3}\frac{e^2\tau}{\hbar}\int \frac{\mathbf{v_k}\,\mathbf{v_k}\,dS_F}{v_\mathbf{k}}\cdot\mathbf{E}. \tag{7.21}$$

We compare this with our standard macroscopic equation,
$$\mathbf{J} = \boldsymbol{\sigma}\cdot\mathbf{E}, \tag{7.22}$$

where $\boldsymbol{\sigma}$ is a tensor. In dyadic notation,
$$\boldsymbol{\sigma} = \frac{1}{4\pi^3}\frac{e^2\tau}{\hbar}\int \frac{\mathbf{v_k}\,\mathbf{v_k}\,dS_F}{v_\mathbf{k}}. \tag{7.23}$$

To derive this from the Kubo formula (7.15), we could define τ as the decay time for a fluctuation in the current $e\mathbf{v_k}$ in the state $\mathbf{k}$. These fluctuations are governed by the Fermi–Dirac distribution in the ensemble average, which replaces $1/kT$ by a density of states factor in (7.23).

We usually deal with crystals having cubic symmetry, in which case the *conductivity tensor* reduces to a scalar. Thinking of the case where $\mathbf{E}$ and $\mathbf{J}$ are both in the x-direction, we find in the integrand
$$(\mathbf{v_k}\,\mathbf{v_k}\cdot\mathbf{E})_x = v_x^2 E, \tag{7.24}$$

which is $\frac{1}{3}$ of the contribution from the square of the total velocity, v^2E. Thus,
$$\begin{aligned}\sigma &= \frac{1}{4\pi^3}\frac{e^2\tau}{\hbar}\frac{1}{3}\int v\,dS_F \\ &= \frac{1}{4\pi^3}\frac{e^2}{\hbar}\frac{1}{3}\int \Lambda\,dS_F,\end{aligned} \tag{7.25}$$

where we introduce the *mean free path*,
$$\Lambda = \tau v. \tag{7.26}$$

This is the basic formula for electrical conductivity.

It is interesting to look (Fig. 122) at the form of the distribution function, $f_\mathbf{k}$, defined by (7.8). As we see from (7.19), $g_\mathbf{k}$ is large only at the Fermi surface. A bit is added on to the side where $\mathbf{v_k}\cdot e\mathbf{E}$ is positive, i.e. on the side where the electrons are accelerated by the field. The same amount is subtracted from the other side.

In fact, we have
$$\begin{aligned}f_\mathbf{k} &= f_\mathbf{k}^0 - \frac{\partial f_\mathbf{k}^0}{\partial \mathscr{E}(\mathbf{k})}\frac{\partial \mathscr{E}(\mathbf{k})}{\partial \mathbf{k}}\cdot\frac{e\tau}{\hbar}\mathbf{E} \\ &= f^0\left(\mathbf{k} - \frac{e\tau}{\hbar}\mathbf{E}\right),\end{aligned} \tag{7.27}$$

by Taylor's theorem. It is as if the whole Fermi surface had been shifted by the amount $(e\tau/\hbar)\,\mathbf{E}$ in $\mathbf{k}$-space. But this is a slightly misleading concept. The states near the bottom of the band, deep within the Fermi sphere, are not really affected by the field. They are prevented, by the Pauli principle, from being accelerated by the field or scattered by impurities.

We notice, however, that the conductivity is independent of the temperature (except perhaps by variation of τ). The same formula holds at $T = 0$, when the edge of the Fermi distribution is perfectly sharp. We may say that the conductivity can be calculated from the displacement of a *hard Fermi surface.*

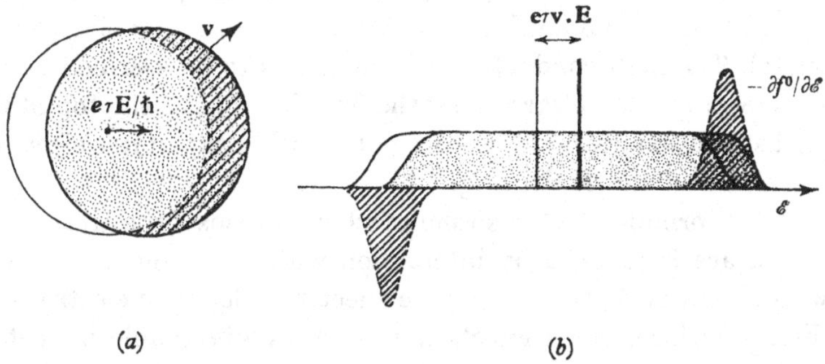

Fig. 122. (*a*) Displaced Fermi surface. (*b*) Displaced Fermi distribution.

Another point to notice is that (7.27) can be written

$$f_{\mathbf{k}} = f^0(\mathscr{E}_{\mathbf{k}} - e\tau\mathbf{v}_{\mathbf{k}}\cdot\mathbf{E}), \tag{7.28}$$

as if the electron in state $\mathbf{k}$ had gained energy of amount

$$\delta\mathscr{E}_{\mathbf{k}} = e\tau\mathbf{v}_{\mathbf{k}}\cdot\mathbf{E}. \tag{7.29}$$

This is just what it would have done classically, if it had moved with velocity $\mathbf{v}_{\mathbf{k}}$ through the field $\mathbf{E}$ for an average time τ. The *kinetic method* of dealing with transport problems is based upon this argument. This extra energy acquired between collisions with impurities is equivalent to a *drift velocity*, $\delta\mathbf{v}$, in the direction of the field, such that

$$\delta\mathbf{v}\cdot\frac{\partial\mathscr{E}}{\partial\mathbf{v}} = e\mathbf{v}\cdot\mathbf{E}\tau, \tag{7.30}$$

i.e.

$$\delta\mathbf{v} = \frac{e\tau v}{mv}\mathbf{E}, \tag{7.31}$$

for a classical particle of mass m.

If there are n particles per unit volume, then

$$\mathbf{J} = ne\delta\mathbf{v}, \tag{7.32}$$

for the net current density, and, by comparison of (7.31), (7.32) and (7.22),

$$\sigma = \frac{ne^2\tau}{m}. \tag{7.33}$$

It can easily be shown that, for a free-electron gas, (7.25) is the same as (7.33): but for a metal the former is much the better formula in principle. From the point of view of linear response theory, it would be even better to write

$$\sigma = \tfrac{1}{3}j_F^2\tau\mathcal{N}(\mathscr{E}_F) \tag{7.34}$$

showing that the conductivity depends only on the properties of the electrons at the Fermi level, not on the total number of electrons in the metal. The high conductivity of metals is to be ascribed to the high current, $j_F = ev_F$, carried by the few electrons at the top of the Fermi distribution, rather than to a high total density of free electrons which can be set slowly drifting.

The basic formula (7.25) also shows what happens if the area of free Fermi surface is reduced by interaction with zone boundaries, and allows for lattice effects reducing the effective velocity of electrons on the Fermi surface. Such effects are certainly observable in metals such as Bi.

On the other hand, the kinetic formula (7.33) is appropriate for a semiconductor, where n would be the number of free carriers. It is customary to write

$$\sigma = n\,|e|\,\mu, \tag{7.35}$$

where

$$\mu = \frac{|e|\,\tau}{m} \tag{7.36}$$

is the *mobility* of the carriers. More generally, one supposes that electrons and holes contribute separately to the current, so we write

$$\sigma = n_e\,|e|\,\mu_e + n_h\,|e|\,\mu_h \tag{7.37}$$

and define *electron mobility* and *hole mobility* separately.

It is not difficult to derive (7.35), say, from (7.20), using a classical distribution function like (4.32) for f^0, and evaluating integrals such as (4.36). One then assumes that τ can be a function of energy, and one finds for its average value, to be put in (7.36)

$$\bar{\tau} = \frac{2}{3kT}\int \mathscr{E}\tau(\mathscr{E})\,e^{-\mathscr{E}/kT}\mathcal{N}(\mathscr{E})\,d\mathscr{E} \Big/ \int e^{-\mathscr{E}/kT}\mathcal{N}(\mathscr{E})\,d\mathscr{E}, \tag{7.38}$$

where $\mathcal{N}(\mathscr{E})$ is the density of states in the particular band of carriers. Thus

$$\mu_e = \frac{|e|\,\bar{\tau}_e}{m_e}, \tag{7.39}$$

where m_e is the effective mass of the electrons—and similarly for holes. These formulae show that the mobility can be a function of temperature; as T increases, the distribution spreads out, and the average relaxation time changes. In a metal the variation of τ with energy makes little difference; only the value $\tau(\mathscr{E}_F)$ is important.

7.3 Calculation of relaxation time

We have still not solved the integral equation (7.16). The most general possible solution for this would be, in fact,

$$g_{\mathbf{k}} = \left(-\frac{\partial f^0}{\partial \mathscr{E}}\right) e\mathbf{E}\cdot\mathbf{\Lambda}(\mathbf{k}), \tag{7.40}$$

where $\mathbf{\Lambda}(\mathbf{k})$ would be a vector defined at each point $\mathbf{k}$ of the Fermi surface. Our elementary solution (7.19) is equivalent to writing

$$\mathbf{\Lambda}(\mathbf{k}) = \tau \mathbf{v}_{\mathbf{k}}, \tag{7.41}$$

showing that $\mathbf{\Lambda}$ is the *vector mean free path* of the electrons. In the general case we might have $\mathbf{\Lambda}(\mathbf{k})$ varying in magnitude, and deviating from the direction of $\mathbf{v}_{\mathbf{k}}$ over the Fermi surface.

It is sometimes assumed that we can write

$$\mathbf{\Lambda}(\mathbf{k}) = \tau(\mathbf{k})\, \mathbf{v}_{\mathbf{k}}, \tag{7.42}$$

where $\tau(\mathbf{k})$ is an *anisotropic relaxation time* varying over the Fermi surface. It is easy to show, however, that this is not always a complete solution of the integral equation—and there is no direct procedure for evaluating the function $\tau(\mathbf{k})$.

The only simple solution is, in fact, the elementary solution (7.19). Suppose we substitute this for $g_{\mathbf{k}}$ in (7.16), and also assume that we have elastic scattering, i.e.

$$Q(\mathbf{k}, \mathbf{k}')\, d\mathbf{k}' = \delta(\mathscr{E} - \mathscr{E}')\, \mathscr{Q}(\mathbf{k}, \mathbf{k}')\, d\Omega' dE, \tag{7.43}$$

where $d\Omega'$ is an element of solid angle for the direction of $\mathbf{k}'$ after scattering (the magnitude of $\mathbf{k}'$ being fixed now by the requirement that it have the same energy, $\mathscr{E}'$, as $\mathbf{k}$).

Eliminating the delta-function of energy on either side, we have

$$\mathbf{v}_{\mathbf{k}}\cdot\mathbf{E} = \tau \int (\mathbf{v}_{\mathbf{k}} - \mathbf{v}_{\mathbf{k}'})\cdot\mathbf{E}\, \mathscr{Q}(\mathbf{k}, \mathbf{k}')\, d\Omega' \tag{7.44}$$

to be evaluated over the Fermi surface. This is a functional relation which imposes conditions on the form of $\mathscr{Q}(\mathbf{k}, \mathbf{k}')$. It is easy to show that it holds when we have a spherical Fermi surface, with $|\mathbf{v}_{\mathbf{k}}|$ constant, and when

$$\mathscr{Q}(\mathbf{k}, \mathbf{k}') = \mathscr{Q}(\theta); \tag{7.45}$$

that is, the relation (7.44) can be satisfied when the scattering probability is a function only of the angle between the two wave-vectors. The proof is simply an exercise in spherical trigonometry which we leave to the reader.

If these conditions are satisfied, then we have at once

$$\frac{1}{\tau} = \int (1-\cos\theta)\,\mathcal{Q}(\theta)\,d\Omega'. \tag{7.46}$$

This shows that the relaxation time is inversely proportional to an integral of the scattering probability over all processes—but weighted, with the factor $(1-\cos\theta)$, in favour of large-angle scattering. This factor comes, as one sees in (7.44), from $(\mathbf{v}_{\mathbf{k}}-\mathbf{v}_{\mathbf{k}'})\cdot\mathbf{E}$; the important fact is not that the electron is scattered but the amount that the component of its velocity along the electric field is changed in the process.

We can express the scattering probability in terms of the differential scattering cross-section $\sigma(\theta)$ of a density N_i of impurities: the mean free path Λ is given by

$$\frac{1}{\Lambda} = N_i\,2\pi\int_0^\pi (1-\cos\theta)\,\sigma(\theta)\sin\theta\,d\theta. \tag{7.47}$$

7.4 Impurity scattering

We now have, in (6.70), (7.25) and (7.47), the material for a calculation, from first principles, of the electrical conductivity of a free electron metal containing charged impurities. We would normally express this as a *resistivity*

$$\begin{aligned}\rho &= \frac{mv_F}{ne^2\Lambda}\\ &= \frac{mv_F}{ne^2}N_i\,2\pi\int_0^\pi (1-\cos\theta)\left(\frac{2mZe^2}{\hbar^2}\right)^2\frac{\sin\theta\,d\theta}{(K^2+\lambda^2)^2}\\ &= \frac{mv_F}{ne^2}N_i\,2\pi\left(\frac{2mZe^2}{\hbar^2\lambda^2}\right)^2\int_0^1\frac{8z^3\,dz}{\{1+(2k_F/\lambda)^2z^2\}^2}.\end{aligned} \tag{7.48}$$

Here we have used the geometry of the scattering, i.e. $K = 2k_F\sin\frac{1}{2}\theta$, and put $z = \sin\frac{1}{2}\theta$ in the integrand.

The integral in (7.48) is easily evaluated as a function of $(2k_F/\lambda)$—we shall not write down the exact formula. The main point to notice is that a charged impurity behaves like a geometrical obstacle of radius

$$R \sim \left(\frac{2mZe^2}{\hbar^2\lambda^2}\right) \tag{7.49}$$

which is, from (5.22) and (3.6), comparable with the radius of an atomic sphere. We also notice that the resistance is independent of the temperature, and should go as the square of the valence difference Z (*Linde's rule*).

In fact, the resistance given by (7.48) is too large. It is better to use the partial wave formula (5.40) for the scattering cross-section. The integration over θ can be done; we get

$$\frac{1}{\Lambda} = N_i \frac{4\pi}{k_F^2} \sum_{l=1}^{\infty} l \sin^2(\eta_{l-1} - \eta_l). \tag{7.50}$$

The phase shifts can then be determined to satisfy the Friedel sum rule (5.43). A large *residual resistivity* must obviously arise from any transition metal impurity with a d-resonance near the Fermi level (§§ 5.5, 6.10), so that $\eta_2 \sim \pi/2$. In the case of magnetic impurities, the spin dependent s–d-interaction also produces a *resistance minimum* or *Kondo effect*, to be discussed in § 10.6.

To calculate the mobility associated with charged impurity scattering in a semiconductor we can use essentially the same argument as in (7.48). In effect the relaxation time is energy dependent:

$$\frac{1}{\tau(\mathscr{E})} = N_i v \,.\, 2\pi \left(\frac{Ze^2}{\epsilon \mathscr{E}}\right)^2 \int_0^1 \frac{\frac{1}{2}z^3\,dz}{\{z^2 + \lambda^2\hbar^2/8m^*\mathscr{E}\}^2}, \tag{7.51}$$

where λ—now given by (5.26)—is supposed to be rather smaller than the range of values of k allowed by thermal excitation.

If we evaluate the integral, we find that it varies only slowly with $\mathscr{E}$: τ depends on the energy of a carrier through the factor $\mathscr{E}^2/v \propto \mathscr{E}^{\frac{3}{2}} m^{*\frac{1}{2}}$. When we put this into (7.38) we find, in effect, that we replace $\mathscr{E}$ by its average value $\boldsymbol{k}T$. Thus, for impurity scattering,

$$\mu \propto T^{\frac{3}{2}} m^{*\frac{1}{2}}. \tag{7.52}$$

This sort of formula is really very crude, but the general qualitative argument that the mobility should increase with temperature is sound. What we are saying is that as the carrier gas gets hotter the average carrier goes faster, and is less easily scattered by a charged impurity.

7.5 'Ideal' resistance

The calculation of the so-called *ideal resistance* of metals—the resistance due to scattering by thermal vibrations of the lattice—is a more complicated problem. We can only sketch out the main lines of the argument.

As a first approximation, we can use the principle of the deformation potential introduced in § 6.14. An electron sees a fluctuating potential of magnitude

$$\delta\mathscr{E} = \tfrac{2}{3}\mathscr{E}_F \Delta, \tag{7.53}$$

where Δ is the local, thermally fluctuating dilatation. The resistance should therefore be proportional to the mean-square density fluctuation, which, by a familiar principle of classical statistical mechanics, is given by

$$\overline{\Delta^2} = NkT\beta, \tag{7.54}$$

where β is the compressiblity.

If we express this in terms of the velocity of sound, s, and the density, D, we have

$$\overline{\Delta^2} = \frac{NkT}{Ds^2} = \frac{kT}{Ms^2}, \tag{7.55}$$

where M is the mass of an ion. This result is more familiar in the general form

$$\rho_i \propto \frac{T}{M\Theta^2} \tag{7.56}$$

showing that the resistance is proportional to the absolute temperature. The variation with atomic mass, and with the Debye temperature, Θ (which is proportional to s, as in § 2.4) is in general agreement with experiment.

To improve on this, we might use (6.87), the effective differential cross-section of each ion in the lattice. We find, from (7.47), a mean free path

$$\frac{1}{\Lambda_i} = N2\pi \int_0^\pi \sigma_a(\theta)\, N\, |\mathbf{K}\cdot\mathbf{U}_\mathbf{q}|^2 (1-\cos\theta) \sin\theta \, d\theta, \tag{7.57}$$

where $\sigma_a(\theta)$ is the differential scattering cross-section of a 'free atom'.

Suppose we consider only N-processes, in which $\mathbf{K}$, the scattering vector, equals $\mathbf{q}$, the phonon wave-vector. In (2.109) and (2.110) the factor $|\mathbf{K}\cdot\mathbf{U}_\mathbf{q}|^2$ was calculated. At high temperatures

$$\begin{aligned} N\,|\mathbf{K}\cdot\mathbf{U}_\mathbf{q}|^2 &\approx \frac{kTK^2}{M\nu_q^2} \\ &\approx \frac{kT}{Ms^2} \\ &\approx \frac{\hbar^2 q_D^2 kT}{Mk^2\Theta^2} \end{aligned} \tag{7.58}$$

in terms of the Debye temperature. In fact, this is the same as $\overline{\Delta^2}$ in (7.54) and (7.55).

Since (7.58) is independent of **K**, i.e. independent of the scattering angle θ—we can perform this integration separately, and write

$$\frac{1}{\Lambda_i} = N\bar{\sigma}_a \frac{\hbar^2 q_D^2 kT}{Mk^2\Theta^2}, \tag{7.59}$$

where
$$\bar{\sigma}_a \equiv 2\pi \int \sigma_a(\theta)\,(1-\cos\theta)\sin\theta\, d\theta, \tag{7.60}$$

the total scattering cross-section of an isolated atom, weighted with the factor $(1-\cos\theta)$ for the change of momentum in the field direction.

This formula can be reduced even further by use of the Lindemann melting formula (2.117). We get

$$\frac{1}{\Lambda_i} \approx N\bar{\sigma}_a \cdot \tfrac{2}{3}x_m^2 \frac{T}{T_m}, \tag{7.61}$$

where x_m is the constant of the order of 0·2 in (2.117), and T_m is the melting temperature. We can go a little further, and assume that the scattering cross-section of an ion is much the same as for a charged impurity. A little juggling with (7.49) gives something like

$$\Lambda_i \sim 50a\frac{T_m}{T}, \tag{7.62}$$

where a^3 is the volume of a unit cell; the mean free path of an electron in a metal should be of the order of 50 times the lattice constant at the melting point. This is not exactly true—but as an order of magnitude estimate it is justifiable.

To get the resistance itself, we should combine (7.59) or (7.62) with (7.33), in the form

$$\rho_i = \frac{mv_F}{ne^2\Lambda_i}, \tag{7.63}$$

which is approximately true for a free-electron gas. These formulae put numbers into our crude functional relation (7.56).

But at low temperatures the situation is more complicated. The structure factor (7.58) was calculated assuming classical statistics. If we go to low temperatures we have to replace $(kT/\hbar\nu_q)$, say, by $\bar{n}_q$, as given by (2.46).† We get

$$N\,|\mathbf{K}\cdot\mathbf{U}_q|^2 \approx \frac{kT}{Ms^2}\left\{\frac{\hbar\nu_q/kT}{\exp(\hbar\nu_q/kT)-1}\right\}. \tag{7.64}$$

† The zero-point energy does not come into this formula; the proof of this requires a detailed analysis of the scattering, distinguishing between phonon emission and absorption processes.

This factor in the integral (7.57) has the effect of a cut-off in the angle variable. The scattering probability falls off rapidly when we need to absorb or emit phonons whose frequency is too high for them to be thermally excited, i.e. we practically exclude scattering for which

$$\hbar\nu_{\mathbf{q}} > kT. \tag{7.65}$$

For a Debye model, and for N-processes on a spherical Fermi surface, this occurs for values of q, and for angles of scatter θ, such that

$$\frac{q}{2k_F} = \sin\tfrac{1}{2}\theta > \frac{q_D}{2k_F}\frac{T}{\Theta}. \tag{7.66}$$

We can calculate the mean free path more or less exactly if we suppose that $\sigma_a(\theta)$ is not a strongly varying function of θ, so that it can be taken as a constant outside the integral. Then (7.57) yields

$$\frac{1}{\Lambda_i} = N\bar{\sigma}_a\frac{\hbar^2 q_D^2 kT}{Mk^2\Theta^2}\left(\frac{T}{\Theta}\right)^4\int_0^{\Theta/T}\frac{4z^4\,dz}{(e^z-1)}, \tag{7.67}$$

where we have written

$$z = \frac{\hbar\nu_{\mathbf{q}}}{kT} = \left(\frac{q}{q_D}\right)\left(\frac{\Theta}{T}\right). \tag{7.68}$$

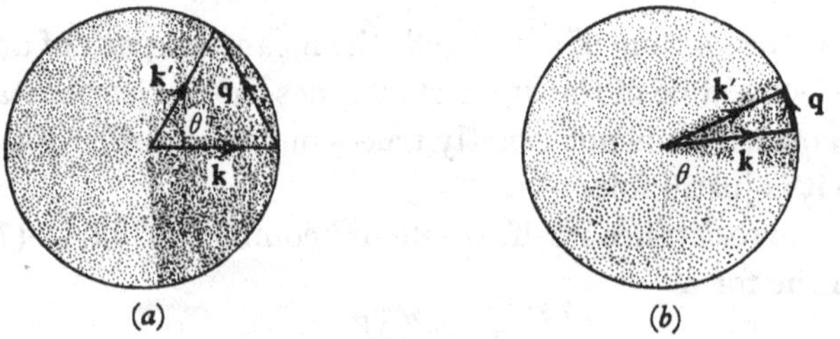

Fig. 123. Electron–phonon N-processes: (*a*) at high temperatures; (*b*) at low temperatures.

At high temperatures, where Θ/T is small, the integral tends to $(\Theta/T)^4$, and we are back at (7.59). At low temperatures the integral tends to a constant, 124·4, and we have a resistance proportional to T^5. This strong temperature dependence of the ideal resistance is a characteristic quantum effect, comparable with the Debye T^3-specific heat law.

We can see what is happening. The effective scattering cross-section, being an integral over angle, with a factor

$$(1-\cos\theta)\sin\theta\,d\theta = 8\sin^3\tfrac{1}{2}\theta\,d(\sin\tfrac{1}{2}\theta) \tag{7.69}$$

is very sensitive to the cut-off suggested by (7.64). If we take (7.66) as the cut-off angle, then we should find a factor proportional to

$$\int_0^{(q_D/2k_F)(T/\Theta)} 8\sin^3(\tfrac{1}{2}\theta)\,d(\sin\tfrac{1}{2}\theta) \propto \left(\frac{T}{\Theta}\right)^4, \tag{7.70}$$

which is just what appears in (7.67). The long-wave phonons are very ineffective as a source of resistance because they can only scatter the electrons through a small angle—an angle which decreases proportionately with the temperature.

The formula obtained by compounding (7.67) with (7.63) is known as the *Bloch–Grüneisen Law*. It is approximately obeyed by many metals at low temperatures—but it should not be taken too literally, for the following reasons:

(i) We have assumed elastic scattering of the electrons. This is not true—but if we make a formal analysis in terms of electron–phonon creation and annihilation processes and allow for the energy lost or gained by the electron, we get almost the same result. The integral in (7.67) becomes

$$\int_0^{\Theta/T} \frac{4z^5\,dz}{(e^z-1)(1-e^{-z})},$$

which has the same asymptotic behaviour as (7.67) at small and large values of T.

(ii) The assumption of a Debye spectrum is only a convenient oversimplification. The separate contributions of longitudinal and transverse modes should be considered.

(iii) The function $\sigma_a(\theta)$ is not a constant, and cannot, therefore, be taken out from under the integral in (7.57). A formula from which to calculate $\sigma_a(\theta)$ is given by (6.89) and (6.93). In particular, $\sigma_a(\theta)$ tends to decrease as θ increases beyond, say, 20–30°—this has its effect in the temperature variation of the resistance.

(iv) The most serious error is in the neglect of U-processes. As shown in § 2.7, if $\mathbf{K}$ goes beyond the boundary of a Brillouin zone, then we should write

$$\mathbf{q} = \mathbf{K} - \mathbf{g}, \tag{7.71}$$

where $\mathbf{g}$ is a reciprocal lattice vector. In our formula (7.57) there is no special reason why such values of $\mathbf{K}$ should not be attained; the scattering cross-section of an atom, as a function of angle, goes round to $\theta = \pi$, which would make $K = 2k_F$, which is greater than the Debye limit q_D.

The difficulty about U-processes in electrical resistance is simply that they give rise to complicated geometrical problems. Even at high temperatures, if we put (7.71) into (7.58), we find that the 'structure factor' is no longer independent of the angle of scatter: it contains an extra factor K^2/q^2 which can be much larger than unity. We can see this in Fig. 124 (*a*), where K is evidently much larger than q. The differential scattering cross-section $\sigma_a(\theta)$ may be small for such a large value of θ—but the contribution to the resistance is heavily weighted by the factor $(1 - \cos\theta)$.

The small value of q required for some U-processes is also important when we consider the temperature variation of the resistance. As shown by (7.66), these processes are not 'frozen out' until one goes to a low temperature. This can be seen more clearly by redrawing the vector triangles of Fig. 124 (*a*) in the repeated zone scheme. The minimum value of q for an Umklapp process is now, obviously, the minimum distance between the Fermi surface in one zone and its repetition in the next.

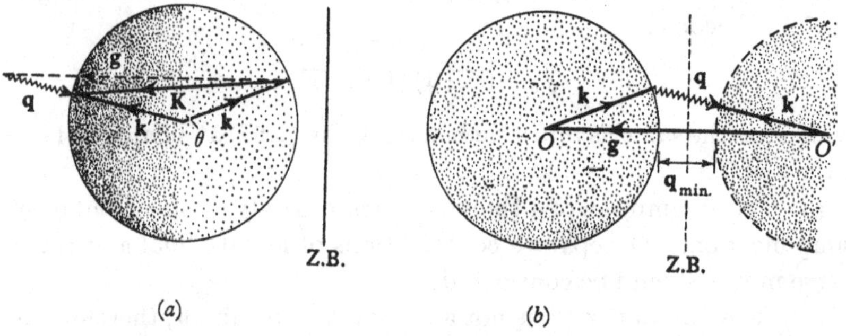

Fig. 124. Electron–phonon U-processes: (*a*) on free-electron sphere; (*b*) in repeated zone scheme.

If the Fermi surface bulges towards the zone boundary, then the minimum value of q decreases—so that contributions to (7.57) with a factor K^2/q^2 will be enhanced. Indeed, it almost looks as though we should get a singularity in ρ_i if the Fermi surface were to touch the zone boundary—but then, as pointed out in § 6.13, for such processes the matrix element of the electron–phonon interaction automatically vanishes.

There is no simple formula for the contribution of U-processes to the electrical resistance. However, the same general factors that give (7.56) and (7.67) are still at work—even though the latter expression has no validity as a precise theoretical result. We can understand all the physical features of the phenomenon, but must rely on tedious numerical computations to get an answer. To mention one point: according to (6.93), the effective cross-section of an atom for large values of K, i.e. for large θ, depends on the magnitude of $\mathscr{V}'$, the 'core pseudo-potential' of a pseudo-atom. This should be reflected in the magnitude of the resistance—but this may be as sensitive to $\mathscr{V}'$ through the way it distorts the Fermi surface, and mixes wave-functions near zone boundaries as through its direct contribution to the matrix element for scattering.

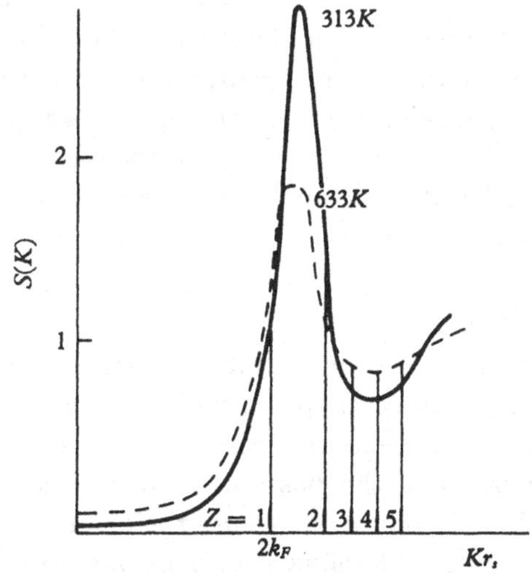

Fig. 125. Typical structure factor of liquid metal, showing $2k_F$ for various valencies, and temperature dependence.

The analogous calculation for a *liquid* metal is actually much easier. It is reasonable to assume an N.F.E. model, where (6.88), (6.91), (7.25) and (7.47) lead us to the formula

$$\rho_L = \frac{3\pi}{4}\frac{1}{\hbar^2 e^2 v_F^2 k_F^4}\frac{1}{N}\int_0^{2k_F} |w_s(K)|^2 S(K)\, K^3 dK. \qquad (7.72)$$

In this expression the active ingredients are the pseudo-atom form factor $|w_s(K)|^2$, and the liquid structure factor, $S(K)$, which may be determined experimentally by neutron or X-ray diffraction. Indeed, the atoms in most liquid metals are arranged so similarly that after scaling for the atomic radius, $S(K)$ can be represented approximately by a single standard curve (Fig. 125). At small K, the structure factor is small because the atomic distribution is nearly uniform in density—here, in fact, the classical formula (7.54) still holds. But the first co-ordination shell of neighbours in the radial distribution function $P(R)$ (cf. (2.103)) gives rise to a peak in $S(K)$, beyond which it oscillates about unity. On this diagram the diameter of the Fermi sphere, $2k_F$, depends on the valence, Z, so that the upper limit of integration in (7.72) does not reach the peak in $S(K)$ in the alkali metals, but falls into the trough beyond for $Z = 2$.

Since there is no long range order to produce Brillouin zones, etc., distinctions between N-processes and U-processes are meaningless. Nevertheless, the geometrical factor K^3 weights the integrand heavily towards values near $2k_F$, so that the 'core' pseudo-potential $\mathscr{V}'$ is once again the predominant effect. The increase of resistance when a metal melts may be ascribed to the change in the structure factor for large K—a change from electron–phonon U-processes, which are still proportional to T, to scattering from around the peak of $S(K)$, which is already of the order of unity and which may actually decrease at higher temperatures.

7.6 Carrier mobility

The scattering of carriers by lattice vibrations in a semiconductor is, in general principle, a much simpler problem. Because the carriers are usually thought of as concentrated in a small region of **k**-space, near a minimum in $\mathscr{E}(\mathbf{k})$, the possible change of **k**-vector in the scattering is small. This allows one to use the deformation potential of § 6.13. The scattering probability is proportional to

$$|\mathscr{M}_{\mathbf{k}'\mathbf{k}}|^2 \mathscr{N}(\mathscr{E}) = |\sum_{ij} \mathsf{W}^0_{ij} \mathscr{E}_{ij}|^2 \mathscr{N}(\mathscr{E}), \tag{7.73}$$

where W^0_{ij} is the amplitude of a component of the strain in an elastic wave, and $\mathscr{E}_{ij}$ is a component of the deformation-potential tensor.

Because $\mathbf{k}' - \mathbf{k}$ is small, we can use classical statistics for the elastic modes: as in (7.54)

$$\overline{|\mathsf{W}^0_{ij}|^2} \propto kT. \tag{7.74}$$

Thus, the relaxation time, which is proportional to the inverse of (7.73), would behave like

$$\frac{1}{\tau(\mathscr{E})} \propto T\mathscr{E}^{\frac{1}{2}} m^{*\frac{3}{2}}, \tag{7.75}$$

because the density of states is a function of energy and of 'effective mass' (cf. (4.33)). Putting this into (7.38) and (7.39), we have

$$\mu \propto T^{-\frac{3}{2}} m^{*-\frac{5}{2}}. \tag{7.76}$$

The premium on small effective mass is very high.

However, this theory is also too simple, even when worked up with matrix elements, etc. Many semiconductors have several different minima for electrons—*many valleys*, as they say (§ 6.6)—and scattering between the valleys, with phonons of large wave-vector, is important. The relation (7.76) is not very satisfactory in practice.

In a semiconductor, the change of carrier energy when a phonon is emitted or absorbed cannot be neglected. The effect on carrier mobility at low fields is not large, but this is the dominant mechanism in *high-field* transport phenomena. Suppose, for example, that an E.M.F. of a few volts is applied across a thin conducting specimen. A carrier travelling between the electrodes must give up this amount of energy as Joule heat—in practice by the emission of a number of phonons. In metals the mean free path between electron–phonon processes is so short (cf. (7.62)) that the specimen itself would be vaporized before this mechanism broke down. In a good semiconductor, however, quite a modest electric field (e.g. 1000 V/cm) can overload the energy transfer processes, so that the electron energy distribution may no longer be in equilibrium at the lattice temperature, as assumed in (7.38). The ensuing *hot electron* phenomena are quite complicated in theory, since they depend in detail on the electron–phonon matrix elements, the spectrum of acoustic and optical modes, and the electronic band structure of the solid. The general effect, however, is departure from Ohm's law towards a condition in which the carrier current is nearly independent of E.

This discussion of electrical conduction also assumes that crystal momentum $\mathbf{k}$ is a good quantum number for electron states. In certain circumstances, as in glasses, amorphous semiconductors, or compensated impurity bands (§§ 4.2, 5.9, 6.7), the carrier may best be described as *localized* at various points in the material. Conduction can then occur only by thermally activated *hopping* from place to place with the emission or absorption of a suitable phonon to permit the transition.

7.7 General transport coefficients

Suppose now that we have a temperature gradient in the specimen, as well as an electric field. Ignoring size and shape effects, the Boltzmann equation (7.14) reads

$$\left(-\frac{\partial f^0}{\partial \mathscr{E}}\right)\mathbf{v}_\mathbf{k}\cdot\left\{\frac{\mathscr{E}(\mathbf{k})-\zeta}{T}(-\nabla T)+e\left(\mathbf{E}-\frac{1}{e}\nabla\zeta\right)\right\} = -\frac{\partial f_\mathbf{k}}{\partial t}\bigg]_{\text{scatt.}} \qquad (7.77)$$

All that we have done is to replace $\mathbf{E}$, in (7.16), by a more complicated vector function of $\mathbf{k}$. If the conditions for the existence of a relaxation time still hold, as discussed in § 7.3, then we know the form of the solution:

$$f_{\mathbf{k}}-f_{\mathbf{k}}^{0}=\left(-\frac{\partial f^{0}}{\partial \mathscr{E}}\right)\tau \mathbf{v}_{\mathbf{k}}\cdot\left[e\left(\mathbf{E}-\frac{1}{e}\nabla\zeta\right)+\frac{\mathscr{E}(\mathbf{k})-\zeta}{T}(-\nabla T)\right]. \quad (7.78)$$

We can put this into an integral like (7.20) to calculate the electric current,

$$\mathbf{J}=\frac{1}{4\pi^{3}}\frac{e^{2}\tau}{\hbar}\iint \mathbf{v}_{\mathbf{k}}\mathbf{v}_{\mathbf{k}}\left(-\frac{\partial f^{0}}{\partial \mathscr{E}}\right)\frac{dS}{v_{\mathbf{k}}}d\mathscr{E}\cdot\left(\mathbf{E}-\frac{1}{e}\nabla\zeta\right)$$

$$+\frac{1}{4\pi^{3}}\frac{e^{2}\tau}{\hbar}\iint \mathbf{v}_{\mathbf{k}}\mathbf{v}_{\mathbf{k}}\left(\frac{\mathscr{E}-\zeta}{T}\right)\left(-\frac{\partial f^{0}}{\partial \mathscr{E}}\right)\frac{dS}{v_{\mathbf{k}}}d\mathscr{E}\cdot(-\nabla T). \quad (7.79)$$

This shows that the thermal gradient, ∇T, acting alone, would give rise to an electric current—evidence of a *thermo-electric effect*.

But the most important effect of a gradient of temperature is to give rise to a current of *heat*. We might think of this as a flux of energy,

$$2\int f_{\mathbf{k}}\mathscr{E}(\mathbf{k})\,\mathbf{v}_{\mathbf{k}}\,d\mathbf{k}.$$

This would not be correct. 'Heat' is 'internal energy' minus 'free energy'. The free energy of an electron is ζ, its chemical potential. The heat carried by the electron in state $\mathbf{k}$ is therefore $\mathscr{E}(\mathbf{k})-\zeta$. A more rigorous analysis confirms that the total flux of heat (per unit volume) must be calculated from

$$\mathbf{U}=2\int f_{\mathbf{k}}\{\mathscr{E}(\mathbf{k})-\zeta\}\,\mathbf{v}_{\mathbf{k}}\,d\mathbf{k}. \quad (7.80)$$

Before writing down an expression for $\mathbf{U}$, we may also note that there is a term in $\nabla\zeta$ in (7.78)—a term arising from the change of chemical potential caused by changing the temperature of the electron gas (cf. (4.23)). But in an ordinary measurement of electrical properties we should include any such gradients of ζ amongst the measured E.M.F.'s in the circuit. We shall just drop this term, on the assumption that any of its effects are subsumed in $\mathbf{E}$, the 'observed' electric field.

From (7.78), (7.79) and (7.80), we can construct the following general transport equations

$$\mathbf{J}=e^{2}\mathbf{K}_{0}\cdot\mathbf{E}+\frac{e}{T}\mathbf{K}_{1}\cdot(-\nabla T), \quad (7.81a)$$

$$\mathbf{U}=e\mathbf{K}_{1}\cdot\mathbf{E}+\frac{1}{T}\mathbf{K}_{2}\cdot(-\nabla T), \quad (7.81b)$$

where the coefficients (actually tensors, but this is only for the sake of abstract generality) are defined by

$$\mathbf{K}_n \equiv \frac{1}{4\pi^3}\frac{\tau}{\hbar}\iint \mathbf{vv}(\mathscr{E}-\zeta)^n\left(-\frac{\partial f^0}{\partial \mathscr{E}}\right)\frac{dS}{v}\,d\mathscr{E}. \tag{7.82}$$

These coefficients can be evaluated by means of our general theorem (4.21) for integrals of the Fermi function over energy;

$$\int \Phi(\mathscr{E})\left(-\frac{\partial f^0}{\partial \mathscr{E}}\right)d\mathscr{E} = \Phi(\zeta)+\tfrac{1}{6}\pi^2(kT)^2\left[\frac{\partial^2\Phi(\mathscr{E})}{\partial \mathscr{E}^2}\right]_{\mathscr{E}=\zeta}+\dots, \tag{7.83}$$

where now $\Phi(\mathscr{E})$ is itself an integral over the energy surface $\mathscr{E}(\mathbf{k}) = \mathscr{E}$. But that presents no special problems.

Just as in (7.21), we have

$$\mathbf{K}_0 = \frac{1}{4\pi^3}\frac{\tau}{\hbar}\int \mathbf{vv}\frac{dS_F}{v}. \tag{7.84}$$

Now imagine $\mathbf{K}_0(\mathscr{E})$ to be the value of (7.84) over some surface of energy $\mathscr{E}$, not at the Fermi level. In the case of $\mathbf{K}_2$ we have to consider $\Phi(\mathscr{E}) = (\mathscr{E}-\zeta)^2\,\mathbf{K}_0(\mathscr{E})$, which vanishes at $\mathscr{E} = \zeta$, and which leaves only

$$\mathbf{K}_2 = \tfrac{1}{3}\pi^2(kT)^2\,\mathbf{K}_0(\zeta) \tag{7.85}$$

in the second term of (7.83). The expression for $\mathbf{K}_1$, however, reduces to

$$\mathbf{K}_1 = \tfrac{1}{3}\pi^2(kT)^2\left[\frac{\partial}{\partial \mathscr{E}}\mathbf{K}_0(\mathscr{E})\right]_{\mathscr{E}=\zeta}, \tag{7.86}$$

which is somewhat more complicated in principle.

The coefficient of $\mathbf{E}$ in (7.81 a) is, of course, the electrical conductivity —the current we should measure under isothermal conditions ($\nabla T = 0$). We merely confirm (7.23); we write

$$\boldsymbol{\sigma} = e^2\mathbf{K}_0, \tag{7.87}$$

and equivalent relations when the tensor, having cubic symmetry, reduces to a scalar. We actually use (7.87) to express $\mathbf{K}_0$ in terms of the known electrical conductivity.

7.8 Thermal conductivity

The coefficient of the temperature gradient in (7.81 b), the equation for the heat current, is what we should call the *thermal conductivity* of our electron gas.

Actually, this is not quite correct. We do not do the experiment in such a way as to keep $\mathbf{E} = 0$. It is easier to put the specimen on open electrical circuit, so that $\mathbf{J} = 0$. From (7.81a), this means that there will be an electric field

$$\mathbf{E} = \frac{1}{e^2}\mathbf{K}_0^{-1}\mathbf{K}_1\frac{e}{T}\cdot\nabla T \tag{7.88}$$

set up along the wire. Putting this into (7.81b) we get

$$\mathbf{U} = \frac{1}{T}\mathbf{K}_1\mathbf{K}_0^{-1}\mathbf{K}_1\cdot\nabla T - \frac{1}{T}\mathbf{K}_2\cdot\nabla T$$

$$= \frac{1}{T}(\mathbf{K}_2 - \mathbf{K}_1\mathbf{K}_0^{-1}\mathbf{K}_1)\cdot(-\nabla T), \tag{7.89}$$

which we should equate with

$$\mathbf{U} = \boldsymbol{\kappa}\cdot(-\nabla T) \tag{7.90}$$

for the thermal conductivity $\boldsymbol{\kappa}$. But the correction is negligible for a metal, so that

$$\boldsymbol{\kappa} = \frac{1}{T}\mathbf{K}_2. \tag{7.91}$$

But, according to (7.85) and (7.87), this can be written

$$\boldsymbol{\kappa} = \frac{\pi^2}{3}\frac{k^2}{e^2}T\boldsymbol{\sigma}, \tag{7.92}$$

a relation known as the *Wiedemann–Franz Law*. It is an old result and very easy to understand. In electrical conduction each electron carries its charge e, and is acted on by the field $e\mathbf{E}$. The current per unit field is proportional to e^2. In thermal conduction each electron carries its thermal energy kT, and is acted on by a thermal force, $k\nabla T$. The heat current per unit thermal gradient is proportional to k^2T. The ratio of these two transport coefficients must be of the order of k^2T/e^2; the factor $\frac{1}{3}\pi^2$ comes from the fact that we are dealing only with the electrons on the Fermi surface, obeying Fermi statistics.

It is easy to show that this law is very general. It would hold if (7.40) were true, that is, if the effect of scattering could only be defined in terms of a vector mean free path varying over the Fermi surface. But it requires that the scattering be *elastic*.

To prove this, let us look at the form of the actual distribution function under thermal conduction conditions. From (7.78)

$$f_{\mathbf{k}} - f_{\mathbf{k}}^0 = \left(-\frac{\partial f^0}{\partial \mathscr{E}}\right)\left(\frac{\mathscr{E}-\zeta}{T}\right)\tau\mathbf{v}_{\mathbf{k}}\cdot(-\nabla T). \tag{7.93}$$

Let us look at a cross-section of the Fermi distribution. As shown in Fig. 126, the effect of adding an expression like (7.93) to f^0 is to spread the distribution out on the side where $\mathbf{v_k}\cdot(-\nabla T)$ is positive, and sharpen it on the other side.

We can express this analytically;

$$f_{\mathbf{k}} = f^0_{\mathbf{k}} - (\tau \mathbf{v_k}\cdot\nabla T)\frac{\partial f^0}{\partial T}$$

$$= f^0(T - \tau \mathbf{v_k}\cdot\nabla T). \tag{7.94}$$

The electrons going in the direction where ∇T is negative, i.e. going down the temperature gradient, are 'hotter' by an amount

$$\delta T = -\tau \mathbf{v_k}\cdot\nabla T. \tag{7.95}$$

Those going in the opposite direction are 'colder' than the average temperature of the electron gas.

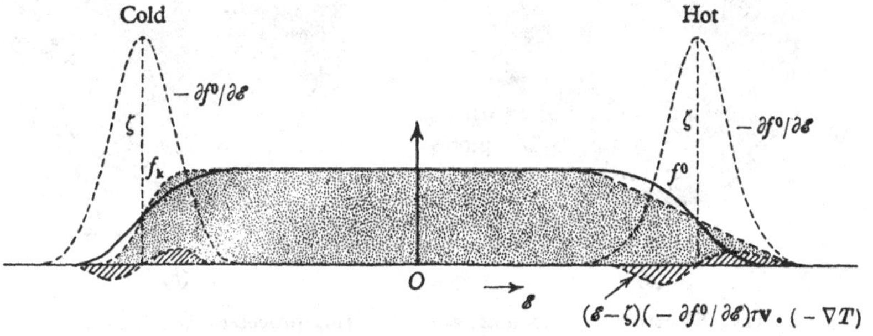

Fig. 126. Fermi distribution in thermal conductivity.

Indeed, this is just what we should argue on kinetic grounds. Electrons arriving in the region of temperature T from the direction $\mathbf{v}$ will have come a distance $\tau\mathbf{v}$. The region they have left—the region where they last suffered a thermalizing collision—will be at a temperature $T+\delta T$. These electrons will thus be 'hotter' by the amount δT.

In a thermal conduction experiment there is thus no net flux of electrons—no net flux of charge. We have a heat current because we have 'hot' electrons travelling one way, and an equal number of 'cold' electrons travelling in the opposite direction. This is sketched in Fig. 127 (a). There are electrons excited above the Fermi level on the right; there are electrons condensed into the Fermi surface on the left.

Now consider ways in which this distribution may relax by scattering. We can scatter a hot electron right round the Fermi sphere reversing its direction. We call this a *horizontal* process. This is the sort of scattering that is also effective in reducing electrical conductivity, as indicated in Fig. 127(*b*). Such processes are *elastic*—and for them the Wiedemann–Franz Law is valid.

But there are also *vertical* transitions, in which a 'hot' electron loses all its extra energy, and falls down below the Fermi level. Such a process has little effect on electrical conductivity—but in thermal conductivity it is just as important in reducing the heat current as scattering through large angles.

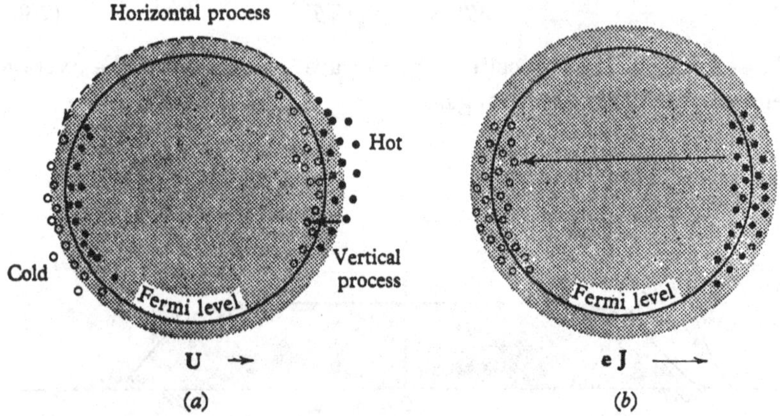

Fig. 127. Electron distributions, and scattering processes: (*a*) in thermal conduction; (*b*) in electrical conduction.

Vertical transitions are, by definition, *inelastic*. They do not occur for impurity scattering, which gives rise to a thermal resistance that obeys the Wiedemann–Franz Law. In other words, since ρ is constant, $\kappa \propto T$. They are also not very important in phonon scattering at high temperatures, where the maximum energy that an electron could gain or lose would be the maximum energy of a phonon, i.e. $k\Theta$. This would be less than the spread, kT, of the Fermi distribution about the Fermi level; it turns out that the Wiedemann–Franz Law also holds for the 'ideal' resistance at the limit of high temperatures. Since ρ_i is proportional to T, we find that κ_i should tend to a constant.

But at low temperatures the average gain or loss of energy by the electron is the average energy of a phonon at that temperature, i.e. it must be of the order of kT. This is just enough to take the electron through the thermal layer—just enough to make a 'hot' electron 'cold'.

The derivation of a formula for this case obviously requires a careful studv of the separate processes in which phonons are created or absorbed. But we can see roughly what happens if we assume that all scattering processes are now equally effective in reducing the heat current. That is, when calculating a 'relaxation time', as in (7.46), we now omit the factor $(1-\cos\theta)$ which allowed for the change in the component of the velocity of the electron along the field direction.

But this factor $(1-\cos\theta)$, at low temperatures, contributed $2\sin^2\frac{1}{2}\theta$ in the integral (7.70). Thus, we should drop a factor $(T/\Theta)^2$ in our 'cut-off' approximation. In a calculation like (7.67), the 'relaxation time for thermal conductivity' would come out as

$$\frac{1}{\tau}\propto\frac{T^3}{M\Theta^4} \tag{7.96}$$

instead of the T^5 law found for electrical conductivity. Putting this into (7.91), (7.85) and (7.84) (i.e. taking in the Wiedemann–Franz factor, proportional to T, which must always be used in comparing a thermal with an electrical conductivity), we find

$$\kappa_i\propto\frac{M\Theta^4}{T^2}. \tag{7.97}$$

This is actually observed—below the Debye temperature, electron–phonon processes produce a *Lorenz ratio* of the form

$$\frac{\kappa_i}{\sigma_i T}\propto\left(\frac{T}{\Theta}\right)^2 \tag{7.98}$$

instead of the constant value $(\frac{1}{3}\pi^2)(\boldsymbol{k}/e)^2$ demanded by the Wiedemann–Franz Law (7.92). But the ratio returns to the *Lorenz number* (7.92) at the very lowest temperatures, where the residual resistance (§ 7.4) is due to elastic scattering by impurities.

Of course, one must allow for the contribution of U-processes to thermal resistivity, but this again is a complicated problem of computation.

7.9 Thermo-electric effects

The general relations (7.81) show that there are interaction effects between electric currents and thermal currents. For example, suppose we establish a temperature gradient in a specimen that is on open circuit. Putting $\mathbf{J}=0$ in (7.81a), we have (slipping into scalar coefficients)

$$\begin{aligned}\mathbf{E}&=\frac{1}{eT}(\mathsf{K}_0^{-1}\mathsf{K}_1)\nabla T\\&=Q\nabla T;\end{aligned} \tag{7.99}$$

an E.M.F. will be observed.

To show this effect, and to measure it, one has to establish a closed circuit of two metals A and B, with junctions at different temperatures T_1 and T_2, and with a voltmeter interposed at some intermediate point, temperature T_0. The E.M.F. round the circuit is defined by the integral of $\mathbf{E}$ along the length of the wire

$$\phi = \int_0^1 E_B\,dx + \int_1^2 E_A\,dx + \int_2^0 E_B\,dx$$

$$= \int_2^1 Q_B \frac{\partial T}{\partial x}\,dx + \int_1^2 Q_A \frac{\partial T}{\partial x}\,dx$$

$$= \int_{T_1}^{T_2} (Q_A - Q_B)\,dT. \tag{7.100}$$

The voltage generated in the circuit looks like a function of the difference in temperature of the two junctions, and of the difference in the *absolute thermo-electric power*, Q, of the two metals. This is the well-known *Seebeck effect*.

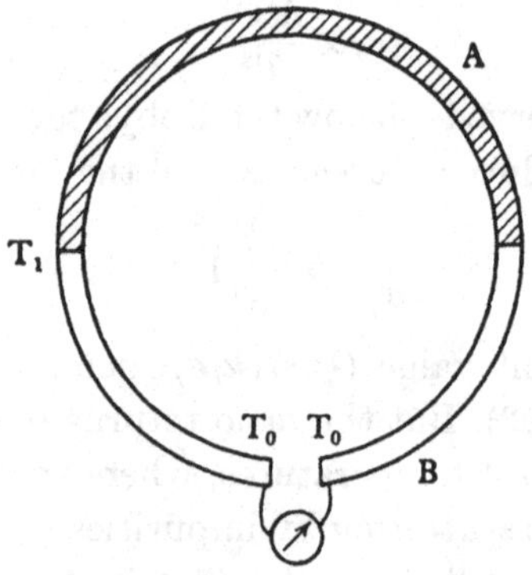

Fig. 128. The Seebeck effect.

But there is another phenomenon, associated with the quantity K_1 that appears in (7.81). Suppose we keep $\nabla T = 0$ round a circuit as in Fig. 129. By (7.81 a), we have

$$\mathbf{U} = e\mathsf{K}_1\mathbf{E}, \quad \mathbf{J} = e^2\mathsf{K}_0\mathbf{E}, \tag{7.101}$$

so that

$$\mathbf{U} = \frac{1}{e}\mathsf{K}_0^{-1}\mathsf{K}_1\mathbf{J}$$

$$= \Pi\mathbf{J}. \tag{7.102}$$

Now drive a current J round the circuit by means of a battery. In the branch A, there will be a heat current $\Pi_A J$. In branch B there will be a different heat current $\Pi_B J$. At the junctions, the balance of heat must be restored; the heat flux $(\Pi_A - \Pi_B)J$ will be emitted at one junction and absorbed at the other. One junction will become warmer, the other colder. This is the *Peltier effect.*

As we see from (7.99) and (7.102), the *Peltier coefficient* is related to the absolute thermo-electric power:

$$\Pi = QT. \tag{7.103}$$

This is no accident. It is one of the *Kelvin relations* of thermo-electricity, and a special case of the *Onsager relations* of the thermodynamics of irreversible processes. In effect, the equations (7.81), if expressed in terms of **J**, and of $\mathbf{U}/T$, must have a symmetric matrix—which explains why K_1 appears twice as a coefficient. There is another

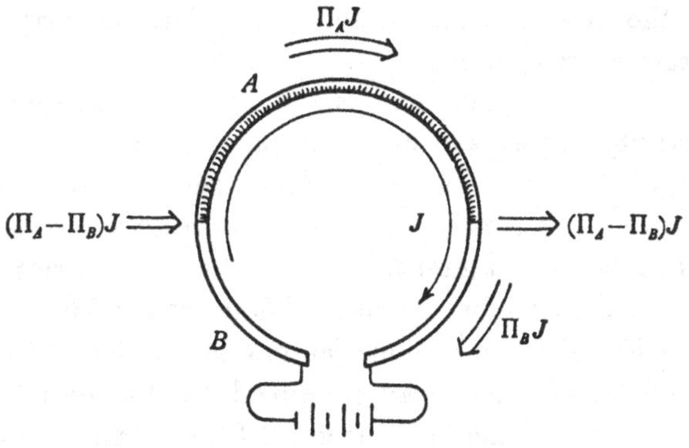

Fig. 129. The Peltier effect.

thermo-electric effect—the *Thomson effect,* which is observed in a circuit bearing both thermal and electric currents—but the coefficient of this is contained already in the general relations (7.81).

For the actual value of the thermo-electric power we refer to (7.86) and (7.87):

$$Q = \frac{\pi^2}{3}\frac{k}{e}kT\left[\frac{\partial \ln \sigma(\mathscr{E})}{\partial \mathscr{E}}\right]_{\mathscr{E}=\zeta}, \tag{7.104}$$

where $\sigma(\mathscr{E})$ means 'the value of the conductivity for a hypothetical metal in which the Fermi level is at $\mathscr{E}$'. It is easy enough in principle (though often complicated and unsuccessful in practice) to apply this formula to the results of calculations of electrical conductivity under various circumstances.

For example, suppose we write

$$\sigma(\mathscr{E}) = n(\mathscr{E})\, e\mu(\mathscr{E}) \tag{7.105}$$

in terms of a density of carriers which is a function of energy, and an effective carrier mobility. We then have

$$Q = \frac{\pi^2}{3}\frac{k}{e}kT\left[\frac{\mathcal{N}(\mathcal{E})}{n}+\frac{\partial \ln \mu(\mathcal{E})}{\partial \mathcal{E}}\right]_{\mathcal{E}=\zeta}. \tag{7.106}$$

The first term is easy to understand. As we saw in § 4.7, the 'specific heat per electron' is

$$C_{\text{el.}} = \frac{\pi^2}{3}k^2T\frac{\mathcal{N}(\mathcal{E}_F)}{n}. \tag{7.107}$$

Thus, the Peltier coefficient, QT, is the ratio of the 'heat per electron' to the electronic charge—exactly as suggested by (7.100).

The derivative of the mobility comes in because we must allow for the way in which the electron current is distributed in energy. Thus, if $\mu(\mathcal{E})$ increases with energy, a high proportion of the current is being carried by the more energetic electrons, and these carry a proportionally larger current of heat.

For a metal, of course, (7.105) is not legitimate; one should express the conductivity in terms of Fermi surface area and mean free path as in (7.25), or density of states, Fermi current and relaxation time, as in (7.34). For a free-electron system the results will be the same either way but the correct interpretation of thermo-electric phenomena in a metal or alloy calls for care in the choice of a model in which the energy variation of these various factors can be correctly assessed. The *giant thermo-electric power* produced by magnetic impurities (§ 10.6) is an extreme effect of this kind. It is worth noting that the charge e is conventionally negative, so that Q ought to be negative—especially since $\mu(\mathcal{E})$ tends to increase with $\mathcal{E}$; faster electrons are scattered less easily.

In a semiconductor this type of formula is not satisfactory. We must go back to the definition of the various coefficients (7.82), and can no longer use the properties of the Fermi function as in (7.83). In particular, the integral for K_1 is of the form

$$\int (\mathcal{E}-\zeta) f_{\mathbf{k}} v_{\mathbf{k}}\, d\mathbf{k}, \tag{7.108}$$

and this will contain the 'heat' corresponding to ζ_e in § 4.6—the energy of a carrier measured, not from the nearest band edge, but from the true Fermi level. Thus, there is a contribution to the thermo-electric power of the form

$$Q_e \sim \frac{1}{T}\frac{\zeta_e}{e} \tag{7.109}$$

as well as a term, usually smaller, coming from the behaviour of $\mu(\mathcal{E})$.

If our semiconductor contains mainly 'holes', then there ought to be a similar result, with ζ_h playing the role of the 'heat per carrier', but now with the charge positive. When both types of carrier are present, we get an average thermo-electric power, balanced between the negative contribution of electrons and the positive contribution of holes, each weighted with its share of the total current.

It is amusing to consider the two alternative procedures that we might use, in the case of a metal, for the consideration of a 'sphere of holes'. As in § 6.6 we could think of it as an electron Fermi surface with e negative but having the peculiar property that the area, velocity, and relaxation time of the electrons are all *decreasing* as the energy increases, so that, in (7.104), $\partial \ln \sigma(\mathscr{E})/\partial \mathscr{E}$ is also negative, and Q is positive. It is more natural to think of a sphere of 'holes', whose number and mobility *increase* as their 'energy' increases outwards from the centre, but which carry positive charge. The result is the same either way.

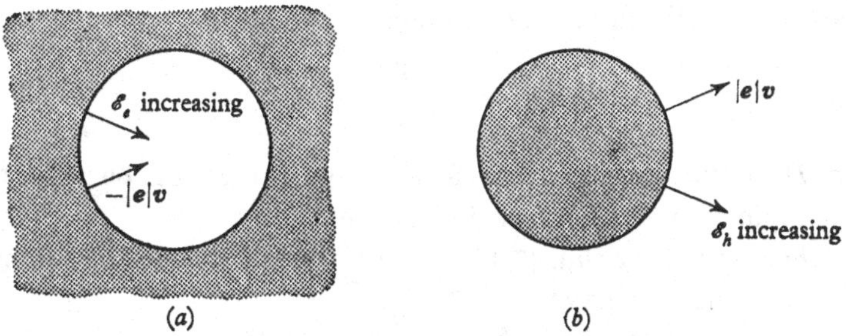

Fig. 130. Alternative models for calculating thermo-electric power: (a) electron Fermi surface; (b) 'hole' Fermi surface.

7.10 Lattice conduction

In electrical insulators and semiconductors there may still be good conduction of heat via the lattice waves. This is a subject of great complexity which we shall only study superficially.

Suppose we think of the phonons as if they were a gas of particles. In elementary kinetic theory it is shown that the thermal conductivity is given by

$$\kappa = \tfrac{1}{3}Cv\Lambda, \tag{7.110}$$

where C is the specific heat, v the velocity, and Λ the mean free path of the carriers.

We may suppose that at high temperatures the phonons have their usual specific heat, $3Nk$, as in (2.60). We may also suppose that they travel with the velocity of sound, s. But to calculate Λ we need to make certain crude approximations. Of these, the simplest is akin to the argument used in § 7.5 to estimate the mean free path of electrons.

Let us recall that a dilatation, Δ, of the lattice changes the local velocity of sound by a fraction

$$\frac{\delta s}{s} = \gamma\Delta, \tag{7.111}$$

where γ is Grüneisen's constant. This was defined in (2.120), in discussion of the theory of thermal expansion.

Local thermal fluctuations of density will cause scattering proportional to the mean square of this quantity. A straightforward comparison with (7.59) suggests

$$\frac{1}{\Lambda} \sim \frac{\gamma^2\overline{\Delta^2}}{a}, \tag{7.112}$$

i.e.

$$\Lambda \sim \frac{Ds^2}{NkT}\frac{a}{\gamma^2}, \tag{7.113}$$

where a is a length of the order of the lattice constant. Thus, with (7.110), and neglecting numerical factors, we have

$$\kappa \sim \frac{Ds^3}{\gamma^2 T}a, \tag{7.114}$$

where D is the density. This formula is interesting because only macroscopic quantities appear in it, except for a.

Another way of writing (7.113) is in terms of the melting temperature: as in (7.62), we have

$$\Lambda \sim \frac{20}{\gamma^2}\frac{T_m}{T}a. \tag{7.115}$$

These formulae are reasonable for functional behaviour and order of magnitude; the dependence of lattice conductivity on the inverse of the absolute temperature is well understood, even if the other factors are very rough.

Nevertheless, this is only valid at high temperatures, above the Debye temperature, and the situation is not nearly as simple as it seems. The fact is that the 'thermal fluctuations of density' are not essentially distinct from the 'phonons carrying the current of heat', and the whole system should be analysed in terms of phonon–phonon interactions as in § 2.10.

We then find a most peculiar result. Suppose we have only phonon–phonon N-processes in which crystal momentum is conserved. Suppose we set up a state with $n_{\mathbf{q}}$ phonons in the mode $\mathbf{q}$—a state with a net crystal momentum

$$\mathbf{P} = \sum_{\mathbf{q}} n_{\mathbf{q}} \hbar\mathbf{q}. \tag{7.116}$$

Then the phonon–phonon N-processes do not change the value of **P**. Thus any net heat current associated with **P** will not be dissipated: the thermal conductivity must be infinite.

In effect, we have a *convection* of phonons down the specimen. Just as in a *flowing* gas, the collisions between the particles allow interchange of momentum and establish some degree of local equilibrium within the gas; they do not, except indirectly by interaction with the walls, slow down the mass transport of heat. In the case of *heat conduction* in a gas, of course the conditions are quite different; we deliberately insist that there be no mass transport of particles: collisions then become important, because 'hot' particles going one way give up their energy to 'cold' particles going in the opposite direction. But in the phonon gas there is no net conservation of 'particles', so that we cannot impose such conditions. Any extra phonons needed to carry the heat are created at the hot end and flow down to the cold end, where they are destroyed.

The finite conductivity of the crystal is due to the existence of U-processes. Since crystal momentum is not conserved in such processes, the quantity (7.116) does not remain constant, and the heat current is dissipated. The calculation of the thermal resistance in such a case is rather complicated, but the result comes out as in (7.114) and (7.115) at high temperatures. This very naïve approach turns out to be justified.

But at low temperatures the resistance due to phonon–phonon U-processes decreases very rapidly to zero. We recall (§ 2.11) the geometrical conditions on the wave-vectors

$$\mathbf{q}+\mathbf{q}' = \mathbf{q}''+\mathbf{g}, \tag{7.117}$$

and the energy condition on the frequencies

$$\nu+\nu' = \nu''. \tag{7.118}$$

Thus, $\mathbf{q}''$ must be large enough for its energy to be equal to the sum of the energies of $\mathbf{q}$ and $\mathbf{q}'$; but the sum of these vectors must go beyond the first Brillouin zone if we are to have U-processes at all, so that $\mathbf{q}''$ cannot be very much less in length than $\frac{1}{2}g$. In a Debye model, this means

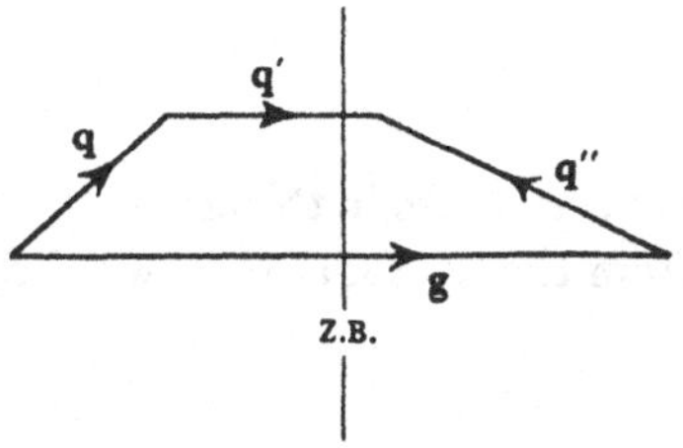

Fig. 131. Phonon–phonon U-processes.

$$q'' \geqslant \beta q_D, \tag{7.119}$$

where β is a fraction, of the order of $\frac{1}{2}$ or $\frac{2}{3}$. In other words,

$$\hbar\nu'' \geqslant \beta k\Theta. \tag{7.120}$$

The probability of the occurrence of U-processes like (7.118) goes as the product of the appropriate occupation numbers of the phonon modes

$$\begin{aligned} n_{\mathbf{q}} n_{\mathbf{q}'} &\approx \exp(-\hbar\nu/kT)\exp(-\hbar\nu'/kT) \\ &\approx \exp(-\hbar\nu''/kT) \\ &\approx \exp(-\beta\Theta/kT) \end{aligned} \tag{7.121}$$

at low temperatures. This falls rapidly to zero. We say that the Umklapp scattering is 'frozen out', and the conductivity tends to infinity as

$$\kappa \sim T^n \exp\left(\frac{\beta\Theta}{T}\right), \tag{7.122}$$

where n is an exponent that depends on the details of the model.

What happens in practice is that the mean free path of the phonons then becomes comparable with the dimensions of the crystal. Suppose that we are dealing with a rod of diameter D. Then in (7.110) we may write

$$\Lambda = D, \tag{7.123}$$

and

$$\kappa \propto T^3 D, \tag{7.124}$$

because of the T^3 specific heat law (§ 2.4) at low temperatures. This *size effect* is well attested experimentally.

There are other mechanisms, however, for the scattering of phonons. Even in a 'perfect' crystal the atoms are not all equivalent; for each chemical element there will usually be a mixture of isotopes of different mass. These variations in mass can scatter phonons. It is easy enough to prove, for example by elementary elastic theory, the *Rayleigh formula* for the scattering cross-section of a point mass δM in a medium of density D for waves of wave-number q;

$$\sigma(q) = \frac{q^4}{4\pi D^2}(\delta M)^2. \tag{7.125}$$

If every atom in the crystal were liable to differ in mass by this amount from the average, then we should have a phonon mean free path

$$\begin{aligned} \Lambda(q) &= \frac{4\pi N M^2}{q^4(\delta M)^2} \\ &= \left(\frac{M}{\delta M}\right)^2 \frac{4\pi a}{(aq)^4}, \end{aligned} \tag{7.126}$$

where $N = 1/a^3$.

The strong variation of $\Lambda(q)$ with q is now a problem. The natural procedure is to generalize (7.110), and write

$$\kappa = \frac{1}{3}\int C(q)\, v(q)\, \Lambda(q)\, \mathbf{dq} \tag{7.127}$$

summing the contributions of the various modes. But this leads to a singularity at $q \to 0$, where the specific heat $C(q)$, and the velocity $v(q)$ are effectively constant, but where $\Lambda(q) \propto q^{-4}$: thus,

$$\kappa \propto \int \frac{q^2\, dq}{q^4} \sim \int_0 \frac{dq}{q^2} \to \infty \tag{7.128}$$

using a Debye spectrum like (2.56) for the distribution of modes of long wavelength.

The resolution of this difficulty is a complicated problem, but the essence of the argument is that there are always strong phonon–phonon N-processes going on in the crystal, even at low temperatures. These cannot, of themselves, produce thermal resistance, but they tend to bring the various modes into equilibrium with one another, and do not allow very large heat currents to be carried by those modes, of long wavelength, which are not scattered strongly by isotopes.

If the phonon–phonon N-processes are very strong indeed, one can argue that the quantity to be averaged in (7.127) should be $1/\Lambda(q)$, which produces a finite integral. Under those circumstances we can estimate the average mean free path to use in the kinetic formula (7.110) by supposing that at the temperature T we have only a 'typical' phonon mode, of wave-vector

$$\bar{q} \sim \frac{T}{\Theta} q_D. \tag{7.129}$$

We should then write $$\Lambda \sim \Lambda(\bar{q}) \propto T^{-4}, \tag{7.130}$$

which would make $$\kappa \propto \left(\frac{M}{\delta M}\right)^2 \frac{1}{T}. \tag{7.131}$$

More detailed analysis, however, using phenomenological models for the phonon–phonon interaction, suggests a more complicated behaviour, in which

$$\kappa \propto \frac{M}{\delta M} \frac{1}{T^{\frac{1}{2}}}. \tag{7.132}$$

This is roughly what is observed.

7.11 Phonon drag

The electrons in a metal are scattered by phonons. Conversely, the phonons are scattered by the electrons. Generally speaking, this scattering is so strong that the lattice conduction is difficult to observe by comparison with the thermal conductivity of the electron gas.

There is, however, one important electrical effect due to phonons in metals and semiconductors. Suppose that we have an electric current flowing in a wire. As shown in (7.27), this can be described as a displacement of the Fermi surface by an amount $(e\tau/\hbar)\mathbf{E}$ in **k**-space. This electron system is interacting with the phonons. If there are only electron–phonon interactions, the phonon system will try to come into equilibrium with the *displaced* electron system, i.e. the phonon system itself will tend to be displaced in its own momentum space.

But such a displacement of the phonon system corresponds to a 'drift' of the phonons in real space—that is, it corresponds to a heat current. Suppose that the drift velocity of the phonons is the same as the drift velocity (7.31) of the electrons. If C_L is the specific heat of the whole phonon system, there will be a lattice heat current

$$\mathbf{U}_L \sim C_L T \boldsymbol{\delta}\mathbf{v}. \tag{7.133}$$

But this is associated with an electric current as in (7.32). Thus, there will be a Peltier coefficient (cf. (7.102))

$$\frac{U_L}{J} \sim \frac{C_L T}{ne}, \tag{7.134}$$

or a *lattice thermo-electric power*

$$Q_L \sim \frac{C_L}{ne}. \tag{7.135}$$

This would be a large effect by comparison with the ordinary electronic contribution (7.106), which has the very small specific heat of the electrons, as in (7.107), in place of the specific heat of the lattice. But of course we have assumed that the phonons are not scattered by any other mechanism except the electron–phonon interaction. This is not the case at ordinary temperatures; there are phonon–phonon U-processes, etc., as discussed in § 7.10, which greatly reduce this *Gurevitch effect*.

However, at low temperatures, where phonon–phonon interaction becomes negligible, the effect of *phonon drag* is easily detected. In that case the lattice specific heat goes as T^3. The formula for a monovalent metal ought to be

$$Q_L \sim \frac{k}{e}\frac{4\pi^4}{5}\left(\frac{T}{\Theta}\right)^3, \tag{7.136}$$

which is a large *negative* thermo-electric power, as fixed by the sign of the electronic charge e.

This is not always what is actually observed, for the following reason. The assumption that the phonon distribution is dragged along at the drift velocity of the electrons depends upon the conservation of crystal momentum in electron–phonon processes. Consider (Fig. 132 (*a*)) an ordinary N-process in which the electron is scattered

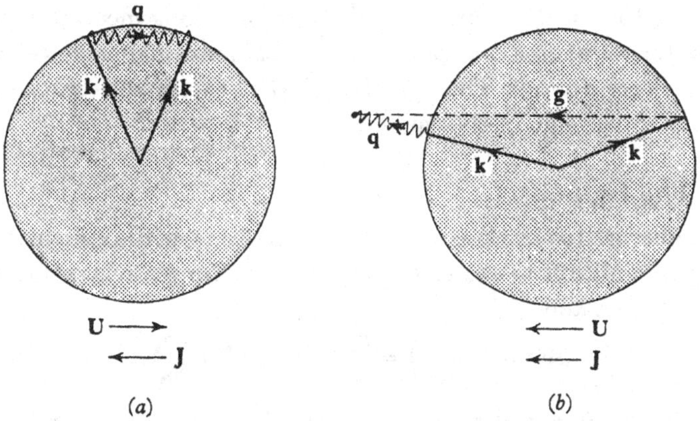

Fig. 132. Phonon drag by electrons: (*a*) N-processes; (*b*) U-processes.

from state **k** to state **k**′ with emission of a phonon. If this process just reverses the component of **k** along the field direction, then the phonon is emitted in that direction. Thus the process by which the electric current is slowed down gives rise to excess phonons moving in the direction of the current carriers.

Now consider what happens in an electron–phonon U-process. As shown in Fig. 132 (*b*), the tendency is for the phonon to be emitted in the *opposite* direction to the carrier current; the vectors **k**, **k**′ and **q** have to add up to give the large reciprocal lattice vector **g**. Thus, the phonon distribution is not 'dragged' with the electrons; it may tend to drift in the opposite direction.

The actual sign of Q_L may thus be positive or negative, depending on whether N-processes or U-processes are the more important in the 'ideal' resistance of the metal. This effect is obviously very sensitive to the shape of the Fermi surface and its distance from the zone boundary, as discussed in § 7.5.

It should be noted that if we had only electron–phonon N-processes the argument of (7.116) would apply: the drag of the phonons would just compensate the drift of the electrons, and there would be no net dissipation of momentum. The electrical conductivity would increase, and in the absence of residual resistance tend to infinity. In real metals this phenomenon is not observed, probably because of the strong electron–phonon U-processes.

The argument that we have given here can also be applied to the carriers in a semiconductor, which 'drag' the phonons with which they interact, as in § 7.6. The effect is larger there, because it is only phonons of small q that scatter carriers and these have long mean free paths. At low temperatures one can even detect a size dependence in the thermo-electric power. The sign of the lattice thermo-electric power is the same as the sign of the carriers.

7.12 The Hall effect

Let us return to our Boltzmann equation (7.14), with now a magnetic field **H** in addition to the electric field. Assuming a relaxation time, this can be written

$$e\mathbf{E}\cdot\mathbf{v}_{\mathbf{k}}\left(-\frac{\partial f^0}{\partial \mathscr{E}}\right) = \frac{g_{\mathbf{k}}}{\tau} + \frac{e}{\hbar c}(\mathbf{v}_{\mathbf{k}}\wedge\mathbf{H})\cdot\frac{\partial g_{\mathbf{k}}}{\partial \mathbf{k}}. \tag{7.137}$$

This is a differential equation for the function $g_{\mathbf{k}}$ in **k**-space. Let us assume we have free electrons, for which

$$\hbar\mathbf{k} = m\mathbf{v}, \tag{7.138}$$

so that we may label each state by its velocity rather than by **k**. We try the following form of solution

$$g_{\mathbf{k}} = \left(-\frac{\partial f^0}{\partial \mathscr{E}}\right)\tau\mathbf{v}_{\mathbf{k}}\cdot e\mathbf{A}, \tag{7.139}$$

where **A** is a vector to be discovered. This is, of course, the analogue of (7.19), where, in the absence of a magnetic field, **A** turns out to be **E**.

Putting (7.138) and (7.139) into (7.137), we get

$$\mathbf{v}\cdot\mathbf{E} = \mathbf{v}\cdot\mathbf{A} + \left(\frac{e\tau}{mc}\right)(\mathbf{v}\wedge\mathbf{H})\cdot\mathbf{A}, \tag{7.140}$$

which is obviously satisfied, for all values of **v**, by

$$\mathbf{E} = \mathbf{A} + \frac{e\tau}{mc}\mathbf{H}\wedge\mathbf{A}. \tag{7.141}$$

This is a vector equation that may be solved for **A**. It is not difficult to prove, by elementary geometry (Fig. 133) that a solution is

$$\mathbf{A} = \frac{\mathbf{E} - \dfrac{e\tau}{mc}\mathbf{H} \wedge \mathbf{E}}{1 + \dfrac{e^2\tau^2}{m^2c^2}H^2}. \tag{7.142}$$

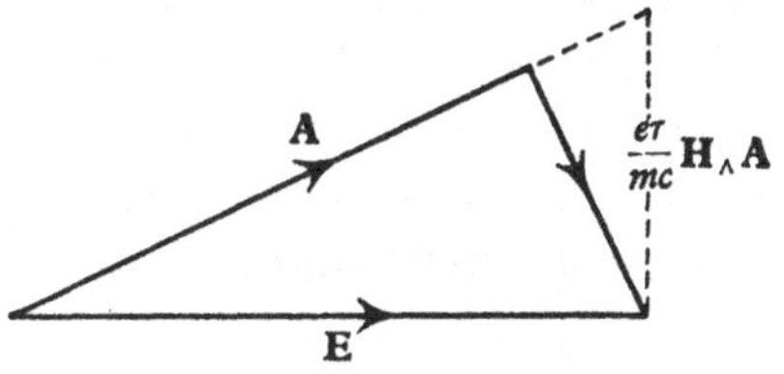

Fig. 133

But we can work directly from (7.141). With g_k given by (7.139), we can easily see that the electric current is simply

$$\mathbf{J} = \sigma_0 \mathbf{A}, \tag{7.143}$$

where σ_0 is the ordinary conductivity of the metal or semi-conductor in the absence of a magnetic field; this follows from (7.19) to (7.23), replacing **E** by **A**. Thus, from (7.141) and (7.143)

$$\mathbf{E} = \frac{1}{\sigma_0}\mathbf{J} + \frac{e\tau}{mc}\mathbf{H} \wedge \frac{1}{\sigma_0}\mathbf{J}$$

$$= \rho_0 \mathbf{J} + \frac{e\tau}{mc}\rho_0 \mathbf{H} \wedge \mathbf{J}, \tag{7.144}$$

where ρ_0 is the ordinary resistivity of the specimen.

This equation indicates that the electric field needed to create the current **J** has two components. Along **J**, i.e. along the length of the wire or strip of metal,

$$E_\parallel = \rho_0 J. \tag{7.145}$$

This tells us that the apparent resistance of the specimen is unaltered by the magnetic field; there is *no magneto-resistance.*

But if **H** is applied perpendicular to **J** there is a transverse electric field

$$E_H = \frac{e\tau}{mc}\rho_0 HJ. \tag{7.146}$$

This is the *Hall effect*. For free electrons the Hall coefficient is given by

$$R = \frac{e\tau}{mc}\rho_0$$

$$= \frac{1}{nec}, \tag{7.147}$$

if we use (7.33) to eliminate the relaxation time.

This result—that the Hall constant is inversely proportional to the density of carriers—is readily comprehensible from kinetic arguments. Thus, if we assume, as in (7.31), that the carriers drift with velocity $\boldsymbol{\delta}\mathbf{v}$, they will be subject to a Lorentz force

$$\frac{e}{c}\boldsymbol{\delta}\mathbf{v}\wedge\mathbf{H} \tag{7.148}$$

across the magnetic field. A transverse electric field

$$\mathbf{E}_H = \frac{1}{c}\mathbf{H}\wedge\boldsymbol{\delta}\mathbf{v}$$

$$= \frac{1}{c}\mathbf{H}\wedge\frac{1}{ne}\mathbf{J} \tag{7.149}$$

is just what is needed to compensate this sideways deflection. The proportionality to $1/n$ arises because, for a given current, the smaller the density of carriers the more rapidly each must travel, and the more it is deflected by the magnetic field.

The kinetic argument also tells us why there is no magneto-resistance in this simple case. The two deflecting effects transverse to the current are in exact balance, so that the carriers travel down the wire, under the influence of the longitudinal electric field, as if there were no magnetic field at all.

The result (7.147) actually holds for any sphere of carriers. Suppose we write

$$\hbar k_F = m^* v_F \tag{7.150}$$

over the Fermi surface, simply defining the parameter m^* by this relation. We should then find

$$R = \frac{e\tau}{m^* c\sigma_0} \tag{7.151}$$

just as in (7.146). But then (cf. (7.25))

$$\sigma_0 = \frac{e^2}{12\pi^3\hbar}\int \tau v_F\, dS_F$$
$$= \frac{e^2\tau}{12\pi^3}\int \frac{k_F\, dS_F}{m^*}$$
$$= \frac{e^2\tau}{m^*}\frac{1}{4\pi^3}\frac{4\pi}{3}k_F^3$$
$$= \frac{ne^2\tau}{m^*}, \tag{7.152}$$

where n is indeed the number of carriers comprised within the sphere of radius k_F. From (7.151) and (7.152), follows (7.147), independently of τ and m^*.

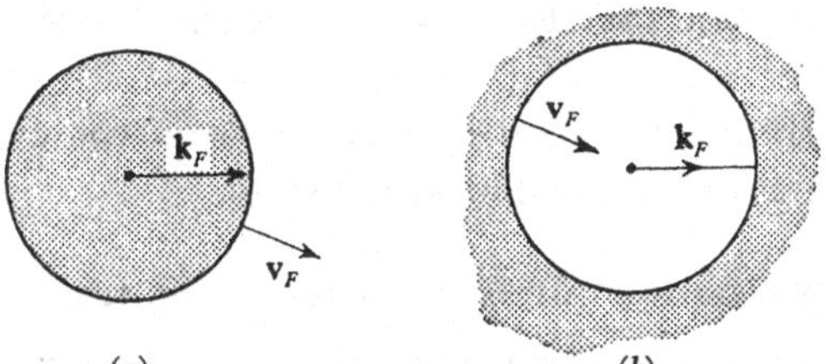

Fig. 134. (*a*) A sphere of 'electrons': m^* positive. (*b*) A sphere of 'holes': m^* negative.

But notice that if we had had a *sphere of holes* we should expect to find e positive, i.e. the Hall field would be of the opposite sign to that found for electrons. This is consistent with the above argument. If we had treated the Fermi surface of *electrons* surrounding the centre of of the sphere, we should still have (7.150), but with m^* negative, because the electron velocity would now point inwards to the centre. This would still be negative in (7.151). In (7.152), however, it would be $|m^*|$ in the denominator because, as in (7.23), we are really calculating

$$\frac{e^2}{12\pi^3\hbar}\int \tau v_F^2 \frac{dS_F}{|v_F|} = \frac{e^2\tau}{12\pi^3}\int \frac{k_F\, dS_F}{|m^*|}. \tag{7.153}$$

Thus, R would contain a factor $|m^*|/m^* = -1$, and the apparent sign of the charge would be reversed.

These formulae do not hold exactly even for a single band of carriers in a semiconductor, because $\tau(\mathscr{E})$ can vary through the occupied states. The correction factor therefore depends on the energy dependence of the scattering mechanisms (7.50) and (7.73), but is seldom far from unity. Combining (7.147) with (7.34) gives the *Hall mobility*

$$\mu_H = |R|\,\sigma c \tag{7.154}$$

of the majority carriers in the specimen, the sign of whose charge is given by the sign of R.

The theory of the Hall effect and other galvanomagnetic phenomena obviously depends heavily upon the Bloch theorem, and the reciprocal lattice representation of the electron states. In a disordered system, where **k** need no longer be a good quantum number, the derivation via the Boltzmann equation is immediately suspect, and we turn to some more fundamental theory such as the Kubo formalism (7.15). The fact is, however, that no exact theory is yet available to replace the simple 'electron sphere' formula (7.147),—which is observed to hold very closely in liquid metals (§ 7.5), even when the electron mean free path is of the order of the interatomic distance. It is doubtful, however, whether a Hall effect should be observable in the extreme case of hopping conduction between localized states.

7.13 The two-band model: magneto-resistance

Suppose now we have two types of carrier, for example, electrons and holes. We find, for each of these separately, the same equations (7.144):

$$\mathbf{E} = \frac{1}{\sigma_1}\mathbf{J}_1 + \beta_1\,\mathbf{H}\wedge\frac{1}{\sigma_1}\mathbf{J}_1, \tag{7.155}$$

where

$$\beta_1 = \frac{e\tau_1}{m_1 c} \tag{7.156}$$

and σ_1 is the conductivity associated with the first type of carrier, which contributes $\mathbf{J}_1$ to the current.

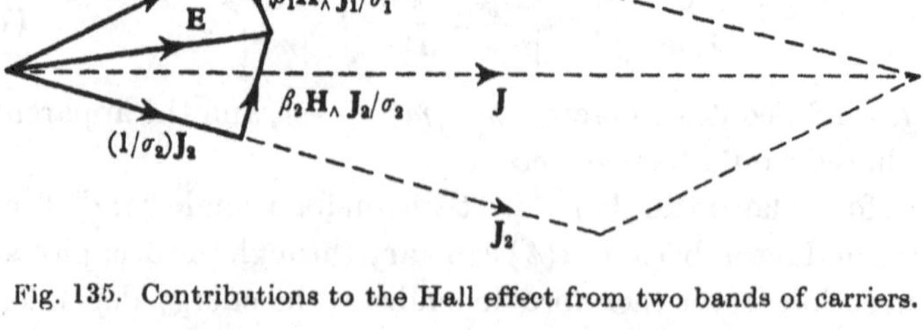

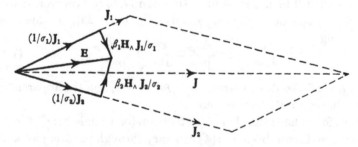

Fig. 135. Contributions to the Hall effect from two bands of carriers.

Similarly, for the second group of carriers,

$$\mathbf{E} = \frac{1}{\sigma_2}\mathbf{J}_2 + \beta_2 \mathbf{H} \wedge \frac{1}{\sigma_2}\mathbf{J}_2 \tag{7.157}$$

represents the relation between their contribution to the current and the applied electric field. We want the total current

$$\mathbf{J} = \mathbf{J}_1 + \mathbf{J}_2. \tag{7.158}$$

Using a solution of the form of (7.142) for each of (7.155) and (7.157), we find

$$\mathbf{J} = \left(\frac{\sigma_1}{1+\beta_1^2 H^2} + \frac{\sigma_2}{1+\beta_2^2 H_2}\right)\mathbf{E} - \left(\frac{\sigma_1 \beta_1}{1+\beta_1^2 H^2} + \frac{\sigma_2 \beta_2}{1+\beta_2^2 H^2}\right)\mathbf{H} \wedge \mathbf{E}, \tag{7.159}$$

a rather complicated formula which is represented geometrically in Fig. 135.

To calculate the Hall coefficient we ought to invert this formula, and express $\mathbf{E}$ in terms of $\mathbf{J}$ and $\mathbf{H} \wedge \mathbf{J}$. But for low magnetic fields the result is easily obtained

$$\begin{aligned} R &= \frac{\sigma_1 \beta_1 + \sigma_2 \beta_2}{(\sigma_1+\sigma_2)^2} \\ &= \frac{\sigma_1^2 R_1 + \sigma_2^2 R_2}{(\sigma_1+\sigma_2)^2}, \end{aligned} \tag{7.160}$$

where R_1 and R_2 would be the Hall constant for each of the corresponding carriers acting alone. Thus, if R_1 and R_2 are of opposite sign, we see a compromise value. In a semiconductor we should express the partial conductivity in terms of the number and mobility of the appropriate group of carriers, i.e. $\sigma_1 = n_1 e \mu_1$, etc. so that the observed Hall mobility (7.154) may not correspond to any particular microscopic quantity unless one band of carriers is decisively in the majority.

The magneto-resistance requires a more careful analysis. We want to compare $\mathbf{J}$ with the component of $\mathbf{E}$ along $\mathbf{J}$. Thus

$$\begin{aligned} \rho &= (\mathbf{J} \cdot \mathbf{E})/J^2 \\ &= \frac{\dfrac{\sigma_1}{1+\beta_1^2 H^2} + \dfrac{\sigma_2}{1+\beta_2^2 H^2}}{\left(\dfrac{\sigma_1}{1+\beta_1^2 H^2} + \dfrac{\sigma_2}{1+\beta_2^2 H^2}\right)^2 + \left(\dfrac{\sigma_1 \beta_1 H}{1+\beta_1^2 H^2} + \dfrac{\sigma_2 \beta_2 H}{1+\beta_2^2 H^2}\right)^2}. \end{aligned} \tag{7.161}$$

We compare this with the resistance in the absence of a magnetic field,

$$\rho_0 = \frac{1}{\sigma_1 + \sigma_2}. \tag{7.162}$$

After a bit of algebra we get

$$\frac{\Delta\rho}{\rho_0} \equiv \frac{\rho-\rho_0}{\rho_0} = \frac{\sigma_1\sigma_2(\beta_1-\beta_2)^2 H^2}{(\sigma_1+\sigma_2)^2+H^2(\beta_1\sigma_1+\beta_2\sigma_2)^2}. \tag{7.163}$$

This formula is not so very important in itself; we are not often in the situation where the carriers can be treated as if they fell into two groups each with such simple properties. Nevertheless, it illustrates the main features of the phenomenon of magneto-resistance.

In the first place, $\Delta\rho$ is essentially positive, and vanishes only if $\beta_1 = \beta_2$. In other words, when the two groups of carriers are deflected by different amounts in the magnetic field, because they have different masses, or different charges, or because they travel different distances before being scattered, it is impossible to find an electric field that will keep both components of the current going in the same direction. The net current, being the vector sum of the components, is reduced; the apparent resistance of the medium is increased.

It is typical, also, that $\Delta\rho$ is proportional to H^2 for small fields, but that it tends to *saturate* at high fields. But this latter effect is related to our choice of closed energy surfaces for the carriers. As we shall see in § 9.4, it is sometimes possible to find special crystalline directions in which $\Delta\rho$ does not saturate.

This sort of formula can easily be generalized to the case where we have many different types of carrier, all contributing separately to the current. In other words, we can deal with complicated Fermi surfaces, with different parts having different values of β. The existence of magneto-resistance in metals is thus evidence for variation of β, i.e. different values of effective mass, or perhaps a variation of τ, over the Fermi surface. But the actual formula is a complicated integral involving various local derivatives of $\mathscr{E}(\mathbf{k})$, and is not very transparent nor easy to use to deduce the shape of the Fermi surface from observed values of $\Delta\rho$.

The simplest case is a 'valley' of carriers in a semiconductor (§ 6.6), whose contribution to galvanomagnetic phenomena can be calculated fairly easily if τ is assumed to be constant on each energy surface. The Lorentz force operator in (7.137) evidently takes a derivative with respect to $\mathbf{k}$ in the direction normal to both $\mathbf{v}$ and $\mathbf{H}$. For a current along the axis of a valley, the 'Hall mass' in a formula such as (7.139) would evidently be the coefficient m_i, say, from (6.44), in this 'transverse' direction. But contributions from the different valleys in different orientations with various off-diagonal components of the inverse mass tensor (6.48), add up to a single Hall coefficient which must be isotropic if the crystal has macroscopic cubic symmetry. This symmetry restriction does not, however, apply to the magnetoresistance, which can vary as the crystal is rotated relative to the electric and magnetic field directions, thus yielding important information about the electronic band structure of the semiconductor.

The above calculation is for the *transverse magneto-resistance*—the magnetic field is perpendicular to the direction of the current. One can also observe *longitudinal magneto-resistance*, when the magnetic field is parallel to the wire carrying the current. Our simple two-band model gives no longitudinal magneto-resistance because it assumes spherical symmetry in each band. The formulae for the longitudinal effect are even more complicated, since they require non-spherical Fermi surfaces. This effect is also observed in metals and in semiconductors. In fact, the change of resistance of a single crystal specimen is a complicated function of the orientations of current and of magnetic field to the crystal axes.

It is not difficult in principle, nor impossible in practice, to invent other transport 'effects' depending upon the ambient magnetic field. The flow of heat along a thermal gradient, for example, produces a transverse *Nernst* electric field analogous to the Hall effect. The classification of these effects, according to the symmetry of the crystal and their connections via the Onsager relations, is a considerable theoretical study in itself, but the coefficients may all be deduced from the Boltzmann equation if the relaxation time approximation is valid.

One further point about (7.163) is worth noting. Suppose that both groups of carriers have the same value of τ. Then $\Delta\rho/\rho_0$ comes out as a function only of τH. But τ itself will then be inversely proportional to the resistivity ρ_0: we can write

$$\frac{\Delta\rho}{\rho_0} = F\left(\frac{H}{\rho_0}\right), \tag{7.164}$$

where F is a function depending on the nature of the metal itself.

This is known as *Kohler's Rule*. It tells us that, in principle, we can plot measurements of magneto-resistance from different specimens on the same basic diagram, or we can investigate the probable effects of very high fields by using very pure specimens at very low temperatures, where ρ_0 is small.

But this rule is obviously only an approximation. It by no means follows that two different types of carrier have the same relaxation time, or even that their relaxation times stand in the same ratio whether they are being scattered by impurities or by phonons, at high temperatures or at low temperatures. All it says is that the deflection of a carrier is proportional to the product of the magnetic field and the time between collisions. Deviations from Kohler's rule are evidence that different types of scattering mechanism have different effects on different groups of carriers.

CHAPTER 8

OPTICAL PROPERTIES

But, soft! what light through yonder window breaks....
Romeo and Juliet

8.1 Macroscopic theory

In this chapter we discuss the propagation of electromagnetic waves into and through solids. It is well known that some solids are *transparent* and others are *opaque*, that some solid surfaces are strongly *reflecting*, whilst others tend to *absorb* the radiation that falls on them. These effects depend on the frequency of the radiation; we therefore include the whole electromagnetic spectrum, from long-wave radio waves to soft X-rays.

Electromagnetic waves are solutions of Maxwell's equations. We shall write these as follows

$$\left.\begin{aligned}\nabla\wedge\mathbf{H} &= \frac{\epsilon}{c}\frac{\partial\mathbf{E}}{\partial t}+\frac{4\pi\sigma}{c}\mathbf{E}, & \nabla\cdot\mathbf{E} &= 0,\\ \nabla\wedge\mathbf{E} &= -\frac{\mu}{c}\frac{\partial\mathbf{H}}{\partial t}, & \nabla\cdot\mathbf{H} &= 0.\end{aligned}\right\} \tag{8.1}$$

In this chapter we shall not consider any magnetic effects: we take $\mu = 1$.

The difference between propagation in free space, and propagation in the solid is expressed by the two coefficients—the *dielectric constant* ϵ and the *conductivity* σ. The former defines the magnitude of displacement currents due to time variation of $\mathbf{E}$: the latter is a measure of the real currents created by the electric field.

The magnetic field may be eliminated between these equations

$$\nabla^2\mathbf{E} = \frac{\epsilon}{c^2}\frac{\partial^2\mathbf{E}}{\partial t^2}+\frac{4\pi\sigma}{c^2}\frac{\partial\mathbf{E}}{\partial t}. \tag{8.2}$$

This represents a wave propagated with dissipation. If we choose the frequency, ω, and write

$$\mathbf{E} = \mathbf{E}_0\exp\{i(\mathbf{K}\cdot\mathbf{r}-\omega t)\}, \tag{8.3}$$

then our wave equation requires

$$-\mathbf{K}^2 = -\epsilon\frac{\omega^2}{c^2} - \frac{4\pi\sigma i\omega}{c^2}, \tag{8.4}$$

i.e.

$$K = \frac{\omega}{c}\left(\epsilon + \frac{4\pi\sigma i}{\omega}\right)^{\frac{1}{2}}. \tag{8.5}$$

Thus, in general the propagation constant K comes out to be a complex number. Now in free space we should have, simply

$$K = \frac{\omega}{c}, \tag{8.6}$$

as for a wave travelling with the velocity of light. In the medium the velocity is modified: we say that the phase velocity is divided by a *complex refractive index*

$$\mathsf{N} = \left(\epsilon + \frac{4\pi\sigma i}{\omega}\right)^{\frac{1}{2}}. \tag{8.7}$$

The whole theory of the optical properties, as observed macroscopically, can be expressed in terms of N. For example, suppose we have a disturbance of frequency ω trying to propagate as a plane wave in the z-direction, and suppose we write

$$\mathsf{N} = \mathsf{n} + i\mathsf{k} \tag{8.8}$$

for the real and imaginary parts of N. The propagation constant becomes

$$K = \frac{\mathsf{n}\omega}{c} + \frac{i\mathsf{k}\omega}{c}, \tag{8.9}$$

so that the wave (8.3) becomes

$$\mathbf{E} = \mathbf{E}_0 \exp\left\{i\omega\left(\frac{\mathsf{n}z}{c} - t\right)\right\}\exp\left(-\frac{\mathsf{k}\omega z}{c}\right), \tag{8.10}$$

the velocity is reduced to c/n, and the wave is *damped*, as it progresses, by a fraction $\exp(-2\pi\mathsf{k}/\mathsf{n})$ per wavelength.

The damping of the wave is, of course, associated with the absorption of electromagnetic energy. To calculate this we should use Maxwell's equations to find the current associated with (8.10). This is the right-hand side of the first equation of (8.1)

$$\mathbf{J} = \left(-\frac{i\omega\epsilon}{c} + \frac{4\pi\sigma}{c}\right)\mathbf{E}$$

$$= -\frac{i\omega}{c}\mathsf{N}^2\mathbf{E} \tag{8.11}$$

by (8.7). The rate of production of Joule heat is the real part of

$$\mathbf{J}\cdot\mathbf{E} = -\frac{i\omega}{c}\mathsf{N}^2E^2. \tag{8.12}$$

Thus the *absorption coefficient*—the fraction of energy absorbed in passing through unit thickness of the material—is given by

$$\eta = \frac{\mathrm{Re}\,(\mathbf{J}\cdot\mathbf{E})}{\mathsf{n}|\mathbf{E}|^2} = \frac{2\mathsf{k}\omega}{c}. \tag{8.13}$$

The other situation which is often studied is where radiation is incident upon a plane surface of the material. In general this is a complicated problem in optics, but we need consider only the case of

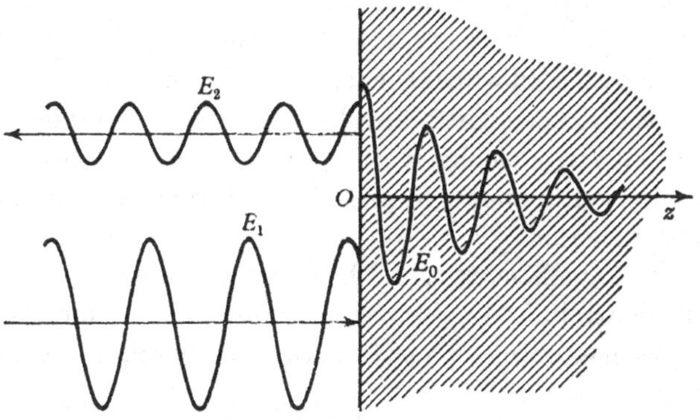

Fig. 136

normal incidence. We want to construct a solution of Maxwell's equations having the form (8.10) inside the medium, but matched to an incident and a reflected wave outside. Thus, for $z > 0$ we write

$$E_x = E_0 \exp\{i\omega(\mathsf{N}z/c - t)\}, \tag{8.14}$$

whilst for $z < 0$, in the free space outside, we write

$$E_x = E_1 \exp\{i\omega(z/c - t)\} + E_2 \exp\{-i\omega(z/c + t)\}, \tag{8.15}$$

corresponding to a wave of amplitude E_1, travelling to the right and a wave of amplitude E_2 travelling to the left. Matching these on the boundary

$$E_0 = E_1 + E_2. \tag{8.16}$$

But there are also magnetic fields associated with these waves, with magnetic vector in the y-direction, say. Using Maxwell's equations to calculate H_y we have

$$-\mathsf{N}E_0 = E_2 - E_1 \tag{8.17}$$

when matched to the wave outside. Thus, the ratio of the complex amplitudes of the reflected and incident waves is

$$\frac{E_2}{E_1} = \frac{1-\mathsf{N}}{1+\mathsf{N}}, \tag{8.18}$$

which corresponds to a real *reflection coefficient*,

$$\mathsf{R} = \left|\frac{1-\mathsf{N}}{1+\mathsf{N}}\right|^2 = \frac{(\mathsf{n}-1)^2+\mathsf{k}^2}{(\mathsf{n}+1)^2+\mathsf{k}^2} \tag{8.19}$$

in terms of the real and imaginary parts (8.8) of the complex refractive index N.

It is evident that independent measurements of the reflection coefficient and of the absorption coefficient are sufficient to fix the values of n and of k. These are, therefore, the optical constants which one usually quotes, and whose theory we shall consider in this chapter. In practice the experiments may be much more complicated—e.g. reflection at an angle to the surface, or transmission through a thin film—but the results ought, in principle, all to be described by these same two coefficients.

However, the coefficients $\mathbf{n}$ and $\mathbf{k}$ are not quite independent of one another. They are linked by *dispersion relations*. A quantity such as N^2 in (8.11) is an example of a *generalized susceptibility*, $\alpha(\omega)$, say, in a relation

$$D(\omega) = \alpha(\omega)\,F(\omega) \tag{8.20}$$

between the Fourier components of a generalized 'displacement', D, and a 'force' F, at some frequency ω. As in elementary A.C. circuit theory, this would have to be a complex quantity

$$\alpha(\omega) = \alpha'(\omega)+i\alpha''(\omega) \tag{8.21}$$

to describe phase differences between D and F.

But (8.20) is merely the Fourier transform of

$$D(t) = \int_{-\infty}^{\infty} A(t-t')\,F(t')\,dt', \tag{8.22}$$

in which the displacement at time t is the resultant *linear response* of the system to forces acting at all other times t'. But there can be no displacement until *after* the application of a force: the response function is subject to the rigid condition of *causality*:

$$A(t-t') \equiv 0 \quad \text{for} \quad t' > t. \tag{8.23}$$

In other words, the Fourier integral for the susceptibility is fully

defined by
$$\alpha(\omega) = \int_{-\infty}^{\infty} A(t)\, e^{i\omega t}\, dt \equiv \int_{0}^{\infty} A(t)\, e^{i\omega t}\, dt. \tag{8.24}$$
without any contributions from negative values of t.

To give meaning to such an integral, it is usual to treat $\alpha(\omega)$ as a function of the *complex* variable ω, with, say, an infinitesimal positive imaginary part $i\epsilon$. The form of (8.24) implies, moreover, that this function has no singularities above the real axis in the complex ω plane, and tends to zero as $|\omega| \to \infty$ in all directions in that region. A little bit of functional analysis, in which Cauchy's theorem is applied to a contour running along the real axis and returning in an infinite semicircle, yields the formula
$$\alpha(\Omega) = \frac{1}{i\pi}\int_{-\infty}^{\infty} \frac{\alpha(\omega)}{\omega - \Omega}\, d\omega \tag{8.25}$$
where the integral takes its principal value and the frequency variables are real.

The definition of the Fourier transforms for 'negative frequencies' implies that $\alpha(-\omega)$ is the complex conjugate of $\alpha(\omega)$. From (8.25) we deduce the *Kramers–Kronig relations* between the real and imaginary parts of the susceptibility (8.21):
$$\alpha'(\Omega) = \frac{2}{\pi}\int_{0}^{\infty} \frac{\omega\alpha''(\omega)\, d\omega}{\omega^2 - \Omega^2} + \text{const.}, \tag{8.26}$$
and
$$\alpha''(\Omega) = -\frac{2\Omega}{\pi}\int_{0}^{\infty} \frac{\alpha'(\omega)\, d\omega}{\omega^2 - \Omega^2}. \tag{8.27}$$

In our case, the real and imaginary parts of N^2 are
$$\mathsf{N}^2 = (\mathsf{n}^2 - \mathsf{k}^2) + 2\mathsf{n}\mathsf{k}i, \tag{8.28}$$
so that we have the conditions
$$\mathsf{n}^2(\omega) - \mathsf{k}^2(\omega) = \frac{2}{\pi}\int_{0}^{\infty} \frac{\omega' 2\mathsf{n}(\omega')\,\mathsf{k}(\omega')\, d\omega'}{\omega'^2 - \omega^2} + \text{const.}, \tag{8.29}$$
and
$$2\mathsf{n}(\omega)\,\mathsf{k}(\omega) = -\frac{2\omega}{\pi}\int_{0}^{\infty} \frac{\{\mathsf{n}^2(\omega') - \mathsf{k}^2(\omega')\}\, d\omega'}{\omega'^2 - \omega^2}. \tag{8.30}$$

From these relations, if we know, say, the absorption coefficient as a function of frequency for all frequencies, then we can evaluate both $\mathsf{n}(\omega)$ and $\mathsf{k}(\omega)$ separately. They are not accidental consequences of some particular model; the proof, which depends solely on the causality condition (8.23), is perfectly general and applies to any linear response function such as the dielectric function (5.16) or the surface impedance (8.106).

8.2 Dispersion and absorption

The simplest model of a solid is a fixed assembly of independent neutral atoms. What is the effect of an electromagnetic wave on such a system? In case this theory is not well known, we sketch it out for a simple case, in which each atom contains only one electron, in a ground-state orbital $\phi_0(\mathbf{r})$, which can be excited to higher orbitals $\phi_j(\mathbf{r})$.

Suppose that the electric field in the wave, near the atom, is given by

$$\mathbf{E}(t) = E_x(e^{i\omega t} + e^{-i\omega t}), \tag{8.31}$$

ignoring the variation with distance as in (8.3). We follow time-dependent perturbation theory, and suppose that the electron wave-function at any particular moment can be written

$$\psi(\mathbf{r},t) = \phi_0 \exp(-i\mathscr{E}_0 t/\hbar) + \sum_j c_j(t)\, \phi_j \exp(-i\mathscr{E}_j t/\hbar). \tag{8.32}$$

We treat $e\mathbf{E}\cdot\mathbf{r}$ as the perturbing potential acting on the electron. It is easy to prove, by putting (8.32) into the time-dependent Schrödinger equation, multiplying through by ϕ_0 and integrating, that the coefficient of the higher state must approximately satisfy the differential equation

$$i\hbar\frac{dc_j}{dt} = \int \phi_j^*\, e\mathbf{E}(t)\cdot\mathbf{r}\, \phi_0\, d\mathbf{r}\; e^{i(\mathscr{E}_j - \mathscr{E}_0)t/\hbar} \tag{8.33}$$

which has the solution

$$\begin{aligned} c_j(t) &= \frac{1}{i\hbar}\int_0^t eE_x x_{j0}(e^{i\omega t} + e^{-i\omega t})\, e^{i(\mathscr{E}_j - \mathscr{E}_0)t/\hbar} \\ &= eE_x x_{j0}\left\{\frac{1 - e^{i(\hbar\omega + \mathscr{E}_j - \mathscr{E}_0)t/\hbar}}{\hbar\omega + (\mathscr{E}_j - \mathscr{E}_0)} - \frac{1 - e^{i(-\hbar\omega + \mathscr{E}_j - \mathscr{E}_0)t/\hbar}}{\hbar\omega - (\mathscr{E}_j - \mathscr{E}_0)}\right\}, \end{aligned} \tag{8.34}$$

where

$$ex_{j0} \equiv \int \phi_j^* ex\, \phi_0\, d\mathbf{r}, \tag{8.35}$$

i.e. it is the matrix element of the dipole moment of the electron in the direction of the electric vector of the field, between states ϕ_0 and ϕ_j.

The usual procedure now, in books of quantum theory, is to study $|c_j(t)|^2$, and to show that it oscillates strongly except for such values of ω as satisfy the condition

$$\hbar\omega = \hbar\omega_j \equiv (\mathscr{E}_j - \mathscr{E}_0). \tag{8.36}$$

One then proves that there will be a quantum jump from state ϕ_0 to state ϕ_j, with probability determined by the square of the matrix element (8.35).

But we now follow another line, and calculate the value of the dipole moment of the atom in this time-dependent state. From (8.32) and (8.34) we have the following

$$\begin{aligned} \langle ex(t)\rangle &= \int \psi^*(\mathbf{r},t)\, ex\, \psi(\mathbf{r},t)\, d\mathbf{r} \\ &= \sum_j \{ex_{0j}\, c_j(t)\, e^{-i\omega_j t} + ex_{j0}\, c_j^*(t)\, e^{i\omega_j t}\} \\ &= E_x \sum_j e^2 |x_{0j}|^2 \frac{1}{\hbar}\left\{\frac{1}{\omega_j-\omega}+\frac{1}{\omega_j+\omega}\right\}(e^{i\omega t}+e^{-i\omega t}) \end{aligned} \tag{8.37}$$

together with terms that oscillate rapidly and out of phase with the applied field.

Thus, the dipole moment of the atom is proportional to the field. We say that the *atomic polarizability* is given by

$$\alpha(\omega) = \sum_j \frac{e^2|x_{0j}|^2}{\hbar}\frac{2\omega_j}{\omega_j^2-\omega^2}. \tag{8.38}$$

To discuss the order of magnitude of such an expression, we may use the *Thomas–Reiche–Kuhn* sum rule—a simpler case of (5.48). This reads

$$\frac{2m}{\hbar^2}\sum_j \hbar\omega_j\, |x_{0j}|^2 = 1; \tag{8.39}$$

it is convenient to define the *oscillator strength* of the jth transition

$$f_j = \frac{2m}{\hbar^2}\hbar\omega_j\, |x_{0j}|^2, \tag{8.40}$$

and write (8.38) in the form

$$\alpha(\omega) = \frac{e^2}{m}\sum_j \frac{f_j}{\omega_j^2-\omega^2}, \tag{8.41}$$

where the numbers f_j add to unity over all transitions.

If now we bring together N such atoms per unit volume, we have a medium whose dielectric constant has a real part

$$\begin{aligned} \epsilon(\omega) &= 1+4\pi N\alpha(\omega) \\ &= 1+\frac{4\pi Ne^2}{m}\sum_j \frac{f_j}{\omega_j^2-\omega^2}. \end{aligned} \tag{8.42}$$

Strictly speaking, we should distinguish between this *transverse* dielectric constant, and the dielectric constant for *longitudinally* polarized waves of the electrostatic field discussed in Chapter 5. In the limit of long wavelengths, however—i.e. for normal 'optical' phenomena—these quantities are the same, since the screening or atomic polarization effects are relatively localized and depend mainly on the strength of the electric field in the immediate vicinity of each ion or atom.

This is the prototype of a dispersion formula. It is interesting to note that it reduces to the case already discussed in § 5.7. If the electrons are 'free', in the sense that all the ω_j are zero and the sum of the f_j are unity, we may write, as in (5.53)

$$\omega_p^2 = \frac{4\pi N e^2}{m} \tag{8.43}$$

to define a plasma frequency for our N electrons per unit volume; (8.42) then reproduces (5.67), the formula leading to the phenomenon of ultra-violet transparency. Notice, however, that there is practically no direct excitation of longitudinal plasmons by transverse photons. To observe *plasmon lines* we need to excite the electron gas with high energy electrons, or look through thin films where *surface plasmon modes* may be observed.

But in general the lowest value of ω_j would correspond to an optical or infra-red frequency, and the variation of ϵ with ω is more complicated. At low frequencies we might find the static dielectric constant

$$\epsilon(0) = 1 + \sum_j f_j \frac{\omega_p^2}{\omega_j^2}, \tag{8.44}$$

which is essentially larger than unity. But $\epsilon(0)$ would not be very large, because ω_p for most solids is comparable with an ordinary optical frequency for the free atom.

Then, as ω increases, $\epsilon(\omega)$ increases until there is a singularity at $\omega = \omega_j$. Thereafter, $\epsilon(\omega)$ becomes negative for a certain range of frequencies before becoming positive again near the next resonance frequency. Eventually, however, $\epsilon(\omega)$ will tend to the formula (5.67);

$$\epsilon(\omega) \to 1 - \frac{\omega_p^2}{\omega^2}, \tag{8.45}$$

for $\omega >$ all ω_j.

Now in the neighbourhood of each ω_j we have a region of *anomalous dispersion*. In particular, if ϵ is negative, then by (8.7) the refractive index N becomes pure imaginary; thus

$$\mathsf{n} = 0, \quad \mathsf{k} = |\epsilon|^{\frac{1}{2}}, \tag{8.46}$$

so that by (8.19) there would be *total reflection* of light from the surface of the solid. In effect, the crystal would seem to remain transparent, although with a very high refractive index, up to $\omega = \omega_j$, would then suddenly become opaque and perfectly reflecting, and would then become transparent again at a higher frequency.

But this would not be consistent with our dispersion relations (8.29) and (8.30). There must also be some absorption, as measured by the product $2\mathsf{nk}\omega$. This, indeed, is the aspect of the time-dependent perturbation that we set aside—the absorption associated with the atomic transition (8.36). This comes as a sharp line at the frequency ω_j—presumably a delta function. In order to satisfy the dispersion relation (8.29) and give the correct dispersion terms in (8.42), we must have

$$2\omega\mathsf{n}(\omega)\,\mathsf{k}(\omega) = \frac{\pi}{2}\frac{4\pi Ne^2}{m}\sum_j f_j\,\delta(\omega-\omega_j); \tag{8.47}$$

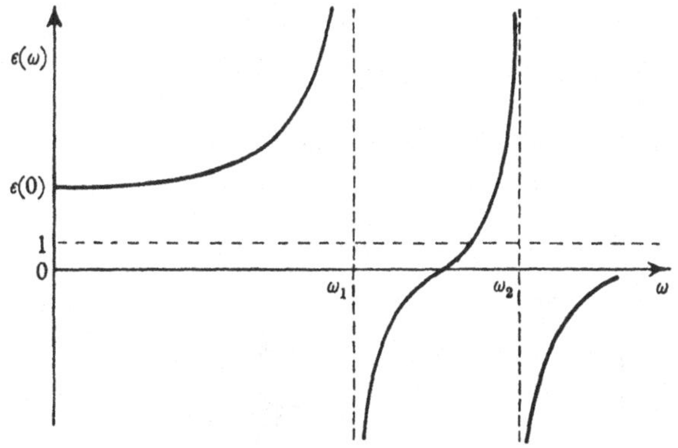

Fig. 137. Dispersion of light.

this must be the imaginary part of the complex dielectric constant which takes the form

$$\epsilon(\omega) = 1+\frac{4\pi Ne^2}{m}\sum_j f_j\left\{\frac{1}{\omega_j^2-\omega^2}+\frac{i\pi}{2\omega}\delta(\omega^2-\omega_j^2)\right\}. \tag{8.48}$$

The large refractive index near ω_j turns into a sharp absorption line at this frequency.

In practice we never see an infinitely sharp line. There is always some broadening, from impurities, etc., or just from the natural radiative relaxation of the levels. As is shown in standard quantum theory, such effects can be incorporated phenomenologically in the analysis by inserting in (8.34), say, a decay factor $\exp(-\frac{1}{2}\Gamma t)$, corresponding to a decay time (in the *square* of the amplitude) of the order of $1/\Gamma$ seconds.

It is easy to follow such a factor through the algebra as the addition

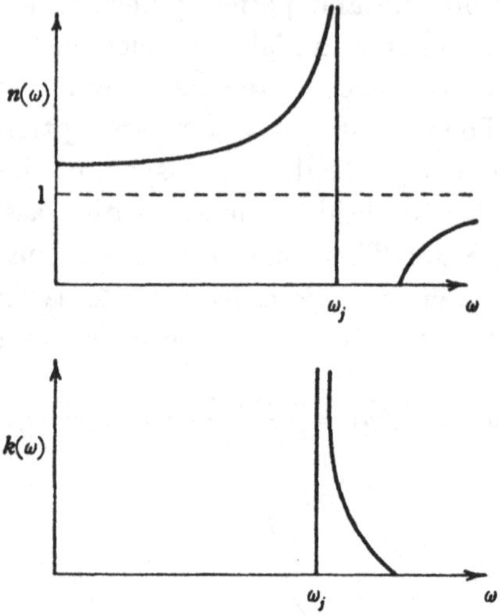

Fig. 138. Real and imaginary parts of the refractive index.

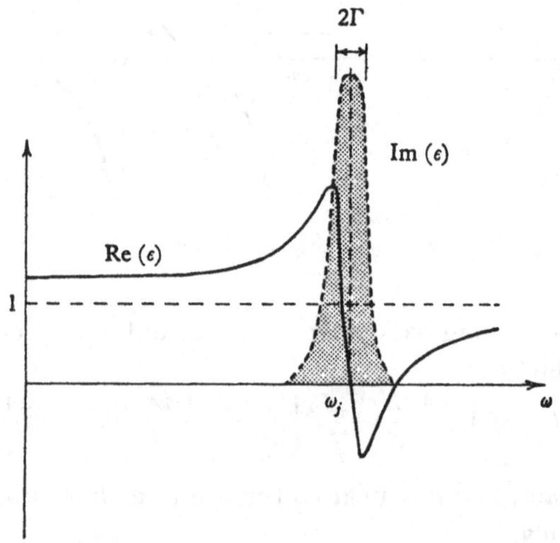

Fig. 139. Broadening of dispersion curve.

of $\frac{1}{2}i\Gamma$ to ω_j, in (8.37) or (8.38). If we ignore Γ^2 relative to ω_j^2, then we get for the real part of the polarizability, and hence in the real part of $\epsilon(\omega)$, terms like

$$f_j \frac{\omega_j^2-\omega^2}{(\omega_j^2-\omega^2)^2+\omega^2\Gamma^2} \tag{8.49}$$

in place of $f_j/(\omega_j^2-\omega^2)$. This has the effect of removing the singularity in n and spreading the dispersion function over a range of frequency of the order of 2Γ in width.

As is well known, the effect of relaxation on the absorption line itself is to spread out the delta function into a finite function of the form

$$\frac{\Gamma/2\pi}{(\omega_j-\omega)^2+\frac{1}{4}\Gamma^2} \approx \frac{2\Gamma\omega^2/\pi}{(\omega_j^2-\omega^2)^2+\Gamma^2\omega^2} \tag{8.50}$$

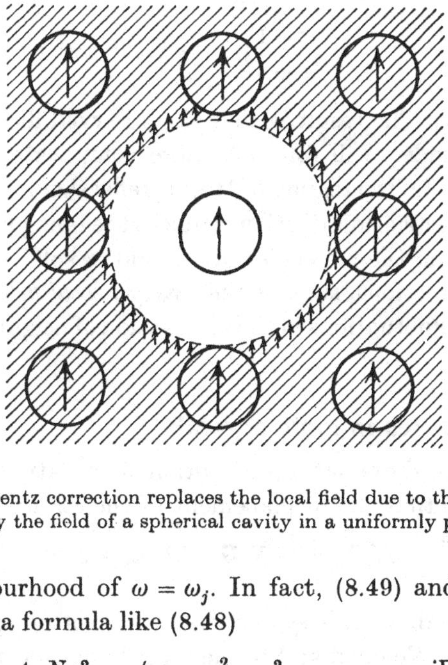

Fig. 140. The Lorentz correction replaces the local field due to the dipoles on neighbouring atoms by the field of a spherical cavity in a uniformly polarized medium.

in the neighbourhood of $\omega = \omega_j$. In fact, (8.49) and (8.50) can be combined into a formula like (8.48)

$$\begin{aligned}\epsilon(\omega) &= 1+\frac{4\pi Ne^2}{m}\sum_j f_j\left\{\frac{\omega_j^2-\omega^2}{(\omega_j^2-\omega^2)^2+\omega^2\Gamma_j^2}+\frac{i\Gamma_j\omega}{(\omega_j^2-\omega^2)^2+\omega^2\Gamma_j^2}\right\}\\ &= 1+\frac{4\pi Ne^2}{m}\sum_j\frac{f_j}{(\omega_j^2-\omega^2)-i\omega\Gamma_j},\end{aligned} \tag{8.51}$$

which is more or less the most general type of dispersion formula. Although calculated for a model of independent atoms, it provides a phenomenological expression for any system whose absorption spectrum is a series of discrete lines.

There is one further correction that should be made in this analysis. The relation (8.42) implies that the *local* field polarizing each atom is the same as the *macroscopic* field applied to the crystal. This is not strictly true; the atom is not situated in an absolutely continuous medium, but occupies a site in a lattice where it is surrounded, but not permeated, by 'other' atoms, similarly polarized (see § 2.3). It is well known, in elementary electrostatics and magnetostatics, that there must be a correction to the local field; if the 'cavity' surrounding the atom is more or less spherical, then we have the *Lorentz correction* to the local field

$$\mathbf{E}_{\text{local}} = \mathbf{E}_{\text{macroscopic}} + \tfrac{4}{3}\pi\mathbf{P}, \tag{8.52}$$

where $\mathbf{P}$ is the average polarization. This leads to the *Clausius–Mosotti relation*

$$\left(\frac{\epsilon - 1}{\epsilon + 2}\right) = \tfrac{4}{3}\pi N\alpha \tag{8.53}$$

in place of (8.42). The result is that the observed dielectric constant and refractive index will follow a more complicated formula than (8.51); for example, the régime of 'total' reflection may be avoided.

When the dielectric polarization is due to the lining up of *permanent* dipole moments on the molecules of a liquid or solid, we must introduce further corrections, such as the *Onsager local field* model, which takes account of fluctuations in the orientation of neighbouring atomic moments.

8.3 Optical modes in ionic crystals

Let us return to the equations of motion of the lattice, as in § 2.1. It was there shown that the displacement of the sth ion in the lth cell satisfies (2.7), i.e.

$$M_s \ddot{\mathbf{u}}_{sl} = -\sum_{s'\mathbf{h}} \mathbf{G}_{ss'}(\mathbf{h}) \cdot \mathbf{u}_{s', l+\mathbf{h}}. \tag{8.54}$$

Now suppose we drive this system with an electromagnetic wave of the form of (8.3). We must add to the right-hand side of (8.54) a force

$$e_s \mathbf{E}_0 \exp\{i(\mathbf{K}\cdot\boldsymbol{l} - \omega t)\}, \tag{8.55}$$

where e_s is the charge of the sth ion in the cell.

We only need a particular integral of the differential equations (8.54) to take account of such an inhomogeneous term. It is obvious that the assumption of (2.8), with $\mathbf{q} = \mathbf{K}$, will remove the space-varying factor; we get an equation for $\mathbf{U}_{s\mathbf{K}}$, the amplitude of the mode of wave-vector $\mathbf{K}$;

$$M_s \ddot{\mathbf{U}}_{s\mathbf{K}} = -\sum \mathbf{G}_{ss'}(\mathbf{K}) \cdot \mathbf{U}_{s'\mathbf{K}} + e_s \mathbf{E}_0 e^{-i\omega t}. \tag{8.56}$$

To see the form of the solution of these equations, let us consider the simple model of the diatomic linear chain as in (2.22). In practice we are thinking of electromagnetic waves of very great wavelength on the atomic scale, so we may take $\cos Ka \approx 1$. If the two atoms are of opposite charge, as in a typical ionic crystal, then we have

$$\left.\begin{aligned} M_1 \ddot{U}_1 &= 2\alpha(U_2 - U_1) + eE_0 e^{-i\omega t}, \\ M_2 \ddot{U}_2 &= 2\alpha(U_1 - U_2) - eE_0 e^{-i\omega t}. \end{aligned}\right\} \tag{8.57}$$

These have the solution

$$(U_1 - U_2) = \frac{eE_0(1/M_1 + 1/M_2)}{\omega_T^2 - \omega^2} e^{-i\omega t}, \tag{8.58}$$

where ω_T is the natural frequency of the system at $q = 0$, as given by (2.25). We may calculate the dipole moment associated with this motion and express it as a polarizability of the lattice

$$4\pi N\alpha(\omega) = \frac{4\pi Ne^2(1/M_1 + 1/M_2)}{\omega_T^2 - \omega^2}. \tag{8.59}$$

We are thus led, by a purely classical argument, to a dispersion formula similar to (8.42). The numerator of (8.59) may be written $\Omega_p'^2$, indicating that it is very like the plasma frequency (6.81) for the assembly of ions. Since the ions are much heavier than electrons, this must be much smaller than ω_p^2. On the other hand, ω_T is just the frequency of a lattice vibration of long wavelength, which is much smaller than the energy of any electronic transition in the atoms or ions. Over a wide frequency range, therefore, we may write

$$\epsilon(\omega) \approx \epsilon(\infty) + \frac{\Omega_p'^2}{\omega_T^2 - \omega^2}, \tag{8.60}$$

where the error arises from local field effects such as (8.53). In this formula, $\epsilon(\infty)$ represents the dielectric constant deduced from observations of the refractive index for $\omega \gg \omega_T$, before electronic transitions are excited. From this we may deduce the *Born equation* for the static dielectric constant

$$\epsilon(0) \approx \epsilon(\infty) + \Omega_p'^2/\omega_T^2. \tag{8.61}$$

In a three-dimensional ionic crystal, ω_T would be the frequency of a *transversely* polarized optical mode, whose dipole moment interacts strongly with a transverse electromagnetic wave. We have here, in fact, *a polariton*—a mixed mode of light and crystal polarization which is the true mode of propagation in the material. When ω passes through ω_T, the dielectric constant becomes negative: just as in (5.67) and (8.46), the radiation would be totally reflected from the crystal surface. The *Reststrahlen effect*—a forbidden region for electromagnetic propagation in the crystal—persists up to the frequency ω_L, say, where (8.60) again becomes positive, i.e.

$$\omega_L^2 = \omega_T^2 + \Omega_p'^2/\epsilon(\infty). \tag{8.62}$$

But the definition of a dielectric constant implies (cf. (5.59)) that a very large electric field variation may then exist within the crystal at this frequency without external excitation. This is the analogue of a plasma oscillation in the ionic lattice, with contributions from both local interatomic forces and long range electrostatic fields: ω_L is the long wave limit of the *longitudinal* optical branch of the lattice spectrum. From (8.61) and (8.62) we derive a companion to (8.60)—the *Lyddane–Sachs–Teller relation*

$$\omega_L^2/\omega_T^2 = \epsilon(0)/\epsilon(\infty) \tag{8.63}$$

which is quite general for any polar material.

In practice, (8.59) and (8.60) should contain an imaginary term in the denominator (cf. (8.51)) corresponding to dissipative effects. For example, energy may be absorbed from the electromagnetic field by the creation of phonons, which would be linked to the optical modes by anharmonicity of the force constants (§§ 2.11, 7.10, 8.4). These effects would be temperature-dependent and would also be observable as a variation of the effective force constant α in the harmonic approximation (2.25) or (8.53). It can even happen that the balance of interatomic forces for some set of lattice displacements is so delicate that this temperature dependence takes ω_T^2 to *negative* values below some critical temperature T_C—an event that we should recognize macroscopically as a transition to a *ferro-electric* phase (§ 10.5) with a permanent built-in dielectric polarization of the lattice structure.

The relatively low frequency of the longitudinal optical modes is important in the theory of the *polaron*. An electron in an ionic crystal carries an electrostatic field that polarizes the lattice around it. For a stationary electron we could use the dielectric constant $\epsilon(0)$ but the response of the lattice to a moving electron is limited dynamically by the frequency ω_L, which describes the natural oscillations of dielectric polarization of the medium.

The situation is very much the same as in the theory of plasmons (§ 5.8). If the energy of the electron exceeds $\hbar\omega_L$ an optical phonon will be produced, and the electron will be decelerated and strongly scattered. Even when the electron is moving slowly, we must take account of the virtual excitation and re-absorption of quanta in the optical modes; in other words, the electron behaves like a particle surrounded by a cloud of virtual phonons. The self-energy of this compound object is a function of the velocity, so that the effective 'mass' of the carrier is considerably increased.

The idealized Hamiltonian of the *large polaron* has been extensively studied as a model of a strongly-interacting field-theoretical system, but it is not an easy entity to observe in practice. The model itself is only valid when the polarization region is somewhat larger than a unit cell of the lattice; otherwise the self-trapped *small polaron* model is more appropriate (§ 6.7).

8.4 Photon–phonon transitions

In the language of field theory, infra-red absorption by a crystal is described as the interaction of a *photon* with one or more phonons. The selection rules for all such processes have already been deduced in §§ 2.7, 2.8, under the guise of *inelastic diffraction*. In the simplest case, for example, the *change* in wave-vector of the diffracted beam had to be equal to the wave-vector of the phonon emitted, i.e.

$$\mathbf{k} - \mathbf{k}' = \mathbf{q} \tag{8.64}$$

and also the change in frequency of the radiation had to be equal to the frequency of the phonon,

$$\omega - \omega' = \nu \tag{8.65}$$

as in (2.101) and (2.102). In special cases it is possible to satisfy these conditions without any resultant diffracted wave; the whole energy and momentum of the photon is transferred to the crystal.

The main point to notice here is that we are dealing with light in the visible and infra-red region, whose wavelength is much larger than the lattice constant of the crystal. The magnitude of $|\mathbf{k} - \mathbf{k}'|$ must therefore be extremely small on the scale of the Brillouin zone. For all ordinary optical phenomena we can take $\mathbf{k} - \mathbf{k}' \approx 0$, and concentrate our attention on the possible changes of energy allowed in the zone scheme; this aspect is complementary to the 'diffraction' point of view appropriate for X-rays and neutrons, where the change in *momentum* is large but the energy differences are not so easy to detect.

The process which we are here considering, the absorption or emission of one optical phonon, is therefore indicated by a nearly vertical line at $\mathbf{q} = 0$, in the reduced zone scheme. If the optical spectrum had several branches, then several different lines ought in principle to be observed, as in Fig. 141 (*b*). But the magnitude of such processes would depend upon the matrix element for the coupling of the electromagnetic wave to the crystal. As we have seen in § 8.3, this coupling is so strong for the transverse optical modes of most polar crystals that a description in terms of weakly interacting photons and phonons is inadequate and we get the Reststrahlen effect. But optical modes of very weak dipole moment are found in some compound semiconductors, where *one-phonon absorption* may be observed in light transmitted through thin films.

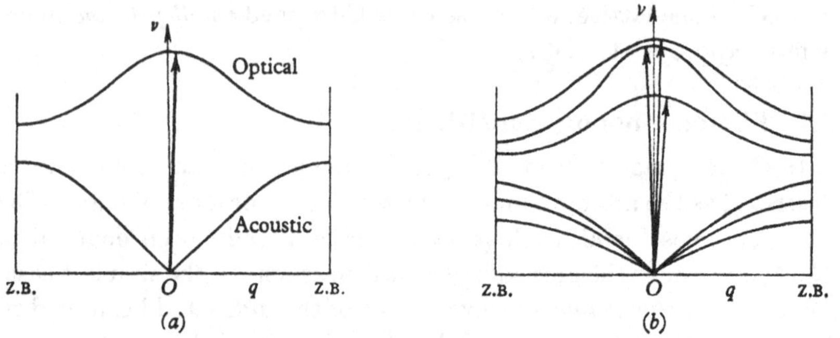

Fig. 141. Absorption by optical modes: (*a*) one-dimensional scheme; (*b*) three-dimensional scheme.

The observation that photon absorption (and emission) processes are 'vertical' in the Brillouin zone suggests that we may have other 'vertical' transitions, as indicated in Fig. 142. This is a transition involving an acoustic mode and an optical mode. The energy of the photon absorbed is evidently made up as follows

$$\hbar\omega = \hbar\nu_{\text{optical}} - \hbar\nu_{\text{acoustic}}. \quad (8.66)$$

In effect, a photon, and an acoustic phonon have disappeared; an optical phonon has been created. The momentum conditions are also satisfied. We may ignore the momentum of the photon, and simply assert that

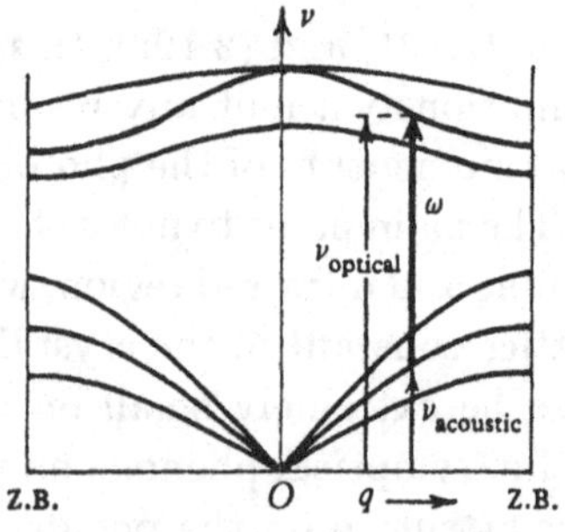

Fig. 142. Multiphonon absorption.

$$\hbar\mathbf{q}_{\text{optical}} = \hbar\mathbf{q}_{\text{acoustic}}, \quad (8.67)$$

which is implied by the statement that the transition must be vertical.

Such processes are allowed by the selection rules, and therefore may be observed. They give rise to whole bands of absorption in the far infra-red. The width and general structure of such bands obviously depends on the variety of possible transitions from acoustic to optical modes and to their spacing. The complete discussion of such processes is obviously rather complicated, since we are dealing with the difference of two functions, each of which is a function of **q** in the Brillouin zone.

This does not exhaust all the possible processes. The above transitions we may describe as '*difference*' processes. It is also possible to have *summation* bands, in which *two* phonons are created out of the photon energy, but of equal and opposite momenta. In that case

$$\hbar\omega = \hbar\nu_{\text{optical}} + \hbar\nu_{\text{acoustic}}, \tag{8.68}$$

with

$$\hbar\mathbf{q}_{\text{optical}} = -\hbar\mathbf{q}_{\text{acoustic}}.$$

To actually calculate the transition probability we need to know the non-linear term in the dipole moment as a function of lattice displacement. We need, for example, the tensor $\mathsf{B}_{ss'}$ in an expression like

$$\mathbf{M}_l = \sum_s \mathsf{A}_s \cdot \mathbf{u}_{sl} + \sum_{ss'} \mathbf{u}_{sl} \cdot \mathsf{B}_{ss'} \cdot \mathbf{u}_{s'l} \tag{8.69}$$

for the dipole moment in the lth cell, due to the displacements $\mathbf{u}_{sl}$. We then use (2.8), (2.10) and (2.129) to express each of the displacements in terms of the annihilation and creation operators for phonons in the various branches of the spectrum. The term in $\mathsf{B}_{ss'}$, corresponding to the product of two such operators, gives rise to the two-phonon processes for which we are looking. The actual probability of a transition will contain the squares of two matrix elements, of the form

$$\left.\begin{aligned} &\langle n_{\mathbf{q}}+1\,|a_{\mathbf{q}}^*|\,n_{\mathbf{q}}\rangle = (n_{\mathbf{q}}+1)^{\frac{1}{2}}, \\ \text{or}\qquad &\langle n_{\mathbf{q}}-1\,|a_{\mathbf{q}}|\,n_{\mathbf{q}}\rangle = n_{\mathbf{q}}^{\frac{1}{2}}, \end{aligned}\right\} \tag{8.70}$$

corresponding to the increase or decrease of the occupation number of the phonon states by one quantum. Since the average value of $n_{\mathbf{q}}$ is a function of the temperature, as in (2.46), these processes will be temperature dependent. In general, however, the non-linear coefficient $\mathsf{B}_{ss'}$ is small, so that these bands are weak by comparison with the one-phonon processes.

It is worth remarking that the function $\nu_{\mathbf{q}}$, for any branch, is a continuous periodic function of **q** with the period of the reciprocal lattice (cf. § 2.2). It follows that the sum or difference of any pair of such functions has the same properties. Thus, van Hove's theorem (§ 2.5) applies to the distribution of frequencies; only the four types of singularity indicated in Fig. 27 can occur in the observed spectrum (unless the matrix elements introduce extra singularities).

In (8.64), (8.65) we wrote down selection rules for the inelastic diffraction of light—i.e. for a photon $|\mathbf{k}, \omega\rangle$ to be transformed into another photon $|\mathbf{k}', \omega'\rangle$ with the emission of a phonon $|\mathbf{q}, \nu\rangle$. This would be observed as a shift in the frequency of some of the light scattered by the crystal. This first-order *Raman effect* is evidently equivalent to one-phonon absorption. The second-order Raman spectrum of a crystal contains bands from two-phonon processes—and so on. Where acoustic modes of finite wave-vector are involved, we refer to *Brillouin scattering*.

The selection rules for momentum (8.64), (8.67), (8.68) depend upon Bloch's theorem (see § 2.7). In the absence of perfect lattice translational symmetry, transitions would be allowed to any point in the Brillouin zone. An impurity or other imperfection of the lattice thus induces infra-red absorption and similar optical phenomena which would otherwise be forbidden. The detailed calculation of such effects is obviously very complicated, but sharp lines from *localized modes* and peaks from resonances (§ 2.12) can thus be observed directly.

8.5 Interband transitions

When the electrons are in Bloch states, forming broad energy bands, the electronic transitions from full to empty states produce a broad absorption band. Again, for optical frequencies, $\mathbf{K}$ is small on the scale of the Brillouin zone, so that one can ignore the wave-vector of the photon, and assume that all transitions are 'vertical', as indicated in Fig. 143.

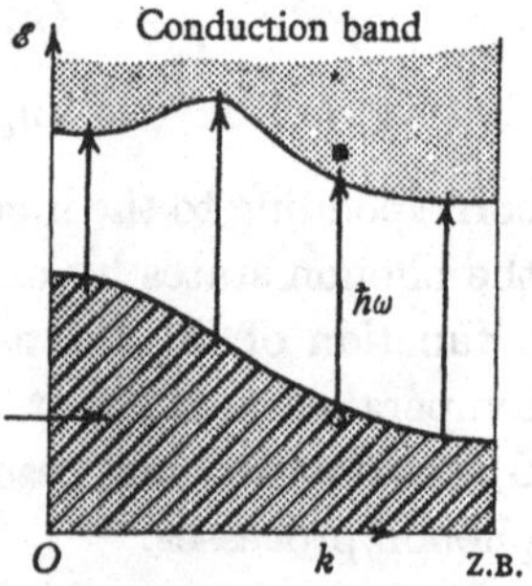

Fig. 143. Vertical interband transitions in a semiconductor.

We have already written down, in (5.45), the formula for the dielectric constant of a semiconductor. In that case we were more interested in the static behaviour at all wavelengths; we are now interested in the frequency dependence of $\epsilon(\mathbf{q}, \omega)$ at long wavelengths. We have from (5.57)

$$\epsilon(\mathbf{q}, \omega) = 1 + \frac{4\pi e^2}{q^2} \sum_{\mathbf{k}} \frac{|\langle \mathbf{k}|\, e^{i\mathbf{q}\cdot\mathbf{r}}\, |\mathbf{k}+\mathbf{q}+\mathbf{g}\rangle|^2\, 2\{\mathscr{E}(\mathbf{k}) - \mathscr{E}(\mathbf{k}+\mathbf{q}+\mathbf{g})\}}{(\hbar\omega)^2 - \{\mathscr{E}(\mathbf{k}) - \mathscr{E}(\mathbf{k}+\mathbf{q}+\mathbf{g})\}^2}$$

$$\approx 1 + \frac{4\pi e^2}{m} \sum_{\mathbf{k}} \frac{f_{\mathbf{k}}}{\omega_{\mathbf{k}}^2 - \omega^2}, \tag{8.71}$$

where $f_{\mathbf{k}}$ is an oscillator strength, as in (8.40), for the transition between $|\mathbf{k}\rangle$ in the valence band and $|\mathbf{k}+\mathbf{g}\rangle$ (directly above it in the reduced zone scheme) in the conduction band, these states differing in energy by the amount $\hbar\omega_{\mathbf{k}}$. But since $\mathbf{k}$ is continuous, the sum becomes an integral, of the form

$$\epsilon(0,\omega) \approx 1 + \frac{4\pi e^2}{m}\int \frac{f(\omega')\mathcal{N}_d(\omega')\,d\omega'}{\omega'^2-\omega^2}, \tag{8.72}$$

where $\mathcal{N}_d(\omega')\,d\omega'$ is the number of levels having a vertical energy difference $\hbar\omega'$, in the range $d\omega'$, and $f(\omega')$ is an oscillator strength—i.e. a number of the order of unity—for transitions in this range.

This formula for the real part of the dielectric constant is not so interesting as the imaginary part. Appeal to the dispersion relations (8.29) and (8.30) shows that the absorption coefficient must be of the form

$$\begin{aligned} 2\omega \mathsf{n}(\omega)\,\mathsf{k}(\omega) &= \frac{\pi}{2}\frac{4\pi e^2}{m}\int f(\omega')\mathcal{N}_d(\omega')\,\delta(\omega-\omega')\,d\omega' \\ &= \frac{2\pi^2 e^2}{m} f(\omega)\mathcal{N}_d(\omega). \end{aligned} \tag{8.73}$$

We could have arrived at the same result by noting that $\epsilon(\mathbf{q},\omega)$ in (5.45) was provided with an imaginary part, governed by the parameter α, which was introduced in (5.1) as the decay constant of the perturbation being screened by the electron system. It is evident that this parameter plays the same role as the line-width parameter Γ_j in (8.51), which is, indeed, essentially of the form of (5.45) and (5.57). Thus, we return to (8.73) by retracing the steps from (8.51) to (8.47)—effectively, by letting the line-width for each transition tend to zero.

The function $\mathcal{N}_d(\omega)$ in (8.73), being the spectrum of the difference of the energies of the valence and conduction bands, each of which is a continuous periodic function in the reciprocal lattice, has only the van Hove singularities noted in § 2.5. The most important of these is the *absorption band edge*, corresponding to the minimum vertical energy difference $\hbar\omega_0$ between the bands. As shown in (2.72), the behaviour of the spectrum in that neighbourhood must be of the form

$$\mathcal{N}_d(\omega) \propto (\omega-\omega_0)^{\frac{1}{2}}. \tag{8.74}$$

It should be noted, however, that $\hbar\omega_0$ is not necessarily the same as the 'energy gap', $\mathscr{E}_{\text{gap}}$, between the top of the valence band and the bottom of the conduction band. These points may not be vertically above one another in $\mathbf{k}$-space: the smallest *vertical* separation of the bands may be somewhat larger.

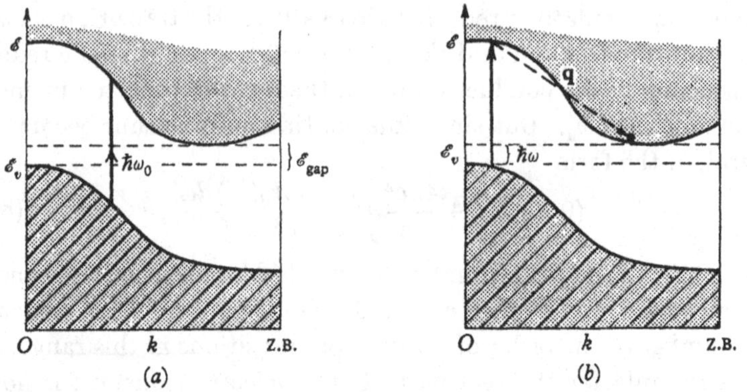

Fig. 144. (*a*) Direct transitions do not necessarily give energy gap. (*b*) Phonon-assisted transition.

However, it is possible to observe optical transitions corresponding to $\hbar\omega \approx \mathscr{E}_{\text{gap}}$, if one allows the simultaneous emission or absorption of a phonon. Such *indirect* or *phonon-assisted* transitions are easily derived by second-order perturbation theory. We have two steps; for example, upwards, vertically by the absorption of a photon, and then to the appropriate minimum in the conduction band by the emission or absorption of a phonon of large enough wave-vector, **q**. We need not worry about energy conservation in the intermediate virtual state; the over-all condition will be

$$\hbar\omega = \mathscr{E}_{\text{gap}} \pm \hbar\nu_{\mathbf{q}}, \tag{8.75}$$

where $\nu_{\mathbf{q}}$ is the frequency of the phonon. Since $\hbar\nu_{\mathbf{q}}$ is a small energy (for example, 0·03 eV.) on the scale of the energy gap, we find that the absorption band seems to start at about $\mathscr{E}_{\text{gap}}$. But the probability of indirect transitions is much smaller than that of direct transitions, and depends on the temperature through the occupation number of the phonon states, as indicated in (8.70).

This discussion of electronic transitions in perfect crystals assumes a simple one-electron model. In reality, however, the final state for a transition such as Fig. 144 contains a hole in the valence band as well as an electron in the conduction band. If these two particles do not immediately escape from one another, they may interact, as in § 6.7, to form bound states—Wannier exciton levels—whose total energy is less than the band gap from which they were created. In other words, the optical spectrum shows *exciton lines* below the fundamental absorption edge. This phenomenon is particularly noticeable in ionic and crystals, where the Frenkel excitons, although 'small' and practically immobile, have large oscillator strengths corresponding to the atomic or molecular excited states from which they derive. In such materials the coupling between the electromagnetic field and the excitonic amplitude may be strong enough to produce polariton modes of propagation (cf. § 8.3).

The optical properties of semiconductors and ionic crystals are also greatly complicated and enriched by the effects of impurities and imperfections. Thus, one can observe a variety of absorption lines associated with transitions between the hydrogen-like levels of charged impurities (§ 6.4).

The theory of optical transitions in colour centres is made somewhat more complicated by the lattice distortion about the imperfection. Since this distortion is affected by electrostatic interaction with the optically active electron, it must depend on the electronic state of the system. The eigenstates of the imperfection must therefore be described by both electronic and atomic co-ordinates. But an optical transition takes place in the periodic time of the frequency absorbed or emitted, whereas the time needed for the lattice to distort into the configuration required by the new electronic state would be the periodic time of a dynamical mode, which is very much longer. The transition must therefore be treated as *non-adiabatic* or *sudden* from the point of view of the lattice (cf. § 6.11).

This is the basis of the *Franck–Condon* principle, illustrated in Fig. 145. Suppose that the distortion of the lattice about the imperfection can be measured by a single co-ordinate u, such as the amplitude of a 'breathing' displacement of the ions, inwards or outwards. When the electron is in state $|a\rangle$, say, the potential energy of the lattice has a minimum at some value u_a, which is not the same as the equilibrium displacement, u_b for a different electron state $|b\rangle$. Optical absorption from $|a\rangle$ to $|b\rangle$ will proceed without change of lattice distortion—i.e. 'vertically' along the line $u = u_a$. On the other hand, emission must occur along the u_b, corresponding to the equilibrium displacement in the initial electron state $|b\rangle$. Evidently, $\hbar\omega_{ab} \neq \hbar\omega_{ba}$. In practice, however, we must allow for thermal excitation of the lattice modes near each potential minimum. Phonon emission will broaden the lines and eventually, at high temperatures, smear out the difference between the two processes.

Interband optical transitions are, of course, also observable in metals—being modified to allow for the fact that the electrons may not completely fill a band, or may fall into several different zones. The theory of such transitions follows, in principle, the argument of this section—but of course the phenomenon may be somewhat masked by

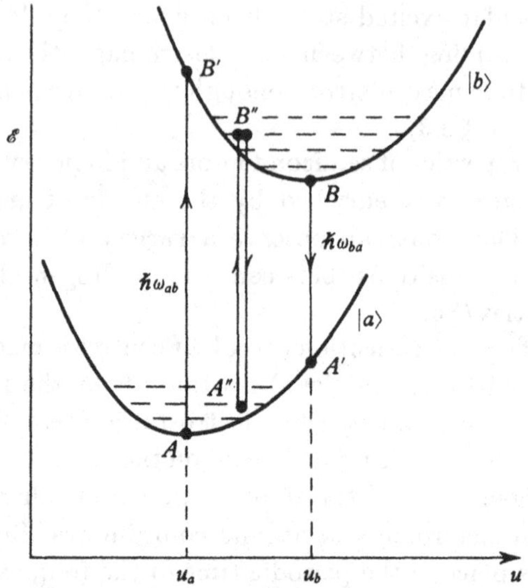

Fig. 145. Optical transitions at F-centre, showing effect of lattice distortions: absorption, AB' requires higher energy than emission, BA'. At high temperatures, phonon levels combine with electronic transitions in both directions, $A''B''$.

the high reflectivity associated with the conductivity of the electron system (see § 8.6).

Another phenomenon which is essentially equivalent to an interband transition is the *emission of soft X-rays*. When an electron is removed from one of the deepest levels in an atom of the solid, X-rays are emitted by the transition of an electron down from various other levels. We are concerned with transitions from the valence bands of a metal—it is intuitively obvious that the spectrum of the emitted radiation will show a band of frequencies reflecting the band of states occupied by the conduction or valence electrons.

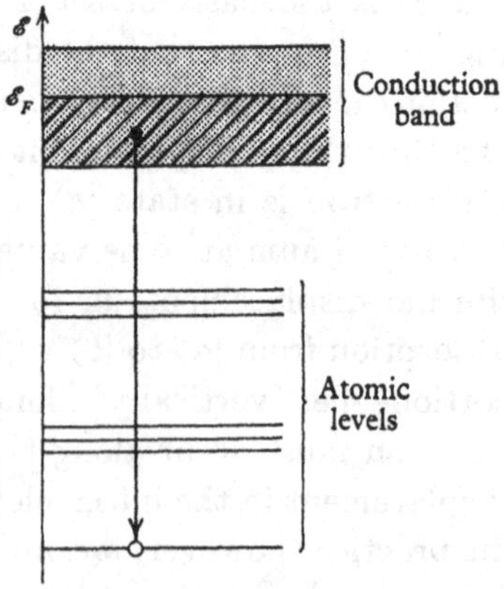

Fig. 146. X-ray emission.

The actual shape of this spectrum will depend on the product of the density of states in the band with the square of the appropriate matrix element for the transition. It is well known that the interaction between an electron and the electromagnetic field proceeds via the momentum operator for the electron (i.e. by the current produced by its motion) so that one must study matrix elements like

$$\int \psi_{\mathbf{k}}^{*}(\mathbf{r})\, e^{i\mathbf{K}\cdot\mathbf{r}}\, \text{grad}\, \phi_a(\mathbf{r})\, d\mathbf{r}, \tag{8.76}$$

where ϕ_a is the atomic orbital from which the electron is missing and $\mathbf{K}$ is the wave-vector of the X-rays. The wavelengths associated with such transitions are still long on the scale of atomic distances, so this wavy factor can be ignored—and, since $\phi_a(\mathbf{r})$ is a localized function there is no selection rule for momentum.

However, this matrix element may vary considerably through the band, according as the level ϕ_a is an s-state or a p-state, and according as $\psi_{\mathbf{k}}(\mathbf{r})$ is more or less s-like or p-like in the interior of the atom. There are also effects due to electron–electron interaction. The exact shape of the emission spectrum is not, therefore, necessarily a direct measure of the density of states, although it should reflect some of the features of that function, especially the sharp cut-off at the Fermi level.

Optical or X-ray absorption by a crystal is accompanied by the excitation of electrons into higher levels. In a semiconductor or insulator these excited electrons are normally mobile, hence giving rise to *photoconductivity*. But carriers produced in this way are very susceptible to *trapping* by impurities and imperfections, as witnessed by a multitude of complex experimental phenomena.

Given sufficient energy, the electron may be *photo-emitted* through the crystal surface. The elementary Einstein formula for the *photo-electric effect* merely tells us that its energy outside the surface cannot exceed

$$\mathscr{E}_{\max} = \hbar\omega - \phi_W \tag{8.77}$$

since it must have surmounted the work-function barrier, ϕ_W (§ 6.9). But photo-electrons of lower energy must have come from levels within the crystal below the Fermi energy. Direct measurements of $n(\mathscr{E}, \omega)$—the energy distribution of emitted electrons for given photon frequency—should give a great deal of information about the probabilities of transitions from occupied to empty band states, similar to those responsible for the absorption of light.

To interpret these observations, we naturally construct an energy level diagram (Fig. 147) and count vertical or *direct* transitions in the manner of Fig. 143. In principle, with control over both $\mathscr{E}$ and ω we should thus obtain somewhat more information about the band structure than is given by, say, the interband spectral density of (8.73). In practice, the photo-emission spectrum for many crystalline solids does not conform to this pattern, but seems to behave like

$$n(\mathscr{E}, \omega) \sim \rho_f(\mathscr{E})\, \rho_i(\mathscr{E} - \hbar\omega), \tag{8.78}$$

as if 'non-direct' transitions were allowed indiscriminately between initial states of energy density $\rho_i(\mathscr{E}-\hbar\omega)$ and final states of energy $\rho_f(\mathscr{E})$. The experiments are technically difficult, but unequivocal: it is not yet clear whether this discrepancy with simple theory is due simply to the breakdown of crystal momentum selection rules near the emitting surface as in L.E.E.D. (§ 6.9), or whether many-electron effects are responsible.

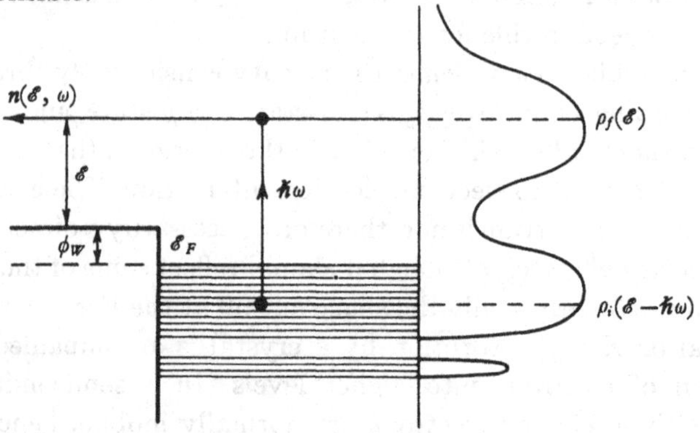

Fig. 147. Photo-emission.

8.6 Interaction with conduction electrons

What happens to the optical properties when the solid is a relatively good conductor? Suppose, for example, that we ignore ϵ in the formula (8.7) for the complex refractive index—an assumption that is valid for metals.

Then we have

$$\mathsf{N}^2 = \frac{4\pi\sigma}{\omega}i, \tag{8.79}$$

and the real and imaginary parts are equal in magnitude;

$$\mathsf{n}+i\mathsf{k} = \left(\frac{2\pi\sigma}{\omega}\right)^{\frac{1}{2}}(1+i). \tag{8.80}$$

The most obvious consequence of this is that the reflecting power of the solid becomes very high. From (8.19),

$$\mathsf{R} \approx 1-2\left(\frac{\omega}{2\pi\sigma}\right)^{\frac{1}{2}}, \tag{8.81}$$

which is known as the *Hagen–Rubens relation*. The deviation from perfect reflectivity is proportional to

$$\left(\frac{\omega}{2\pi\sigma}\right)^{\frac{1}{2}} \sim \left(\frac{2\omega}{\omega_p^2\tau}\right)^{\frac{1}{2}}, \tag{8.82}$$

where τ is the relaxation time in the classical formula (7.33) for the conductivity, and ω_p is the ubiquitous plasma frequency of the electron gas. We know that this ratio will be quite small even when ω approaches infra-red frequencies.

But this elementary macroscopic theory assumes that σ is independent of the frequency. This will not be true when the field oscillates so rapidly that the electrons are not given time to make collisions, i.e. when

$$\omega\tau > 1. \tag{8.83}$$

To investigate this region we must return to our transport equation (7.14).

For completeness, and later reference, let us write down the Boltzmann equation for a distribution that may vary in space and time:

$$e\mathbf{E}\cdot\mathbf{v}_\mathbf{k}\left(-\frac{\partial f^0}{\partial \mathscr{E}}\right) = \frac{g_\mathbf{k}}{\tau} + \mathbf{v}_\mathbf{k}\cdot\frac{\partial g_\mathbf{k}}{\partial \mathbf{r}} + \frac{\partial g_\mathbf{k}}{\partial t}. \tag{8.84}$$

We assume a relaxation time as in (7.17). The variation of the 'out of balance' distribution $g_\mathbf{k}$ with time may be included explicitly; as we see from (7.7), this is what is left over when all the other contributions to $\partial f_\mathbf{k}/\partial t$ have been accounted for.

Now suppose

$$\mathbf{E} = \mathbf{E}_0 \exp\{i(\mathbf{K}\cdot\mathbf{r} - \omega t)\}, \tag{8.85}$$

as in (8.3), and that $g_\mathbf{k}$ varies, in space and time, in the same manner, by 'sympathy'. Thus, let us write

$$g_\mathbf{k} = \left(-\frac{\partial f^0}{\partial \mathscr{E}}\right)\Phi(\mathbf{k})\exp\{i(\mathbf{K}\cdot\mathbf{r} - \omega t)\}. \tag{8.86}$$

Substituting in (8.84), we have

$$e\mathbf{E}_0\cdot\mathbf{v}_\mathbf{k} = \frac{\Phi(\mathbf{k})}{\tau} + i\mathbf{K}\cdot\mathbf{v}_\mathbf{k}\,\Phi(\mathbf{k}) - i\omega\Phi(\mathbf{k}), \tag{8.87}$$

which has the solution

$$\Phi(\mathbf{k}) = \frac{e\tau\mathbf{v}_\mathbf{k}\cdot\mathbf{E}_0}{1 - i\omega\tau + i\tau\mathbf{K}\cdot\mathbf{v}_\mathbf{k}}. \tag{8.88}$$

The linearity of our Boltzmann equation makes this elementary solution possible.

For the conductivity we have, as in (7.20)–(7.24),

$$\begin{aligned} \boldsymbol{\sigma}\cdot\mathbf{E}_0 &= \frac{e}{4\pi^3}\int \Phi(\mathbf{k})\,\mathbf{v}_\mathbf{k}\frac{dS_F}{\hbar v_\mathbf{k}} \\ &= \frac{e^2}{4\pi^3}\int \frac{\tau\mathbf{v}_\mathbf{k}\mathbf{v}_\mathbf{k}}{1 - i\tau(\omega - \mathbf{K}\cdot\mathbf{v}_\mathbf{k})}\frac{dS_F}{\hbar v_\mathbf{k}}\cdot\mathbf{E}_0. \end{aligned} \tag{8.89}$$

Thus the conductivity itself has real and imaginary parts; the real part of $\boldsymbol{\sigma}$ will contribute to the imaginary part of N^2, whilst the imaginary part of $\boldsymbol{\sigma}$ will look like part of the real dielectric constant.

For an ordinary electromagnetic wave, in which the phase velocity is greater than the Fermi velocity of an electron, we can drop the term $\mathbf{K}\cdot\mathbf{v_k}$, which is much smaller than ω. For formal simplicity assuming that the metal has cubic symmetry, we get from (8.89)

$$\sigma(\omega) = \frac{e^2}{12\pi^3}\int \frac{\tau v(1+i\omega\tau)}{(1+\omega^2\tau^2)}\,dS_F$$

$$= \sigma(0)\frac{1+i\omega\tau}{1+\omega^2\tau^2}, \tag{8.90}$$

where $\sigma(0)$ is the ordinary static conductivity of the metal.

This formula should be put into (8.7) and (8.19). If we take $\epsilon = 1$, i.e. if we ignore any internal polarizability of the ions, then we have, from (8.7), (8.8), (7.33), and (5.53), for the real and imaginary parts of N^2

$$\mathsf{n}^2 - \mathsf{k}^2 = 1 - \frac{4\pi\sigma(0)\,\omega\tau}{\omega(1+\omega^2\tau^2)}$$

$$= 1 - \frac{\omega_p^2\,\tau^2}{1+\omega^2\tau^2} \tag{8.91}$$

and

$$2\mathsf{nk} = \frac{4\pi\sigma(0)}{\omega(1+\omega^2\tau^2)}$$

$$= \frac{\omega_p^2\,\tau}{\omega(1+\omega^2\tau^2)}, \tag{8.92}$$

where, once more, the plasma frequency of the electron gas plays a most important role.

There are three different frequency regions covered by this *Drude theory*.

(i) $\omega \ll 1/\tau$. This is the ordinary low-frequency region. The imaginary part of N^2 is much larger than the real part, so that the metal is strongly reflecting, and the *Hagen–Rubens* relation (8.81) is valid. The absorption coefficient is more or less independent of ω, and is proportional to the conductivity. The real part of N^2 is negative, and much larger than unity in magnitude.

(ii) $1/\tau \ll \omega \ll \omega_p$. This is the *relaxation region* where $\omega^2\tau^2$ takes over in the denominators of (8.91) and (8.92). The absorption coefficient falls rapidly, proportionately to $1/\omega^2$—and, strangely, is *inversely* proportional to the conductivity. The imaginary part of N^2 becomes less than the real part—but this is still large and negative, being now of the form

$$n^2 - k^2 = 1 - \frac{\omega_p^2}{\omega^2} \tag{8.93}$$

as in (5.67) and (8.45). Thus, the metal is still strongly reflecting, with

$$R \approx 1 - \frac{1}{\sqrt{\pi\sigma_0\tau}} \approx 1 - \frac{2}{\omega_p\tau}. \tag{8.94}$$

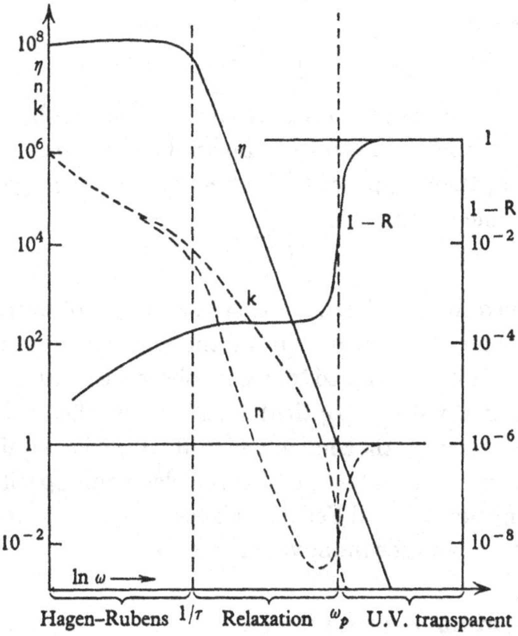

Fig. 148. Schematic behaviour of optical properties of metals, showing real and imaginary parts of dielectric constant, reflection coefficient and absorption coefficient. Note logarithmic scale of frequency.

(iii) $\omega_p \ll \omega$. The real part of N^2 becomes positive, and the reflecting power falls to zero. The metal should now appear more or less transparent, with an absorption coefficient

$$\frac{2\omega n(\omega)\, k(\omega)}{c} \approx \frac{\omega_p^2}{\omega^2\tau c}. \tag{8.95}$$

The formulae (8.91) and (8.92) imply certain relations between n and k, as functions of ω, which are not, in fact, always obeyed. But this can, to some extent, be understood if one supposes that the relaxation time, τ, in the integral (8.89), is not a constant, but varies over the Fermi surface. Thus, we can, in principle, distinguish three different integrals

$$\int \tau v \, dS_F, \quad \int \tau^2 v \, dS_F, \quad \int \frac{v}{\tau} dS_F, \tag{8.96}$$

corresponding to three different types of average of $\tau(\mathbf{k})$ over the Fermi surface. There is no reason why these different averages should all be equal.

We also notice that the real part of N^2 at high frequencies depends upon the integral

$$\int v \, dS_F, \tag{8.97}$$

which is, for free electrons, proportional to the number of electrons and inversely proportional to their mass. In the density of states, on the other hand, there appears a different type of average of the velocity—a harmonic mean

$$\int \frac{dS_F}{v} \tag{8.98}$$

as in (4.6). The difference between these two types of average can give information about the anisotropy of the electron velocity over the Fermi surface. It is not wise to 'correct' observed values of (8.97) by a factor associated with an 'optical mass', which then turns out to be very different from the 'thermal mass' that would be required to make (8.98) fit the free-electron formula for the electronic specific heat. The implicit assumption that all Fermi surfaces are spheres often turns out to be unjustified even for monovalent metals.

8.7 The anomalous skin effect

The Hagen–Rubens relation (8.81) may break down in another way. The absorption coefficient given by (8.79) is very high and the electromagnetic wave is very rapidly attenuated as it enters the metal. The damping distance, as derived from (8.10) and (8.80), is of the order

$$\begin{aligned} \delta &= \frac{c}{\mathsf{k}\omega} \\ &= \frac{c}{(2\pi\sigma\omega)^{\frac{1}{2}}}. \end{aligned} \tag{8.99}$$

This phenomenon is well known, classically, as the *skin effect*. We refer to δ, as given by (8.99), as the *classical skin depth*, $\delta_{\text{cl.}}$.

At high frequencies this may become very small. If we have a metal of great purity at low temperatures, we find that the skin depth can become much less than the electron mean free path, Λ. The ordinary theory of the electrical conductivity is no longer valid; the effective field acting on a carrier is varying rapidly over the distance the carrier moves between collisions.

To deal with this case we must try to solve the Boltzmann equation with $g_{\mathbf{k}}$ varying in space. If we ignore the time variation of $\mathbf{E}$ and $g_{\mathbf{k}}$ (and this phenomenon can easily be observed at frequencies well below the onset of the relaxation régime discussed above), then we have, as in (8.84),

$$e\mathbf{E}(\mathbf{r})\cdot\mathbf{v}_{\mathbf{k}}\left(-\frac{\partial f^0}{\partial \mathscr{E}}\right) = \frac{g_{\mathbf{k}}}{\tau} + \mathbf{v}_{\mathbf{k}}\cdot\frac{\partial g_{\mathbf{k}}}{\partial \mathbf{r}} \tag{8.100}$$

as the differential equation to be solved for $g_{\mathbf{k}}$.

The most direct procedure for the solution of (8.100)—a procedure that is applicable to a variety of problems involving surfaces, thin films, wires, etc.—is to use the *Chambers formula*,

$$g(\mathbf{v},\mathbf{r}) = \frac{e}{v}\left(-\frac{\partial f^0}{\partial \mathscr{E}}\right)\int^{\mathbf{r}} \mathbf{v}\cdot\mathbf{E}(\mathbf{r}')\, e^{-s'/\tau v}\, ds'. \tag{8.101}$$

It is not very difficult to prove, quite formally, that this is a solution of the differential equation; it is simply a generalization of the usual formula for the solution of an equation of the form $dy/dx + \alpha y = f(x)$.

What it means, physically, is that the electron current at the point $\mathbf{r}$, in the direction $\mathbf{v}$, depends upon the previous history of the electrons that contribute to that current. The integral is not over all space, but simply over the distance s' back along the trajectory passing through $\mathbf{r}$ in the direction $\mathbf{v}$. Thus, at the point $\mathbf{r}'$ the electrons are accelerated by the force $e\mathbf{E}(\mathbf{r}')$. But this contribution decays with time, because of relaxation processes; the exponential factor measures this effect as a function of distance along the trajectory. This semi-classical formula is closely related in principle to the Kubo formula (7.15) where we similarly integrate over the previous history of the system.

From (8.101), one can calculate the total electric current by the usual formula (7.20). But we notice that this current will be a function of $\mathbf{r}$—and also of the boundary conditions on the function $g(\mathbf{v},\mathbf{r})$. Thus, there is an integral equation defining $\mathbf{J}(\mathbf{r})$ as a functional of the distribution of electric field $\mathbf{E}(\mathbf{r})$. On the other hand, these two quantities are linked by Maxwell's equations in a form more general than (8.2), i.e.

$$\begin{aligned}\nabla^2\mathbf{E} &= \frac{4\pi}{c^2}\frac{d\mathbf{J}}{dt}\\ &= \frac{4\pi i\omega}{c^2}\mathbf{J}(\mathbf{r})\end{aligned} \tag{8.102}$$

(ignoring the ordinary displacement current). Thus, one can solve the integro-differential equation, derived from (8.101) and (8.102), and obtain the distribution of $\mathbf{J}(\mathbf{r})$ and $\mathbf{E}(\mathbf{r})$ in the neighbourhood of the surface.

The exact solution of this equation is rather elaborate, and depends upon the boundary conditions—it depends on whether the electrons are specularly reflected at the surface, or whether they are randomly scattered there. The actual formulae are not so important as their physical interpretation. The fact is that not all the electrons are participating in the absorption and reflection of the electromagnetic wave. Only those that are running inside the skin depth for most of a mean free path Λ are capable of picking up much energy from the

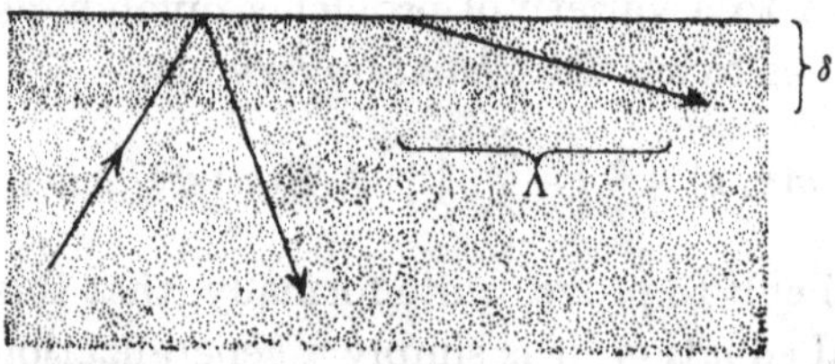

Fig. 149. Only electrons in the skin depth are 'effective'.

electric field. If, for example, the skin depth is δ', then one may suppose that only a fraction δ'/Λ of the electrons are *effective* in the conductivity. We may write

$$\sigma' = \tfrac{3}{2}\beta\frac{\delta'}{\Lambda}\sigma_0 \tag{8.103}$$

for the apparent conductivity—the number β being just a 'fudge factor'.

But then, a surface with this conductivity would, according to (8.99), itself have a skin depth

$$\begin{aligned}\delta' &= \frac{c}{(2\pi\sigma'\omega)^{\frac{1}{2}}}\\ &= \frac{c}{\left(2\pi\,\tfrac{3}{2}\beta\dfrac{\delta'}{\Lambda}\sigma_0\,\omega\right)^{\frac{1}{2}}},\end{aligned} \tag{8.104}$$

which can be solved for δ', and for σ'. Thus

$$\sigma' = \left(\frac{9\beta^2}{8\pi}\right)^{\frac{1}{3}}\frac{1}{\omega^{\frac{1}{3}}}\left(\frac{c\sigma_0}{\Lambda}\right)^{\frac{2}{3}}; \tag{8.105}$$

the effective conductivity will appear to behave as $\omega^{-\frac{1}{3}}$, but will be independent of the electron mean free path, which cancels with the ordinary static conductivity σ_0. We are then in the conditions of the *anomalous skin effect*. What one does, in practice, is to make the surface of the metal part of a resonant cavity, and then observe the effect upon the resonance properties of the cavity. For examples, one measures the *surface impedance*, which is just

$$\mathsf{Z}(\omega) = -4\pi i\omega \frac{E(z)}{(\partial E/\partial z)}\Big]_{z=0} = \frac{4\pi c}{\mathsf{N}} \tag{8.106}$$

in the notation of § 8.1.

There are various other changes in the theory of the apparent optical properties of metals in the anomalous limit which can be worked out, especially when one gets into the relaxation region of frequencies. For example, there is an additional absorption of power from a reflected wave due to the actual scattering of electrons from the surface, if this is 'rough'—a contribution that is as large as the effects calculated in the previous section where such scattering was ignored.

The most interesting property of the anomalous skin effect is, however, its dependence on the geometry of the Fermi surface. As we saw above, only the electrons travelling nearly parallel to the surface are 'effective' in the conductivity. If our specimen is a single crystal, then those electrons derive from a narrow belt running round the Fermi surface. If the Fermi surface is highly anisotropic, we should see different values of the surface impedance according as we measure on different cuts of the metal crystal, and at different orientations of the field.

The geometrical property that is being measured can be derived by an extension of the *ineffectiveness concept*, (8.103). We assume that we need only consider electrons whose velocity vectors lie within an angle $\pm\beta\delta'/\Lambda$ of parallelism with the (x, y)-plane, which defines the metal boundary (we take the electric field in the x-direction).

At some general point, P, on the belt of effective electrons, the velocity $\mathbf{v}$ makes an angle θ with $\mathbf{E}$. The width of the belt there will be $2\beta\delta'\,|\rho|/\Lambda$, where $|\rho|$ is the radius of curvature in the plane of $\mathbf{v}$ and of the z-axis. An element of the circumference of the belt, of length ds, will thus correspond to an area of Fermi surface

$$dS = \frac{2\beta\delta'\,|\rho|}{\Lambda}\,ds, \tag{8.107}$$

and will evidently contribute to the conductivity in the x-direction, by the usual formula (7.23),

$$d\sigma'_{xx} = \frac{e^2}{4\pi^3\hbar}\tau v_x\, dS_x$$

$$= \frac{e^2}{4\pi^3\hbar}\Lambda\cos\theta\frac{2\beta\delta'\,|\rho|}{\Lambda}\cos\theta\, ds. \qquad (8.108)$$

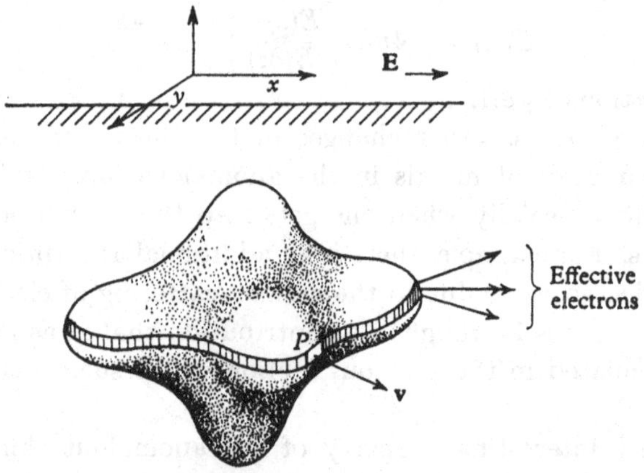

Fig. 150. Belt of effective electrons on a Fermi surface.

Thus, the total apparent conductivity of the metal will be

$$\sigma' = \frac{e^2\beta\delta'}{2\pi^3\hbar}\int|\rho|\cos^2\theta\, ds$$

$$= \frac{e^2\beta\delta'}{2\pi^3\hbar}\oint|\rho_y|\, dk_y, \qquad (8.109)$$

where ρ_y means the radius of curvature, on the belt, of a section of the Fermi surface parallel to the (x, z)-plane, and we integrate over the range of the momentum component k_y around the belt. This is the quantity we put into (8.104), and then solve for δ' and σ' separately. A more complete analysis shows that

$$\beta = \frac{8\pi}{3\sqrt{3}} \qquad (8.110)$$

is the proper value for this arbitrary 'fudge factor'.

The beauty of this result is that the surface impedance does not depend upon the mean free path, or velocity, of the electrons, but only on the local curvature integrated round a well-defined belt. The integral (8.109) is especially sensitive to regions of large radius of curvature—that is, to relatively flat portions of the Fermi surface. It thus provides information that is to some extent complementary to other 'effects', which often depend on small details and little bits of disconnected Fermi surface. It is not, however, easy to invert the relation between a given surface and the values of a set of integrals like (8.109) for belts drawn in different orientations; a tedious procedure of trial and error is usually necessary.

8.8 Ultrasonic attenuation

Although not strictly an 'optical' property, the theory of the attenuation of high-frequency elastic waves by interaction with the conduction electrons in a metal comes most naturally in this chapter. A sound wave in a solid gives rise to electric fields which accelerate electrons in much the same way as an electromagnetic wave. But the velocity of sound is so much less than the velocity of light—less even than the Fermi velocity of the electrons—that certain special effects are observed.

In the ordinary low-frequency region, the attenuation can be calculated by appeal to elementary kinetic theory. A gas of electrons, of mass m, number density n, average velocity v_F and mean free path Λ would have a viscosity

$$\eta = \tfrac{1}{3}nm\Lambda v_F. \tag{8.111}$$

This viscosity can be included as a term in the ordinary classical equation for the forces acting on an element of the medium. The attenuation constant for longitudinal elastic waves of frequency ω comes out as

$$\alpha = \frac{4}{5}\frac{\omega^2}{Ds^3}\eta, \tag{8.112}$$

where D is the mass density and s is the velocity of sound. These may be put together into various forms, for example

$$\alpha = \frac{4}{15}\frac{\omega^2 m v_F^2}{e^2 D s^3}\sigma, \tag{8.113}$$

in terms of the conductivity (7.33).

If the metal is very pure, and the measurement is made at low temperatures, the relaxation time of the electrons is greatly increased. Yet it is still difficult, with ultrasonic waves, to get into the relaxation region where $\omega\tau > 1$. However, because the velocity of sound is so much less than the Fermi velocity, it is possible to get the mean free path longer than the wavelength of the sound, i.e. to have

$$q\Lambda > 1. \tag{8.114}$$

To discuss this case, we return to (8.89), where we constructed a formula for the solution of the Boltzmann equation in an electric field varying sinusoidally in space and time. In effect, we found a generalized conductivity

$$\sigma(\mathbf{q},\omega) = \frac{e^2}{4\pi^3}\int \frac{\tau\mathbf{v}\mathbf{v}}{1-i\tau(\omega-\mathbf{q}\cdot\mathbf{v})}\frac{dS_F}{\hbar v}. \tag{8.115}$$

When discussing this formula in § 8.6, we took $q = 0$, on the grounds that the velocity of an electromagnetic wave is so large that its wave-number may be taken to be nearly zero. But when the electric field is

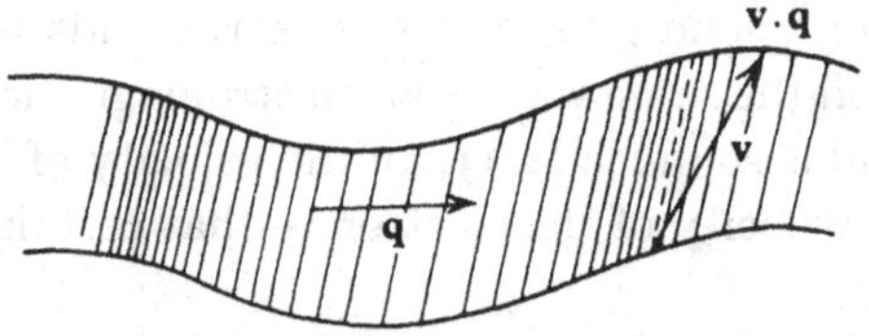

Fig. 151. Surf-riding resonance.

generated by an acoustic wave, which travels so much more slowly than the electrons, we must retain this term. The real part of the conductivity then comes predominantly from regions of the Fermi surface where $\omega \sim \mathbf{q}\cdot\mathbf{v}$. This is another way of saying that the component of the electron velocity in the direction of propagation of the wave is equal to the velocity of sound.

This is known as the *surf-riding resonance*. An electron that happens to be on the crest of the wave can continue to draw energy from the field by travelling in a direction nearly normal to the propagation vector of the wave. It is evident that only a few electrons on the Fermi surface can satisfy this condition. If we neglect ω completely, it defines a belt round the Fermi surface where the electron velocity is normal to $\mathbf{q}$—a belt of exactly the same form as the one defined in the theory of the anomalous skin effect (§ 8.7).

It is not difficult to evaluate (8.115) in the limit $q\Lambda \gg 1$. The effective width of the belt depends upon the rate at which $\tau\mathbf{q}\cdot\mathbf{v}$ increases as we go away from the exact resonance position. The larger the value of τ, the narrower does it become. Just as in (8.107), the element of 'effective' area is proportional to the local radius of curvature of the surface, and inversely proportional to Λ. For example, one can prove by elementary differential geometry the exact formula for the limit $q\Lambda \gg 1$

$$\sigma_{xx}(\mathbf{q}, 0) = \frac{e^2}{4\pi^3\hbar q}\oint |\rho_y|\, dk_y, \tag{8.116}$$

where the directions x and y are normal to the direction of propagation, $\mathbf{q}$, and the other symbols are exactly as in (8.109).

The interesting feature of this formula is that it is independent of Λ. It is essential that there should be some mechanism by which the electrons are finally scattered, but in this limit the strength of the

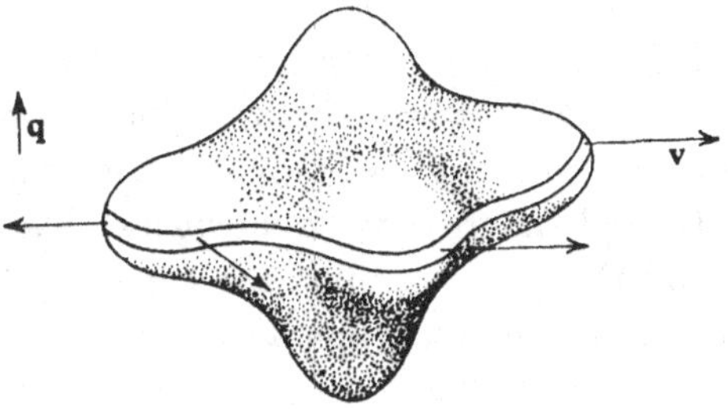

Fig. 152. Belt of electrons interacting with phonon $\mathbf{q}$.

scattering does not matter. This is the situation we have already encountered in § 8.7. Indeed, one can use (8.116) as the starting point for a complete derivation of the formula (8.109) for the anomalous skin effect, including the value of the fudge factor β. It is a matter, essentially, of expressing the relation (8.102), derived from Maxwell's equations, in Fourier components, and also putting in boundary conditions that allow for the behaviour of the electrons at the metal surface.

It would be very agreeable if we could use (8.116) by itself in a theory of ultrasonic attenuation. The experiment of measuring the attenuation of waves propagating in various directions through a single crystal is obviously much easier, technically, than cutting crystal surfaces in various orientations, and keeping them clean and perfect whilst measuring the surface impedance.

Unfortunately, this will not give direct information about the shape of the Fermi surface. The attenuation does, indeed, tend to a value independent of the mean free path of the electrons—(8.113) is no longer valid—but the actual magnitude depends upon the details of the coupling of the electrons to the lattice wave. In other words, the electric field seen by an electron as a result of lattice displacements or lattice strains cannot be calculated directly, and varies over the Fermi surface. One may think of having to insert, in the integrand of (8.116) a *deformation potential* of the sort discussed in § 6.14—but much more complicated, much less easily expressed by a few parameters, on a multiply connected Fermi surface, say, than in a simple free-electron metal.

Ultrasonic attentuation by free carriers can also occur in a semiconductor. An ionic material without a centre of symmetry—e.g. a III–V or II–VI compound (§ 4.2)—is usually *piezo-electric*. A local strain $W_{ij}(\mathbf{r})$ gives rise to a local electric *field*

$$\delta E_k(\mathbf{r}) = \sum_{ij} \beta_{ijk} W_{ij}(\mathbf{r}) \tag{8.117}$$

in addition to the deformation potential (6.95). This travelling wave of electric field carried by a macroscopic sound wave affects the carriers, and is the dominant mechanism of interaction until one reaches relatively high microwave frequencies. The details of the phenomenon would then depend upon the components of the piezo-electric tensor β_{ijk} projected on the polarization vector of the sound wave.

In a medium of relatively low conductivity, the *acousto-electric field* produced by the beam of ultrasonic phonons transferring momentum to the carriers may be directly observed. This is obviously the inverse of the phonon drag effect (§ 7.11). A closely related phenomenon *is acousto-electric amplification* of sound by the application of a D.C. electric field along the ultrasonic beam. This may be regarded as a form of maser action: energy from the electric field unbalances the carrier distribution, which is then stimulated into the coherent emission of phonons. We may note the analogy with Cerenkov radiation: the electric field gives the carrier a drift velocity (7.31) that exceeds the velocity of sound, hence permitting the direct creation of phonons with conservation of energy and momentum.

We have based the above derivation upon a quasi-classical treatment of the Boltzmann equation. But the same results may be obtained, some people might say more rigorously, from quantum theory. For example, consider the energy condition (2.101) for the absorption of a phonon of frequency ω in the transition from state $\mathbf{k}$ to state $\mathbf{k}+\mathbf{q}$. This gives us

$$\begin{aligned} \hbar\omega &= \mathscr{E}(\mathbf{k}+\mathbf{q}) - \mathscr{E}(\mathbf{k}) \\ &\approx \mathbf{q}\cdot\frac{\partial \mathscr{E}(\mathbf{k})}{\partial \mathbf{k}} \\ &= \hbar\mathbf{q}\cdot\mathbf{v}_{\mathbf{k}}, \end{aligned} \tag{8.118}$$

essentially the same as (8.115). In other words, the 'surf-riding resonance' is no more than an elementary process of the electron–phonon interaction, as discussed in §§ 2.8, 6.13.

Again (8.115) can be derived from quite a different source. Let us write the denominator in the form

$$1 - i\tau(\omega - \mathbf{q}\cdot\mathbf{v}) = \frac{i\tau}{\hbar}\{\mathscr{E}(\mathbf{k}+\mathbf{q}) - \mathscr{E}(\mathbf{k}) - \hbar\omega - i\hbar\alpha\}, \qquad (8.119)$$

where $\alpha = -1/\tau$. We recognize the denominator of the sum occurring in (5.16) and elsewhere in Chapter 5. With a little juggling, and use of (8.118), we find that we may write

$$\epsilon(\mathbf{q}, \omega) \approx 1 + \frac{4\pi i \sigma_l(\mathbf{q}, \omega)}{\omega} \qquad (8.120)$$

when q and ω are small. Here σ_l means the longitudinal conductivity—the component of $\boldsymbol{\sigma}$ in the direction of propagation of the phonon.

In other words, we have rederived (8.7). The complex dielectric constant of the electron gas, as calculated in § 5.1, already contains the conductivity as its imaginary part. In (5.1) we took $-\alpha$ to be an arbitrary decay constant for the perturbing field—but the identification with $1/\tau$ is obviously appropriate. There is an extraordinary amount of physics packed up in (5.16).

CHAPTER 9

THE FERMI SURFACE

Thou com'st in such a questionable shape. *Hamlet*

9.1 High magnetic fields

Generally speaking, the electronic properties of metals at high temperatures are complicated averages of velocities, wave-functions, transition probabilities, etc., over the whole Fermi surface. Apart from vague and fragmentary information derived from the Kohn effect (§ 5.4) and from positron annihilation (§ 5.8), procedures for actually delineating that surface, from experimental observations, depend upon the use of pure specimens, at low temperatures where the scattering of the electrons does not blur the phenomena. It is doubtful even whether a Fermi surface can properly be said to exist when the relaxation time of the electrons is very short. For example, an electron with mean free path Λ has an uncertainty in momentum of the order of $\hbar/\Lambda$; at high temperatures this might well be as large as some of the detailed structure of the energy surfaces in reciprocal space.

We have already discussed the anomalous skin effect (§ 8.7). Almost all other studies of the Fermi surface depend upon the use of high magnetic fields. The effect of a magnetic field on an electron state is given by (6.40);

$$\dot{\mathbf{k}} = \frac{e}{c\hbar}\mathbf{v}\wedge\mathbf{H}. \tag{9.1}$$

This means that the change in the vector **k** is

(i) normal to the direction of **H**,

(ii) normal to **v**, which is itself normal to the energy surface.

Thus, **k** must be confined to the *orbit* defined by the intersection of the Fermi surface with a plane normal to **H**. The magnetic field simply drives the representative point round this orbit without change of energy.

If the electron is not scattered, it makes a circuit in the period

$$\frac{2\pi}{\omega_H} = \frac{c\hbar}{eH}\oint\frac{dk}{v_\perp}, \tag{9.2}$$

where $v_\perp$ is the component of **v** in the plane normal to **H** at the point **k**.

The corresponding frequency, ω_H, is called the *cyclotron frequency.* For free electrons we have, by elementary geometry,

$$\oint \frac{dk}{v_\perp} = \frac{m}{\hbar}\oint \frac{dk}{k_\perp} = \frac{2\pi m}{\hbar}, \tag{9.3}$$

so that

$$\omega_H = \frac{eH}{mc}. \tag{9.4}$$

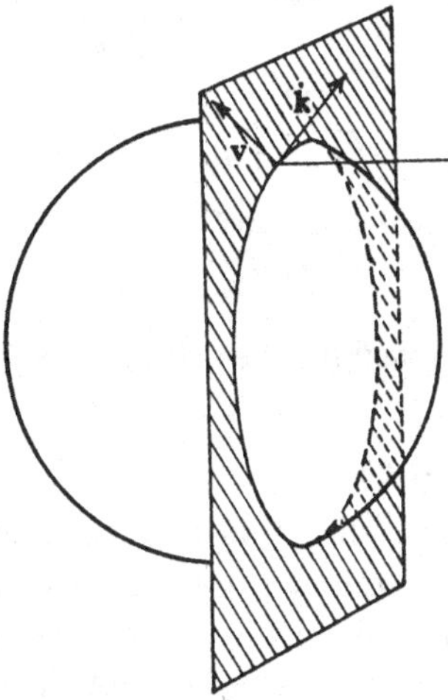

Fig. 153. 'Orbit' of electron in a magnetic field.

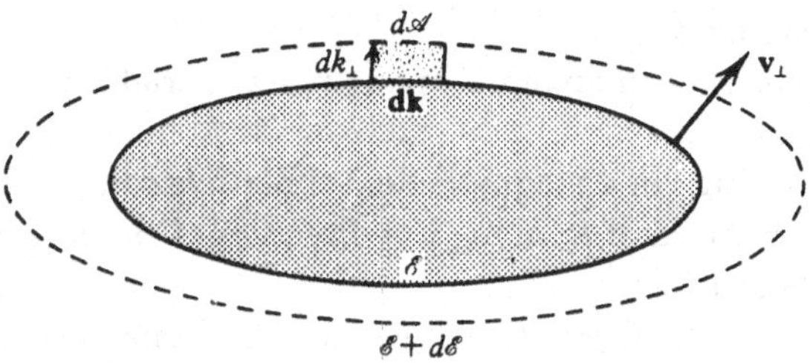

Fig. 154

It is customary to define yet another 'effective mass'—the *cyclotron mass*, m_H^*, such that

$$\omega_H = \frac{eH}{m_H^* c}. \tag{9.5}$$

It should be noted that this is not the same as the dynamical mass of the electron. It is a property of an orbit, not of a particular electronic state.

A useful geometrical definition of the cyclotron mass comes from the fact, implicit in (6.2), that

$$v_\perp = \frac{1}{\hbar}\frac{d\mathscr{E}}{dk_\perp}, \tag{9.6}$$

where $dk_\perp$ is an increment of k, in the plane of the orbit, normal to the Fermi surface. From (9.2) and (9.5) we have (Fig. 154)

$$\begin{aligned} m_H^* &= \frac{eH}{c}\frac{1}{2\pi}\frac{c\hbar}{eH}\oint\frac{dk}{v_\perp} \\ &= \frac{\hbar^2}{2\pi}\oint\frac{dk_\perp}{d\mathscr{E}}dk \\ &= \frac{\hbar^2}{2\pi}\frac{\partial\mathscr{A}}{\partial\mathscr{E}}, \end{aligned} \tag{9.7}$$

where $\mathscr{A}$ is the area enclosed by the orbit in the plane normal to **H**.

9.2 Cyclotron resonance

It is natural to suggest that this periodic motion of the electron could be detected by resonance with an electromagnetic field of suitable frequency. The only condition is that the electron should make at least one circuit of the orbit before being scattered. It is easy enough to construct a formal theory for the response of the system to a circularly polarized electromagnetic wave travelling parallel to the axis of the magnetic field.

For example, let us map out **k**-space in new variables. A point **k** is specified as follows:

(i) Energy $\mathscr{E}(\mathbf{k})$; in a magnetic field this stays constant.

(ii) Component k_H of **k**, along the direction of the magnetic field; this also stays constant.

(iii) A 'phase variable', ϕ, defined by the same sort of integral round the orbit as appears in (9.2), i.e.

$$\phi = \omega_H\frac{c\hbar}{eH}\int^{\mathbf{k}}\frac{dk}{v_\perp}. \tag{9.8}$$

This is a convenient variable because it increases automatically at a constant rate under the influence of the magnetic field:

$$\dot{\phi} = \omega_H, \tag{9.9}$$

and $\phi = 2\pi$ for a complete circuit.

We can use these variables as a framework of co-ordinates for the Boltzmann equation. We can express the 'out-of-balance' part $g_{\mathbf{k}}$ in (7.10) as a function of $(\mathscr{E}, k_{\mathbf{H}}, \phi)$. If we assume a relaxation time, as in (8.84), we need only add a term

$$\left.\frac{\partial f}{\partial t}\right]_{\text{mag.}} = \dot{\phi}\frac{\partial g}{\partial \phi} = \omega_H \frac{\partial g}{\partial \phi} \tag{9.10}$$

to include the effect of the magnetic field on the electron distribution. This is equivalent to, but more compact than, the expression used in (7.137), where the magnetic field was supposed to be small.

Let us take a time-varying field, of frequency ω, and ignore the space variation of $\mathbf{E}$. Let us assume, also, by analogy with (8.86), that

$$g(\mathscr{E}, k_{\mathbf{H}}, \phi) = \left(-\frac{\partial f^0}{\partial \mathscr{E}}\right)\Phi(k_{\mathbf{H}})\, e^{i(\phi-\omega t)}, \tag{9.11}$$

i.e. that the electron distribution (if not each individual electron) is actually driven round the orbit at the frequency of the applied field. Then our Boltzmann equation

$$e\mathbf{E}\cdot\mathbf{v}\left(-\frac{\partial f^0}{\partial \mathscr{E}}\right) = \frac{g}{\tau} + \omega_H \frac{\partial g}{\partial \phi} + \frac{\partial g}{\partial t} \tag{9.12}$$

has the solution

$$\Phi(k_{\mathbf{H}}) = \frac{e\tau\mathbf{v}\cdot\mathbf{E}}{1+i(\omega_H-\omega)\tau}. \tag{9.13}$$

The deviation is, as before, proportional to the strength of the field, but follows round the orbit, out of phase, unless the applied frequency is the same as the natural cyclotron frequency of the system.

Just as in (8.90), the metal will behave as if it had the conductivity

$$\sigma(\omega) = \sigma(0)\frac{1-i(\omega_H-\omega)\tau}{1+(\omega_H-\omega)^2\tau^2}. \tag{9.14}$$

Observation of the surface resistance, reflecting power, absorption, etc., of the material will show a *cyclotron resonance* line, of width $1/\tau$, at the frequency $\omega = \omega_H$. The analogy of this result with the theory of optical resonance, in (8.51), is obvious. It is not difficult to get this answer for free electrons by a simple analysis of the equation of motion—but the result we have proved here is somewhat more general, although not quite exact for an arbitrary Fermi surface where the assumption (9.11) ignores higher harmonics in the variable ϕ.

This formula is for a circularly polarized beam rotating *with* the natural cyclotron motion of the electrons in the magnetic field. It is obvious, from consideration of the geometry, that the formula for a beam polarized in the opposite sense, will have $+\omega$ in place of $-\omega$; so that there would be no resonance. But consider the *imaginary* part of $\sigma(\omega)$, which contributes to the *real* part of the complex refractive index (8.7), and hence alters the phase velocity of the electromagnetic wave. For a linearly polarized beam propagated along the direction of the magnetic field, the two circularly polarized components travel with different velocities, so that the plane of polarization continually rotates as the wave passes through the crystal. This is the well-known *Faraday effect*.

For semiconductors this is almost all the theory one needs to describe a cyclotron resonance. The carriers, whether electrons or holes, occupy little pockets of the $\mathscr{E}(\mathbf{k})$ function, at minima or maxima of the band. In these neighbourhoods the energy surfaces are usually ellipsoidal (cf. § 2.5); it can be shown that ω_H is the same on all parallel sections of an ellipsoid, so that the line predicted by (9.14) is observed. Of course ω_H varies with orientation, thus providing information about the axes of the ellipsoid. The situation is more complicated near a maximum of the valence band at the centre of the zone, where the levels are split by spin-orbit interaction (§ 3.9) but the analysis still follows essentially the above lines.

For metals there are two difficulties which seem to bar the way. In order to resolve the resonance line, we need to be working at high frequencies, with $\omega\tau \gg 1$. In these conditions the skin depth is very small—indeed, we are in the anomalous region of § 8.7. Thus the radius of the helical trajectory of an electron in the magnetic field, in real space, will be much larger than the distance to which the electric field will penetrate.

This suggests that one ought to apply the magnetic field normal to the metal surface, so that the resonating electrons are the 'effective' ones in the skin depth, and can absorb power. But if one makes an analysis of this situation one finds that the 'resonance' is smeared out until it is scarcely detectable. One can see, in principle, how this happens by putting (9.14) into (8.7). Just as in (8.91), the complex refractive index has a large imaginary part, chiefly because the *real* part of N^2 is large and negative. The metal is too highly conducting; all that we can measure is the slight imperfection of its reflecting power, as in (8.94). We should have to go above the plasma frequency ω_p, which would require enormous magnetic fields, to see the same sort of resonance line as may readily be detected in a semiconductor.

On the other hand, suppose that we have 'good' cyclotron material such as very pure Na at 4 °K, in which $\omega_H \tau \gg 1$ in a magnetic field of a few thousand oersteds. If we work at a very low frequency—no more than a few cycles per second—we can make the refractive index (8.7) nearly real and observe electromagnetic propagation through a slab of the metal. This follows from (9.14); putting $\omega \ll \omega_H$, we get

$$\mathsf{N} = \left(\epsilon + \frac{4\pi i\sigma(\omega)}{\omega}\right)^{\frac{1}{2}}$$

$$\approx \left(\frac{4\pi\sigma(0)}{\omega\omega_H\tau}\right)^{\frac{1}{2}} = \left(\frac{4\pi nec}{H\omega}\right)^{\frac{1}{2}} = \frac{\omega_p}{(\omega\omega_H)^{\frac{1}{2}}}. \qquad (9.15)$$

This turns out to be a very large number, of the order of 10^9. The wave therefore travels through the metal at a velocity of only a few centimetres per second—a phenomenon on the 'human' scale of length and time. In (9.15) it is important that ω and ω_H should have the same sign; the convention established in (9.8) and (9.11) implies that we are dealing with a circularly polarized wave rotating in phase with the cyclotron motion of the electrons. A *helicon* mode thus combines cyclotron and plasmon excitations of the electron gas. Unfortunately the information obtainable from this phenomenon is limited since (9.15) does not depend to first order simply on the geometry of the Fermi surface. In a compensated semicondutor (§ 4.6), the electron and hole helicon terms cancel one another, leaving higher order terms describing the propagation of *Alfvén waves*—another magneto-plasma phenomenon familiar in astrophysics.

It turns out that a strong resonance can be detected when the magnetic field is *parallel* to the surface. This *Azbel'–Kaner resonance*, arises as follows. The electron circles round in a helical orbit, whose axis, in the x-direction, say, is parallel to the surface. At each revolution it comes into the skin depth, and 'sees' the oscillating electric field. If it comes back in phase with the field each time, then it will gain energy from the field, and a 'resonance' will be observed.

It is not difficult to derive a formula to describe this phenomenon. Let us write the general kinetic expression (8.101) in the form

$$\mathbf{J}(\mathbf{r}, t) = \frac{e^2}{4\pi^3\hbar}\int \mathbf{v}\left\{\int_{-\infty}^{t} \mathbf{v}(\mathbf{r}', t')\cdot\mathbf{E}(\mathbf{r}', t')\, e^{-(t-t')/\tau}\, dt'\right\}\frac{dS_F}{v}. \qquad (9.16)$$

Here we have integrated over the Fermi surface, to get the whole current. But we still retain explicitly the main feature of the argu-

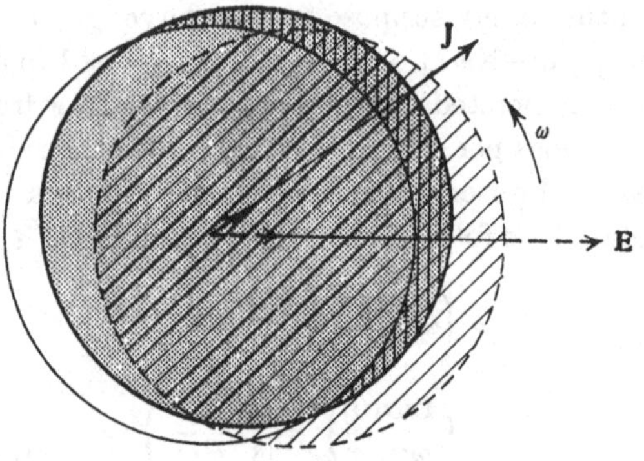

Fig. 155. The electric field displaces the Fermi distributions. The displacement is driven round the Fermi surface, decaying with time.

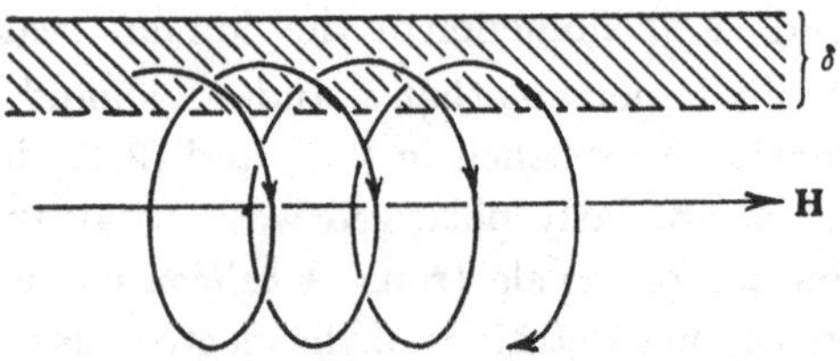

Fig. 156. Azbel'–Kaner resonance.

ment—that the contribution to the current at **r**, at time t, from the electron of velocity **v**, depends upon the impulse given by the field at a previous point **r**′ through which the electron passed at a previous time t'; but the memory of this impulse fades with time, being governed by a decay factor with relaxation time τ. This description of electrical conduction is very close in spirit to the Kubo formula (7.15).

Now the trajectories are helical. The electron reaches **r** along a curved path, along which the decay with time is governed by the same exponential factor. In one period this factor will be $\exp(-2\pi/\omega_H \tau)$. The electric field only acts for the brief time that the electron is in the skin depth; in each revolution the electron will gain or lose in phase by an amount $(2\pi\omega/\omega_H)$ on the oscillating field. Thus, we may reduce the time integral in (9.16) to the result of the first passage through the skin depth, multiplied by the sum of the decay factors for all subsequent passages, i.e.

$$(1+e^{-w}+e^{-2w}+\ldots) = \frac{1}{1-e^{-w}}, \tag{9.17}$$

where
$$e^{-w} = \exp\left(-\frac{2\pi}{\omega_H \tau} - i\frac{2\pi\omega}{\omega_H}\right). \tag{9.18}$$

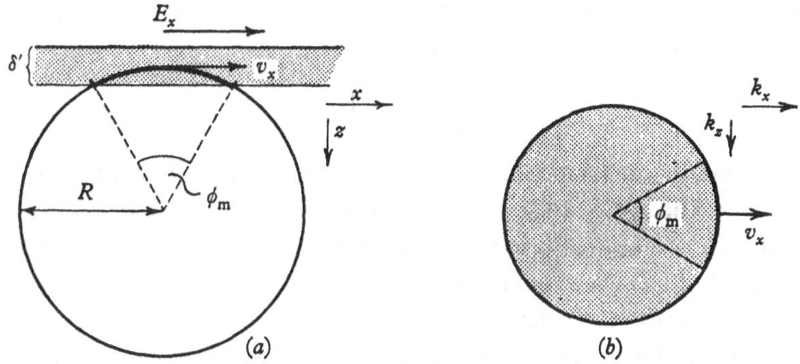

Fig. 157. Electrons are 'effective' in the angle ϕ_m: (a) on the helical path in real space; (b) on the orbit on the Fermi surface.

We can estimate the 'effectiveness' of the electrons by noting that each electron only stays in the skin depth for a time of the order of ϕ_m/ω_H, such that (Fig. 157)

$$\phi_m^2 \approx \frac{\delta'}{R}$$

$$\approx \frac{\delta'\omega_H}{v_x}, \tag{9.19}$$

where R is the radius of the orbit in real space, and v_x is the velocity of the electron, in the plane of section normal to the magnetic field, parallel to the surface.

Then again, most electrons passing through the skin depth will hit the surface of the metal, where they are scattered and lost. Only those which are moving parallel to the surface inside the skin depth are effective. By the sort of argument used in (8.107), we count only those belonging to the elementary area of Fermi surface

$$dS = 2\beta\phi_m |\rho_y|\, dk_y, \tag{9.20}$$

where, again, ρ_y is the radius of curvature of the orbit in the slice of thickness dk_y at the point where the velocity is parallel to the metal surface.

Putting these arguments together, we get a result essentially like (8.109), i.e.

$$\sigma' = \frac{e^2\beta\delta'}{2\pi^3\hbar}\int \frac{|\rho_y|\, dk_y}{1-e^{-w}}. \tag{9.21}$$

The surface resistance can now be calculated by putting (9.21) into (8.104) and (8.105).

For our present purposes it is the resonance factor (9.17) which is of interest. This oscillates, with peaks when

$$\omega = n\omega_H. \tag{9.22}$$

In practice, one keeps the frequency ω constant, and varies the magnetic field. Thus, from (9.5),

$$\frac{1}{H} = n\frac{e}{\omega m_H^* c} \tag{9.23}$$

is the condition for peaks to occur as H is varied. It is important to note that the *Azbel'–Kaner resonance* is not a single peak as in ordinary cyclotron resonance (9.14). Energy will still be fed in if the frequency is such that the electromagnetic field goes through an integral number of cycles between the times that the electron makes its appearance at the surface. The sharpness of each resonance obviously depends on the strength of the real part of w in (9.18); we want $\omega_H \tau \gg 1$ for a strong effect.

There is still one further complication. The value of ω_H (or, equivalently, of m_H^*) depends upon the orbit. For a general Fermi surface, it is *not* the same on different orbits, even when these belong to a set of parallel slices normal to the same magnetic field. Thus, the variable w in (9.21) is itself a function of k_y; contributions from different slices will not necessarily stay in phase.

The exact calculation of the shape of the resonance curve now becomes very complicated. But one can argue—and the argument can be backed by formal mathematics—that the main contributions will come from sections where m_H^* is stationary, either a maximum or a minimum, as a function of k_y. If one expands m_H^* as a function of δk_y, measured from each such point, the constant term provides a resonance peak, which is not removed by the next term in the expansion (of order $(\delta k_y)^2$). The contributions to the integral from other regions of the Fermi surface, where m_H^* varies linearly with δk_y, get out of phase and interfere to cancellation. We have the usual argument of wave optics, where the Cornu spiral is a useful geometrical construction for the solution.

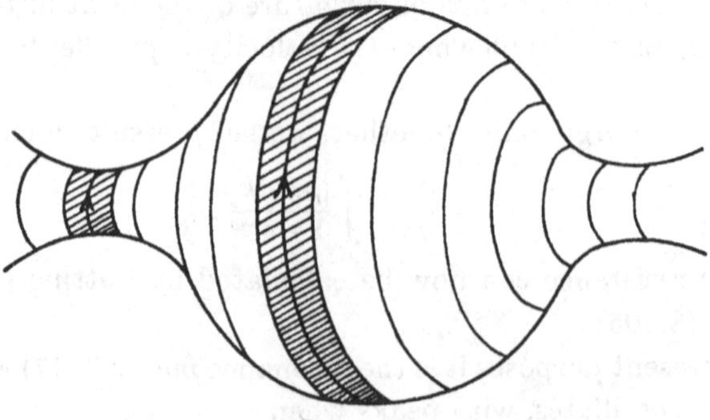

Fig. 158. Extremal orbits.

9.3 High-field magneto-resistance

The theory of *galvanomagnetic phenomena* presented in § 7.12 is only valid in the case of carriers on a spherical energy surface. When $\omega_H \tau > 1$, the representation of electron states by the Cartesian components of $\mathbf{k}$ is mathematically clumsy, and leads to formulae that are physically difficult to interpret. It is simpler to derive the transport properties from the Boltzmann equation expressed in terms of the variables $(\mathscr{E}, k_H, \phi)$. We are looking now for a steady-state solution, in constant electric field. Thus, (9.12) reduces to

$$e\mathbf{E}\cdot\mathbf{v}\left(-\frac{\partial f^0}{\partial \mathscr{E}}\right) = \frac{g}{\tau} + \omega_H \frac{\partial g}{\partial \phi}. \tag{9.24}$$

By analogy with (8.101), this has the integral

$$g(\mathscr{E}, k_{\mathbf{H}}, \phi) = \frac{e}{\omega_H}\left(-\frac{\partial f^0}{\partial \mathscr{E}}\right)\int_{-\infty}^{\phi} \mathbf{v}(\mathscr{E}, k_{\mathbf{H}}, \phi)\exp\{(\phi'' - \phi)/\omega_H \tau\}\, d\phi''\cdot\mathbf{E}. \tag{9.25}$$

As in (7.27), one can say that the displacement of the Fermi surface at the point whose phase angle is ϕ is the sum of the displacements created by the electric field at other points on the orbit, which are then driven round the orbit by the magnetic field, decaying with the characteristic relaxation time τ. In the low-field case the decay is so rapid that we only see a slight rotation of the direction of the displacement away from the electric field direction, giving rise to the Hall effect (cf. (7.141)).

The conductivity tensor can now be calculated from the current

$$\begin{aligned}\mathbf{J} &= 2\int e\mathbf{v}_{\mathbf{k}} g_{\mathbf{k}}\, \mathbf{dk} \\ &= \frac{e}{4\pi^3}\iiint \mathbf{v}(\mathscr{E}, k_{\mathbf{H}}, \phi)\, g(\mathscr{E}, k_{\mathbf{H}}, \phi)\frac{m_H^*}{\hbar^2}\, d\mathscr{E}\, dk_{\mathbf{H}}\, d\phi,\end{aligned} \tag{9.26}$$

where we have used (9.7) and (9.8) to define an element of area on the plane $k_{\mathbf{H}} = $ constant. Substituting (9.25), and using the properties of the derivative of the Fermi function to keep to the Fermi surface, we find for the conductivity tensor, in Cartesian components,

$$\sigma_{\alpha\beta} = \frac{1}{4\pi^3}\frac{e^2}{\hbar^2}\int \frac{m_H^*}{\omega_H}\left\{\int_0^{2\pi}\int_0^{\infty} v_\alpha(\phi, k_{\mathbf{H}})\, v_\beta(\phi - \phi', k_{\mathbf{H}})\, e^{-\phi'/\omega_H \tau}\, d\phi\, d\phi'\right\} dk_{\mathbf{H}}. \tag{9.27}$$

This is the *Shockley tube-integral formula*—a universal formula for the conductivity tensor in a magnetic field. This explicit derivation of a 'Kubo-type' expression (cf. (7.15)) is quite simple—yet it is a formula of considerable power.

In low fields, we may take $\omega_H \tau \ll 1$, and write

$$v_\beta(\phi-\phi') = v_\beta(\phi) - \phi' \frac{\partial v_\beta}{\partial \phi} + \ldots, \tag{9.28}$$

since the exponential factor prevents ϕ' from growing large. Evidently the terms in $\sigma_{\alpha\beta}$ that depend on H come from derivatives of the electron velocity tangential to the orbit, as in (7.137). It is a good exercise to deduce the more familiar free electron formulae in this way.

In high fields, where $\omega_H \tau \gg 1$, we may use the fact that the velocities are periodic functions of ϕ and ϕ'. The range of integration of ϕ', from 0 to ∞, can be cut up into stretches, each of length 2π. Thus

$$\begin{aligned}\int_0^\infty e^{-\phi'/\omega_H\tau} f(\phi')\,d\phi' &= \sum_n e^{-2\pi n/\omega_H\tau} \int_0^{2\pi} e^{-\phi'/\omega_H\tau} f(\phi')\,d\phi' \\ &= \frac{1}{1-e^{-2\pi/\omega_H\tau}} \int_0^{2\pi} e^{-\phi'/\omega_H\tau} f(\phi')\,d\phi' \\ &\approx \frac{\omega_H\tau}{2\pi} \int_0^{2\pi} e^{-\phi'/\omega_H\tau} f(\phi')\,d\phi'.\end{aligned} \tag{9.29}$$

In (9.27) a factor ω_H may be cancelled:

$$\sigma_{\alpha\beta} = \frac{1}{4\pi^3}\frac{e^2}{\hbar^2}\int \frac{m_H^*\tau}{2\pi}\left\{\int_0^{2\pi}\int_0^{2\pi} v_\alpha(\phi)\, v_\beta(\phi-\phi')\, e^{-\phi'/\omega_H\tau}\, d\phi\, d\phi'\right\} dk_{\mathbf{H}} \tag{9.30}$$

(the velocities, cyclotron mass and cyclotron frequency, depend, of course, on $k_{\mathbf{H}}$, but we need not show this explicitly here).

With now the range of ϕ' limited, we can expand the exponential

$$e^{-\phi'/\omega_H\tau} = 1 - \frac{\phi'}{\omega_H\tau} + \frac{1}{2}\left(\frac{\phi'}{\omega_H\tau}\right)^2 - \ldots. \tag{9.31}$$

The other functions in (9.30) are independent of ω_H, i.e. independent of the magnetic field. Thus, putting (9.31) into (9.30) provides a series expansion of the conductivity tensor in powers of $1/H$.

The behaviour of the magneto-resistance and Hall effect can be worked out when one has identified the first non-vanishing term in this expansion for each component of $\boldsymbol{\sigma}$. We shall follow the usual convention and take the magnetic field to be along the z-axis. For the leading term, we have, from (9.8),

$$\begin{aligned}\int v_x(\phi')\,d\phi' &= \int v_\perp \cos\theta \frac{\hbar}{m_H^*}\frac{dk}{v_\perp} \\ &= -\frac{\hbar}{m_H^*}\int dk_y,\end{aligned} \tag{9.32}$$

which will vanish when we go round a closed orbit.

In this case, there will be no term independent of H, in $\sigma_{\alpha\beta}$, if either of α or β refers to a direction in the plane normal to **H**.

For the next term in the expansion we may use (9.32) and integrate by parts;

$$\int_0^{2\pi} \phi' v_x(\phi-\phi')\,d\phi' = \frac{2\pi\hbar}{m_H^*}k_y(\phi) - \frac{\hbar}{m_H^*}\int_0^{2\pi} k_y(\phi-\phi')\,d\phi' \quad (9.33)$$

of which the second term will vanish by centro-symmetry of the orbit. But the first term contributes to the component of $\boldsymbol{\sigma}$ that governs the Hall effect: using (9.32) again

$$\begin{aligned}\sigma_{yx} &= \frac{1}{4\pi^3}\frac{e^2}{\hbar^2}\int \frac{m_H^*\tau}{2\pi}\frac{1}{\omega_H\tau}\int_0^{2\pi} v_y(\phi)\frac{2\pi\hbar}{m_H^*}k_y(\phi)\,d\phi\,dk_z \\ &= \frac{1}{4\pi^3}e^2\int \frac{1}{m_H^*\omega_H}\oint k_y\,dk_x\,dk_z \\ &= \frac{ec}{H}\frac{1}{4\pi^3}\int \mathscr{A}(k_z)\,dk_z, \end{aligned} \quad (9.34)$$

where $\mathscr{A}(k_z)$ is the area of the orbit on the slice k_z.

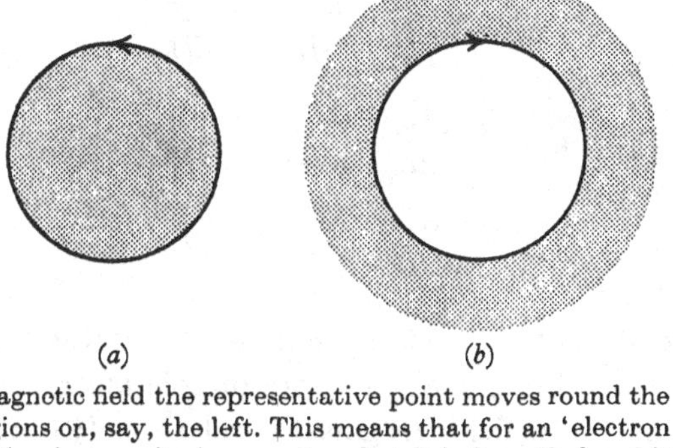

Fig. 159. In a magnetic field the representative point moves round the Fermi surface keeping filled regions on, say, the left. This means that for an 'electron orbit' (*a*), the direction of rotation is opposite in sense to what it is for a 'hole orbit' (*b*).

This integral is just the volume of the Fermi surface that has been sliced into orbits. In other words,

$$\sigma_{yx} = \frac{ec}{H}n, \quad (9.35)$$

where n is the number of electrons enclosed within this Fermi surface. Comparison with (7.147) will show that we verify this relation at very high fields, for an arbitrary Fermi surface.

But what happens if we have a 'hole' surface? One must look back carefully through the analysis, and note that the direction of traversal of the orbit governs the sign of (9.32). Thus, a closed volume of holes will contribute to (9.34) with the opposite sign. We should really write

$$\sigma_{yx} = \frac{ec}{H}(n_e - n_h). \tag{9.36}$$

This is the proper generalization of the argument of § 7.13; at high fields the relative mobilities of the two types of carrier do not come into the argument. But this only applies when we have closed surfaces of electrons and holes.

Returning to (9.30) and (9.31), it is evident that the terms of order $(1/\omega_H \tau)$ will vanish in σ_{xx} and σ_{yy}; this can easily be checked by using (9.32) and (9.33). We need to go to the next term, of order $(1/\omega_H \tau)^2$, in (9.31), before we find a non-vanishing component.

For the z-components of $\boldsymbol{\sigma}$, the same sort of argument holds. Thus σ_{xz} and σ_{yz} vanish in first order, but retain a term of order $1/H$; σ_{zz} contains a term that is independent of H.

Putting all these results together, we see that the form of the conductivity tensor in high fields is as follows:—

$$\boldsymbol{\sigma} = \begin{pmatrix} \dfrac{A_{xx}}{H^2} & \dfrac{-A_{yx}}{H} & \dfrac{-A_{zx}}{H} \\ \dfrac{A_{yx}}{H} & \dfrac{A_{yy}}{H^2} & \dfrac{-A_{zy}}{H} \\ \dfrac{A_{zx}}{H} & \dfrac{A_{zy}}{H} & A_{zz} \end{pmatrix}, \tag{9.37}$$

where the coefficients A_{xx}, etc., are non-zero, and independent of H. Notice that the off-diagonal elements of this matrix are skew symmetric: in a magnetic field the Onsager relations (§ 7.9) take the form

$$\sigma_{xy}(\mathbf{H}) = \sigma_{yx}(-\mathbf{H}) \tag{9.38}$$

because inversion of the axes has the effect of reversing the sign of the magnetic field.

A preliminary inspection of this matrix suggests that the magnetoresistance, measured normal to the field, ought to increase indefinitely as H^2; does not σ_{xx} decrease as $1/H^2$? This is not the case, as an elementary argument will indicate. Our experiment consists in measuring the E.M.F. along a wire carrying a current. As in (7.161), we must look at the appropriate component of the *resistivity tensor*, which is the inverse of (9.37).

For example, let us look at ρ_{xx}—the transverse *resistivity*. By elementary matrix algebra, we have

$$\rho_{xx} = \frac{1}{\Delta}\left(\frac{A_{yy}A_{zz}+A_{zy}^2}{H^2}\right)$$

$$= \frac{\left(\dfrac{A_{yy}A_{zz}+A_{zy}^2}{H^2}\right)}{\left(\dfrac{A_{zz}A_{yx}^2}{H^2}+\dfrac{A_{xx}A_{yy}A_{zz}+A_{xx}A_{zy}^2+A_{yy}A_{zx}^2}{H^4}\right)}$$

$$\to \frac{A_{yy}A_{zz}+A_{zy}^2}{A_{zz}A_{yx}^2} \tag{9.39}$$

to the lowest power in $1/H$. Thus, *the transverse magneto-resistance tends to a constant in high fields.* We say that there is *saturation* of this component. This is characteristic of all the diagonal components of $\boldsymbol{\rho}$. We have already noted this phenomenon in the simple two-band model of § 7.13.

What happens is that, although the conductivity σ_{xx} tends to zero, this does not mean that no current can pass. An E.M.F. in the x-direction is associated with a Hall current in the y-direction. The E.M.F. in the y-direction needed to annul this current gives, in its turn, a Hall current in the x-direction, and this is the current we observe.

For the Hall coefficient itself, we get, in the same limit,

$$\rho_{xy} \to \frac{H}{A_{yx}}, \tag{9.40}$$

which, with (9.36), gives

$$R = \frac{1}{(n_e - n_h)\,ec}. \tag{9.41}$$

In the special case of a *compensated* metal—i.e. with equal numbers of electron and hole states—this expression would become infinite. Turning back again to (9.36), we see that the linear term in $1/H$ in σ_{yx} vanishes: all components of the resistivity tensor, including the Hall coefficient itself, should then increase quadratically in magnetic field, without saturation.

9.4 Open orbits

Throughout the above argument, from (9.33) on, we have tacitly assumed that we have closed energy surfaces, of electrons or of holes, or both. But this is not the only possibility. Consider, for example, a

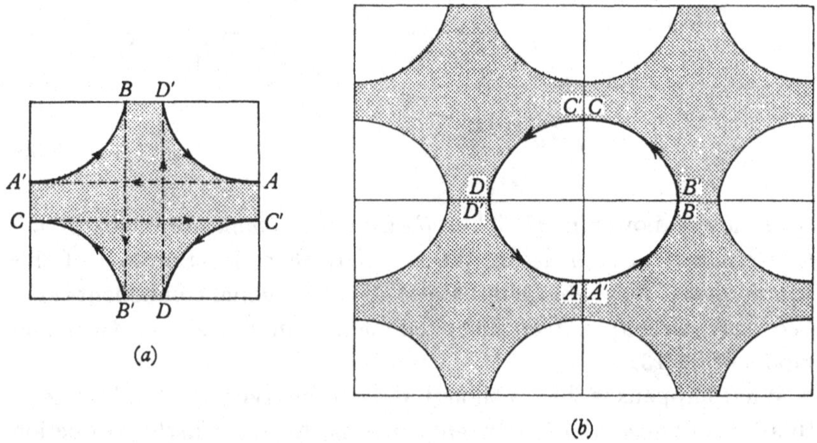

Fig. 160. Orbit: (a) in reduced zone; (b) in repeated zone.

Fermi surface that touches the zone boundary, as in Fig. 160(*a*). Consider an electron starting out from D', on this plane section. What happens when it reaches A? As shown in § 3.3, this point in **k**-space is equivalent to the point A', reached by the translation of a reciprocal lattice vector. From A' the representative point goes on to B, then is translated to the equivalent point B' and so on, to C, C', D and D'.

This construction of an 'orbit' seems artificial, as if there were discontinuities at the zone boundaries. But A and A' really do represent exactly the same wave-function. To make this clearer, we draw the *repeated* zone scheme, as in §§ 3.3 and 6.8. Instead of shifting the representative point through a reciprocal lattice vector each time it crosses a zone boundary, we provide another cell of the reciprocal lattice on the other side. The orbit is continuous across the zone boundary, and the bits join up as shown. We should call this a *hole orbit*, because it encloses an empty region of **k**-space. It will have its characteristic cyclotron frequency, just as in § 9.1.

In practice, our reciprocal lattice is three-dimensional, as in Fig. 161. It is evident that there may be both electron orbits and hole orbits on

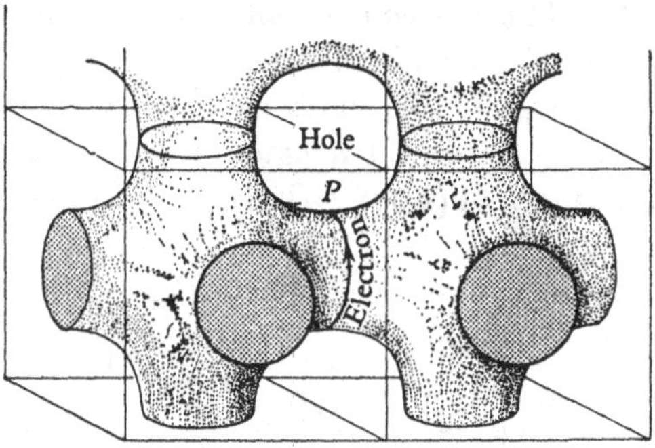

Fig. 161. On a multiply connected Fermi surface there may be 'electron' orbits and 'hole' orbits. At a point such as P a carrier may belong to either type, depending on the field direction.

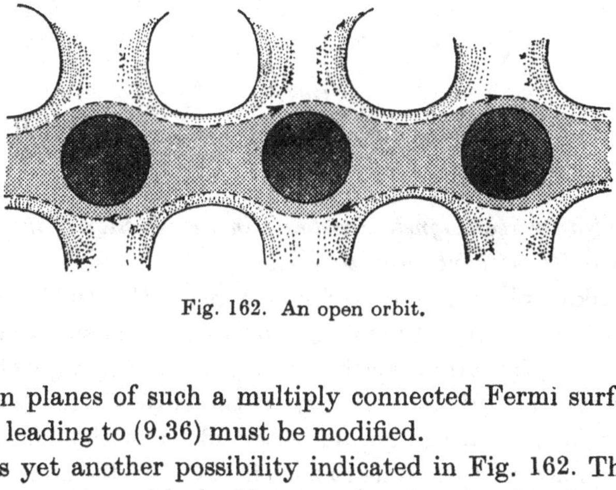

Fig. 162. An open orbit.

the section planes of such a multiply connected Fermi surface; the argument leading to (9.36) must be modified.

There is yet another possibility indicated in Fig. 162. There is a plane whose section with the Fermi surface is not a closed curve. The 'orbits' in this plane are *open*. A magnetic field does not bring the representative point back to where it started. The **k**-vector of the state continues onwards to infinity in the repeated zone scheme.

The existence of open orbits profoundly alters the argument of the last section. Suppose, for example, that there is an open orbit in the y-direction. Then the proof no longer holds that the integral (9.32),

$$\int v_x(\phi')\,d\phi' = -\frac{\hbar}{m_H^*}\int dk_y, \tag{9.42}$$

must vanish. Indeed, one can see by inspection that this integral along the path may contain a positive definite integrand, and therefore must be finite.† Thus, all components of $\boldsymbol{\sigma}$ referring to the x-direction will receive contributions from this type of integral; in particular, σ_{xx} will not go as $(1/\omega_H \tau)^2$, but will come out independent of H.

Again, the effect of this on the magneto-resistance has to be calculated properly. In place of (9.37) we have

$$\boldsymbol{\sigma} = \begin{pmatrix} B_{xx} & \dfrac{-A_{yx}}{H} & -B_{zx} \\ \dfrac{A_{yx}}{H} & \dfrac{A_{yy}}{H^2} & \dfrac{-A_{zy}}{H} \\ B_{zx} & \dfrac{A_{zy}}{H} & A_{zz} \end{pmatrix} \tag{9.43}$$

for the terms of lowest order in $1/H$. We then find that ρ_{xx} behaves as in (9.38), tending to saturation; but, for the y-component of resistance we get

$$\begin{aligned} \rho_{yy} &= \frac{1}{\Delta}(B_{xx}A_{zz}+B_{zx}^2) \\ &= \frac{B_{xx}A_{zz}+B_{zx}^2}{\{A_{zz}(A_{yx}^2+B_{xx}A_{yy})+B_{xx}A_{yx}^2+A_{yy}B_{zx}^2\}/H^2} \\ &\propto H^2. \end{aligned} \tag{9.44}$$

Thus, *the transverse magneto-resistance in the direction of the open orbit increases as* H^2, *without saturation.*

This striking phenomenon is of importance in the study of the Fermi surface. An inspection of the magneto-resistance of single crystals, as a function of orientation relative to the magnetic and electric fields, provides evidence of the topology of the surface. For example, if directions of non-saturation are observed, then the Fermi surface must be connected from zone to zone in the repeated zone scheme; it cannot consist of closed regions of electrons and holes unless these are of exactly equal volume (cf. (9.41)).

Fig. 162 shows a typical open orbit. At first sight one would imagine that all open orbits are of this nature, running along 'axes' of the multiply connected Fermi surface. Thus one expects the directions of non-saturation of magneto-resistance to be exact symmetry directions of the crystal.

This is not true, as may be verified by the following argument. Suppose a plane cuts a multiply connected Fermi surface at such an

† Strictly speaking, we should redefine the 'phase variable' ϕ for open orbits. There is no difficulty about this; it is only a matter of defining a variable to be along the path of the electron in **k**-space.

angle that it contains both closed electron orbits and closed hole orbits (Fig. 163). Then there will be an open orbit, or at least an extended orbit, separating the region of the plane that contains electron orbits from the region containing hole orbits.

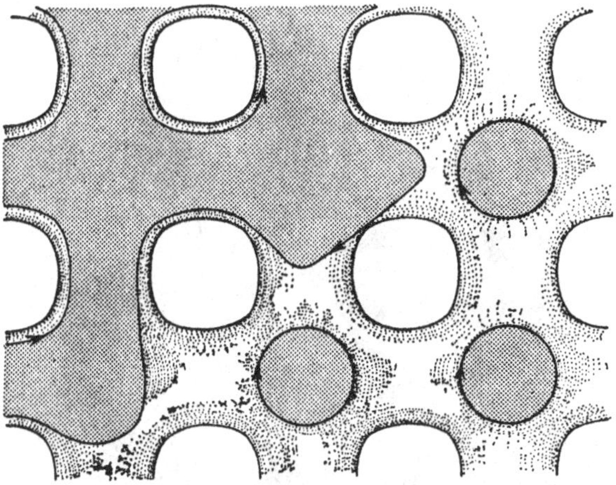

Fig. 163. Open orbit separating regions of 'electron' orbits and 'hole' orbits.

The proof is intuitive. One cannot draw closed loops shaded inside, representing electron orbits, and closed loops shaded outside, representing hole orbits, without putting in an extra boundary between shaded and unshaded regions. Of course this boundary may itself form a large closed loop—an *extended orbit*—but it is not difficult to construct cases where it continues indefinitely in the same general direction through the reciprocal lattice. This is best understood by inspection of simple models. One finds that there are often *aperiodic* open orbits of this kind running in directions that make quite a substantial solid angle round the symmetry axes. This is the case, for example, in the noble metals Cu, Ag, Au, whose Fermi surfaces are sketched in Fig. 164.

9.5 Magneto-acoustic oscillations

As we saw in § 8.8, the attenuation of ultrasonic waves does not give a direct measure of the shape of the Fermi surface. But if a magnetic field is applied along with the ultrasonic wave, then phenomena are observed that do depend directly on the geometry of the surface.

These phenomena are exceedingly complex, in theory and in practice; there are so many different ways of arranging the directions of the magnetic field, of the propagation vector of the ultrasonic

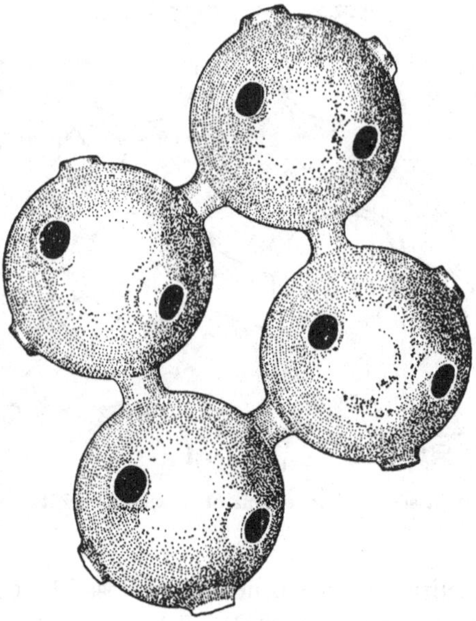

Fig. 164. Fermi surface of copper.

waves, and of the polarization vector of the waves. In order to make a complete analysis, one must construct a generalized conductivity $\boldsymbol{\sigma}(\mathbf{q}, \omega)$, similar to (8.115), but expressed in terms of cyclotron coordinates as in (9.14) or (9.21). This can be done—but it is inevitably rather complicated by the geometry of the situation.

However, the following simple case exemplifies the effect. Suppose we have a magnetic field in the z-direction and a transverse ultrasonic wave, with polarization vector in the y-direction, propagating in the x-direction. Consider the actual path of an electron in real space. For free electrons this would be a helix, i.e. a circle projected on the (x, y)-plane. For an 'orbit' in **k**-space the projection is easily determined. In general,

$$\dot{\mathbf{k}} = \frac{e}{c\hbar}\mathbf{H} \wedge \dot{\mathbf{r}}, \tag{9.45}$$

as in (6.40). It follows, by an elementary time integration, that

$$\mathbf{k} = \frac{e}{c\hbar}\mathbf{H} \wedge \mathbf{r}. \tag{9.46}$$

The orbit of $\mathbf{k}$ in $\mathbf{k}$-space is the same as the x-y projection of the path of $\mathbf{r}$ in real space, except that it is multiplied by $eH/c\hbar$ and rotated through a right angle about $\mathbf{H}$.

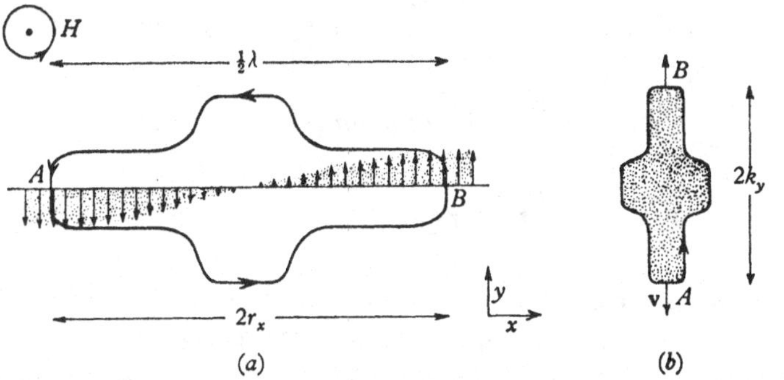

Fig. 165. (*a*) Trajectory of electron in real space. (*b*) Orbit on Fermi surface.

Now consider the electric field set up by the acoustic wave. For all practical purposes we may treat this as a stationary oscillating electric field transverse to the direction of propagation. In practice, to make $\omega_H \tau \gg 1$, we need such a strong magnetic field that $\omega_H \gg \omega$; the electron makes many circuits of its orbit before being scattered or before the electric field changes. The important factor then is the variation of electric field seen by the electron as it goes round the path. It is evident from Fig. 165 that it is possible for this field to be so arranged that it accelerates the electron in the same sense round the orbit at both points where $\mathbf{E}$ and $\mathbf{v}$ are parallel. This is a *geometrical resonance*; the condition is, obviously, that the diameter of the circuit should be an odd number of half waves of the field:

$$2r_x = \frac{\hbar c}{eH} 2k_y = (n + \tfrac{1}{2})\lambda. \tag{9.47}$$

For fixed wavelength, the absorption will oscillate, as H is varied, with a definite period in $1/H$. This period should give the diameter of the Fermi surface, as measured between points where the velocity of the electrons is parallel to the electric vector produced by the sound wave.

However, this resonance condition (9.47) is not very exact. One can see this by considering the case of a circular orbit, where, say

$$v_y(t) = v_0 \sin \omega_H t, \tag{9.48}$$

and where
$$x(t) = r_x \sin \omega_H t, \tag{9.49}$$

so that in the field $E_y \exp(iqx)$ the energy absorbed per cycle will be

$$\begin{aligned}\int \mathbf{E}\cdot\mathbf{v}\,dt &= E_y v_0 \int_0^{2\pi/\omega_H} \exp(iqr_x \sin \omega_H t) \sin \omega_H t\, dt \\ &= \frac{E_y v_0}{\omega_H} \int_0^{2\pi} \exp(iqr_x \sin \xi) \sin \xi\, d\xi \\ &= \frac{E_y v_0}{\omega_H} 2\pi J_1(qr_x). \end{aligned} \tag{9.50}$$

The 'resonances' would thus come at the maxima of the Bessel function $J_1(qr_x)$. Instead of $(n+\frac{1}{2})$ in (9.47) we get numbers like 1.22, 2.23, 3.24...$(n+\frac{1}{4})$. Care must therefore be taken in the interpretation of the effect. Furthermore, the relationship between the strain field of the elastic wave and the forces on the electrons comes into the calculation and affects the positions of the resonances.

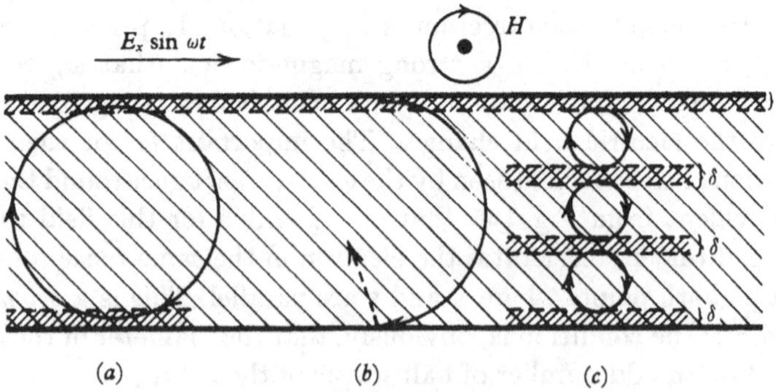

Fig. 166. R.f. size effect (*a*) extremal trajectory; (*b*) scattered trajectory; (*c*) chain of trajectories induced by internal skin layers.

The caliper diameter of a set of electron orbits may be determined even more directly by the *Gantmakher* or *r.f. size effect*. The surface impedance of a thin slab of very pure metal, with magnetic field parallel to the surface, is measured at an arbitrary radio frequency $\omega \ll \omega_H$. An electron accelerated in the skin depth follows a closed trajectory whose diameter increases as H decreases. When this trajectory is large enough, the electron hits the lower face of the slab, and is scattered. The surface impedance, being sensitive to the 'effective' electrons (§§ 8.7, 9.2), changes when $2r_x$ in (9.47) equals the thickness of the specimen (Fig. 166).

This phenomenon makes no use of cyclotron resonance conditions, and is sharper than the magneto-acoustic effect because the surface is a genuine spatial discontinuity. But oscillations with constant period in $1/H$ are again observed, with interference between contributions from various groups of extremal orbits. This is because the effective electrons from the skin depth of the top face of the slab combine to produce a thin current sheet at the caliper diameter below this face. This new 'skin layer' acts as a source for another set of trajectories—and so on. If a chain of such trajectories can just be fitted into the thickness of the slab, so as to produce a 'skin depth' region on the bottom face, then the specimen will appear much more transparent to the r.f. field than we would have expected. But as soon as electrons in the bottom trajectory hit this surface, the impedance changes sharply. With the additional possibilities of combining different groups of extremal trajectories, the observations acquire considerable complexity. But these and similar non-resonant *magnetomorphic* phenomena are capable of yielding a lot of detailed information about the Fermi surface.

9.6 Quantization of orbits

The cyclotron frequency, ω_H, is like the frequency of a simple harmonic oscillator. We expect, therefore, to find the energy quantized in units of $\hbar\omega_H$. This is true—although a perfectly general proof cannot yet be given.

But consider free electrons in a magnetic field. They satisfy the Schrödinger equation

$$\frac{1}{2m}\left(\frac{\hbar}{i}\nabla - \frac{e}{c}\mathbf{A}\right)^2 \psi = \mathscr{E}\psi, \tag{9.51}$$

where $\mathbf{A}$ is the vector potential. We want the stationary states of this system. We choose the gauge where

$$\mathbf{A} = (0, Hx, 0), \tag{9.52}$$

whose curl gives the field H in the z-direction.

With (9.52) in (9.51) we must solve

$$\frac{\partial^2\psi}{\partial x^2}+\left(\frac{\partial}{\partial y}-\frac{ieH}{\hbar c}x\right)^2\psi+\frac{\partial^2\psi}{\partial z^2}+\frac{2m\mathscr{E}}{\hbar^2}\psi=0. \tag{9.53}$$

This obviously has a solution of the form

$$\psi(x,y,z)=\exp\{i(\beta y+k_z z)\}\,u(x), \tag{9.54}$$

where $u(x)$ has to satisfy the equation

$$\frac{\partial^2 u}{\partial x^2}+\left\{\frac{2m\mathscr{E}'}{\hbar^2}-\left(\beta-\frac{eH}{\hbar c}x\right)^2\right\}u=0, \tag{9.55}$$

with
$$\mathscr{E}'=\mathscr{E}-\frac{\hbar^2}{2m}k_z^2. \tag{9.56}$$

The motion in the z-direction—that is, parallel to the field—is exactly as for a free electron, and the contribution to the kinetic energy is the same. But for the motion in the (x, y)-plane we have to solve a new eigenvalue equation, (9.55), which we can write

$$-\frac{\hbar^2}{2m}\frac{\partial^2 u(x)}{\partial x^2}+\tfrac{1}{2}m\left(\frac{eH}{mc}x-\frac{\hbar\beta}{m}\right)^2 u(x)=\mathscr{E}'u(x). \tag{9.57}$$

This is no more than the one-dimensional Schrödinger equation for the wave-function $u(x)$ of a simple harmonic oscillator, of frequency

$$\omega_H=\frac{eH}{mc}, \tag{9.58}$$

centred on the point
$$x_0=\frac{1}{\omega_H}\frac{\hbar\beta}{m}. \tag{9.59}$$

Thus,
$$\mathscr{E}'=(n+\tfrac{1}{2})\hbar\omega_H, \tag{9.60}$$

and
$$\mathscr{E}=(n+\tfrac{1}{2})\hbar\omega_H+\frac{\hbar^2}{2m}k_z^2. \tag{9.61}$$

This is the result that we are after: the energy of the electron states is expressed as the sum of a translational energy along the magnetic field, together with the quantized energy of the cyclotron motion in the plane normal to the field.

The interesting question is—how do we count states? Suppose we have a box, as in § 1.7, of sides L_x, L_y, L_z. Obviously, k_z is quantized, as usual, in units of $2\pi/L_z$. Again, from the form of (9.54), β is quantized in units of $2\pi/L_y$. But notice that the energy is independent of β, so that for a given value of n we could, one might suppose, have any value of β out of an infinite set.

This is not the case. As shown in (9.59), the function u depends on β by being centred at the point

$$x_0 = \frac{1}{\omega_H}\frac{\hbar\beta}{m} = \frac{v_y}{\omega_H}. \tag{9.62}$$

This means, in effect, that if the electron starts off in the y-direction with velocity v_y it will move in a circular path in the magnetic field, with centre x_0. This path must not be too big. We must have x_0 inside the box, so we have

$$0 < x_0 < L_x \tag{9.63}$$

as a restriction on the allowed position of this centre.

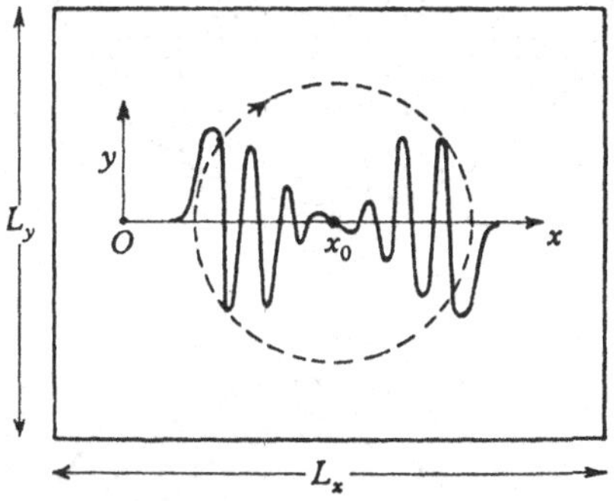

Fig. 167. Solution of Schrödinger equation for electron in a magnetic field.

But from (9.62) this puts a restriction on the range of values allowed for β. Not only is this variable quantized, in units of $2\pi/L_y$, but it must satisfy

$$0 < \beta < \frac{m\omega_H}{\hbar}L_x = \frac{eH}{c\hbar}L_x. \tag{9.64}$$

Thus, there are only

$$p = \frac{L_y}{2\pi}\frac{m\omega_H}{\hbar}L_x \tag{9.65}$$

different values of β. Each level of (9.61), corresponding to a particular choice of n and k_z, is p-fold degenerate.

The magnetic field has, of course, broken down our original quantization scheme. The variables k_x, k_y, k_z, which were our co-ordinates in **k**-space are no longer 'good quantum numbers'; the wave-functions (9.54) no longer each correspond to fixed values of a set of these parameters. Nevertheless, let us use this same space, and let us represent the p new levels having some value of $\mathscr{E}$ given by (9.61) by the surfaces corresponding to this energy in our original scheme. Ignoring the z co-ordinate, these surfaces are circles in the (x, y)-plane. The new states are not really located at any point on this circle; they will be

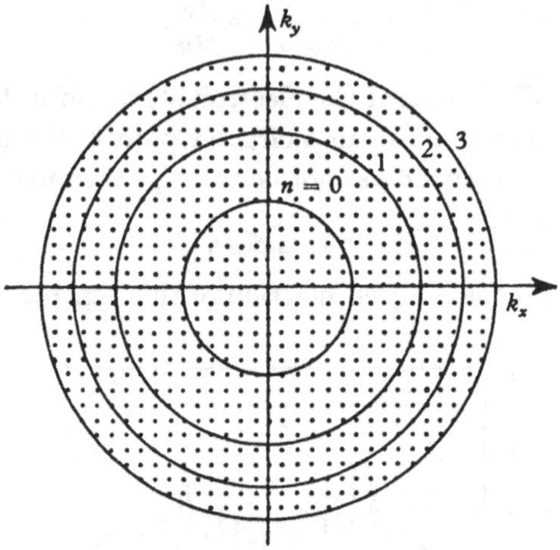

(a)

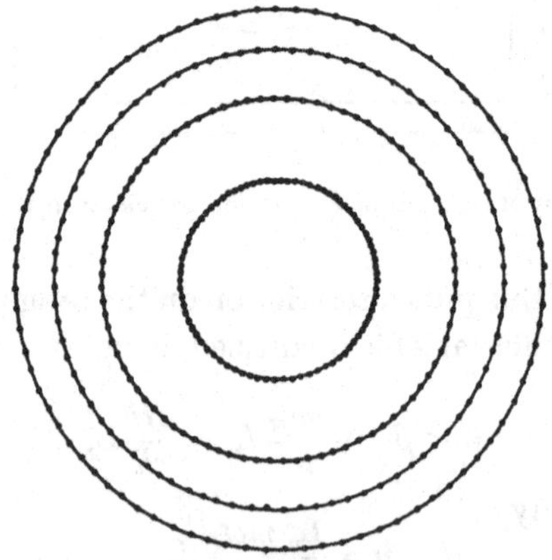

(b)

Fig. 168. Quantization scheme for free electrons: (a) without magnetic field; (b) in a magnetic field.

rotating round it with frequency ω_H. But we can classify the various levels in the magnetic field by naming the circles on which they lie.

In fact we can check that the total number of levels associated with a given macroscopic volume of **k**-*space is the same in the new scheme as it* *was before*. We can use (9.7); the area between two energy surfaces separated by the energy $\delta\mathscr{E}$ is

$$\delta\mathscr{A} = \frac{2\pi m_H^*}{\hbar^2}\,\delta\mathscr{E}. \tag{9.66}$$

Suppose that $\delta\mathscr{E}$ is a quantum of cyclotron frequency, $\hbar\omega_H$. Recall that the density of 'allowed states' in the conventional quantization scheme is $(L_x L_y)/(2\pi)^2$ per unit area of the (k_x, k_y)-plane (cf. § 1.6). Thus, the number of states between two quantized orbits is

$$\begin{aligned}\frac{L_x L_y}{(2\pi)^2}\,\delta\mathscr{A} &= \frac{2\pi m_H^*}{\hbar^2}\,\hbar\omega_H\,\frac{L_x L_y}{(2\pi)^2}\\ &= p\end{aligned} \tag{9.67}$$

by (9.65).

This is the result we are after. It shows that the effect of the magnetic field is, so to speak, to create these quantized orbits in **k**-space, and to cause the free-electron states to 'condense' on to the nearest such orbit. The number of states on each orbit is exactly the number available amongst the 'allowed states' in the annulus in which it lies.

What happens in the general case, when we are dealing with electrons in a crystal? For a single band the quasi-classical theory of § 6.5 is valid. The Schrödinger equation (6.30) for the equivalent Hamiltonian has solutions in a magnetic field that will satisfy the Correspondence Principle limit at large quantum numbers, and will be quantized according to the Bohr phase-integral formula

$$\oint \mathbf{p}\cdot d\mathbf{r} = (n+\gamma)\,2\pi\hbar, \tag{9.68}$$

where n is an integer, γ is a phase correction (typically, $\frac{1}{2}$), and **p** and **r** are conjugate variables representing the momentum and position of the particle as it traces out its orbit.

We can use (9.45) to calculate **r**, the position of the electron on its path in real space, and we can assume the usual momentum rule

$$\mathbf{p} = \hbar\mathbf{k} + \frac{e}{c}\mathbf{A}. \tag{9.69}$$

The result, by trivial geometry, is that the *area of the orbit in* **k**-*space is quantized*

$$\mathscr{A}_n = \frac{2\pi e H}{c\hbar}(n+\gamma). \tag{9.70}$$

It is obvious from (9.61), (9.66) and (9.5) that this is the result already obtained for free electrons.

The description of the electron levels in a magnetic field now requires the following construction in **k**-space. Choose a value of n. On each plane of section of the Fermi surface normal to the magnetic field, draw the energy contour of area $\mathscr{A}_n$. Join these contours into a continuous tube, with axis running parallel to **H**, and of constant area of cross-section. Draw, similarly, tubes for other values of n. Assume that the 'allowed points' in the conventional quantization scheme have all condensed on to the nearest tube. This will give the number of states, degenerate in energy, that must now be assumed to be circulating round each tube with the cyclotron frequency appropriate to their orbit.

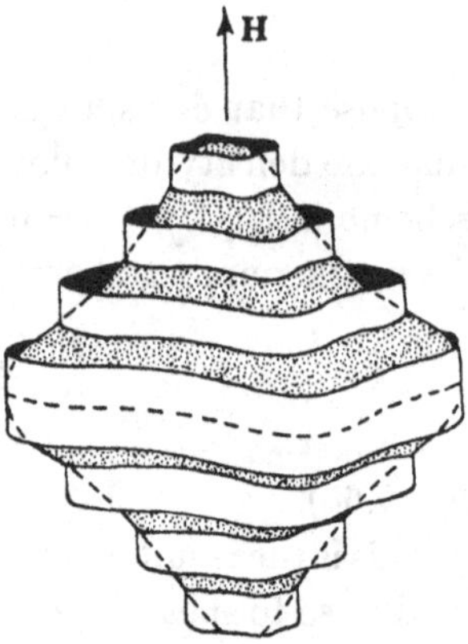

Fig. 169. Tubes of quantized magnetic levels.

9.7 The de Haas–van Alphen effect

This quantization has some interesting and striking effects. The energy of the electron gas, as a whole, depends upon the strength of the magnetic field. But the magnetic susceptibility, which is basically diamagnetic (we ignore spin effects here) and independent of temperature, oscillates as the magnetic field changes. This is the *de Haas–van Alphen effect.*

To calculate this effect, for a general shape of Fermi surface, we may proceed as follows. Let us write down the free energy of the system; for a Fermi–Dirac assembly this is given by

$$F = N\zeta - kT \sum_i \ln(1 + e^{(\zeta - \mathscr{E}_i)/kT}), \tag{9.71}$$

where we sum over all possible states, of energy $\mathscr{E}_i$, but we fix ζ, the Fermi potential, by the number of occupied states (as in (4.9), for example).

For our system of quantized *Landau levels*, the energy is given by

$$\mathscr{E}_i = \mathscr{E}(n + \gamma, k_z), \tag{9.72}$$

where n is the orbital quantum number as in (9.61)—but of course each such level is p-fold degenerate, as in (9.65). Take a cube of crystal, with unit sides; by (9.67) or (9.70), there will be

$$\frac{1}{4\pi^3}d^3k = \frac{eH}{2\pi^2\hbar c}dk_z \tag{9.73}$$

states belonging to the length dk_z (measured in the direction of **H**, as usual) of each tube in **k**-space. Thus

$$F = N\zeta - kT\int_{-\infty}^{\infty}\left\{\frac{eH}{2\pi^2\hbar c}\sum_{n=0}^{\infty}\ln\left(1+\exp\left[\{\zeta-\mathscr{E}(n+\gamma, k_z)\}/kT\right]\right)\right\}dk_z. \tag{9.74}$$

To evaluate this beastly integral, we use a trick of pure mathematics —the *Poisson summation formula*. Consider $f(x)$, an arbitrary function. In the range $n < x < n+1$ we can write

$$f(x) = \sum_{s=-\infty}^{\infty} e^{-2\pi ixs}g_s, \tag{9.75}$$

where
$$g_s = \int_n^{n+1} f(x)\,e^{2\pi ixs}dx. \tag{9.76}$$

This is just a Fourier series, as in (1.7). Now we can write

$$\begin{aligned} f(n+\tfrac{1}{2}) &= \sum_{s=-\infty}^{\infty} e^{-2\pi is(n+\frac{1}{2})}g_s \\ &= \sum_{s=-\infty}^{\infty} e^{-i\pi s}g_s \\ &= \sum_{s=-\infty}^{\infty}(-1)^s\int_n^{n+1} f(x)\,e^{2\pi ixs}dx. \end{aligned} \tag{9.77}$$

Now summing this over all intervals of the variable x, we have

$$\sum_{n=0}^{\infty} f(n+\tfrac{1}{2}) = \int_0^{\infty} f(x)\,dx + 2\sum_{s=1}^{\infty}(-1)^s\int_0^{\infty} f(x)\cos 2\pi xs\,dx. \tag{9.78}$$

We apply this formula to the sum over n in (9.74)

$$\begin{aligned} F = N\zeta - kT\int_{-\infty}^{\infty}&\left\{\frac{eH}{2\pi^2\hbar c}\int_0^{\infty}\ln\left(1+\exp\left[\{\zeta-\mathscr{E}(x+\gamma, k_z)\}/kT\right]\right)dx\right\}dk_z \\ - 2kT\sum_{s=1}^{\infty}\int_{-\infty}^{\infty}&\left\{\frac{eH}{2\pi^2\hbar c}(-1)^s\int_0^{\infty}\ln\left(1+\exp\left[\{\zeta-\mathscr{E}(x+\gamma, k_z)\}/kT\right]\right)\right. \\ &\left.\times\cos 2\pi xs\,dx\right\}dk_z. \end{aligned} \tag{9.79}$$

Let us concentrate our attention, for the moment, upon integrals of the form

$$\int_0^\infty \ln\left(1+\exp\left[\{\zeta-\mathscr{E}(x+\gamma,k_z)\}/\boldsymbol{k}T\right]\right)\cos 2\pi xs\,dx$$
$$= \frac{1}{4\pi^2 s^2 \boldsymbol{k}T}\left[f^0(\mathscr{E})\frac{\partial\mathscr{E}}{\partial x}\right]_{x=0} + \int_0^\infty \frac{\cos 2\pi xs}{4\pi^2 s^2 \boldsymbol{k}T}\left[f^0(\mathscr{E})\frac{\partial^2\mathscr{E}}{\partial x^2}+\frac{\partial f^0}{\partial x}\frac{\partial\mathscr{E}}{\partial x}\right]dx. \tag{9.80}$$

Here we have used the fact that

$$\boldsymbol{k}T\frac{\partial \ln\left(1+e^{(\zeta-\mathscr{E})/\boldsymbol{k}T}\right)}{\partial\mathscr{E}} = f^0(\mathscr{E}), \tag{9.81}$$

the Fermi function defined in (4.8), and we have integrated twice by parts. We shall ignore the first term in (9.80), because we are only interested in the oscillatory part of the free energy. Using (9.61), we have, by definition,

$$\mathscr{E}(x+\gamma,k_z) = (x+\gamma)\,\hbar\omega_H + f(k_z), \tag{9.82}$$

so that $\partial\mathscr{E}/\partial x = \hbar\omega_H$ and $\partial^2\mathscr{E}/\partial x^2 = 0$. Again, the lower limit of integration might as well be $-\infty$, since the only important contributions to the integral come from near the Fermi level where $\mathscr{E} = \zeta$. Thus, the contribution to the free energy from this slice of the Fermi distribution is proportional to

$$I_s(k_z) = \frac{\hbar\omega_H(\zeta,k_z)}{4\pi^2 s^2 \boldsymbol{k}T}\int_{-\infty}^{\infty} \cos 2\pi xs\,\frac{\partial f^0}{\partial x}\,dx. \tag{9.83}$$

This integral, although nominally of the form of (4.13), cannot be evaluated by that formula because the integrand oscillates too rapidly inside the thickness of the Fermi layer. But we may use the fact that $-\partial f^0/\partial x$ has a maximum at the point X where

$$\mathscr{E}(X,k_z) = \zeta: \tag{9.84}$$

we can say that

$$X = \frac{c\hbar}{2\pi e H}\mathscr{A}(\zeta,k_z), \tag{9.85}$$

where $\mathscr{A}(\zeta,k_z)$ is the area of cross-section of the Fermi surface, at the actual Fermi level, at the slice k_z. We expand the integral about this point, recalling that $\partial f^0/\partial x$ is an even function of x, and get

$$I_s(k_z) \approx \frac{\hbar\omega_H}{4\pi^2 s^2 \boldsymbol{k}T}\int_{-\infty}^{\infty}\cos 2\pi s\left(X+\frac{\boldsymbol{k}T}{\hbar\omega_H}\eta\right)\frac{\partial f^0}{\partial\eta}\,d\eta$$
$$= \frac{\hbar\omega_H}{4\pi^2 s^2 \boldsymbol{k}T}\cos 2\pi sX\int_{-\infty}^{\infty}\cos\left(\frac{2\pi s\boldsymbol{k}T}{\hbar\omega_H}\eta\right)\frac{\partial f^0}{\partial\eta}\,d\eta$$

$$= -\frac{\hbar\omega_H}{4\pi^2 s^2 kT}\cos\left(\frac{sc\hbar\mathscr{A}(\zeta, k_z)}{eH}\right)\left\{\frac{2\pi^2 skT/\hbar\omega_H}{\sinh(2\pi^2 skT/\hbar\omega_H)}\right\}$$

$$= -g(s, k_z)\cos\left(\frac{sc\hbar\,\mathscr{A}(\zeta, k_z)}{eH}\right), \quad \text{say.} \tag{9.86}$$

At this moment, perhaps we ought to pause for breath, and try to see what is happening. We are considering a particular slice of the Fermi surface. The Fermi level, fixed by the volume of the whole

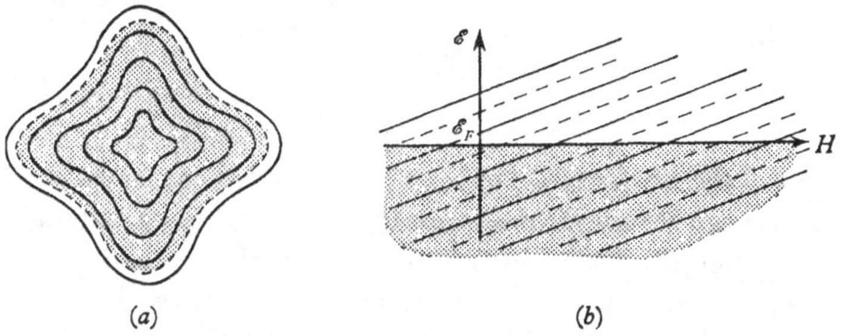

Fig. 170. (*a*) Fermi surface need not coincide with quantized orbit. (*b*) As magnetic field changes, quantized levels pass through Fermi level.

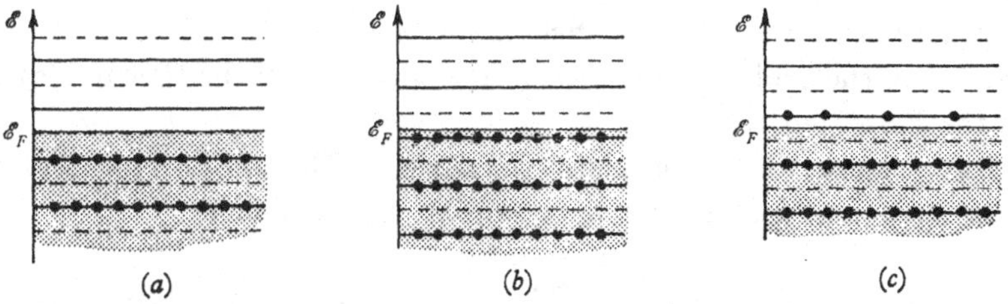

Fig. 171. Occupation of magnetic levels as magnetic field changes.

figure, does not necessarily coincide with a quantized magnetic orbit. As the field changes, quantized orbits are drawn in and out through the Fermi level, as indicated in Fig. 170.

Let us suppose we are at $T = 0$, so that the Fermi distribution is perfectly sharp. What happens as a quantized orbit passes through ζ? Suppose, for example, that the field is such that ζ lies half way between two orbits (Fig. 171 (*a*)). Then the number of states below the Fermi level will be exactly as if there were no magnetic levels—but the total energy of the electron gas will be less than in the absence of a magnetic field—by an amount of the order of $\frac{1}{2}\hbar\omega_H$ per electron at the Fermi level. Now, as H increases, these electrons will be drawn up to the Fermi level (Fig. 171(*b*)) so that their free energy increases to a maximum. But when the magnetic level passes through the Fermi level, it begins to empty (Fig. 171(*c*)) and the average energy of the electrons drops again, reaching a minimum when ζ again lies half way between two quantized orbits. Thus, the free energy of the electron gas oscillates regularly, with period determined by the interval between coincidences of a quantized orbit with the Fermi level. This is the meaning of (9.86); the period comes from the condition

$$\frac{sc\hbar\mathscr{A}(\zeta, k_z)}{eH} = 2n\pi. \tag{9.87}$$

Most of the algebra from (9.74) to (9.81) was a device for analysing the oscillations as a Fourier series, since the change of energy as each level passes through the Fermi surface is not a simple sinusoidal function. This explains the index s, leading to higher harmonics.

We have also allowed for the fact that the Fermi surface is not perfectly sharp when $T \neq 0$; indeed, the oscillations will all be smoothed away if

$$kT \gg \hbar\omega_H. \tag{9.88}$$

In other words, the magnetic field must be strong enough to make the spacing of the orbits of the order of the thickness of the thermal layer on the Fermi surface.

But we have still not completed the analysis. This was for a particular, cross-section at k_z; we must sum over all different slices

$$F = 2kT\sum_{s=1}^{\infty}(-1)^s\int_{-\infty}^{\infty}\frac{eH}{2\pi^2\hbar c}g(s, k_z)\cos\left(\frac{sc\hbar\mathscr{A}(\zeta, k_z)}{eH}\right)dk_z. \tag{9.89}$$

This is a Fresnel-type integral; it is well known that the major contributions come from regions where the phase is stationary. Let us suppose that

$$\frac{\partial\mathscr{A}(\zeta, k_z)}{\partial k_z} = 0 \tag{9.90}$$

at $k_z = k_0$. Expand about this point: $k_z = k_0 + k'$, and

$$\mathscr{A} = \mathscr{A}_0 \pm \tfrac{1}{2}k'^2\mathscr{A}_0'' \dots \tag{9.91}$$

(the variation of $g(s, k_z)$ can be ignored—and we are also implicitly assuming that each part of the Fermi surface has a centre of symmetry). Thus,

$$F = 2kT \sum_{s=1}^{\infty} (-1)^s \frac{eH}{2\pi^2\hbar c} g(s, k_0) \int_{-\infty}^{\infty} \cos\left\{\frac{s c\hbar}{eH}(\mathscr{A}_0 \pm \tfrac{1}{2}k'^2 \mathscr{A}_0'' \ldots)\right\} dk'$$

$$\approx 2kT \sum_{s=1}^{\infty} (-1)^s \left(\frac{eH}{2\pi^2\hbar c}\right) g(s, k_0) \left(\frac{2eH}{s c\hbar\, |\mathscr{A}_0''|}\right)^{\frac{1}{2}} \cos\left(\frac{s c\hbar \mathscr{A}_0}{eH} \pm \frac{\pi}{4}\right)$$

$$= 2kT \sum_{s=1}^{\infty} (-1)^s \left(\frac{eH}{2\pi s c\hbar}\right)^{\frac{3}{2}} \frac{1}{\sinh(2\pi^2 s kT/\hbar\omega_H)} |\mathscr{A}_0''|^{-\frac{1}{2}} \cos\left(\frac{s c\hbar}{eH}\mathscr{A}_0 \pm \frac{\pi}{4}\right), \tag{9.92}$$

by the standard properties of the Fresnel integral. By a double differentiation with respect to H, this can be represented as a contribution to the magnetic susceptibility of the metal.

It looks a very complicated formula—but again the interpretation is simple enough. Instead of thinking of a single section, we watch the whole length of the magnetic tubes growing outwards as the magnetic field increases. As we saw above, there is a change in energy as each tube passes through the Fermi level. But as a given tube grows, all that happens is that its intersection with the Fermi surface moves up or down it (see Fig. 169) so there is little effect on the energy, except when the tube leaves the Fermi surface altogether, at a maximum or minimum area of cross-section. This explains the appearance of $\mathscr{A}_0$ in the periodic factor. For the first harmonic,

$$\frac{c\hbar}{eH}\mathscr{A}_0 = 2\pi(n+\gamma), \tag{9.93}$$

where γ is a phase correction. Thus, *the period of the oscillations, when the magnetic moment is plotted as a function of* $1/H$, *gives directly the area of a maximal or minimal cross-section of the Fermi surface normal to the magnetic field.*

Actually, if there are different maxima and minima of cross-section for a complicated Fermi surface, each will contribute its oscillatory term, and the resulting magnetic moment may show very complex behaviour as a function of field. The amplitude of each oscillation will depend on $\mathscr{A}_0''$—the local curvature of the Fermi surface round the extremal cross-section. It will also depend on the temperature, approximately as $\exp(-kT/\hbar\omega_H)$ so that the cyclotron frequency on the stationary orbit can be estimated. But if there is impurity scattering, the amplitude will be reduced further by a factor like

$\exp(-1/\omega_H \tau)$, i.e. as if the system could not be cooled below a fixed temperature $T_0 = \hbar/k\tau$.

The above analysis has been for the free energy, and magnetic moment of the electron gas. Other observable properties of the system, such as electrical and thermal conductivities, also show a similar oscillatory effect in strong magnetic fields. The formal analysis of such behaviour is much more complicated—but it depends essentially on the variation of the effective density of states at the Fermi level as the field varies. The oscillatory electrical conductivity is called the *de Haas–Shubnikov effect.*

9.8 Magneto-optical absorption

In a semiconductor the Landau levels can be detected directly by optical methods. Consider, for example, a simple parabolic conduction band of electrons with effective mass m^*. In the magnetic quantization scheme (9.61) the energy levels lie on a series of parabolas in the variable k_z, spaced by the magnetic quantum $\hbar\omega_H$ (Fig. 172).

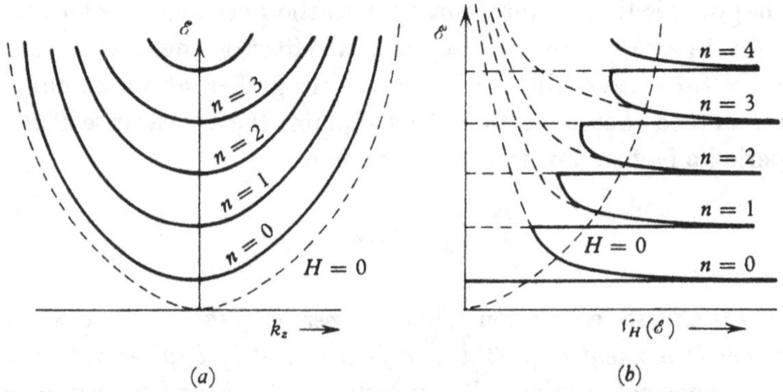

Fig. 172. (*a*) Magnetic quantization scheme for parabolic band. (*b*) Density of states.

The number of states to be assigned to each segment of a parabola is given by (9.67). For each value of the magnetic quantum number n, we have a simple density of states problem in one dimension. Adding contributions from all parabolas below the energy $\mathscr{E}$, we get a total density of states function

$$\mathscr{N}_H(\mathscr{E}) = \frac{1}{4\pi^2}\left(\frac{2m^*}{\hbar^2}\right)^{\frac{3}{2}} \hbar\omega_H \sum_n \{\mathscr{E} - (n+\tfrac{1}{2})\hbar\omega_H\}^{-\frac{1}{2}}. \tag{9.94}$$

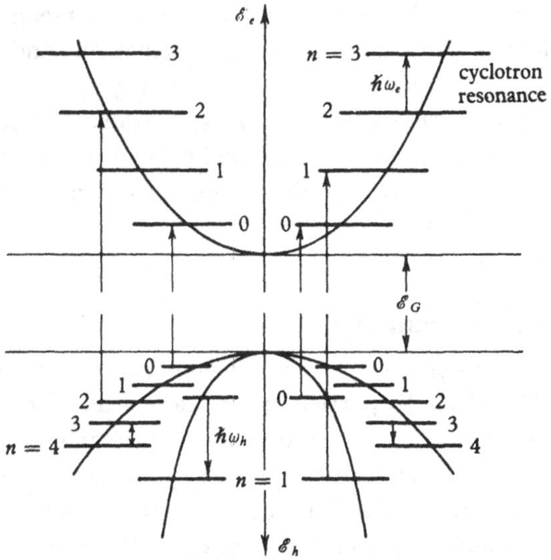

Fig. 173. Magneto-optical transitions in semiconductor with two bands of holes.

As $\omega_H \to 0$, this formula integrates correctly to the standard parabolic density of states function (4.33). But the magnetic field produces strong van Hove singularities (§ 2.5) of the type $\mathscr{E}^{-\frac{1}{2}}$ at each Landau level, in agreement with the general argument for the de Haas–van Alphen effect in § 9.7.

Transitions between such levels would obey the selection rule $\Delta n = \pm 1$; this is just another way of describing cyclotron resonance (§ 9.2). But the magnetic structure in the density of states is also sufficiently broad to be observed in the spectrum of infra-red transitions from other bands, as discussed § 8.5. This is the basic *magneto-optical effect*.

To interpret the actual observations, one must allow for quantization of the valence band states, with different values of ω_H and m^* (Fig. 173). For a typically strong transition that would occur in the absence of a magnetic field, we may take the matrix element (8.40) to be independent of H, and impose a magnetic selection rule $\Delta n = 0$. The joint density of states (8.73) then takes the same form as (9.94), with singularities of the form $(\omega - \omega_n)^{-\frac{1}{2}}$ appearing as ω coincides with and interband resonance frequency

$$\hbar\omega_n = \mathscr{E}_G + (n + \tfrac{1}{2})\,\hbar(\omega_e + \omega_h). \tag{9.95}$$

In practice, however, these singularities are broadened by collisions.

This phenomenon thus provides a very precise means of mapping out the cyclotron frequencies ω_e and ω_h of electrons and holes, over a range of energies on either side of the energy gap. In the general anisotropic case, with valence band degeneracy and non-parabolic dependence on $|\mathbf{k}|$, the magneto-optical spectrum may be very complicated, but provides invaluable evidence concerning the electronic band structure of the crystal.

9.9 Magnetic breakdown

Throughout this chapter we have been concerned with electrons in high magnetic fields. In particular, we have studied the phenomena that arise when the electron makes many cyclotron revolutions before it is scattered. We have seen that a new quantization scheme is now appropriate, in which the magnetic levels absorb the separate allowed states of the Bloch scheme. Of course, if there is scattering, the magnetic levels will themselves be broadened, but so long as $\omega_H \tau \gg 1$ this effect can be ignored.

But the quantization scheme that we have proposed, with (9.70) satisfied for each orbit, is not the ultimate scheme that we should use in a truly enormous magnetic field. The quasi-classical theory involved in § 9.6 depends upon the principle of the equivalent Hamiltonian, of § 6.4; *it is only valid when we can ignore transitions between bands.* If the magnetic field is very large, then such transitions must occur, by tunnelling, if you like, as in the Zener effect (§ 6.8), and the *Onsager Scheme* (9.70) will break down.

To understand this effect, it is easiest to start from the very high field end, where the electron wave-functions are essentially those of particles moving freely in a magnetic field, with the crystal potential of the lattice as a perturbation. Confining ourselves to the two-dimensional space normal to the magnetic field, we have a circular orbit in the (k_x, k_y)-plane.

Now introduce a perturbation of the form

$$\mathscr{V}(x) = \sum_g \mathscr{V}_g e^{igx} \tag{9.96}$$

—a series of planes spaced $2\pi/g$ apart. When the orbit passes through the zone boundary, i.e. when its effective wave-vector k_x in the x-direction is equal to $\pm\frac{1}{2}g$, then there is the possibility of Bragg diffraction. Instead of continuing along AB, say, the orbit may switch to the direction AC, and so on.

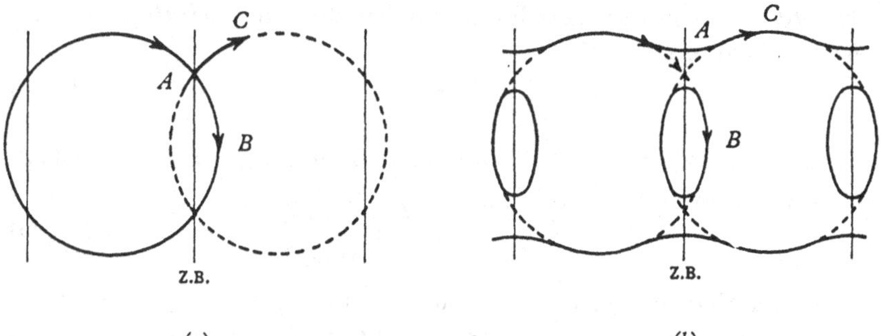

Fig. 174. (*a*) Free electron orbit in a magnetic field. (*b*) In the periodic potential of the lattice, the orbits are reconnected at the zone boundary; but in a strong magnetic field the orbit may jump back to the free-electron path.

If we increase the strength of the perturbation, then the trajectories at A will split apart in energy, and the route AC will be preferred. The electron is now moving on an ordinary (open) orbit in the conventional repeated zone scheme. The part B of the circle in Fig. 174(*a*) has now been joined up into a separate branch of the Fermi surface, traversed quite separately.

If we now increase the magnetic field, we shall go back towards the circular orbit scheme. Instead of the electron always going along AC, it may *break through* the energy gap, or 'tunnel through' the region separating the two orbits in reciprocal space, and find itself going around B.

Indeed, we can use the formula (6.65) for Zener breakdown to estimate the transition probability. An electric field E can cause tunnelling through an energy gap $\mathscr{E}_{\text{gap}}$ if

$$\frac{eEa\mathscr{E}^0}{(\mathscr{E}_{\text{gap}})^2} > 1, \tag{9.97}$$

where $\mathscr{E}^0$ is the kinetic energy of the electron at the gap—effectively the Fermi energy $\mathscr{E}_F$—and a is the corresponding lattice spacing.

Now an electron approaching A in the repeated zone scheme has velocity

$$v \sim \frac{\hbar k_F}{m}, \tag{9.98}$$

where k_F is the Fermi radius. This will not be strictly true at the zone boundary, because the energy surfaces will be distorted by the existence of the energy gap, but we are assuming that this is small.

The motion of the electron, with this velocity, across the magnetic field gives rise to a Lorentz force, which is the same as if there were an electric field of strength

$$E \sim \frac{vH}{c} \tag{9.99}$$

at right angles to **v**. This field can cause tunnelling if (9.97) is satisfied, i.e. if the parameter

$$\frac{ev\mathbf{H}}{c}\frac{a\mathscr{E}^0}{(\mathscr{E}_{\text{gap}})^2} \approx \frac{e\hbar}{mc}H\frac{k_F a\mathscr{E}_F}{(\mathscr{E}_{\text{gap}})^2} \tag{9.100}$$

is greater than unity. By (3.3), the product $k_F a$ is a number of the order of unity. Using (9.4), the condition for *magnetic breakdown* becomes

$$\frac{\hbar\omega_H \mathscr{E}_F}{(\mathscr{E}_{\text{gap}})^2} > 1. \tag{9.101}$$

This condition is far less restricting than, shall we say, that the separation of the magnetic orbits shall be greater than the energy gap. For some metals it becomes important in fields of the order of 100 kG., and can cause the appearance, in the de Haas–van Alphen effect and in other phenomena, of new orbits of a variety of different areas.

CHAPTER 10

MAGNETISM

All things began in order, so shall they end, and so shall they begin again; according to the ordainer of order and mystical mathematics of the city of heaven. SIR THOMAS BROWNE

10.1 Orbital magnetic susceptibility

In this chapter we shall be concerned with the magnetic susceptibility of solids, as distinct from the effects of magnetic fields upon their other properties, such as the electrical conductivity. This is a large subject, which we cannot treat in detail. We shall, for example, ignore most *magnetic resonance* phenomena. We also omit the theory of the technically important macroscopic properties of permanently magnetizable materials, such as *hysteresis*, *coercivity*, *remanent magnetism*, etc., that depend essentially upon the specimen becoming spontaneously magnetized in *domains* whose magnetic moments are oriented in different directions. We shall not even consider the elegant theory of the *Bloch wall* that separates domains of different magnetization, and which can be made to move by the application of a field.

We first consider *non-magnetic solids*, where there is no suspicion that the atoms or ions carry localized magnetic moments. Let us also ignore electron spin, or assume that all spins are tightly paired. There still remain contributions to the magnetic susceptibility from the orbital motion of the electrons.

The simplest case is where each atom, or ion consists of closed shells of electrons, and where the energy of excitation to a higher state is large. It is well known that this gives rise to *diamagnetism*; the susceptibility is negative and is given approximately by

$$\chi = -\frac{Ze^2N}{6mc^2}\overline{r^2}, \tag{10.1}$$

where Z is the total number of electrons in the atom, N is the number of atoms per unit volume, and $\overline{r^2}$ is the mean square radius of the electron charge cloud about each atom. This is the sort of formula we should use for a rare-gas solid, or for the contributions of the ions in an ionic crystal.

The above result derives, essentially, from the first-order perturbation energy

$$\left\langle 0 \left| \frac{e^2}{2mc^2} A^2 \right| 0 \right\rangle, \tag{10.2}$$

where **A** is the vector potential of the magnetic field, in the ground state $|0\rangle$ of each atom. This arises automatically when we use the standard expression for the Hamiltonian in a magnetic field by writing the kinetic energy in the form

$$T = \frac{1}{2m}\left(\frac{\hbar}{i}\nabla - \frac{e}{c}\mathbf{A}\right)^2 \tag{10.3}$$

as in (9.51).

But if there are excited states of the atom, $|n\rangle$, say, whose orbital moment is not zero, these may be mixed into the ground state by the term in (10.3) that is linear in **A**—the term which gives the operator

$$\frac{e}{2mc}\mathbf{H}\cdot\mathbf{L} = \mathbf{H}\cdot\boldsymbol{\mu}_L, \tag{10.4}$$

where **L** is the orbital angular momentum operator for the electrons and $\boldsymbol{\mu}_L$ is the usual magnetic moment operator to which it is proportional. In second-order perturbation this contributes to the energy

$$\delta\mathscr{E} = \sum_n \frac{|\langle 0|\mathbf{H}\cdot\boldsymbol{\mu}_L|n\rangle|^2}{\mathscr{E}_0 - \mathscr{E}_n} \tag{10.5}$$

which is positive and quadratic in H. This is the source of the constant *van Vleck paramagnetism* of the atom, ion, or molecule.

The application of (10.2) and (10.5) to actual solids turns out to be extremely complicated in practice; that is why we only quote these formulae in principle. But we notice, in (10.5), that the contribution should increase when the energy of excitation becomes small—as, for example, when there is a small gap above a filled energy band. What happens when this gap goes to zero—when we have free carriers?

Again the theory becomes very complicated, but there is an exact solution for free electrons. In § 9.6 we derived the energy levels of free electrons in a magnetic field; it is not difficult to calculate the average change in energy of the whole electron gas, by following through the formalism of § 9.7, which led to the de Haas–van Alphen effect. Instead of looking for oscillatory terms in the free energy, we concentrate on the terms that are monotonic functions of H. In (9.80), for example, we dropped

$$\frac{1}{4\pi^2 s^2 kT}\left[f^0(\mathscr{E})\frac{\partial\mathscr{E}}{\partial x}\right]_{x=0} = \frac{1}{4\pi^2 s^2 kT}\,\frac{e\hbar H}{mc}\,[f^0(\mathscr{E})]_{x=0} \tag{10.6}$$

(using (9.82) and (9.58)).

But this was a term in (9.79), which contained a sum over all values of s, and integration over all slices, k_z, of the Fermi sphere. The total contribution to the free energy (excluding the first two terms in (9.79), which are independent of H) is then

$$\Delta F_L = -2\boldsymbol{k}T\sum_{s=1}^{\infty}(-1)^s\int_{-\infty}^{\infty}\frac{eH}{2\pi^2\hbar c}\frac{1}{4\pi^2 s^2\boldsymbol{k}T}\frac{e\hbar H}{mc}[f^0(\mathscr{E})]_{x=0}\,dk_z$$

$$= -\frac{1}{4\pi^4}\left(\frac{eH}{c}\right)^2\frac{1}{m}\int_{-\infty}^{\infty}[f^0(\mathscr{E})]_{x=0}\,dk_z\sum_{s=1}^{\infty}\frac{(-1)^s}{s^2}. \tag{10.7}$$

The integral over k_z is essentially a measure of the length of the filled region in **k**-space along the axis of the Fermi sphere, and the sum over s yields $-\frac{1}{12}\pi^2$. Expressed as a susceptibility, we get the *Landau diamagnetism*

$$\chi_L = -\frac{e^2 k_F}{12\pi^2 mc^2}. \tag{10.8}$$

The result is negative because the bunching of the energy levels tends to increase the total energy of the system. We can see this in Fig. 171. Only when the Fermi level lies exactly half way between two quantized levels can the electrons 'condense' on to their appropriate tube without change in their average energy. As a quantized level approaches the Fermi level it draws electrons up with it; that is why ΔF_L is positive. The susceptibility is independent of temperature, because this is an average effect and does not require that the spacing of the levels, $\hbar\omega_H$, be greater than $\boldsymbol{k}T$.

It is not difficult to calculate (10.7) for the more general case of an arbitrary Fermi surface, following the lines of the argument of § 9.7. But the result is not correct. The Landau diamagnetism for electrons in a periodic lattice requires a more complicated analysis, which we shall not give here. There are also cross-terms between the van Vleck and Landau contributions.

10.2 Spin paramagnetism

The degeneracy between the electrons of opposite spin that share the same orbital state is resolved by a magnetic field. In a metal this causes a redistribution of electrons between the two spin orientations, and hence gives rise to a magnetic moment. The calculation of this effect is elementary. Suppose we distinguish between electrons of + and − spin (relative to the direction of **H**). Then their energies will be

$$\left.\begin{aligned}\mathscr{E}_{\mathbf{k}+} &= \mathscr{E}(\mathbf{k}) - \mu_0 H,\\ \mathscr{E}_{\mathbf{k}-} &= \mathscr{E}(\mathbf{k}) + \mu_0 H,\end{aligned}\right\} \tag{10.9}$$

where $\mathscr{E}(\mathbf{k})$ is the energy in the absence of the magnetic field, and μ_0 is the magnetic moment of the electron.

The number of electrons in each state will be given by two distinct Fermi–Dirac distributions, like (4.8), with the *same* chemical potential ζ. Thus, as we see in Fig. 175, the + spin states hold more electrons than previously—a total of

$$n_+ = \int_0^\infty \tfrac{1}{2}\mathscr{N}(\mathscr{E}) f^0(\mathscr{E}_{\mathbf{k}+})\, d\mathscr{E}, \tag{10.10}$$

since the density of states at the original energy $\mathscr{E}$ is equally divided between electrons of + and − spin.

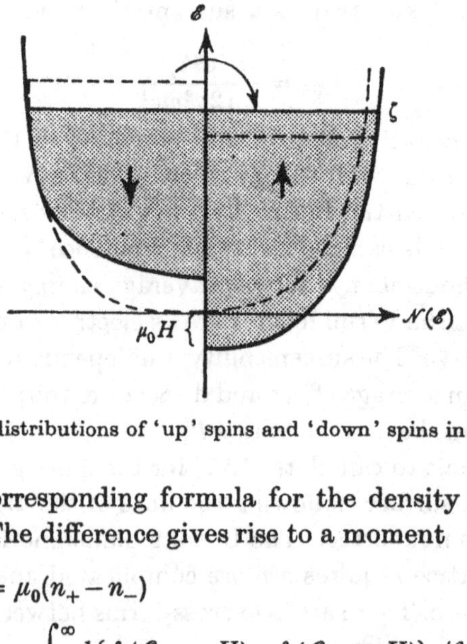

Fig. 175. Fermi distributions of 'up' spins and 'down' spins in a magnetic field.

There is a corresponding formula for the density of electrons of negative spin. The difference gives rise to a moment

$$\begin{aligned} M &= \mu_0(n_+ - n_-) \\ &= \mu_0 \int_0^\infty \tfrac{1}{2}\{f_0(\mathscr{E} - \mu_0 H) - f_0(\mathscr{E} + \mu_0 H)\}\mathscr{N}(\mathscr{E})\, d\mathscr{E} \\ &\approx \mu_0^2 H \int_0^\infty \left(-\frac{\partial f^0}{\partial \mathscr{E}}\right)\mathscr{N}(\mathscr{E})\, d\mathscr{E} \\ &= \mu_0^2 H \mathscr{N}(\mathscr{E}_F) \end{aligned} \tag{10.11}$$

by (4.18).

This positive magnetic moment gives rise to the *Pauli paramagnetism*. It is independent of the temperature and measures directly the density of states at the Fermi level. In principle, then, it ought to correlate directly with the electronic specific heat (§ 4.7), although when one makes corrections for many-electron interactions

(§ 5.8) the exact equivalence is destroyed. The difficulty about making comparisons of this sort, in practice, is that the diamagnetic correction, (10.8) is not negligible, and is not known very well. But one can measure the spin susceptibility directly, in some cases, by observing the electron-spin resonance of the conduction electrons.

For a free-electron gas it is easy to show that

$$\chi_P = \mu_0^2 \mathcal{N}(\mathscr{E}_F) = \frac{e^2 k_F}{4\pi^2 mc^2}, \tag{10.12}$$

which is exactly three times the magnitude of the Landau term (10.8). But the effects of the lattice on these two types of magnetic susceptibility are quite different. Suppose, for example, that the Fermi surface is spherical, but that the 'effective mass' becomes m^*. It is easy to prove that χ_P being dependent on the density of states, will be directly proportional to m^*, whereas χ_L is determined by the orbital motion of the electrons, without the intervention of the Bohr magneton, and so will be *inversely* proportional to m^*.

Another phenomenon associated with the spin susceptibility is the *Knight shift*. In a magnetic field $\mathbf{H}$ the electron spins are polarized to an average magnetic moment per electron

$$\langle \boldsymbol{\mu}_e \rangle = \frac{1}{n}\chi_P \mathbf{H}. \tag{10.13}$$

An electron at $\mathbf{r}$ will interact with the magnetic moment $\boldsymbol{\mu}_I$ of a nucleus at $\mathbf{R}$ by the 'contact interaction'

$$\mathscr{H}_{\text{int.}} = \tfrac{8}{3}\pi(\boldsymbol{\mu}_I \cdot \boldsymbol{\mu}_e)\,\delta(\mathbf{r}-\mathbf{R}), \tag{10.14}$$

an effect that is well known in the theory of nuclear resonance. Putting (10.13) into (10.14), we find an average energy

$$\langle \mathscr{H}_{\text{int.}} \rangle = \tfrac{8}{3}\pi \frac{\chi_P}{n} H\,|\psi(0)|^2 \mu_I = \Delta\boldsymbol{\mu}_I \cdot \mathbf{H}, \tag{10.15}$$

say, where $|\psi(0)|^2$ is the amplitude of the wave-functions at the nucleus. This energy is interpreted as a change, $\Delta\mu_I$, in the apparent moment of the nucleus—a shift of its resonance frequency in a magnetic field.

It is evident that the Knight shift should be proportional to χ_P. But there is the factor $|\psi(0)|^2$ which is not given by any independent experiment. In principle this refers only to states on the Fermi surface, so that one says that it 'measures the amount of s-like character in the wave functions at the Fermi level'. The Knight shift is a nice quantity to determine, but it is not very easy to interpret the results.

10.3 The Curie–Weiss Law and ferromagnetism

Let us now suppose that each atom behaves like a small magnet of moment $\boldsymbol{\mu}$, such as might arise from an unfilled shell of d- or f-electrons in an ion of a transition metal or rare earth. In the magnetic field $\mathbf{H}$ each little magnet will have the energy $-\boldsymbol{\mu}\cdot\mathbf{H}$. Suppose that each atom is independent of its neighbours; because it is localized, it satisfies Boltzmann statistics. The fraction of atoms in which the moment is $\boldsymbol{\mu}$ will be proportional to

$$n(\boldsymbol{\mu}) = e^{\boldsymbol{\mu}\cdot\mathbf{H}/kT}. \tag{10.16}$$

If we assume that each magnet can rotate freely, then the average moment must be

$$\langle\boldsymbol{\mu}\rangle = \int \boldsymbol{\mu}\, e^{\boldsymbol{\mu}\cdot\mathbf{H}/kT}d\Omega \Big/ \int e^{\boldsymbol{\mu}\cdot\mathbf{H}/kT}d\Omega, \tag{10.17}$$

where $d\Omega$ is the element of solid angle for the rotation of $\boldsymbol{\mu}$. An assembly of N such atoms per unit volume will have susceptibility

$$\begin{aligned}\chi = N\frac{\partial\langle\mu\rangle}{\partial\mathbf{H}} &= \int (N/kT)\,\boldsymbol{\mu\mu}\, e^{\boldsymbol{\mu}\cdot\mathbf{H}/kT}d\Omega \Big/ \int e^{\boldsymbol{\mu}\cdot\mathbf{H}/kT}d\Omega \\ &\approx \frac{N}{kT}\langle\boldsymbol{\mu\mu}\rangle = \frac{1}{3}\frac{N\langle\mu^2\rangle}{kT}\end{aligned} \tag{10.18}$$

when H is so small that the exponential can now be neglected, and where we have compared the average of the dyadic tensor $\boldsymbol{\mu\mu}$ with the average of μ^2, the square of the length of the vector $\boldsymbol{\mu}$.

It is more natural for us, nowadays, to relate each magnet to a spin angular momentum S which is quantized along the magnetic field:

$$\boldsymbol{\mu}\cdot\mathbf{H} = \beta H S_z, \tag{10.19}$$

where S_z goes from S to $-S$ by integer jumps and β is twice the Bohr magneton. Suppose, for example, that $S = \frac{1}{2}$. In (10.17) we have a sum instead of an integral

$$\begin{aligned}\langle\mu\rangle &= \beta S\left[\frac{e^{\beta SH/kT} - e^{-\beta SH/kT}}{e^{\beta SH/kT} + e^{-\beta SH/kT}}\right] \\ &= \beta S \tanh\frac{\beta SH}{kT}.\end{aligned} \tag{10.20}$$

In the limit where H is small, this gives us the same result as (10.18) provided that we note that, by the rules for the magnitude of the total angular momentum,

$$\langle \mu^2 \rangle = S(S+1)\beta^2. \tag{10.21}$$

The result (10.18) is the well-known *Curie Law* for paramagnetic susceptibility. It is obeyed approximately by certain types of solid, especially those where the magnetic atoms or ions are well separated from one another in the crystal, so that the assumption of independence is valid.

In many substances, however, there is interaction between the neighbouring spins. To take account of this, approximately, one may introduce the *Weiss field*,

$$\mathbf{H}_I = \lambda N \langle \boldsymbol{\mu} \rangle. \tag{10.22}$$

This field is supposed to be due to the average environment of magnetization in which each atom finds itself—but the strength of the interaction, as measured by the parameter λ, we leave arbitrary.

The total field acting on an atom is now $\mathbf{H}+\mathbf{H}_I$. Putting this into (10.20) in place of $\mathbf{H}$, we have, approximately,

$$N\langle \boldsymbol{\mu} \rangle = \frac{1}{3}\frac{N\langle \mu^2 \rangle}{kT}(\mathbf{H}+\lambda N\langle \boldsymbol{\mu} \rangle), \tag{10.23}$$

which gives

$$\chi \equiv N\left\langle \frac{\partial \boldsymbol{\mu}}{\partial \mathbf{H}} \right\rangle$$

$$= \frac{N\langle \mu^2 \rangle}{3k(T-\theta)}, \tag{10.24}$$

where

$$\theta = \frac{\lambda N\langle \mu^2 \rangle}{3k}. \tag{10.25}$$

This is known as the *Curie–Weiss Law*, and describes the behaviour of the susceptibility of many solids.

The parameter θ evidently has the dimensions of a temperature. What happens if $T < \theta$ in (10.24)? The susceptibility would seem to become infinite, then negative. We must return to (10.20), and solve the equation

$$N\langle \mu \rangle = N\beta S \tanh\left\{\frac{\beta S}{kT}(\lambda N\langle \mu \rangle + H)\right\}. \tag{10.26}$$

When $H = 0$, the usual solution for this is $\langle \mu \rangle = 0$; but when $T < \theta$ we find that there is another root with $\langle \mu \rangle \neq 0$. This root is easily located by a graphical construction; at temperature $T = 0$ it obviously tends to

$$M(0) = N\beta S. \tag{10.27}$$

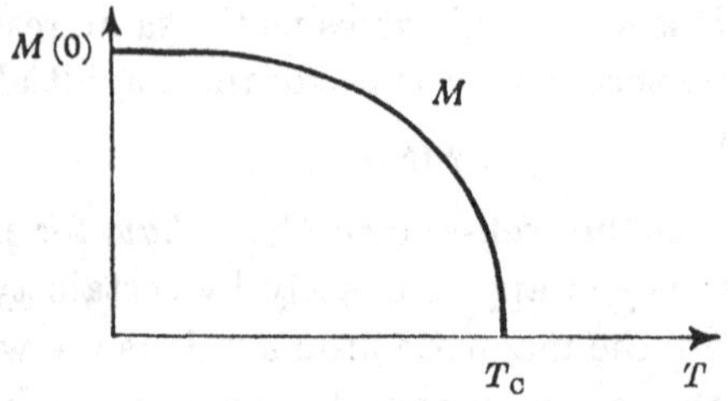

Fig. 176. Spontaneous magnetization in a ferromagnet.

The system behaves as if every magnetic moment were aligned parallel; we say that the solid is *ferromagnetic*, and observe that it shows a large spontaneous magnetization even in the absence of a magnetic field. As we warm the crystal, spontaneous magnetization decreases, until it vanishes at a well-defined temperature, the *Curie temperature*

$$T_C = \theta. \tag{10.28}$$

10.4 Exchange interaction

As measured by the Curie temperature (10.25), the Weiss internal field is very large—much larger than the ordinary internal dipole field of an array of magnets. The coefficient λ, in many ferromagnetic substances, is not $\frac{4}{3}\pi$, as it might be for the Lorentz field (8.52), but is of the order of 1000.

It is usual to ascribe this apparent field to an *exchange interaction* between the spins of the electrons on neighbouring atoms. For the strictly localized model of Heisenberg and Dirac, where we may assign a total spin operator S_l to the atom on the lth lattice site, we write down a Hamiltonian

$$\mathscr{H} = -\sum_{ll'} J_{ll'} \mathbf{S}_l \cdot \mathbf{S}_{l'} - \beta \mathbf{H} \cdot \sum_l \mathbf{S}_l. \tag{10.29}$$

The *exchange integral* $J_{ll'}$ is a function of the relative positions of the sites l and l', but is large only when $l - l'$ is one or two lattice spacings.

We may write this Hamiltonian in the form

$$\mathscr{H} = -\sum_l \{\sum_{l'} J_{ll'} \mathbf{S}_{l'} + \beta \mathbf{H}\} \cdot \mathbf{S}_l. \tag{10.30}$$

If we think what happens at very low temperatures, when all the spins are aligned, so that, for all l'

$$\langle \beta \mathbf{S}_{l'} \rangle \approx \langle \boldsymbol{\mu} \rangle, \tag{10.31}$$

then we can identify the first term in (10.30) as the internal field

$$\beta \mathbf{H}_I = \sum_{l'} J_{ll'} \mathbf{S}_{l'}$$

$$= \frac{1}{\beta} (\sum_{l'} J_{ll'}) \langle \boldsymbol{\mu} \rangle, \tag{10.32}$$

i.e.

$$\lambda = \frac{1}{N\beta^2} (\sum_{l'} J_{ll'}) \tag{10.33}$$

or, in terms of the Curie temperature (10.25),

$$k\theta = \tfrac{1}{3} S(S+1) \sum_{l'} J_{ll'}. \tag{10.34}$$

This gives the magnitude of the exchange interaction, in terms of the temperature at which the transition to the ferromagnetic state occurs.

The properties of the *spin Hamiltonian* (10.29) will be the major topic of the remainder of this chapter. But is it justified as a basic assumption? This is a question to which many different answers have been given, and which is at the heart of the mystery of ferromagnetism as a physical phenomenon.

It can readily be shown, from group theoretical arguments, that the form of the interaction, a scalar product of the spin operators, is at least the simplest possible, although there may be other terms, of greater complexity, such as the dipole–dipole interaction,

$$\mathscr{H}_{\text{dip.-dip.}} = \beta^2 \left\{ \frac{\mathbf{S}_1 \cdot \mathbf{S}_2}{R^3} - \frac{3(\mathbf{S}_1 \cdot \mathbf{R})(\mathbf{S}_2 \cdot \mathbf{R})}{R^5} \right\}, \tag{10.35}$$

for two spins at a distance $\mathbf{R}$ (a vector) apart.

It is also well known that electron spins on the same atom, or on adjacent atoms tend to be coupled by the exchange effect—a consequence of the Pauli principle. For example, if ϕ_a and ϕ_b are two wave-functions into which we 'put two electrons' then there are two types of state that we can construct, according as the two electrons have parallel or antiparallel spins. These states are the symmetric and antisymmetric combinations

$$\left.\begin{aligned} \Phi_S(\mathbf{r}_1, \mathbf{r}_2) &= 2^{-\frac{1}{2}} \{\phi_a(\mathbf{r}_1)\phi_b(\mathbf{r}_2) + \phi_a(\mathbf{r}_2)\phi_b(\mathbf{r}_1)\}, \\ \Phi_A(\mathbf{r}_1, \mathbf{r}_2) &= 2^{-\frac{1}{2}} \{\phi_a(\mathbf{r}_1)\phi_b(\mathbf{r}_2) - \phi_a(\mathbf{r}_2)\phi_b(\mathbf{r}_1)\}, \end{aligned}\right\} \tag{10.36}$$

of which Φ_S is associated with antiparallel spins (*singlet* state) and Φ_A goes with parallel spins (*triplet* state). Now, if we calculate the averages of the Coulomb energy $e^2/|\mathbf{r}_1 - \mathbf{r}_2|$ in these two states, we shall find them differing by an amount

$$2\iint \phi_a^*(\mathbf{r}_1)\,\phi_b^*(\mathbf{r}_2)\,\frac{e^2}{|\mathbf{r}_1-\mathbf{r}_2|}\,\phi_a(\mathbf{r}_2)\,\phi_b(\mathbf{r}_1)\,d\mathbf{r}_1\,d\mathbf{r}_2, \qquad (10.37)$$

which is the *exchange integral*. It is easy to rewrite this difference so that it represents the difference between the values of $2J\mathbf{S}_1\cdot\mathbf{S}_2$ for $\mathbf{S}_1$ and $\mathbf{S}_2$ parallel and antiparallel, respectively. Thus, the sign and magnitude of J depends upon the sign and magnitude of this integral.

There are two well-known cases. Suppose the two electrons belong to the *same* atom, but do not complete a closed shell. The form of the Coulomb interaction and of the atomic orbitals ϕ_a and ϕ_b is then such that J is positive; the electron spins tend to line up to make the maximum total spin consistent with the number of independent states to be filled in the shell. This is *Hund's rule*, which explains why the electrons in an incomplete d-shell of an ion of a transition metal tend to combine to give a large permanent magnetic moment to the ion (§ 5.5). The study of the paramagnetism and magnetic resonance properties of such ions, when situated in various environments in complex crystals, is an important branch of physics.

But we are concerned here with the interaction between the spins of electrons on *different* ions—and it turns out, then, that when one calculates J it almost always comes out negative, favouring *antiparallel* spins on neighbouring sites. The simplest case of this is the *Heitler–London model* of the hydrogen molecule, where the *bonding state* has electron spins paired. We took this for granted in § 4.2, when dealing with covalent bonds.

On this basis, therefore, it is difficult to explain the fact that many metals, usually 'transition' elements, are ferromagnetic. We can understand why the d-electrons in each ion tend to combine to a total spin which is as large as $\frac{5}{2}$ in the case of Fe; we cannot understand why the moments on adjacent ions should interact in favour of a parallel orientation. The problem is further complicated by the fact (§ 4.1) that these so-called d-electrons are not strictly localized on particular ions but lie in states that overlap from atom to atom to form a narrow band. This band, in turn, hybridizes with the ordinary s-band, where electrons conduct very freely. The *Heisenberg model* Hamiltonian (10.29), despite its success as a phenomenological theory of ferromagnetism in metals, demands serious reconsideration from first principles. Some aspects of this subtle problem—one of the most important fundamental questions in the theory of solids—are discussed in §§ 10.5–10.6.

10.5 Band ferromagnetism

Going to the opposite extreme from the localized spins of the Heisenberg model, let us put all the magnetic electrons into a band of Bloch states. Let $n_{\mathbf{k}+}$ be the occupation number of the state $|\mathbf{k}\rangle$ with + spin. In the *collective electron model* of Stoner, we add to the electron Hamiltonian a term such as

$$\mathscr{H}_{\text{int.}} = \frac{U}{N}\sum_{\mathbf{k},\mathbf{k}'} n_{\mathbf{k}+} n_{\mathbf{k}'-}. \tag{10.38}$$

Every pair of electrons of opposite spins thus contribute a positive 'exchange' energy, of magnitude U/N, arising perhaps from their occasional simultaneous residence in the d-shell of the same atom, but not necessarily represented precisely by an integral such as (10.37). Contributions from electrons of the *same* spin need not be counted explicitly, since they can be included in the definition of the zero of energy.

Although this term may be interpreted as giving rise to an internal field like (10.22), we are now dealing with a Fermi–Dirac distribution, and may follow the argument of § 10.2. Let n_σ be the total number of electrons per atom of spin σ. Instead of (10.9), for the energy of an electron of + spin, we have

$$\mathscr{E}_{\mathbf{k}+} = \mathscr{E}(\mathbf{k}) - \mu_0 H + U n_-, \tag{10.39}$$

and similarly for those of − spin.

The chemical potential of the two electron assemblies is still the same, so that we have to satisfy the following equations:

$$n\langle\mu\rangle = \mu_0 \int_0^\infty \tfrac{1}{2}\{f^0(\mathscr{E}_{\mathbf{k}+}) - f^0(\mathscr{E}_{\mathbf{k}-})\}\mathscr{N}(\mathscr{E})\,d\mathscr{E}, \tag{10.40}$$

and

$$n = \int \tfrac{1}{2}\{f^0(\mathscr{E}_{\mathbf{k}+}) + f^0(\mathscr{E}_{\mathbf{k}-})\}\mathscr{N}(\mathscr{E})\,d\mathscr{E}. \tag{10.41}$$

In the limit $H \to 0$, $T \to 0$, these equations have a solution analogous to (10.11). The susceptibility takes the form

$$\chi = \frac{\mu_0^2 \mathscr{N}(\mathscr{E}_F)}{1 - \frac{1}{2}U\mathscr{N}(\mathscr{E}_F)} \tag{10.42}$$

as if we had simply enhanced the Pauli susceptibility (10.12) by a factor $1/\{1 - \frac{1}{2}U\mathscr{N}(\mathscr{E}_F)\}$. The exchange interaction (10.38), by favouring parallel spins, makes the system more easily polarizable.

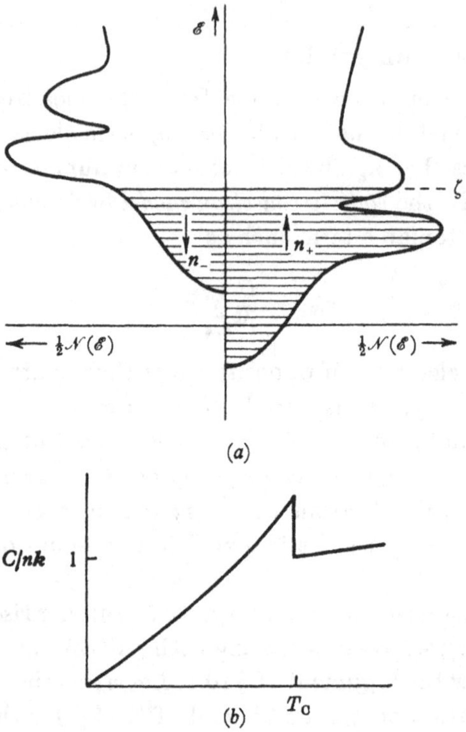

Fig. 177. (a) Band ferromagnetism. (b) Electronic specific heat of ferromagnetic metal.

But when the exchange field is strong enough to make

$$\tfrac{1}{2}U\mathcal{N}(\mathcal{E}_F) > 1, \tag{10.43}$$

this formal solution is obviously unstable. This represents a transition to ferromagnetism. For consistency of (10.39)–(10.41) we must make n_+ (say) much larger than n_-, so that the system has a permanent magnetic moment. The equations can then be solved numerically for $\langle \mu \rangle$ as a function of temperature. The result is not dissimilar, qualitatively, to the result of (10.26). The system exhibits a Curie temperature below which the magnetic moment saturates to a nearly constant value. Above the Curie point, also, something like the Curie–Weiss Law (10.24) is predicted.

In this model it is not difficult to calculate explicitly the specific heat associated with the transition. The energy (cf. §4.7) can be evaluated as a function of T and differentiated to give the specific heat. It turns out that there is a discontinuity in $C_{\text{el.}}$ at the Curie point, with a sharp peak just below. This, again, is an observed feature of the transition to ferromagnetism, although the theory does not reproduce the exact shape of the peak in all detail.

The *itinerant electron* picture of ferromagnetism thus explains most of the observed properties of ferromagnetic metals quite as well as the localized spin model. It is also consistent with general band theory (§ 3.10), with transport theory, and with Fermi surface studies, where one may, for example, observe two distinct band systems containing 'up' spin and 'down' spin electrons respectively. On the other hand, the precise nature of the 'exchange' energy U is not at all clear, and there is independent evidence for some degree of apparent localization of moments (§ 10.6, § 10.11). The best opinion favours the band model for transition metals with d-bands, whereas the spins in partially filled f-states in rare earth metals seem always to behave as if localized in the Heisenberg manner.

As noted in § 8.3, some substances are *ferro-electric*; they can carry a permanent electric moment in the absence of an electric field. This phenomenon is macroscopically akin to ferromagnetism, but arises from a much more complex and unusual microscopic effect. It is due to the capacity of certain crystal structures of distorting themselves slightly so as to give every cell of the lattice a dipole moment.

10.6 Magnetic impurities

Atoms of a transition or rare earth element dissolved in an ordinary metal often retain localized magnetic moments. This is the source of a number of interesting experimental phenomena and theoretical concepts. The Friedel interpretation in the language of phase shifts (§ 5.5) is complementary to the conventional description through the *Anderson Hamiltonian*, which permits semi-quantitative estimates of many features of these phenomena in terms of a small number of phenomenological parameters.

Let us assume, first of all, that the 'exchange' or 'correlation' repulsion between electrons of opposite spin comes into force only when they lie in the same atomic d-level, of energy $\mathscr{E}_d$. In the simplified Hartree–Fock approximation already used in (10.38), we write

$$\mathscr{H}_{dd} = U n_{d+} n_{d-} \tag{10.44}$$

where $n_{d\sigma}$ means the average number of d-electrons of spin σ. If these atomic levels remained perfectly sharp when the transition element is dissolved in the ordinary metal, they would undoubtedly be polarized, with, say, the value n_{d+} as large as possible, and n_{d-} as small as possible—an exaggerated case of the type discussed in § 10.5. In this configuration, the effective energies of the levels are split apart, i.e. (Fig. 178),

$$\mathscr{E}_{d\sigma} = \mathscr{E}_d + Un_{d\sigma}. \tag{10.45}$$

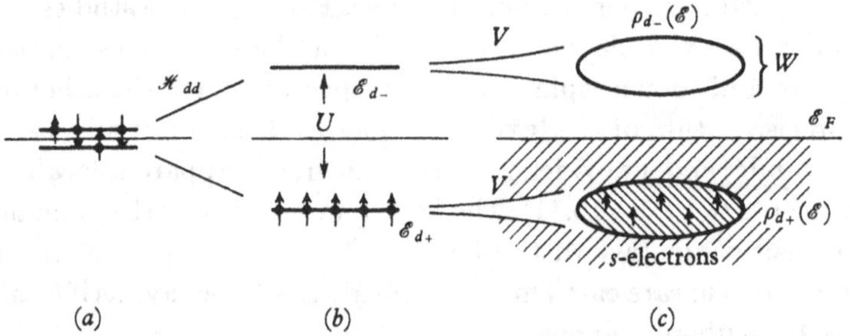

Fig. 178. Atomic d-levels, (a), are polarized and split by the d–d-interaction (b), and then broadened into resonances by hybridization with s-electrons (c).

Now we introduce an interaction, of strength V, between these localized d-states and the Bloch states of the conduction electrons ('s-electrons') of the solvent metal. This interaction is not necessarily spin-sensitive; in principle, it is much the same as an s–d-hybridization matrix element in the model Hamiltonian formulation for the band structure of a transition metal (§ 3.10) which might have been derived from an L.C.A.O. 'overlap integral' such as (3.27). We assume, therefore, that this interaction only links s- and d-states of the same spin, and we ignore the small exchange interaction that may exist among the s-electrons themselves.

Considered as a perturbation on each d-level, however, this interaction gives rise to transitions to and from the s-band at a rate

$$W_\sigma/\hbar = (\pi/\hbar)\, V^2 \mathscr{N}_s(\mathscr{E}_{d\sigma}) \tag{10.46}$$

In this 'Golden Rule' formula, the properties of the solvent metal appear only through the density of states in its s-band at the energy (10.45) of the d-state under consideration. But this shortened life time may be interpreted as an energy broadening of the d-level by an amount W_σ. In place of a sharp level at $\mathscr{E}_{d\sigma}$, we find a spectral density function of the typical Lorentzian form

$$\rho_{d\sigma}(\mathscr{E}) = \frac{2l+1}{\pi} \frac{W_\sigma}{(\mathscr{E}-\mathscr{E}_{d\sigma})^2 + W_\sigma^2}. \tag{10.47}$$

In principle, the centre of this line will also be shifted a bit, but this is not a significant physical effect.

This formula is, in fact, exactly what we should get by differentiating the Friedel sum (5.40), as a function of energy, near a resonance such as (3.81). The present interpretation of the width W_σ in terms of an s–d matrix element V in (10.46) is not, therefore, an essential feature of the physics, although these equations are valid when this interaction is not too strong.

But now, having broadened the d-levels on the impurity, we can no longer be sure that they will remain magnetically polarized. The situation is, indeed, exactly as in (10.39)–(10.42), with (10.47) in place of $\frac{1}{2}\mathcal{N}(\mathscr{E})$. We thus arrive at a condition for a permanent moment on the impurity ion; as in (10.43), if

$$U\rho_{d+}(\mathscr{E}_F) > 1 \tag{10.48}$$

the polarized state is stable energetically against the transfer of an electron from the n_{d+} distribution to the n_d-distribution (cf. § 5.2).

Various observable properties of dilute alloys of transition and rare earth elements, with plausible and consistent values of the parameters U and V, follow from this theory. For example, the iron group elements from V to Co, which are magnetic when dissolved in Cu, Ag and Au, do not show magnetic moments when they occur as impurities in Al. The higher concentration of conduction electrons in Al makes $\mathcal{N}_s$ too large in (10.46), hence broadening the d-levels, and lowering the peak of $\rho_{d\sigma}(\mathscr{E})$ in (10.47), so that the magnetization condition (10.48) can no longer be satisfied.

The conduction electrons of the host metal are not themselves indifferent to the presence of the impurity. But the effect is spin-dependent. Suppose that the magnetic ion is fully polarized, with total spin S. If all the 'up' spin d-states are full, they cannot be entered by s-electrons of the same spin. In this case, therefore, only the 'down' spin s-electrons benefit. By ordinary perturbation theory, we estimate that virtual transitions, via the matrix element V, into the d-levels at an energy U above the Fermi level, changes the energy of each of these electrons by an amount

$$J/N \approx -V^2/U. \tag{10.49}$$

The system thus behaves as if there were a spin-dependent interaction energy of the form

$$\mathscr{H}_{sd} = -(J/N)\,\mathbf{s}\cdot\mathbf{S}_i\,\delta(\mathbf{r}-\mathbf{R}_i) \tag{10.50}$$

between the localized spin moment $\mathbf{S}_i$ of the impurity at $\mathbf{R}_i$ and the spin $\mathbf{s}$ of the conduction electron at $\mathbf{r}$. Note that the sign of this interaction favour antiparallel spins (see § 10.7) whatever the sign of V.

The s–d interaction (10.50) is a strongly localized perturbation on the free electron gas. To appreciate its effect, let us calculate the magnetic 'response' of the system, as a function of wave-number, in much the same way as we calculated the electrostatic response function in § 5.1. The present problem is, indeed much simpler, since we are neglecting spin–spin interactions within the gas, and may treat the whole effect as being due to the 'external' perturbation. 'Up' spin and 'down' spin s-electrons, each of magnetic moment $g\beta\mathbf{s}$, are treated independently; by the arguments leading to (5.16), we arrive at a generalized susceptibility (cf. § 8.1):

$$\chi_0(q) = 2(g\beta s)^2 \sum_{\mathbf{k}} \frac{f^0(\mathbf{k}) - f^0(\mathbf{k}+\mathbf{q})}{\mathscr{E}(\mathbf{k}+\mathbf{q}) - \mathscr{E}(\mathbf{k})}, \tag{10.51}$$

where the sum is over electron states of both spins.

In the free-electron model, this sum can be evaluated, as in (5.35), as a simple analytic function with a second-derivative singularity at $q = 2k_F$. A Fourier transformation (cf. (5.17)) then tells us the spatial distribution of magnetization in the neighbourhood of the localized moment (10.50). The singularity gives rise to an oscillatory term, of wave-number $2k_F$, falling off as $1/r^3$ at large distances (Fig. 179). This is the R.K.K.Y. (*Ruderman–Kittel–Kasuya–Yosida*) *effect*, which can be detected by small variations in the Knight shifts (§ 10.2) of various ions situated at various distances from the impurity.

In the language of § 5.5 we could have described the magnetically polarized impurity as a region into which 'up' spin electrons are drawn by a resonance (3.81) in their d-phase shifts. The corresponding resonance for 'down' spin electrons comes at a much higher energy, so that the two distributions have a big difference in phase shift at the Fermi level. From (5.41), this gives rise to oscillations of magnetic polarization, entirely analogous to the Friedel oscillations of charge about an electrostatic impurity. This description is equivalent to the Anderson model, but does not immediately give us the magnitude of J in terms of V and U as in (10.49).

What is more important, perhaps, is that this phenomenon provides a mechanism for the coupling of the moments on two adjacent magnetic ions. The excess of 'down' spin s-electrons in the immediate vicinity of an 'up' spin impurity induces a corresponding 'up' spin

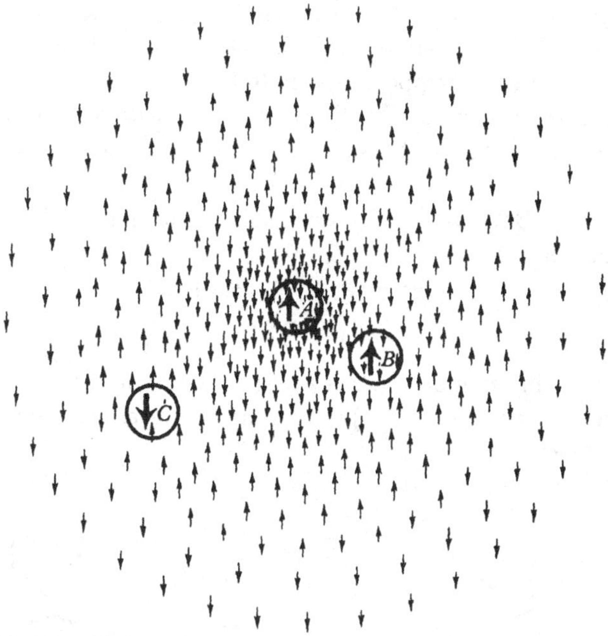

Fig. 179. Friedel oscillations of spin polarization of the *s*-electrons about the impurity *A* lead to a ferromagnetic interaction with the neighbouring impurity *B*, but may favour antiferromagnetism at a greater distance, *C*.

on any other impurity in that neighbourhood. We thus have the machinery for a Heisenberg exchange model of ferromagnetism (§ 10.4) in which the moments localized in the *d*-shells of the transition metal ions are encouraged into mutual parallelism by their tendency to polarize antiparallel to the gas of conduction electrons in which they are bathed.

This sort of interaction gives rise to complex magnetic and thermal effects in dilute alloys, where the magnetic ions are distributed at random on the solvent lattice. In the extreme case, where the non-magnetic solvent is removed, leaving a pure transition metal, this is the *Vonsovski–Zener model* for ferromagnetism in metals such as Fe. Notice that it is a much more 'dynamical' model than is suggested by the Heisenberg–Dirac Hamiltonian (10.29). The *d*-shell moments are not made up by integral numbers of perfectly localized '*d*-electrons', but represent average contributions in Bloch states where the *s*- and *d*-bands have been substantially hybridized. This model, therefore, is not inconsistent with the simpler version of band ferromagnetism discussed in § 10.5, although the parameter U would need to be reinterpreted.

The spin-dependent s–d interaction (10.50) is also responsible for a peculiar effect associated with the electrical resistivity due to a magnetic impurity. Suppose, as in (6.90), we try to calculate the differential cross-section of such an object, for scattering of an 'up' spin s-electron in the state $|\mathbf{k},+\rangle$ into the state $|\mathbf{k}',+\rangle$ of the same spin.

In the first Born approximation, the scattering amplitude is just the matrix element

$$\begin{aligned} t^{(1)} &= \langle \mathbf{k},+|\mathcal{H}_{sd}|\,\mathbf{k}',+\rangle \\ &= -(J/N)\,\mathrm{S}_i^{(z)}. \end{aligned} \tag{10.52}$$

This term, being non-zero when squared, and additional to valence-charge scattering, etc., (§ 7.4) would normally satisfy our curiosity on the subject.

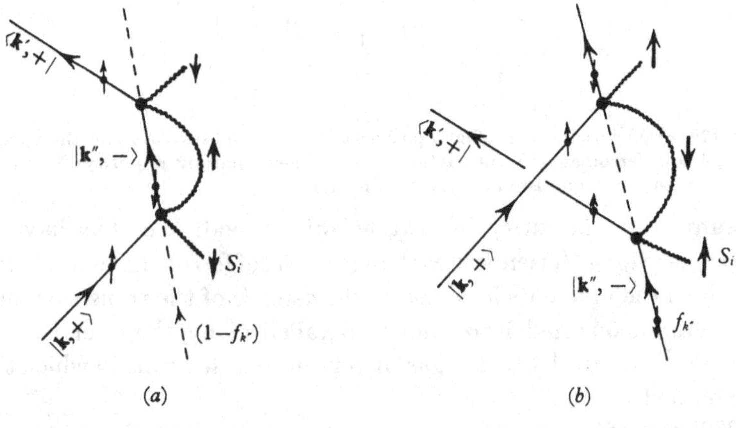

Fig. 180. Direct (*a*) and exchange (*b*) processes in second-order *s*–*d*-scattering. Note relative orientations of *s*- and *d*-spins for spin-flip transitions.

But, in a spirit of pedantry, let us look at the higher terms in the perturbation series. In elementary texts on quantum mechanics we are taught that the second Born approximation takes the form

$$t^{(2)} = \sum_{\mathbf{k}'',\sigma} \frac{\langle \mathbf{k},+|\mathcal{H}_{sd}|\,\mathbf{k}'',\sigma\rangle\langle \mathbf{k}'',\sigma\,|\mathcal{H}_{sd}|\,\mathbf{k}',+\rangle}{\mathscr{E}(\mathbf{k})-\mathscr{E}(\mathbf{k}'')}, \tag{10.53}$$

where the 'intermediate' state $|\mathbf{k}'',\sigma\rangle$ is linked to both the initial and final states by the interaction Hamiltonian $\mathcal{H}_{sd}$. This expression is not, however, quite correct. In the full time-dependent perturbation theory, we have to consider two types of 'graph' (Fig. 180). In the *direct* process (a) the incoming electron goes first into the state $|\mathbf{k}'', \sigma\rangle$ (which must therefore be empty) before being emitted into the final state: in the *exchange* process (b) we suppose that an electron already in $|\mathbf{k}'', \sigma\rangle$ goes off in $|\mathbf{k}', +\rangle$ before the 'initial' electron is accepted into the intermediate state. The rules tell us that we should rewrite (10.53) in the form

$$t^{(2)} = \sum_{\mathbf{k}'', \sigma} \frac{1}{\mathscr{E}(\mathbf{k}) - \mathscr{E}(\mathbf{k}'')} \{(1 - f_{\mathbf{k}''}) \langle \mathbf{k}, + |\mathscr{H}_{sd}| \mathbf{k}'', \sigma\rangle \langle \mathbf{k}'', \sigma |\mathscr{H}_{sd}| \mathbf{k}', +\rangle$$
$$+ f_{\mathbf{k}''} \langle \mathbf{k}'', \sigma |\mathscr{H}_{sd}| \mathbf{k}', +\rangle \langle \mathbf{k}, + |\mathscr{H}_{sd}| \mathbf{k}'', \sigma\rangle\}, \quad (10.54)$$

where $f_{\mathbf{k}''}$ is the occupation number of the intermediate state.

For normal potential scattering, the product of the two matrix elements in (10.54) does not depend on their order, so that $f_{\mathbf{k}''}$ cancels out and we are back at (10.53), whose contribution to the scattering cross-section would be a small correction to the first-order term (10.52). But for spin scattering we must remember that $\mathscr{H}_{sd}$ contains terms like $\mathrm{S}_i^{(x)} s_x$ and $\mathrm{S}_i^{(y)} s_y$ which 'flip' the spin of the conduction electron at the expense of the polarization of the impurity (see (10.104)). The operators $\mathrm{S}_i^{(x)}$ and $\mathrm{S}_i^{(y)}$ do not commute, so that the order of the terms in (10.54) becomes significant. From the point of view of the spin moment of the impurity, it really does matter which of the two transitions occurs first. It is obvious, for example, that the direct process is forbidden if the impurity already has maximum 'up' spin.

The upshot is that in the algebraic reduction of (10.54) we are left with a term of the form

$$t_{\mathrm{K}} \approx 2(J/N)^2 \mathrm{S}_i^{(z)} \sum_{\mathbf{k}''} \frac{f_{\mathbf{k}''}}{\mathscr{E}(\mathbf{k}) - \mathscr{E}(\mathbf{k}'')}, \quad (10.55)$$

where the factor $f_{\mathbf{k}''}$ has not been compensated by contributions from other graphs. Since $f_{\mathbf{k}''}$ changes abruptly at the Fermi energy, this sum becomes large as $\mathscr{E}(\mathbf{k})$ approaches $\mathscr{E}_F$. Assuming, for example, that the density of states in the conduction band is of the order of $\mathscr{N}(\mathscr{E}_F)$ down to some energy $(\mathscr{E}_F - D)$ we can evaluate (10.55) approximately in the form

$$t_{\mathrm{K}} \approx 2(J/N)^2 \mathrm{S}_i^{(z)} \mathscr{N}(\mathscr{E}_F) \ln \frac{D}{|\mathscr{E}_F - \mathscr{E}(\mathbf{k})|}. \quad (10.56)$$

The *residual resistance* produced by ordinary charged impurities in metals is independent of temperature (§§ 7.2–7.4). But when the total probability of scattering from a magnetic impurity is integrated through the 'thermal layer' of thickness $\boldsymbol{k}T$ on the Fermi surface, the singularity in (10.56) is not smoothed out. The resistivity should vary logarithmically with temperature, more or less as

$$\rho(T) \sim \rho_0\{1-2(J/N)\mathscr{N}(\mathscr{E}_F)\ln D/\boldsymbol{k}T\}. \tag{10.57}$$

Since J is essentially negative, this formula predicts a rapid increase in the residual resistivity as the temperature falls. Combined with the ideal resistance (§ 7.5) of the host metal, which varies at T^5 at low temperatures, this explains the mysterious *resistance minimum* which had long been observed in dilute alloys containing transition elements. The strong energy dependence of (10.56) in a formula such as (7.104) explains the *giant thermo-electric power* also observed in such materials.

But the complete theory of the *Kondo effect* must take into account the mutual polarization of neighbouring impurities, higher-order corrections to the perturbation expansion for t, the nature of the ground state of a system with an *s–d* interaction, and many other subtleties. This field is now one of the major exercise grounds for advanced quantum-mechanical methods in the physics of solids.

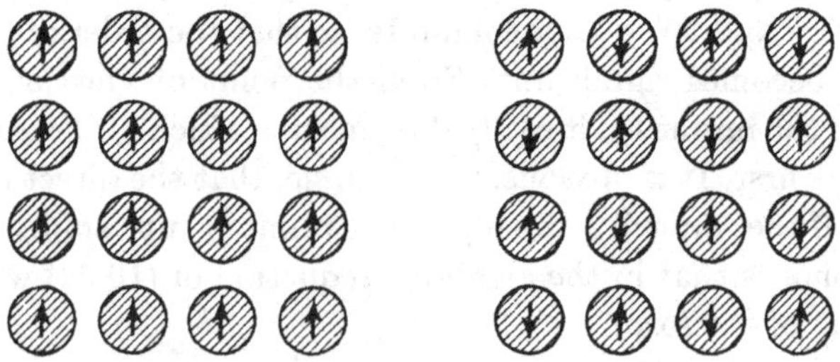

(*a*) (*b*)

Fig. 181. (*a*) Ferromagnetic ordering. (*b*) Antiferromagnetic ordering.

10.7 Antiferromagnetism

The general condition for ferromagnetism in the Heisenberg model is that the exchange integral $J_{ll'}$ in (10.29) should be positive. The internal field is then parallel to the alignment of the spins, and the whole system prefers to take up the configuration of Fig. 181 (*a*).

But suppose, as the calculation of $J_{ll'}$ often suggests, that the sign of the integral is negative. The internal field (10.32) at any site would be opposite to the mean direction of magnetization of neighbouring spins. The arrangement of Fig. 181 (*b*) is favoured. We call such a system *antiferromagnetic*.

To see what happens, let us refer separately to the two sublattices on which the spins would be oriented in opposite directions. Suppose that μ_+ is the average magnetization of an atom on the ⊕ sublattice. The exchange interaction will give rise to an internal field on the ⊖ sublattice

$$\mathbf{H}^- = \frac{1}{\beta^2}(\Sigma J)\boldsymbol{\mu}_+$$
$$= \lambda N\boldsymbol{\mu}_+, \qquad (10.58)$$

say. The sum is supposed to be only over nearest neighbours, because $J_{ll'}$ is supposed to be of very short range. The essential point is that λ will now be *negative*.

Again, because of the same interaction, there will be an internal field at every ⊕ site

$$\mathbf{H}^+ = \lambda N\boldsymbol{\mu}_-. \qquad (10.59)$$

Now suppose that our magnets are classical, so that (10.17) holds for the *average* value of a moment $\boldsymbol{\mu}$ in the field $\mathbf{H}$. Let us write (10.17) in the form

$$\langle\boldsymbol{\mu}\rangle = \mathscr{L}\{(\boldsymbol{\mu}\cdot\mathbf{H})/\boldsymbol{k}T\}, \qquad (10.60)$$

using the symbol $\mathscr{L}$ to signify the *Langevin function* defined by the right-hand side of (10.17). We now have two equations to solve

$$\left.\begin{aligned}\boldsymbol{\mu}_- &= \mathscr{L}\{\boldsymbol{\mu}\cdot(\mathbf{H}+\lambda N\boldsymbol{\mu}_+)/\boldsymbol{k}T\},\\ \boldsymbol{\mu}_+ &= \mathscr{L}\{\boldsymbol{\mu}\cdot(\mathbf{H}+\lambda N\boldsymbol{\mu}_-)/\boldsymbol{k}T\}.\end{aligned}\right\} \qquad (10.61)$$

Now at high temperatures, when the approximation (10.18) is valid, these two equations can be added together to give (10.23). The result is almost exactly as before, in (10.24):

$$\chi = \frac{N\langle\mu^2\rangle}{3\boldsymbol{k}(T+\theta)}; \qquad (10.62)$$

with λ negative it is convenient to define θ by $-\lambda N\langle\mu^2\rangle/3\boldsymbol{k}T$ as would be consistent with (10.25). The paramagnetic susceptibility is reduced by the magnetic moments tending to align themselves antiparallel, instead of being drawn into the same direction by the magnetic field.

But at low temperatures there is a solution even in the absence of the magnetic field. If we write

$$\boldsymbol{\mu}_- = -\boldsymbol{\mu}_+ \qquad (10.63)$$

(as is intuitively the case in the ordered antiferromagnetic array of Fig. 181 (*b*)) then the negative sign of λ is compensated, and we have the equation

$$\boldsymbol{\mu}_+ = \mathscr{L}\{(\boldsymbol{\mu}\cdot|\lambda|\,N\boldsymbol{\mu}_+)/\boldsymbol{k}T\} \qquad (10.64)$$

to be solved. This is exactly the same equation as gives rise to ferromagnetism in § 10.3—for example, (10.26), which contains the 'quantum' equivalent of the Langevin function.

Thus, from the point of view of a single spin, ferromagnetism and antiferromagnetism seem similar; the system is paramagnetic at high temperatures, but becomes strongly polarized below a temperature which in the antiferromagnetic case is called the *Néel temperature*,

$$T_N = \theta. \tag{10.65}$$

Macroscopically, however, the behaviour is very different. The antiferromagnetic array does not have any net polarization, so that the system does not show any permanent magnetization. All that is observed is a change in the susceptibility, which can be calculated by solving (10.61) in the presence of a small magnetic field H.

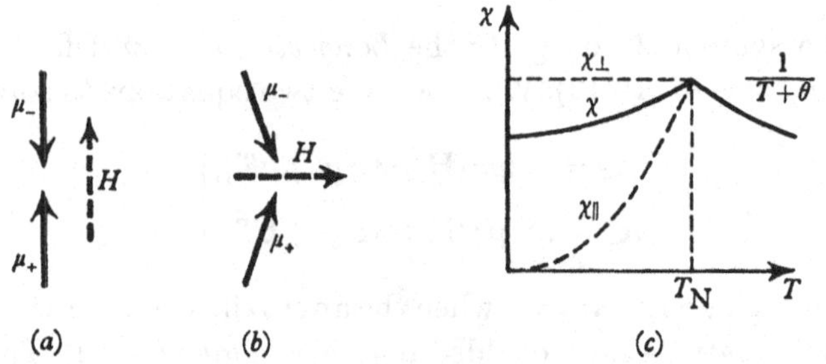

Fig. 182. Susceptibility of antiferromagnet: (a) $\chi_\parallel$ tends to zero; (b) $\chi_\perp$ remains constant; (c) $\chi = \frac{1}{3}\chi_\parallel + \frac{2}{3}\chi_\perp$.

This calculation is elementary in principle, though a little tedious in practice. The result is interesting, in that it depends on the orientation of **H** relative to the direction of magnetization of the two sublattices. When **H** is parallel to $\boldsymbol{\mu}_+$ the susceptibility $\chi_\parallel$ goes to zero at $T = 0$. This is because the internal field is so strong that it prevents any spins in the $\ominus$ sublattice from turning over. On the other hand, $\chi_\perp$ at $T = 0$ is the same as the value it would have at T_N; the polarization of the sublattices normal to the applied field exerts only a feeble opposition to their magnetization being slightly turned in the field direction.

In practice, a macroscopic sample will contain numerous domains, each with a different axis of polarization, and we can write for the average susceptibility

$$\chi = \tfrac{1}{3}\chi_\parallel + \tfrac{2}{3}\chi_\perp. \tag{10.66}$$

The discontinuity of slope of χ at T_N is easily observable, and is associated with a specific heat anomaly as the sublattices become polarized.

It is worth remarking that, because $\chi_\perp > \chi_\parallel$, a magnetic field should be capable of pulling all domains into the perpendicular configuration. This does not occur—presumably because of the anisotropy of the spin-spin interaction, which tends to favour special crystal directions.

The model that we have considered here is only a simple case out of a wide range of systems showing a bewildering variety of phenomena. The argument may be elaborated in different directions. For example, we may have a complex crystal structure, in which the magnetic ions do not form a simple cubic network, and where it may not be perfectly obvious just how the system will fall into sublattices. Again, the interaction may be of longer range than between nearest neighbours, or may be antiferromagnetic for some pairs of spins and ferromagnetic

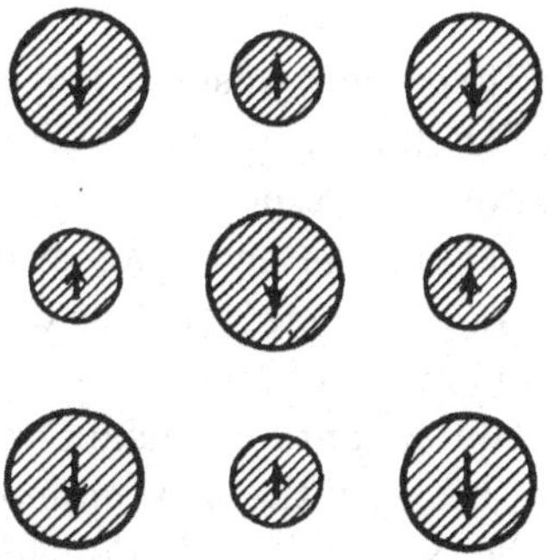

Fig. 183. Ferrimagnetism.

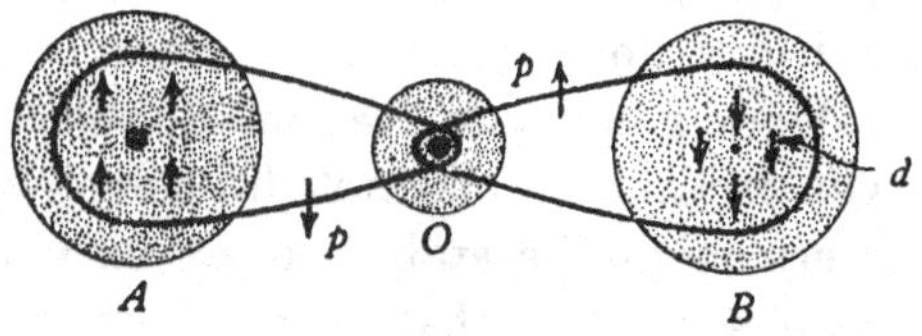

Fig. 184. Super-exchange.

for others. One result of this (which can easily be dealt with, inside our formalism, by adding a term like $\lambda' N \mu_-$ to (10.58), and $\lambda' N \mu_+$ to (10.59)) is to change the Néel temperature, which is no longer the same as the parameter θ in (10.62).

Another case is where there are two different types of magnetic ion in the crystal, and where these do not have the same magnetic moment. We may then have the situation sketched in Fig. 183, where the A and B atoms have their spins oppositely oriented, but where the system will show a permanent moment because the magnitude of the moment on one sublattice is larger than that on the other. This is called *ferrimagnetism*, being characteristic of ferrites; macroscopically, it looks very like ferromagnetism.

In many antiferromagnetic substances the magnetic ions are quite a long way apart, so that it becomes difficult to believe that they can be linked by an exchange integral such as (10.37). This does not mean that the crystal lattice is very attenuated; it means that there are many non-magnetic ions filling the spaces between the magnetic ions. For example, these may be oxygen ions, as in MnO. We then have the mechanism known as *super-exchange*, where the d-electrons in the Mn ion are coupled to electrons in the oxygen, which then interact, by exchange, with the d-electrons in a second Mn ion.

Schematically, the situation may be as in Fig. 184. If atom A has $+$ spin, there is room for an electron of $-$ spin to come in from the oxygen p-state. If atom B has $-$ spin, then it can take in the other electron of the pair from the oxygen. Thus, the interaction between the spins of the p-electrons on the oxygen is transformed into an interaction between the spins of the d-shells in the magnetic ions. The actual calculation of this effect is somewhat subtle, but it explains, in principle, the observed phenomena.

In dilute magnetic alloys, the *indirect exchange* mechanism of § 10.6 can be antiferromagnetic. As indicated in Fig. 179, this would occur when the second impunity stands at a distance corresponding to an 'up' spin peak in the R.K.K.Y. oscillations about the first magnetic ion. But to understand antiferromagnetism in pure metals such as Cr, we should go back to the itinerant electron picture of § 10.5.

Quantitatively, it is not difficult to see how an antiferromagnetic pattern might be stabilized by exchange amongst the conduction electrons. Suppose that magnetic order of the type of Fig. 181 (*b*) had already been established. An 'up' spin electron travelling through the crystal will see an effective field depending upon the local polarization through the exchange interaction U. This field would have the periodicity of the magnetic superlattice, and hence would give rise to an energy discontinuity at a 'sub-zone' boundary inside the usual Brillouin zone (cf. § 2.6). This feature in the band structure would be reflected as a superlattice of electron density for 'up' spin electrons. For electrons of the opposite spin, the sign of the perturbation would be reversed, so that they would tend to occupy the other superlattice. If these distributions happened to agree with the magnetization pattern we first thought of, we should have found a self-consistent antiferromagnetic ordering of the electrons without forcing them to be localized.

To give mathematical substance to this type of argument, suppose we generalize the uniform susceptibility (10.42) into a q-dependent magnetic response function. It is quite easy, following the lines of § 5.1, to include an exchange interaction of the form of (10.38) in a calculation of the type leading to the susceptibility $\chi_0(\mathbf{q})$ of (10.51). The result is, in fact, recognizably similar to (10.42), i.e.

$$\chi(\mathbf{q}) = \frac{\chi_0(\mathbf{q})}{1-\frac{1}{2}U(g\beta s)^{-2}\chi_0(\mathbf{q})}. \tag{10.67}$$

Now if the denominator of this expression should vanish, the assumed non-magnetic state of the metal is not stable. The *Overhauser condition* for magnetic ordering is

$$\tfrac{1}{2}U(g\beta s)^{-2}\chi_0(\mathbf{q}) > 1 \tag{10.68}$$

of which the condition (10.43) for ferromagnetism is the special case where $q = 0$.

The question of the occurrence of stable *spin density waves* in ordinary metals as a result of the usual exchange interaction between free electrons has not been firmly resolved. In a transition metal, however, U is known to be quite large, and the electron energy states $\mathscr{E}(\mathbf{k})$ to be put into the formula (10.51) for $\chi_0(\mathbf{q})$ are dominated by the band structure. It is obvious from § 5.4, for example, that a wave-vector $\mathbf{q}$ spanning parallel flat regions of Fermi surface may induce a large value of $\chi_0(\mathbf{q})$, so that (10.68) may be satisfied. In such a case the antiferromagnetic ordering may be exceedingly complicated, the variations in spin density being equivalent to a helical pattern of spin moments upon a cone, the repeat distance being unrelated to the underlying crystal lattice.

10.8 The Ising model

In discussing ferromagnetism and antiferromagnetism we have used a very crude approximation for the states of the Heisenberg Hamiltonian (10.29). In effect, we have supposed that every spin except the one we are dealing with has its average value; we have ignored the fluctuations and correlations between the spins on neighbouring sites. Moreover, we have ignored the fact that the $\mathbf{S}_l$ are operators, which cannot be replaced by their average expectation values.

The distribution of the eigenvalues of the spin Hamiltonian is, in reality, very complex and the attempt to give an exact discussion of the thermodynamics of the system is one of the major enterprises of the theory of solids. There are two directions into which the theory has proliferated. We can assume that the system is near its ground state (ferromagnetic or antiferromagnetic) and we can then discuss the low-energy excitations. This is the theory of *spin waves* which is outlined in § 10.11.

The other type of approximation is to ignore the off-diagonal elements of the spin operators, and to consider only their components along some fixed direction—usually the direction of **H**. We assume, in other words, that the energy levels of the system are given by

$$\mathscr{E} = -\sum_{ll'} J_{ll'}\sigma_l\sigma_{l'} - \beta H \sum_l \sigma_l, \tag{10.69}$$

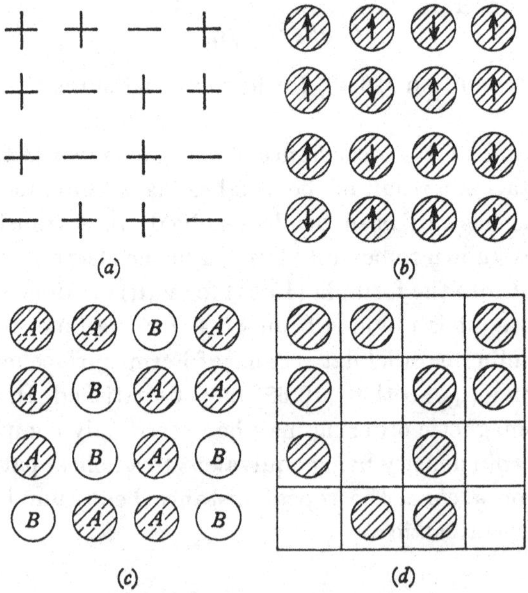

Fig. 185. A configuration of the Ising model, (*a*), may represent: (*b*) an arrangement of spins; (*c*) an arrangement of atoms in a binary alloy; (*d*) a configuration of a 'lattice gas'.

where on each site l we define a 'spin' quantum number σ_l which can only be $+1$ or -1. In most cases we also assume that $J_{ll'}$ acts only between nearest neighbours in the lattice.

This is called the *Ising model*. It cannot be an exact description of the ferromagnetic or antiferromagnetic system, but by short-circuiting the very difficult problem of finding the exact eigenfunctions of an assembly of spins it allows one to concentrate on the combinatorial problem of the statistical distribution of the states of such a system.

The Ising model does, indeed, describe more exactly another type of system—an alloy where two elements, A and B, may be substituted for one another on a set of fixed lattice sites. We may say that $\sigma_l = +1$ corresponds to an A atom on the lth site, and $\sigma_l = -1$ prescribes a B atom on that site. The 'exchange integral' J is then to be interpreted as

$$J = \tfrac{1}{4}\{\mathscr{V}_{AB} - \tfrac{1}{2}(\mathscr{V}_{AA} + \mathscr{V}_{BB})\}, \tag{10.70}$$

where $\mathscr{V}_{AB}$ is the energy of interaction between an A atom and a B atom on neighbouring sites and where we subtract the average energy of an AA pair, or a BB pair.

Such alloy systems can be studied experimentally, and it is found that they do show a transition temperature analogous to a Curie temperature or a Néel temperature. Above the transition point the alloy is disordered. At lower temperatures it tends to be ordered, either by segregating out nearly pure A and nearly pure B material (this would be for $J > 0$, corresponding to ferromagnetism), or by forming regions where the A and B atoms fall regularly on interpenetrating sublattices—the analogue of antiferromagnetism. The difference between the *order-disorder transition* and a ferromagnetic or antiferromagnetic transition is that the total number of A type and B type atoms is fixed in the former, whereas 'up' and 'down' spins can be transformed freely into one another. This difference can be allowed for, formally, in the theory, so that many of the results hold generally for both types of system.

Another problem to which the Ising model can be applied is that of the 'lattice gas', or 'lattice liquid', where one puts atoms, which can interact with one another over short distances, on to a set of sites, leaving vacant sites. This is formally equivalent to the alloy problem. It is doubtful, however, whether it really applies very closely to the thermodynamics of the liquid-gas transition which it is sometimes supposed to describe.

The importance of the Ising model is not, however, in the description of particular physical effects; it is a mathematically tractable model of a system that should exhibit *co-operative phenomena* and phase transitions. In that sense it belongs more to the general theory of statistical mechanics than to the theory of solids. Yet there are certain principles which are important in the general theory of crystal lattices, and which we shall now try to explain.

10.9 Combinatorial method

The advantage of the Ising model is that the calculation of its thermodynamic properties can be reduced to a combinatorial problem. For example, suppose we ask for the *partition function*

$$Z = \Sigma\, e^{-\mathscr{E}/kT}, \tag{10.71}$$

where the sum is over all 'configurations', that is, over all arrangements of + spins and − spins on the N sites of the lattice. This sum looks very complicated, but in a formal sense it need involve only two variables, for example, N_A, the number of 'up' spins [or A atoms] and N_{AB}, the number of neighbouring antiparallel spins [or the number of AB pairs].

To see this, let us express the energy (10.69) in terms of the total numbers of links of various kinds

$$\mathscr{E} = J(N_{AB} - N_{AA} - N_{BB}) - \beta H(N_A - N_B). \tag{10.72}$$

But these variables are not independent. It is obvious that

$$N_A + N_B = N.$$

It is also obvious that there can only be pN_A links ending on A sites, where p is the *co-ordination number* of the lattice (i.e. the number of nearest neighbours of a given site). Each of these links would need to be counted twice in N_{AA} and once in N_{AB}. Thus $2N_{AA} + N_{AB} = pN_A$; similarly $2N_{BB} + N_{AB} = pN_B$. Putting in these conditions, we find that (10.72) becomes

$$\mathscr{E} = -\tfrac{1}{2}pNJ + 2N_{AB}J - (2N_A - N)\beta H. \tag{10.73}$$

Now suppose that $g(N; N_A, N_{AB})$ represents the total number of ways of putting N_A 'up' spins on to the lattice with exactly N_{AB} pairs of antiparallel neighbours. All these configurations have the same energy; we may express the partition function (10.71) in the form

$$\begin{aligned} Z &= y^{\frac{1}{2}N} z^{-\frac{1}{4}pN} \sum_{N_A, N_{AB}} g(N; N_A, N_{AB})\, y^{N_A} z^{N_{AB}} \\ &= y^{\frac{1}{2}N} z^{-\frac{1}{4}pN} \Lambda_N(y, z), \end{aligned} \tag{10.74}$$

where the variables y and z are related to the magnetic field and to the interaction J:

$$y \equiv e^{2\beta H/kT}; \quad z \equiv e^{-2J/kT}. \tag{10.75}$$

All the interesting properties of the system now reside in the function Λ_N from which can be derived formulae applying to the 'ferromagnetic' and 'antiferromagnetic' systems or to the theory of alloy phases and *regular solutions*.

But the problem of calculating the combinatorial factor in (10.74) is not soluble in general; only approximate formulae are known. It is not difficult to calculate the total number of configurations for a given value of N_A. By elementary algebra this must be the total number of arrangements of N_A objects in N places

$$\sum_{N_{AB}} g(N;\, N_A, N_{AB}) = \frac{N!}{N_A!\,(N-N_A)!}. \tag{10.76}$$

The difficulty is to partition this total into separate numbers for each value of N_{AB}.

A very crude approximation is to replace the separate summation over N_{AB}, by the assumption that this has its average value. That is, we distribute our N_A A atoms and N_B B atoms at random over the sites of the lattice, and calculate the probability that any one of the $\frac{1}{2}pN$ links has an A atom at one end and a B atom at the other. The result is, of course,

$$\begin{aligned}\langle N_{AB}\rangle &= 2\frac{N_A}{N}\frac{N_B}{N}\tfrac{1}{2}pN \\ &= \frac{pN_AN_B}{N}.\end{aligned} \tag{10.77}$$

Putting (10.76) and (10.77) into (10.74), we get

$$\Lambda_N \approx \sum_{N_A} \frac{N!}{N_A!\,N_B!} y^{N_A} z^{pN_A N_B/N}. \tag{10.78}$$

We can calculate from this a free energy, taking out the largest term in the logarithm and expanding asymptotically by Stirling's formula. Thus

$$\begin{aligned}F &= -kT\ln\Lambda_N \\ &\approx kT\left\{-N\ln N + N_A\ln N_A + N_B\ln N_B - N_A\ln y - p\frac{N_A N_B}{N}\ln z\right\}.\end{aligned} \tag{10.79}$$

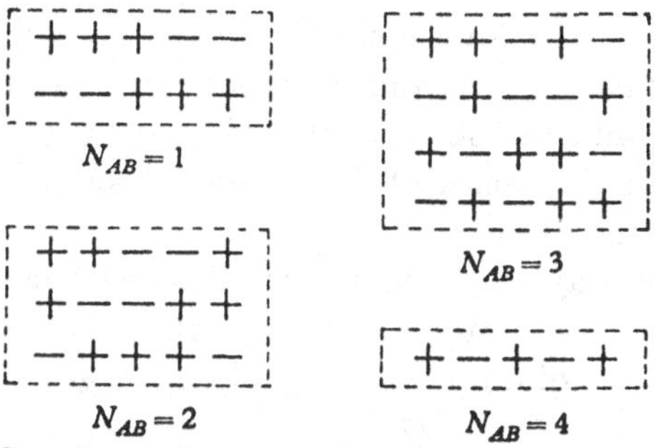

Fig. 186. Counting configurations for a chain with $N = 5$, $N_A = 3$. $\langle N_{AB}\rangle = 2{\cdot}4$ for this system.

The condition for minimum free energy, or a maximum term in the sum, follows by differentiating with respect to N_A (of course, putting $N_B = N - N_A$). Using (10.75) we get

$$\ln N_A - \ln N_B - \frac{2\beta H}{kT} - \frac{(N_A - N_B)}{N}\frac{2pJ}{kT} = 0,$$

i.e.
$$\frac{N_A}{N_B} = \exp\left\{\frac{2\beta H}{kT} + \left(\frac{N_A - N_B}{N}\right)\frac{2pJ}{kT}\right\}, \tag{10.80}$$

or
$$\left(\frac{N_A - N_B}{N}\right) = \tanh\left\{\frac{\beta H}{kT} + \left(\frac{N_A - N_B}{N}\right)\frac{pJ}{kT}\right\}. \tag{10.81}$$

If we identify $(N_A - N_B)/N$ with $\langle\mu\rangle$, the average net polarization of the system, and express pJ as an internal field, as in (10.33), then we are back at the Curie–Weiss condition (10.26) for a ferromagnetic transition. In the theory of alloys this is known as the *Bragg–Williams approximation.*

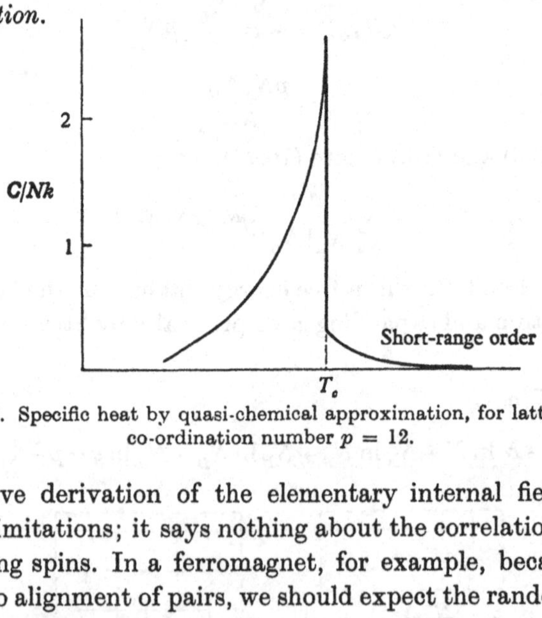

Fig. 187. Specific heat by quasi-chemical approximation, for lattice with co-ordination number $p = 12$.

The above derivation of the elementary internal field formula shows its limitations; it says nothing about the correlations between neighbouring spins. In a ferromagnet, for example, because of the tendency to alignment of pairs, we should expect the random average (10.77) to be an overestimate of the number of AB pairs in the actual equilibrium configuration.

An interesting way of improving on this is Guggenheim's *quasi-chemical approximation.* Suppose we work out the other averages like (10.77), on the basis of random mixing. Then we get

$$\frac{\langle N_{AB}\rangle^2}{\langle N_{AA}\rangle\langle N_{BB}\rangle} = 4. \tag{10.82}$$

But if we had been dealing with the *chemical* equilibrium between molecules, as expressed by the chemical reaction

$$AA + BB \rightleftharpoons 2AB, \tag{10.83}$$

then we know that the equilibrium concentrations of the different species would be given by

$$\frac{\langle N_{AB}\rangle^2}{\langle N_{AA}\rangle\langle N_{BB}\rangle} = 4\frac{(e^{-\mathscr{E}_{AB}/kT})^2}{e^{-\mathscr{E}_{AA}/kT}\,e^{-\mathscr{E}_{BB}/kT}}, \tag{10.84}$$

where $\mathscr{E}_{AB}$, etc., is the energy of an AB molecule, etc. In the notation of (10.75), this suggests that the right-hand side of (10.84) should, on the average, equal $4z^2$. With a little bit of juggling with (10.73), one can calculate $\langle N_{AB}\rangle$, and put this into (10.78) in place of (10.77).

We shall not follow this calculation further, except to remark that the system shows spontaneous polarization (ferromagnetism or anti-ferromagnetism) below a transition temperature defined by

$$\frac{kT_c}{pJ} = \frac{2/p}{\ln(1-2/p)}, \tag{10.85}$$

which would tend to (10.28) (in our present notation, $k\theta = pJ$) in the limit of large co-ordination number p. We can also calculate, approximately, the specific heat near the transition temperature; it shows a sharp singularity, with a discontinuity on the high temperature side, as in Fig. 187.

As it happens, the quasi-chemical approximation is equivalent to what seems to be quite a different method, due to Bethe and Peierls. This is an attempt to generalize the internal field argument. One takes a cluster of spins—the central one at $l = 0$, say, and its p neighbours. One takes for the energy of this cluster the following terms out of (10.69)

$$\mathscr{E}_p = -\sum_{j=1}^{p} J\sigma_0\sigma_j - \beta H\sigma_0 - \beta H_I\sum_{j=1}^{p}\sigma_j. \tag{10.86}$$

In other words, we suppose that the central spin sees the actual external field H, and is also linked explicitly to the spins of its neighbours, whereas for the outer sites of the cluster we suppose that we can treat the interaction with the environment as if it were the effect of an internal field H_I.

It is easy to write down a complete partition function for this cluster, since it contains only p terms. One can then calculate the mean orientation of the atom at the centre of the cluster. But this is not a

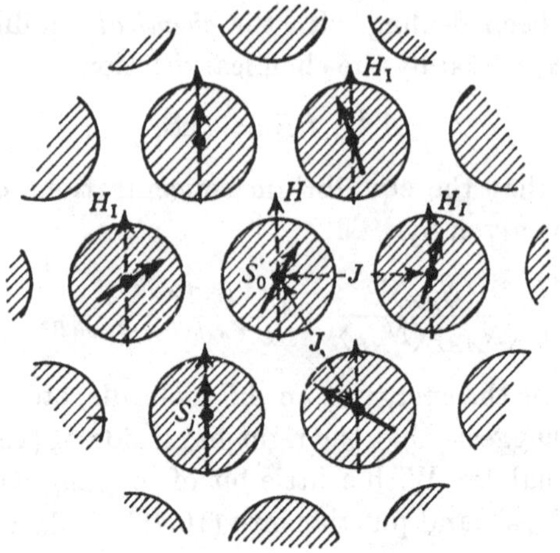

Fig. 188. Bethe–Peierls cluster.

privileged site in the lattice; $\langle\sigma_0\rangle$ must be the same as $\langle\sigma_j\rangle$, the mean moment of one of the spins on the edge of the cluster, which can also be written down as a function of H_I. This gives us a self-consistency condition on the internal field H_I, which can be non-zero, even when $H = 0$, at temperatures below the transition temperature T_c defined by (10.75).

The Bethe method is attractive, as a physically intuitive procedure, and can even be made to yield results for a more realistic model, where proper spin operators are introduced in place of the 'Ising spin variables' of (10.86), and the partition function is calculated for the energy levels of this cluster Hamiltonian. Or we can introduce further shells surrounding the simple cluster of first neighbours, and solve the problem to a higher approximation.

These higher-order approximations are really attempts to determine more accurately the coefficients of a power series expansion of $\Lambda_N(y, z)$. It can be shown that each coefficient, whether of the expansion in powers of z, which is valid at low temperatures, or in powers of

$$w = \tanh\frac{J}{kT}, \tag{10.87}$$

which is valid at high temperatures, depends upon the enumeration of the polygons of a given number of sides that can be traced out on the 'links' of the lattice, rather as in the Mayer cluster-sum expansion of the theory of imperfect gases, or as in the 'diagrams' of perturbation theory. This enumeration, easy enough in principle, is extremely laborious in practice.

Moreover, such expansions cannot quite convince one that there really is a singularity in the thermodynamic properties of the system at some definite temperature, and the closed analytical formulae of the Bragg–Williams and Bethe methods do not fill this gap. These methods *assume* that an internal field exists, or that the system has a net polarization which is the same, on the average, at each lattice site. In other words, they assume that the system has *long-range order*—that if we know the direction of the spin at one site we can guess at its direction at any other site, however distant.†

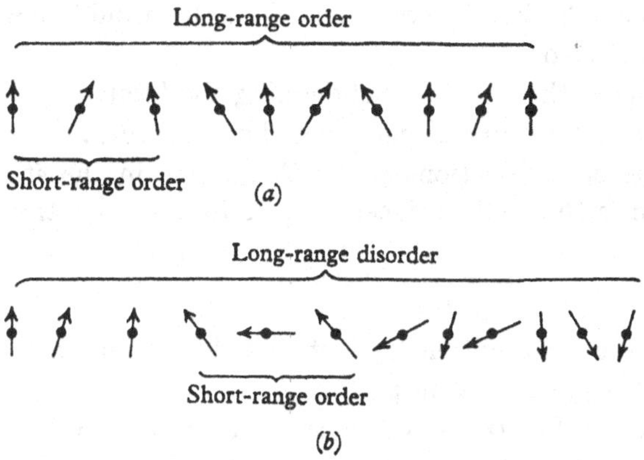

Fig. 189. (*a*) Long-range order implies short-range order. (*b*) Short-range order does not imply long-range order.

We have already remarked that the Bragg–Williams method does not even give any *short-range order*, except such as is implied by the assumed long-range order: above the Curie temperature the distribution is assumed to be quite random, with each spin independent of the detailed behaviour of its neighbours. The Bethe method will give short-range order above T_c, and is, therefore, more realistic. Both methods are quite good at describing the general properties of the long-range ordered state, once it is well established, but they cannot be used to prove that it exists.

† In the ferromagnetic case we need to know one spin, to discover the net direction of polarization of the whole lattice. In the antiferromagnetic case the direction of spin on the distant site will depend, of course, on whether or not it belongs to the same sublattice as the site at the origin.

10.10 Exact solutions of the Ising problem

It turns out that the Ising problem can be solved exactly in two special cases. One of these cases is trivial, and the proof is simple; the other case is not difficult to formulate, but the proof of the final formula for the partition function is much too difficult to present here.

The trivial case is that of the linear chain, whose solution we shall give because it demonstrates a general technique. Let us consider a lattice consisting of N 'layers'. In the one-dimensional case, such a 'layer' is a single site. In two-dimensions this would be a whole row of atoms, and so on.

Now suppose that ν_j is a label defining the 'state' of the jth layer, for example, in two dimensions ν_j would be a symbol like $(+ + - + \ldots)$ defining the spin directions of all the atoms in the jth row. Our assumption is that only adjacent layers interact, i.e. the energy can be written

$$\mathscr{E} = \sum_{j=1}^{N} \mathscr{V}(\nu_j, \nu_{j+1}) + \sum_{j=1}^{N} \mathscr{V}(\nu_j) \tag{10.88}$$

(including the 'internal energy' of each layer, and also closing the ends so that $\nu_{N+1} \equiv \nu_1$, as in § 1.6).

The partition function of our system can now be written

$$Z = \sum_{\nu_1}\sum_{\nu_2}\ldots\sum_{\nu_N}\left\{\prod_{j=1}^{N} \exp\{-[\mathscr{V}(\nu_j, \nu_{j+1}) + \tfrac{1}{2}\mathscr{V}(\nu_j) + \tfrac{1}{2}\mathscr{V}(\nu_{j+1})]/\boldsymbol{k}T\}\right\}. \tag{10.89}$$

But this is the same thing as writing down the product of a successsion of matrices of which the typical element is

$$P_{\nu_j\nu_{j+1}} = \exp\{-[\mathscr{V}(\nu_j, \nu_{j+1}) + \tfrac{1}{2}\mathscr{V}(\nu_j) + \tfrac{1}{2}\mathscr{V}(\nu_{j+1})]/\boldsymbol{k}T\}, \tag{10.90}$$

recognizing that ν_j is a label capable of running through two or more values, according to the complexity of the layer. Indeed, all that we have in (10.89) is

$$\begin{aligned} Z &= \text{trace}\,(P^N) \\ &= \sum_i \lambda_i^N, \end{aligned} \tag{10.91}$$

where the λ_i are the eigenvalues of the matrix P in (10.90). In practice, we are assuming that N is large—that our chain is nearly infinite in length—so that only the largest eigenvalue of P is at all important. We have the result

$$\begin{aligned} F &= -\boldsymbol{k}T \ln Z \\ &\to -N\boldsymbol{k}T \ln \lambda_{\text{max.}}, \end{aligned} \tag{10.92}$$

showing that the free energy is an extensive property.

Now, in the one-dimensional chain, the labels ν_j are just the values $+1$ and -1 of σ_j. The matrix (10.90) can be written

$$P = \begin{pmatrix} y^{-\frac{1}{2}}z^{-\frac{1}{2}} & z^{\frac{1}{2}} \\ z^{\frac{1}{2}} & y^{\frac{1}{2}}z^{-\frac{1}{2}} \end{pmatrix} \tag{10.93}$$

in the notation of (10.75). The eigenvalues of this matrix are

$$\lambda_{\pm} = \tfrac{1}{2}\{(1+y) \pm \surd[(1-y)^2+4yz^2]\}\, y^{-\frac{1}{2}}z^{-\frac{1}{2}}; \tag{10.94}$$

putting λ_+ into (10.92) our problem is solved.

The interesting point about this result is that F is a continuous function both of y and of z—that is, it is a continuous function of H and of T without singularities. The specific heat at zero field can be calculated; it is just

$$C_v = Nk\left(\frac{J}{kT}\right)^2 \operatorname{sech}^2\left(\frac{J}{kT}\right), \tag{10.95}$$

which has a smooth maximum in the neighbourhood of $kT = J$, but exhibits no phase transition.

Furthermore, there is no solution corresponding to spontaneous magnetization at $H = 0$. *A linear chain is not ferromagnetic.* This would not be apparent from the Bragg–Williams formula (10.81), but does show up in the Bethe method where (10.85) would make $T_c \to 0$ as the co-ordination number p tends to 2. The same result also holds, of course, for antiferromagnetism.

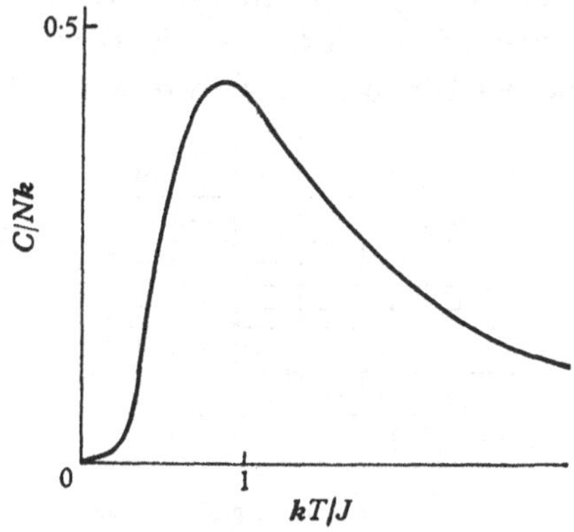

Fig. 190. Specific heat of linear chain.

It is important to realize that this is a *topological* theorem; the fact that $p = 2$ cannot be made to give a transition temperature in the Bethe method may well be accidental. The point is that a linear chain is too easily broken at any point. By introducing a single break, and hence destroying long-range order, we increase the energy by $2J$. But a single break may occur at any one of N sites, so that we gain $k \ln N$ in entropy. The change in free energy would be

$$F = 2J - kT \ln N, \tag{10.96}$$

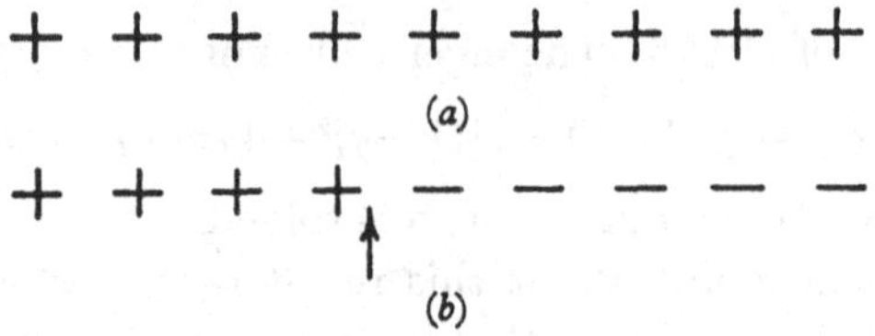

Fig. 191. Long-range order in a linear chain, (*a*), is destroyed by a single break, (*b*).

which can always be made negative, however low the temperature, if N is sufficiently large. From this argument—or from a more sophisticated discussion of the eigenvalues of a matrix such as (10.93), which is of finite order and has all its elements positive—one can deduce that any lattice which is infinite in only one dimension will have no singularities. This is a characteristic exemplification of the maxim that *one chain does not make a crystal*, which should be kept firmly in mind by all students of the condensed state.

It is not difficult to show that in a two-dimensional lattice the above objection does not hold. Suppose, for example, that we make a connected region of reversed spins as in Fig. 192. This region will be bounded by a polygon of length L—that is, there will be L links of type AB, in the border between the two regions. The energy of the

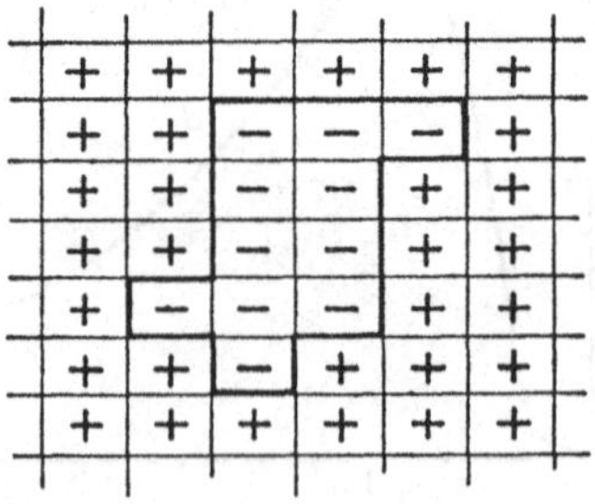

Fig. 192. A region of reversed spins.

system will thus be increased by $2LJ$. But what is the entropy associated with this energy? Generally speaking, at each node of the boundary there are 3 choices of direction, so that there are something like 3^L ways of laying out the boundary. In fact, this is an overestimate because the polygon must close again—but when L is large this error will not be important. Thus, the contribution to the free energy will be

$$F \approx 2LJ - kT \ln 3^L$$

$$= L(2J - kT \ln 3). \tag{10.97}$$

All such contributions will be positive if

$$kT < \frac{2J}{\ln 3}. \tag{10.98}$$

The ordered state should be stable below this temperature.

This crude argument can be supported by an exact calculation, since the other soluble case of the Ising model is the two-dimensional square lattice. Onsager's proof that this system is 'ferromagnetic' (or 'antiferromagnetic') below a temperature

$$kT_c = \frac{2J}{\ln(1+\sqrt{2})} \tag{10.99}$$

requires algebraic theorems that are far beyond our present scope (although the starting point is, in fact, the relation quoted in (10.92)).

The most interesting feature of this result is that although the free energy itself is a continuous function of temperature, the specific heat

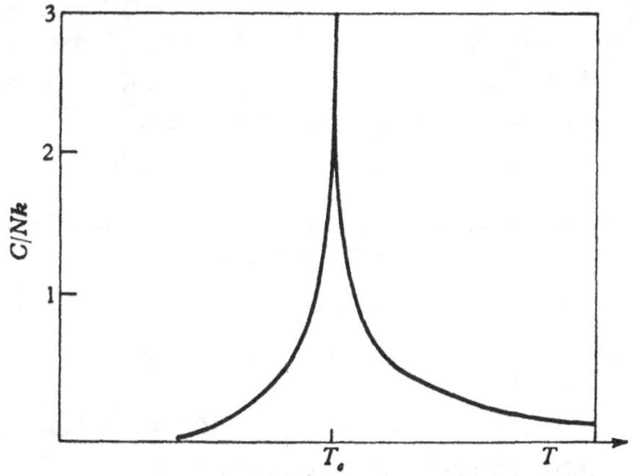

Fig. 193. Specific heat of quadratic layer lattice: Onsager formula.

is logarithmically infinite on both sides of the Curie temperature. Such a singularity is not actually observed in any solid system, but this may be because we cannot easily achieve a two-dimensional assembly of Ising spins.

The Onsager method can be applied to a few other two-dimensional lattices, which show similar properties, and there are also some elegant topological theorems by which the transition temperature can be located in other cases. There is no exact solution, however, for a two-dimensional Ising ferromagnet in a finite magnetic field.

For three-dimensional lattices there are no exact solutions of the Ising problem at all. One can feel sure that they will show long-range order, since the topological conditions are even more favourable to co-operative effects than they are in two dimensions, and with the aid of series expansions the Curie points can be located fairly accurately.

One important principle arises from this discussion. According to the Bethe method, the order-disorder transition should depend only on the co-ordination number, p (cf. (10.75)). Thus, the two-dimensional triangular lattice and the three-dimensional simple cubic lattice, both with $p = 6$, should have identical properties. In fact, they behave quite differently, with different specific heat curves and different transition temperatures. The dimensionality of the lattice is fundamental to co-operative phenomena, and, indeed, to many properties of solids.

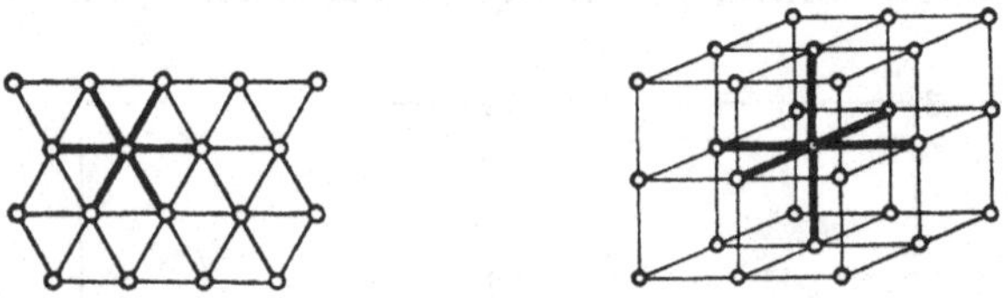

(a) (b)

Fig. 194. Triangular layer lattice, (a), has same co-ordination number as simple cubic lattice, (b).

10.11 Spin waves

Let us return to our spin Hamiltonian (10.29), and assume a ferromagnetic exchange integral. In zero magnetic field

$$\mathscr{H} = -\sum_{ll'} J_{ll'} \mathbf{S}_l \cdot \mathbf{S}_{l'}. \tag{10.100}$$

Suppose that S is the total spin of each ion. We can classify the state of spin on the lth ion by the eigenstates of $\mathbf{S}_l^z$, the z-component of the spin. Thus

$$\mathbf{S}_l^z \,|S'\rangle_l = S' \,|S'\rangle_l \tag{10.101}$$

for the state with component S' along the z-direction (we take $\hbar = 1$ in this section).

The assumption of the general theory of ferromagnetism is that the Hamiltonian $\mathscr{H}$ has a ground state in which all the spins have maximum alignment along z. We assume that this state can be written

$$|0\rangle = |S\rangle_1 |S\rangle_2 |S\rangle_3 \ldots |S\rangle_l \ldots |S\rangle_N, \qquad (10.102)$$

in which S^z has its maximum value, S, at each site of the lattice.

It is easy to check that this is an eigenstate of $\mathscr{H}$. Let us define two new operators for the spin on some site (whose label l we drop for the moment)

$$\mathsf{S}^+ \equiv \mathsf{S}^x + i\mathsf{S}^y, \quad \mathsf{S}^- \equiv \mathsf{S}^x - i\mathsf{S}^y. \qquad (10.103)$$

These are *spin-deviation operators*. Consider, for example, the effect of operating with S^+ on one of the eigenstates of S^z; the result is another eigenstate of S^z. Thus, using the notation of (10.101),

$$\begin{aligned}
\mathsf{S}^z\{\mathsf{S}^+ |S'\rangle\} &= \mathsf{S}^z(\mathsf{S}^x + i\mathsf{S}^y)\,|S'\rangle \\
&= \{(\mathsf{S}^x\mathsf{S}^z + i\mathsf{S}^y) + i(\mathsf{S}^y\mathsf{S}^z - i\mathsf{S}^x)\}\,|S'\rangle \\
&= \mathsf{S}^+\mathsf{S}^z\,|S'\rangle + \mathsf{S}^+\,|S'\rangle \\
&= \mathsf{S}^+(S'+1)\,|S'\rangle \\
&= (S'+1)\{\mathsf{S}^+\,|S'\rangle\},
\end{aligned} \qquad (10.104)$$

using the familiar commutation relations of spin operators. Thus $\mathsf{S}^+\,|S'\rangle$ is an eigenstate of S^z with eigenvalue $S'+1$—in other words, it is the state $|S'+1\rangle$. S^+ has the effect of increasing the component of the spin in the z-direction by one unit. Similarly, S^- decreases S' by one quantum.

Fig. 195. Operator $\mathsf{S}_l^+\mathsf{S}_{l'}^-$ exchanges spin deviations.

It is easy enough to substitute from (10.103) into the Hamiltonian, which we may write

$$\mathscr{H} = -\sum_{ll'} J_{ll'}\{\mathsf{S}_l^z\,\mathsf{S}_{l'}^z + \tfrac{1}{2}(\mathsf{S}_l^+\,\mathsf{S}_{l'}^- + \mathsf{S}_l^-\,\mathsf{S}_{l'}^+)\}. \qquad (10.105)$$

When $\mathscr{H}$ operates on the state (10.102), only the z-components contribute:

$$\mathscr{H}\,|0\rangle = -\sum_{ll'} J_{ll'}\,S^2\,|0\rangle. \qquad (10.106)$$

The operator S_l^+ gives zero when operating on the state $|S\rangle_l$ because we cannot increase the eigenvalue of S_l^z beyond its maximum value S. The ordered state $|0\rangle$ is thus an eigenstate of the total Hamiltonian.

But now consider the excitations of the system. The natural suggestion is that these correspond to states in which a single spin is deviated, as in Fig. 195(*a*). However, such a state is not an eigenstate of $\mathscr{H}$; the operator $S_l^+ S_{l'}^-$ acting on this state will change it into another state such as Fig. 195(*b*), where the spin deviation has been exchanged with its neighbour. The classification of states used in the Ising model (e.g. Fig. 185) will not do; spin deviations keep moving about through the lattice.

This exchange effect is rather like the process by which electrons are handed on from one atom to the next in the tight-binding model of Bloch states (§ 3.4), and it turns out to have very similar mathematical properties. The mathematical formulation of this effect is as follows.

Let us suppose that S is rather a large number (such as $\frac{5}{2}$). The effect of S^- is to 'create a spin deviation'. Instead of using the classification (10.101), let us write

$$n = S - S', \tag{10.107}$$

and define the state $|n\rangle$ as having n quanta of spin deviation upon it. So long as $n < 2S$, we have freedom to create and annihilate these quanta, by operating with S^- and S^+. Thus, these operators have some of the properties of the annihilation and creation operators of the conventional theory of the simple harmonic oscillator (cf. § 2.11).

To show this formally, we may calculate the commutator

$$\begin{aligned}[S^+, S^-] &= i[S^y, S^x] - i[S^x, S^y] \\ &= -2i(iS^z) \\ &= 2S^z.\end{aligned} \tag{10.108}$$

Thus, if we write

$$a = \frac{S^+}{(2S^z)^{\frac{1}{2}}}, \quad a^* = \frac{S^-}{(2S^z)^{\frac{1}{2}}}, \tag{10.109}$$

then a and a^* have the standard commutation relation

$$[a, a^*] = 1, \tag{10.110}$$

and are complex conjugates of one another

The definition (10.109) leaves uncertain the exact order in which S^+ and $[S^z]^{-\frac{1}{2}}$ are to be taken—since these operators do not commute this could be serious. But if S is large enough, we can ignore the error, and replace the operator $2S^z$ by the number $2S$, since we are only going to consider states which are nearly perfectly aligned. To this approximation the spin deviation states $|n\rangle$ are eigenstates of a^*a, and we can write

$$a|n\rangle = n^{\frac{1}{2}}|n-1\rangle, \quad a^*|n\rangle = (n+1)^{\frac{1}{2}}|n+1\rangle. \tag{10.111}$$

The definition of n also implies that we can write

$$S^z = S - a^*a. \tag{10.112}$$

Let us now substitute for (10.109) and (10.112) into the Hamiltonian (10.105). This becomes, approximately,

$$\begin{aligned}\mathscr{H} &\approx -\sum_{ll'} J_{ll'}\{(S-a_l^* a_l)(S-a_{l'}^* a_{l'}) + \tfrac{1}{2}.2S(a_l^* a_{l'} + a_l a_{l'}^*)\} \\ &\approx -\sum_{ll'} J_{ll'}\{S^2 + S(a_l a_{l'}^* + a_l^* a_{l'} - a_l^* a_l - a_{l'}^* a_{l'})\},\end{aligned} \tag{10.113}$$

where we have dropped a term of order $a_l^* a_l a_{l'}^* a_{l'}$.

The first term in (10.113) is the energy of the ground state (10.116): the remaining terms are operators expressing explicitly the handing on of spin deviations from one site to the next. This is rather like the situation in lattice dynamics, where a lattice displacement is passed on from site to site by the elastic forces, or, as we have remarked, like the tight-binding formula for electron waves. We can find the eigenstates of this Hamiltonian by the same device as in § 2.1; substitute

$$a_l = N^{-\frac{1}{2}} \sum_{\mathbf{q}} a_{\mathbf{q}} e^{i\mathbf{q}\cdot\mathbf{l}}, \quad a_l^* = N^{-\frac{1}{2}} \sum_{\mathbf{q}} a_{\mathbf{q}}^* e^{-i\mathbf{q}\cdot\mathbf{l}}, \tag{10.114}$$

for the annihilation and creation operators on each site. This is equivalent to (2.8); the wave-vectors $\mathbf{q}$ satisfy the same conditions as in lattice dynamics, and the new operators $a_{\mathbf{q}}$ and $a_{\mathbf{q}}^*$ satisfy the commutation relations (10.110).

Putting (10.114) into (10.113), and using the fact that $J_{ll'}$ depends only on $\mathbf{h} = \mathbf{l} - \mathbf{l}'$ (as in (2.6)), we have

$$\mathscr{H} = \mathscr{H}_0 + \sum_{\mathbf{q}}\{\sum_{\mathbf{h}} 2SJ(\mathbf{h})(1-e^{-i\mathbf{q}\cdot\mathbf{h}})\} a_{\mathbf{q}}^* a_{\mathbf{q}}, \tag{10.115}$$

where $\mathscr{H}_0$ is independent of the operators $a_{\mathbf{q}}$. But $\mathscr{H} - \mathscr{H}_0$ is the Hamiltonian of a set of simple harmonic oscillators, one for each wave-vector $\mathbf{q}$, each of which has the energy levels

$$\mathscr{E}_{\mathbf{q}} = n_{\mathbf{q}} \sum_{\mathbf{h}} 2SJ(\mathbf{h})(1-e^{-i\mathbf{q}\cdot\mathbf{h}}). \tag{10.116}$$

In other words, the operators $a_{\mathbf{q}}$, $a_{\mathbf{q}}^*$ are annihilation and creation operators for *spin waves*, which are the elementary excitations of the system. The energy in the $\mathbf{q}$th spin-wave mode is quantized in units

$$\hbar\omega_{\mathbf{q}} = \sum_{\mathbf{h}} 2SJ(\mathbf{h})(1-e^{-i\mathbf{q}\cdot\mathbf{h}}). \tag{10.117}$$

By analogy with photons, phonons, and plasmons, such an elementary excitation is sometimes called a *magnon*.

The form of (10.117), at small wave-vectors, for a simple lattice with only nearest-neighbour interaction is easily calculated;

$$\hbar\omega_{\mathbf{q}} \approx 2SJq^2a^2, \tag{10.118}$$

where a is the lattice constant. The energy of a magnon is thus proportional to the square of its wave-vector—rather like the energy of an electron. Indeed, one can define an 'effective mass' m^*, such that

$$\hbar\omega_{\mathbf{q}} \approx \frac{\hbar^2 q^2}{2m^*}. \tag{10.119}$$

For a typical ferromagnet with a Curie temperature of a few hundred degrees Kelvin (this gives the magnitude of J, from (10.34)), we find m^* to be about 10 times the mass of a free electron.

The form of the excitation spectrum has certain simple consequences. For example, since the magnons obey Bose–Einstein statistics one can use (2.46) to calculate the average occupation number of each mode:

$$\bar{n}_{\mathbf{q}} = \frac{1}{e^{\hbar\omega_{\mathbf{q}}/kT}-1}. \tag{10.120}$$

If we accept (10.118), and calculate the total number of magnons in all modes excited at the temperature T, we get

$$\begin{aligned}\bar{n} &= \frac{1}{8\pi^3}\int_0^\infty \frac{4\pi q^2\,dq}{\exp(2SJa^2q^2/kT)-1} \\ &= \frac{N}{4\pi^2}\left(\frac{kT}{2SJ}\right)^{\frac{3}{2}}\int_0^\infty \frac{x^{\frac{1}{2}}\,dx}{e^x-1}.\end{aligned} \tag{10.121}$$

But every magnon corresponds to the excitation of one spin deviation; it reduces the total spin NS of the system by one unit (as can be checked from (10.112) and (10.114)). Thus, the excitation of $\bar{n}$ quanta in the spin-wave modes reduces the total spin by $\bar{n}$; the magnetization should decrease as the temperature increases according to the formula

$$M(T) = M(0)\left\{1-\gamma\left(\frac{kT}{2SJ}\right)^{\frac{3}{2}}\right\}, \tag{10.122}$$

where $M(0)$ is the saturation magnetization at $T = 0$, and γ is a number that can be calculated from the value of the definite integral (10.121).

Another simple consequence of the dispersion formula (10.118) is that there should be a contribution from the magnons to the specific heat. It is obvious that putting $\hbar\omega_{\mathbf{q}}$ under the integral in (10.121) (cf.

(2.49)) will introduce a further power of T in the evaluation of $\mathscr{E}$. This extra power of T will be taken out by differentiation to get the specific heat, which thus also varies as $T^{\frac{3}{2}}$.

Spin waves in ferromagnetic media have a number of other interesting properties. They can, for example, interact with phonons, and can give inelastic scattering of neutrons in much the same way as lattice waves. The general arguments of § 2.8 apply, and give selection rules for energy and crystal momentum.

From a general theoretical point of view, all that we have done in this section is to verify the Bloch theorem (§ 1.4) for the states of the spin-Hamiltonian (10.100). This satisfies the conditions for lattice translational invariance, so that the states must be classifiable according to a wave-vector and must have a spectrum satisfying the laws of § 2.5. Nevertheless, the reduction of the Hamiltonian to the form (10.113) necessitated the dropping of numerous terms, especially those involved in the definition of the operators a and a^* from the spin operators S^+ and S^-. These terms give interactions between the spin waves, yielding further terms in the expansion (10.122) in powers of T. The fact is that the number of spin deviations that can be created at a single site is limited to $2S$. If a high density of magnons is created, then there is a tendency for this limit to be exceeded, and the assumptions of superposibility and independence are no longer valid. In fact, the ordered state breaks down, and we go through the Curie point; the whole approximation scheme falls in ruins.

Our derivation of the ferromagnon spectrum started from the Heisenberg model of § 10.4. But physically equivalent phenomena occur in the itinerant electron model of § 10.5, even though the spins are not localized on the lattice sites. An energy of the order of the exchange energy, U, is obviously needed to construct the determinantal wave function $\Psi(\mathbf{k}, \mathbf{q})$ in which an electron from some state $|\mathbf{k}, +\rangle$ has been put into some other state $|\mathbf{k}+\mathbf{q}, -\rangle$ with reversed spin. But the electron gas is really a many-body system, with collective modes (cf. § 5.7) as well as single-particle excitations. To describe a spin wave, we make a linear combination of these determinants,

$$\Psi_{\mathbf{q}} = \sum_{\mathbf{k}} A_{\mathbf{k}} \Psi(\mathbf{k}, \mathbf{q}) \qquad (10.123)$$

and choose the coefficients $A_{\mathbf{k}}$ so that $\Psi_{\mathbf{q}}$ is an approximate eigenstate of the total Hamiltonian of the system. The result looks exactly like (10.118); the energy goes to zero as $q \to 0$. According to the *Goldstone theorem* on the effects of broken symmetry this is a necessary consequence of the invariance of the Heisenberg Hamiltonian (10.29) under rotations of the direction of the magnetic field.

Macroscopically, this is obvious: in a spin wave of long wavelength, the medium is locally polarized ferromagnetically, without abrupt reversals of relative spin direction at neighbouring points. A magnon is thus a collective excitation in which the spins of electrons are strongly correlated in space. For this reason, the general theory of the magnon–neutron interaction does not depend on whether the electrons are localized or itinerant, since diffraction phenomena depend only on such two-particle space and time correlations (§ 2.8). The spin correlations implicit in the spin-density wave theory of antiferromagnetism (§ 10.7) similarly yield the correct neutron diffraction results for the analysis of the magnetic order pattern in the crystal (§ 2.6).

10.12 The antiferromagnetic ground state

It does not, at first sight, seem difficult to apply the method of § 10.11 to a Heisenberg antiferromagnet. Let us suppose that we have a lattice of Ising spins, which would have the ordered arrangement defined by the variable σ_l, which would be $+1$ on a 'spin-up' site, and -1 on a 'spin-down' site. We can write our Hamiltonian almost exactly as in (10.105).

$$\mathscr{H} = -\sum_{ll'} J_{ll'}\{\mathsf{S}_l^z \mathsf{S}_{l'}^z + \tfrac{1}{2}(\mathsf{S}_l^+ \mathsf{S}_{l'}^- + \mathsf{S}_l^- \mathsf{S}_{l'}^+)\} - \beta \sum_l (\sigma_l H_A \mathsf{S}_l^z + \mathbf{H}\cdot\mathsf{S}_l). \quad (10.124)$$

In this expression we have included, in addition to the external magnetic field $\mathbf{H}$, an 'anisotropy field' H_A, which is always parallel to the local ordered spin direction. Thus H_A is 'up' on an 'up' site, and 'down' on a 'down' site. This field provides the direction of the z-axis of quantization (which is not necessarily the direction of the applied field), but is otherwise arbitrary; we shall see the need for it shortly.

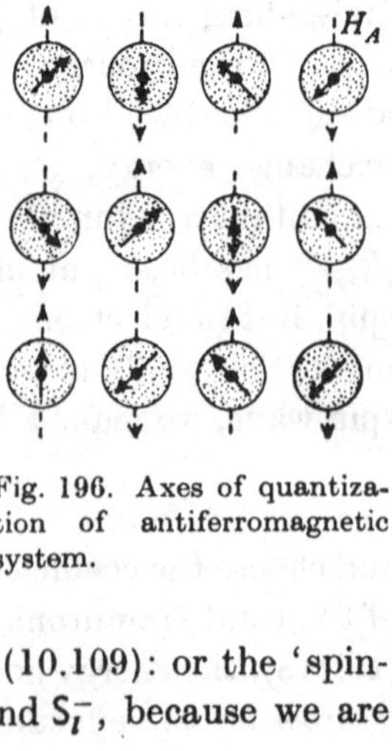

Fig. 196. Axes of quantization of antiferromagnetic system.

Now we define our local annihilation and creation operators in terms of deviations from the ordered 'Ising' state. On the sites where σ_l is $+1$, the definition is the same as (10.109): or the 'spin-down' sites we interchange the role of S_l^+ and S_l^-, because we are measuring from the state $|-S\rangle_l$ of the lth ion. Thus, on every site, both 'up', and 'down', the equations

$$\left.\begin{aligned} S_l^x &\approx (\tfrac{1}{2}S)^{\frac{1}{2}}(a_l+a_l^*),\\ S_l^y &\approx -i\sigma_l(\tfrac{1}{2}S)^{\frac{1}{2}}(a_l-a_l^*),\\ S_l^z &= \sigma_l(S-a_l^*a_l), \end{aligned}\right\} \tag{10.125}$$

define operators a_l and a_l^* which destroy and create spin deviations. (In some expositions one defines different operators on the two different sublattices; but since a_l and a_l^* commute if $l \neq l'$, this is only a matter of labels.)

It is easy enough to substitute (10.125) in (10.124). The first term is the 'Ising' energy of the state

$$\mathscr{E}_0 = -\sum_{ll'} J_{ll'}\sigma_l\sigma_{l'}S^2 - N\beta SH_A \tag{10.126}$$

which we can drop. The remaining terms, retaining products up to the second order in a_l, a_l^*, are complicated, but can easily be written down.

A Fourier transformation like (10.114) introduces new annihilation and creation operators, $b_{\mathbf{q}}$ and $b_{\mathbf{q}}^*$, such that

$$a_l = N^{-\frac{1}{2}}\sum_{\mathbf{q}} b_{\mathbf{q}}e^{i\mathbf{q}\cdot\mathbf{l}}, \quad a_l^* = N^{-\frac{1}{2}}\sum_{\mathbf{q}} b_{\mathbf{q}}^* e^{-i\mathbf{q}\cdot\mathbf{l}}. \tag{10.127}$$

In addition to the 'Ising' energy (10.126), we find that the Hamiltonian can be written

$$\mathscr{H} = \tfrac{1}{2}\sum_{\mathbf{q}}\{A_{\mathbf{q}}(b_{\mathbf{q}}^*b_{\mathbf{q}}+b_{-\mathbf{q}}^*b_{-\mathbf{q}})+B_{\mathbf{q}}(b_{\mathbf{q}}b_{-\mathbf{q}}+b_{\mathbf{q}}^*b_{-\mathbf{q}}^*)\}+\mathbf{H}\cdot\mathbf{M}, \tag{10.128}$$

where the coefficients are similar to (2.12) or (10.116):—

$$\left.\begin{aligned} A_{\mathbf{q}} &= \beta H_A + \sum_{\mathbf{h}} 2SJ(\mathbf{h})\{\sigma_{\mathbf{h}}-\tfrac{1}{2}(1+\sigma_{\mathbf{h}})\cos\mathbf{q}\cdot\mathbf{h}\},\\ B_{\mathbf{q}} &= -\sum_{\mathbf{h}} SJ(\mathbf{h})(1-\sigma_{\mathbf{h}})\cos\mathbf{q}\cdot\mathbf{h}. \end{aligned}\right\} \tag{10.129}$$

In these expressions, $\sigma_{\mathbf{h}}$ means the sign of the site at $\mathbf{h}$ relative to the sign of the central site. We can include cases where $J(\mathbf{h})$ is of longer-range than nearest-neighbour interaction, and may even be ferromagnetic at some distances; it is only necessary that σ_l should describe the ground state of the Ising model correctly.

In (10.128) the external magnetic field occurs linked to a complicated operator $\mathbf{M}$ which we shall not attempt to define in general, although in simple cases one can see that it corresponds to a moment that seems to alternate in sign (along the z-axis) at sites on the two sublattices.

This is a result of our arbitrary definition of local axes of quantization in (10.125)—we shall not pursue this term further.

The other effect of our 'alternating' state is that there are terms in (10.128) linking $b_{\mathbf{q}}$ with $b_{-\mathbf{q}}$. These arise when we put (10.125) into (10.124), through products like $a_l a_{l'}$ which do not arise in the ferromagnetic state (10.113). The imposition of a background system of Ising spins reduces the basic translational symmetry of the system, so that the Hamiltonian is not automatically diagonalized by the Fourier transformation (10.127). We ought, perhaps, to take two spins per unit cell, and make a rather different Fourier analysis.

However, the problem is really no more difficult than the reduction of the interaction matrix (2.13) in the analogous problem of lattice dynamics. We introduce new variables $c_{\mathbf{q}}, c_{\mathbf{q}}^*$ by a canonical transformation

$$\left.\begin{aligned} b_{\mathbf{q}} &= c_{\mathbf{q}} \cosh\theta_{\mathbf{q}} + c_{-\mathbf{q}}^* \sinh\theta_{\mathbf{q}}, & b_{\mathbf{q}}^* &= c_{\mathbf{q}}^* \cosh\theta_{\mathbf{q}} + c_{-\mathbf{q}} \sinh\theta_{\mathbf{q}}, \\ b_{-\mathbf{q}} &= c_{\mathbf{q}}^* \sinh\theta_{\mathbf{q}} + c_{-\mathbf{q}} \cosh\theta_{\mathbf{q}}, & b_{-\mathbf{q}}^* &= c_{\mathbf{q}} \sinh\theta_{\mathbf{q}} + c_{-\mathbf{q}}^* \cosh\theta_{\mathbf{q}}, \end{aligned}\right\} \quad (10.130)$$

which conserves the commutator

$$[c_{\mathbf{q}}, c_{\mathbf{q}}^*] = \cosh^2\theta_{\mathbf{q}} - \sinh^2\theta_{\mathbf{q}} = 1. \quad (10.131)$$

The parameter $\theta_{\mathbf{q}}$ represents a rotation through an imaginary angle in the space of these four operators. It is worth noting that the conjugate of $c_{\mathbf{q}}$ is $c_{-\mathbf{q}}^*$; we have to reverse the direction of motion if we want it to look like the physical conjugate of a given wave.

If one substitutes from (10.130) into (10.128), then one gets a set of 'off-diagonal' terms

$$\tfrac{1}{2}\sum_{\mathbf{q}} \{2A_{\mathbf{q}} \cosh\theta_{\mathbf{q}} \sinh\theta_{\mathbf{q}} + B_{\mathbf{q}}(\cosh^2\theta_{\mathbf{q}} + \sinh^2\theta_{\mathbf{q}})\} (c_{\mathbf{q}} c_{-\mathbf{q}} + c_{\mathbf{q}}^* c_{-\mathbf{q}}^*), \quad (10.132)$$

which can be made to vanish for each value of $\mathbf{q}$ if we choose $\theta_{\mathbf{q}}$ so that

$$\begin{aligned} \frac{B_{\mathbf{q}}}{A_{\mathbf{q}}} &= -\frac{2\cosh\theta_{\mathbf{q}} \sinh\theta_{\mathbf{q}}}{\cosh^2\theta_{\mathbf{q}} + \sinh^2\theta_{\mathbf{q}}} \\ &= -\tanh 2\theta_{\mathbf{q}}. \end{aligned} \quad (10.133)$$

The diagonal terms then yield the following

$$\mathscr{H} = \sum_{\mathbf{q}} [(A_{\mathbf{q}}^2 - B_{\mathbf{q}}^2)^{\frac{1}{2}} (c_{\mathbf{q}}^* c_{\mathbf{q}} + \tfrac{1}{2}) - A_{\mathbf{q}}], \quad (10.134)$$

where we have used the commutation relations to put the operators in this order. The sum now is over all values of $\mathbf{q}$, positive and negative.

This result is very similar to (10.115). The eigenstates of $c_{\mathbf{q}}^* c_{\mathbf{q}}$ are $|n_{\mathbf{q}}\rangle$, say, corresponding to $n_{\mathbf{q}}$ quanta excited in this mode. The energy of an *antiferromagnon* in simple cases will be

$$\begin{aligned}\hbar\omega_{\mathbf{q}} &= (A_{\mathbf{q}}^2 - B_{\mathbf{q}}^2)^{\frac{1}{2}} \\ &= \{(\beta H_A - 2zSJ)^2 - (2pSJ\langle\cos qa\rangle)^2\}^{\frac{1}{2}},\end{aligned} \tag{10.135}$$

where $\langle\cos qa\rangle$ is the average value of this quantity over the p nearest neighbours, distant a from the origin, which we assume to interact through the exchange integral J. If we write

$$\hbar\omega_A = \beta H_A, \quad \hbar\omega_J = -2pSJ \tag{10.136}$$

(recalling that J is negative for antiferromagnetism), then

$$\omega_{\mathbf{q}} = \{(\omega_A + \omega_J)^2 - (\omega_J\langle\cos qa\rangle)^2\}^{\frac{1}{2}} \tag{10.137}$$

is the spin-wave spectrum.

This looks simple enough. For example, suppose that $\omega_A \ll \omega_J$. Then

$$\begin{aligned}\omega_{\mathbf{q}} &\approx \omega_J\{1 - \langle\cos qa\rangle^2\}^{\frac{1}{2}} \\ &\approx \frac{1}{\sqrt{3}}\omega_J . qa,\end{aligned} \tag{10.138}$$

corresponding to a linear dependence of frequency on wave-number, just as in the lattice spectrum (2.21). The specific heat contribution of the magnons should go as T^3, just as in (2.57).

Let us try to follow (10.122) and calculate the magnetization of each sublattice as a function of temperature. The average spin deviation is obviously

$$\begin{aligned}\langle a_l^* a_l\rangle &= \frac{1}{N}\sum_{\mathbf{q}}\langle b_{\mathbf{q}}^* b_{\mathbf{q}}\rangle \\ &= \frac{1}{N}\sum_{\mathbf{q}}\langle\cosh^2\theta_{\mathbf{q}}\, c_{\mathbf{q}}^* c_{\mathbf{q}} + \sinh^2\theta_{\mathbf{q}}\, c_{\mathbf{q}} c_{\mathbf{q}}^*\rangle \\ &= \frac{1}{N}\sum_{\mathbf{q}}\left\{(n_{\mathbf{q}} + \tfrac{1}{2})\frac{A_{\mathbf{q}}}{\hbar\omega_{\mathbf{q}}} - \tfrac{1}{2}\right\},\end{aligned} \tag{10.139}$$

using the transformations (10.127) and (10.130). In this expression $n_{\mathbf{q}}$ is the number of magnons excited in the $\mathbf{q}$th mode; we naturally calculate it as a function of temperature using the Bose–Einstein distribution (10.120).

But even at absolute zero, the average spin deviation will not vanish. If $n_{\mathbf{q}} = 0$ in (10.139) we still have

$$\langle a_l^* a_l \rangle_{T=0} = \frac{1}{2N} \sum_{\mathbf{q}} \left(\frac{A_{\mathbf{q}}}{\hbar\omega_{\mathbf{q}}} - 1 \right)$$

$$\approx \frac{1}{2N} \sum_{\mathbf{q}} \left\{ \frac{1}{(1 - \langle \cos qa \rangle^2)^{\frac{1}{2}}} - 1 \right\} \qquad (10.140)$$

in the simple case described by (10.138). This sum can be calculated; it behaves like the integral

$$\int \frac{1}{q} \, \mathbf{dq} \qquad (10.141)$$

which converges only if **dq** is over a two-dimensional or three-dimensional manifold. Thus, we confirm that a one-dimensional antiferromagnet does not have a stable ordered state, even at $T = 0$. This goes beyond the result (10.96), which showed that the one-dimensional Ising model is unstable against thermal fluctuations; in the antiferromagnetic case, the fluctuations associated with the exchange of spins destroy the supposed ordered ground state of the system.

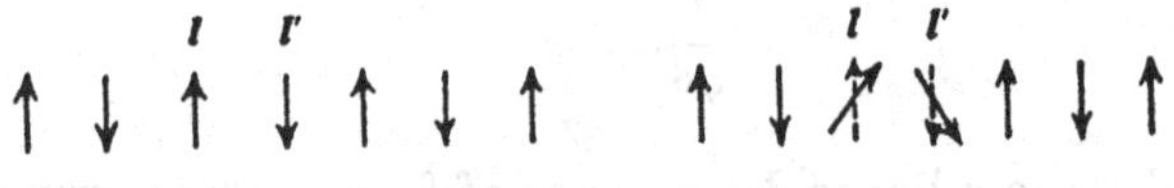

(a) (b)

Fig. 197. In the antiferromagnetic ordered state, (a), the exchange operator $S_l^+ S_{l'}^-$ creates two spin deviations, (b).

Even in two and three dimensions, the moment of each sublattice does not reach its assumed maximum. In fact, the alternating state defined by the σ_l in (10.125) is not really an eigenstate of the Hamiltonian. We can try to check this by the method of (10.106); as noted there, the term $S_l^+ S_{l'}^-$ tends to hand on a spin deviation from one site to the next. In the antiferromagnetic case, it creates spin deviations simultaneously on two adjacent sites—so that the ground state must contain an admixture of all the states one can get out of this and similar configurations.

The fact that the antiferromagnetic ground state is not fully ordered in the case of a three-dimensional system (with $S = \frac{1}{2}$, each sublattice achieves perhaps 90% of its maximum moment) is reflected in the expectation value of the energy. As we see from (10.126), (10.129) and (10.134), this is

$$\begin{aligned}\langle \mathscr{H} \rangle + \mathscr{E}_0 &= \sum_{\mathbf{q}} (n_{\mathbf{q}} + \tfrac{1}{2}) \hbar\omega_{\mathbf{q}} - A_{\mathbf{q}} + \mathscr{E}_0 \\ &= -2NpS(S+1)J + \sum_{\mathbf{q}} (n_{\mathbf{q}} + \tfrac{1}{2}) \hbar\omega_{\mathbf{q}}, \qquad (10.142)\end{aligned}$$

which even at $T = 0$, when $n_{\mathbf{q}} = 0$, is greater than the diagonal value one would calculate for the 'Ising' state. The 'zero-point energy' in the magnon modes must be included; it increases the energy of the ground state, which lies somewhere between $-2NpJS(S+1)$ and $-2NpJS^2$.

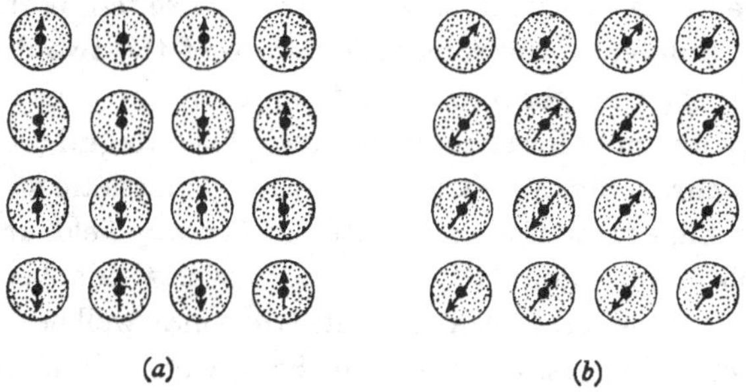

Fig. 198. In the absence of an anisotropy field, the spins of an antiferromagnetic array (*a*) can rotate without change of energy (*b*).

There is a further point to notice about (10.139). According to (10.138), $\omega_{\mathbf{q}} = 0$ at $q = 0$, so that the contribution of this term in the calculation of the average spin deviation would be infinite. We need the anisotropy field, H_A, to make

$$\omega_{\mathbf{q}} \rightarrow \sqrt{(2\omega_A \omega_J)} \quad \text{as} \quad q \rightarrow 0 \qquad (10.143)$$

and thus to avoid the singularity. The fact is that our system, in its ground state, of total spin zero, is unstable against a homogeneous swivelling of all the spins, as in Fig. 198. The anisotropy field is needed to stabilize the direction of quantization. This field can be very small—the limit of (10.141) can be finite at $q = 0$, without H_A—but in principle the anisotropy field must exist, and in practice it does.

Actually, one can observe the frequency (10.143) as the *antiferromagnetic resonance* frequency of the system. This is a high frequency, even if H_A is only a few hundred gauss, but it can be brought into resonance by the application of a strong external magnetic field.

It is not difficult to check the susceptibility rules for the antiferromagnetic ordered state, and there are, of course, various calculations for magnon–phonon interaction, etc. Perhaps the most interesting point is, however, that if we use (10.139) to calculate the sublattice polarization at a finite temperature we find an expression that behaves like

$$\sum_{\mathbf{q}} \frac{\bar{n}_{\mathbf{q}} A_{\mathbf{q}}}{\hbar\omega_{\mathbf{q}}} \approx \sum_{\mathbf{q}} \frac{kTA_{\mathbf{q}}}{(\hbar\omega_{\mathbf{q}})^2}$$

$$\propto \int \frac{1}{q^2} d\mathbf{q} \tag{10.144}$$

and diverges logarithmically at $q = 0$ in the two-dimensional case.

Thus, the exchange effect, when thermally excited, is sufficient to destroy the ordered antiferromagnetic state of the two-dimensional Ising model; the Onsager result is not valid for real spins in this case (although it does hold for ferromagnetism). The integral (10.144) does converge in three dimensions, so the antiferromagnetic state is then probably stable—there is always a little anisotropy field, arising from dipole-dipole terms like (10.35). But this has not been proven rigorously; the best demonstration of its truth may well be the neutron diffraction experiments, mentioned in § 2.6, which show, physically, the two magnetic sublattices.

CHAPTER 11

SUPERCONDUCTIVITY

Since 'tis Nature's law to change
Constancy alone is strange. EARL OF ROCHESTER

11.1 The attraction between electrons

Superconductivity was long considered the most extraordinary and mysterious of the properties of metals; but the theory of Bardeen, Cooper and Schrieffer—the *BCS theory*—has explained so much that we can say that we now understand the superconducting state almost as well as we do the 'normal' state. We shall not attempt here

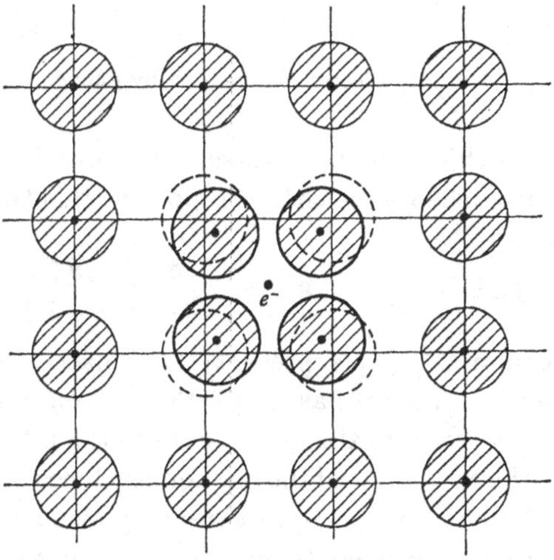

Fig. 199. Polarization of lattice near electron.

to cover the whole of a large and rapidly advancing subject; the emphasis will be on the atomic processes from which so many unusual macroscopic phenomena arise.

The whole effect springs from a small force of attraction between any two electrons which have nearly the same energy. We usually assume that free electrons *repel* one another, through their Coulomb interaction, although, as shown in § 5.8, this field is considerably reduced at long distances by the screening effect of the 'other' electrons. But in a lattice an electron tends to pull towards itself the positive ions, so that it is surrounded by a region where the lattice is slightly denser than usual. Another electron coming into the vicinity will be drawn towards this region; it will look as if it were attracted towards the first electron. The two particles can, so to speak, gain by sitting close together in the same depression of the mattress.

The simplest way of calculating this force is to describe it as the emission of a 'virtual' phonon by one electron, and its absorption by the other. Consider the process defined by Fig. 200 (*a*). An electron in state $\mathbf{k}$ emits a phonon, and is scattered into state $\mathbf{k}-\mathbf{K}$. The electron

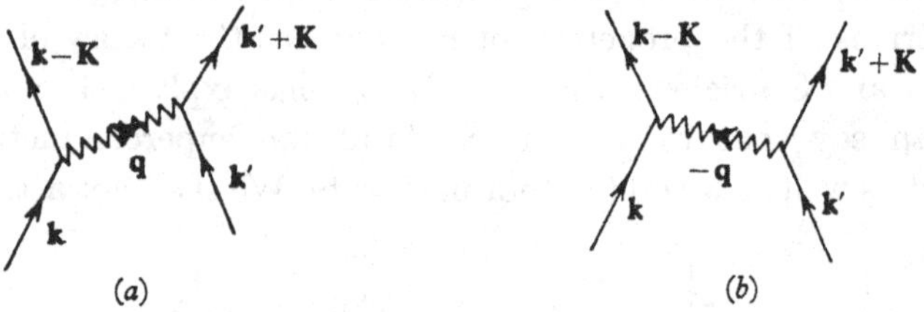

Fig. 200. Electron–electron interaction by phonon exchange.

in state $\mathbf{k}'$ absorbs this phonon, and is scattered into $\mathbf{k}'+\mathbf{K}$. In second-order perturbation theory we should represent this transition by the matrix element

$$\langle \mathbf{k}-\mathbf{K}, \mathbf{k}'+\mathbf{K} \,|\, V \,|\, \mathbf{k}, \mathbf{k}' \rangle = \frac{\mathscr{M}_{\mathbf{k},\mathbf{k}-\mathbf{K}} \mathscr{M}^*_{\mathbf{k}',\mathbf{k}'+\mathbf{K}}}{\mathscr{E}(\mathbf{k}) - \mathscr{E}(\mathbf{k}-\mathbf{K}) - \hbar\nu_{\mathbf{q}}}. \qquad (11.1)$$

In this expression $\mathscr{M}_{\mathbf{k},\mathbf{k}-\mathbf{K}}$ represents the matrix element of the electron–phonon interaction, as in § 6.13. The energy denominator represents the change from the initial state to the intermediate state, which includes the energy $\hbar\nu_{\mathbf{q}}$ of the phonon produced; for N-processes, $\mathbf{q} = \mathbf{K}$, but electron–phonon U-processes may also be included.

There is another process contributing to the over-all scattering, as represented in Fig. 200 (*b*), where the phonon of wave-vector $-\mathbf{q}$ is emitted by $\mathbf{k}'$, before being absorbed by $\mathbf{k}$. This contributes a similar expression to (11.1), with a slightly different denominator. Combining these two processes, and assuming that the electron–phonon matrix elements are essentially the same, we have

$$\langle \mathbf{k}-\mathbf{K}, \mathbf{k}'+\mathbf{K} \,|\, V \,|\, \mathbf{k}, \mathbf{k}' \rangle = \frac{|\mathscr{M}_{\mathbf{k},\mathbf{k}-\mathbf{K}}|^2 \, 2\hbar\nu_{\mathbf{q}}}{\{\mathscr{E}(\mathbf{k}) - \mathscr{E}(\mathbf{k}-\mathbf{K})\}^2 - (\hbar\nu_{\mathbf{q}})^2}. \qquad (11.2)$$

The effect of this on the electrons is as if there were a direct interaction between them, whose Fourier coefficient, $V(\mathbf{K})$, is given by

(11.2). Usually this is positive, but for the narrow range of energies where

$$|\mathscr{E}(\mathbf{k})-\mathscr{E}(\mathbf{k}-\mathbf{K})| < \hbar\nu_{\mathbf{q}} \tag{11.3}$$

the interaction is negative, corresponding to an attractive force. In the range (11.3), this force has Fourier components

$$V(\mathbf{K}) \approx -\frac{2\,|\mathscr{M}_{\mathbf{k},\mathbf{k}-\mathbf{K}}|^2}{\hbar\nu_{\mathbf{q}}}. \tag{11.4}$$

This is not, perhaps, a very strong force. We are working at very low temperatures so we may take the phonon emission processes to be induced by the zero-point motion of the lattice. By (2.109), (6.86) and (6.93), we get

$$\begin{aligned}|\mathscr{M}_{\mathbf{k},\mathbf{k}-\mathbf{K}}|^2 &= |\mathscr{U}(\mathbf{K})|^2\,|\mathbf{K}\cdot\mathbf{U}_{\mathbf{q}}|^2\\ &= |\mathscr{U}(\mathbf{K})|^2\frac{\mathbf{K}^2\hbar}{2NM\nu_{\mathbf{q}}},\end{aligned} \tag{11.5}$$

where $\mathscr{U}(\mathbf{K})$ is the Fourier transform of the screened potential of an ion. If we have only N-processes, and if the velocity of sound is given correctly by (6.84), we have

$$\begin{aligned}V(\mathbf{K}) &\approx -\frac{|\mathscr{U}(\mathbf{K})|^2 q^2}{NM\nu_q^2}\\ &= -\frac{|\mathscr{U}(\mathbf{K})|^2}{D_0 s^2} = -\frac{3}{2}\frac{|\mathscr{U}(\mathbf{K})|^2}{NZ\mathscr{E}_F}.\end{aligned} \tag{11.6}$$

Thus, the effective interaction does not decrease rapidly in the range of values of **K** allowed by N-processes. For larger values of K, the U-processes will increase $V(\mathbf{K})$ somewhat, although the mixing of plane waves in the Bloch states of the electron, as discussed in §§ 6.13 and 7.5, prevents a catastrophe at $\mathbf{K} = \mathbf{g}$. It is evident that the attraction is of short range—it has large matrix elements out to the maximum value of the momentum exchanged between the two electrons.

There is, of course, also the ordinary Coulomb repulsion between the electrons. The corresponding matrix element would be, as in (5.32),

$$V_{\text{Coul.}}(\mathbf{K}) = \frac{4\pi e^2}{K^2+\lambda^2}, \tag{11.7}$$

where λ is the screening parameter (5.22), or some more complicated expression as in (5.36). We take the sum of (11.4) and (11.7) as our estimate of the *Fröhlich interaction*. The criterion for superconductivity is that this be negative. This criterion does not have any very simple form, in terms of the electron–phonon interaction, although there is evidence that the strength of the 'pseudo-potential part' $\mathscr{V}'(K)$ in the Bardeen formula (6.93) is the important factor.

The above argument is simple enough; to justify it properly one needs to investigate the various contributions to the energy of the joint electron–phonon system, including the renormalization of the velocity of sound in § 6.11. It should be noted that we are dealing only with electrons in a narrow range of relative energies, as in (11.3); outside of this range the lattice moves too slowly to follow the electron motion. But because the most important contributions to the net interaction come from fairly large values of K, corresponding to large values of q, this range is of the order of $\boldsymbol{k}\Theta$ where Θ is the Debye temperature. Since superconductivity occurs only at very low temperatures, this is a far wider range of energies than the thickness $\boldsymbol{k}T$ of the 'thermal layer' on the Fermi surface. Thus, the approximation (11.4) is justified for the electrons that can really take part in the superconducting transition.

11.2 Cooper pairs

To appreciate the effect of the attractive force produced by the exchange of phonons between electrons, let us now calculate the total scattering amplitude t, i.e. the matrix element whose square gives the probability of the process in which two electrons in states $\mathbf{k}_1$ and $\mathbf{k}_2$ are scattered into two other states $\mathbf{k}_1'$ and $\mathbf{k}_2'$. As in Fig. 201 (*a*) we shall assume that these states lie just outside the Fermi sphere.

The most obvious contribution to the scattering will be the direct interaction $V(\mathbf{K})$ defined by (11.2). This is rather too complicated for our present calculation, so let us abstract the attractive part, and assume that it is a constant. Thus, our first-order scattering amplitude will be assumed to be

$$t_1 = V(\mathbf{K}) = V, \tag{11.8}$$

where V is a negative constant, independent of $\mathbf{k}_1'$ and $\mathbf{k}_1$. But this attractive potential only holds for electrons of nearly the same energy; to put the condition (11.3) into the theory, we shall say that $V(\mathbf{K})$ is zero unless

$$|\mathscr{E}(\mathbf{k}_1) - \mathscr{E}(\mathbf{k}_2)| < w, \tag{11.9}$$

where w is a constant. From (11.3) it is evident that w ought to be of the order of $\boldsymbol{k}\Theta$.

Usually, in scattering problems, the approximation (11.8) is sufficient. Nevertheless, in a perturbation expansion of t there will be further terms, corresponding to successive virtual processes in which the electrons start in $\mathbf{k}_1$ and finish in $\mathbf{k}_1'$. The next term, for example, might be the process indicated in Fig. 201 (*b*), where we interpose between our initial and final states the virtual state in which $\mathbf{k}_1$ has gone to $\mathbf{k}_1+\mathbf{K}$ and $\mathbf{k}_2$ has gone to $\mathbf{k}_2-\mathbf{K}$, before making a second transition to $\mathbf{k}_1'$ and $\mathbf{k}_2'$.

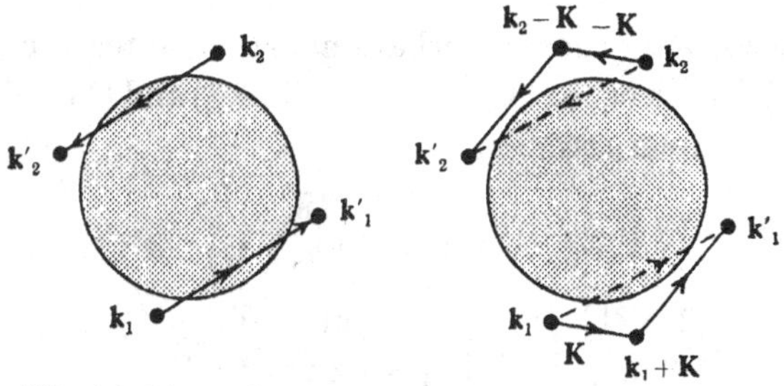

Fig. 201. (*a*) Direct electron scattering. (*b*) Second-order term.

The contribution of such a process to the scattering amplitude is given by the usual perturbation expression

$$t_2(\mathbf{K}) = \frac{V^2}{\mathscr{E}(\mathbf{k}_1)+\mathscr{E}(\mathbf{k}_2)-\mathscr{E}(\mathbf{k}_1+\mathbf{K})-\mathscr{E}(\mathbf{k}_2-\mathbf{K})}. \tag{11.10}$$

But this term only exists under very limited conditions. The intermediate states must not lie below the Fermi surface, because of the Pauli principle, and the energy difference in each transition must be less than w, by (11.9). Thus, we have rules

$$\mathscr{E}_F < \mathscr{E}(\mathbf{k}_1+\mathbf{K}) < \mathscr{E}(\mathbf{k}_1)+w, \quad \mathscr{E}_F < \mathscr{E}(\mathbf{k}_2-\mathbf{K}) < \mathscr{E}(\mathbf{k}_2)+w, \tag{11.11}$$

etc., to be satisfied if (11.10) is to contribute at all to t_2. These conditions are analogous to the selection rules for virtual spin-flip transitions in the Kondo effect (§ 10.6); the occupation numbers of the various states do *not* cancel from the second-order terms, taking account of both direct and exchange processes.

These rules are so restrictive that we can usually neglect t_2. But when $\mathbf{k}_1$ and $\mathbf{k}_2$ are nearly in opposite directions the two conditions (11.11) coalesce, and do not bear so harshly upon the range of values of $\mathbf{K}$ for which scattering is possible. In fact, t_2 can then become quite large.

To demonstrate this, it is convenient to measure energies from the Fermi level. We define

$$\begin{aligned}\xi(\mathbf{k}) &\equiv \mathscr{E}(\mathbf{k})-\mathscr{E}_F \\ &\approx \hbar v_F(k-k_F)\end{aligned} \tag{11.12}$$

for each state, if k is nearly equal to k_F. We also introduce two new auxiliary variables, having the dimensions of energy,

$$\Delta = \xi(\mathbf{k}_1)+\xi(\mathbf{k}_2), \quad \eta = \hbar v_F\,|\mathbf{k}_1+\mathbf{k}_2| \tag{11.13}$$

which measure the total energy and total momentum of our pair of electrons.

To get the second-order perturbation term we integrate (11.10) over all values of $\mathbf{K}$ subject to (11.11). Using (11.12) and (11.13) this can be reduced to a simpler formula,

$$\begin{aligned} t_2 &= V^2\int\frac{d\mathbf{K}}{\mathscr{E}(\mathbf{k}_1)+\mathscr{E}(\mathbf{k}_2)-\mathscr{E}(\mathbf{k}_1+\mathbf{K})-\mathscr{E}(\mathbf{k}_2-\mathbf{K})} \\ &= \frac{2\pi V^2k_F^2}{8\pi^3\hbar v_F}\int_0^w d\xi\int_{-1}^1\frac{d(\cos\theta)}{\Delta+\eta\cos\theta-2\xi}. \end{aligned} \tag{11.14}$$

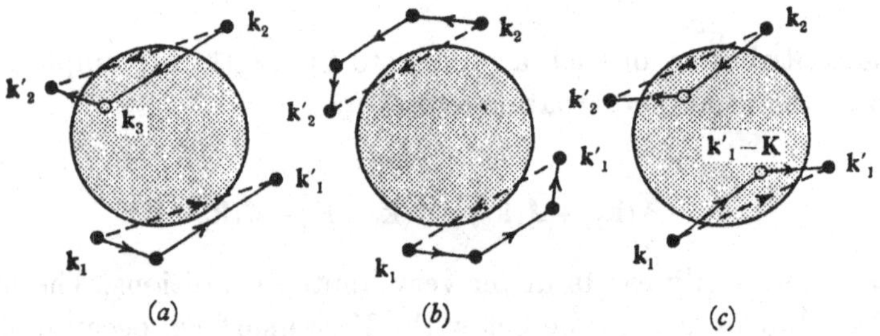

Fig. 202. (a) Process in which hole is created at $\mathbf{k}_3$, and subsequently filled by $\mathbf{k}_2$. (b) Successive pair scattering. (c) Pair scattering with holes.

When Δ and η are much less than w, there is a singularity in the integrand which makes the integral behave logarithmically (cf. (10.56)):

$$t_2 \approx -\frac{V^2}{4\pi^2}\frac{k_F^2}{\hbar v_F}\ln\frac{2w}{\max(\Delta,\eta)}. \tag{11.15}$$

Usually, this term is much smaller than t_1, and would be neglected. But if Δ and η happen to be very small, then (11.15) can become large—and we may have to consider further terms in the perturbation series for t.

Such terms may be very complicated, since we may consider virtual processes such as Fig. 202(a) in which an electron from inside the Fermi surface interacts with one of the electrons in the pair, leaving a hole which eventually gets filled up in a later process. But such processes tend to give much smaller contributions in the perturbation series. The important ones are those in which the process indicated in Fig. 201(b) is repeated—the pair scatters off itself again, as in

Fig. 202(*b*). Each such virtual scattering introduces a factor like (11.15), including the large logarithmic integral if Δ and η are small.

To get the correct answer we also have to include the processes in which a pair is scattered from out of the Fermi sea, leaving a pair of holes which can receive the electrons in $\mathbf{k}_1$ and $\mathbf{k}_2$ (Fig. 202(*c*)). This gives the same contribution to t_2 as (11.15). Adding up all these terms, we see that the scattering amplitude must behave like the infinite series

$$t \approx V + \gamma V^2 + \gamma^2 V^3 + \ldots, \tag{11.16}$$

where

$$\gamma = -\frac{1}{2\pi^2}\frac{k_F^2}{\hbar v_F}\ln\frac{2w}{\max(\Delta, \eta)}. \tag{11.17}$$

But the series (11.16) can be summed:

$$t \approx \frac{V}{1-\gamma V}. \tag{11.18}$$

Thus, the scattering amplitude behaves as if it had a pole. This is a well-known sign that the assumption of free particles scattered from one another is inadequate; it is the sign of a *bound state* of the two electrons. Suppose we take $\eta = 0$, so that $\mathbf{k}_1 = -\mathbf{k}_2$. The energy at which the pole occurs is determined by

$$\begin{aligned} 1 &= \gamma V \\ &= -\frac{V}{2\pi^2}\frac{k_F^2}{v_F}\ln\frac{2w}{\Delta}. \end{aligned} \tag{11.19}$$

In other words, when the total energy of the pair of electrons is equal to

$$\Delta_0 = 2w\exp\left(-\frac{2\pi^2\hbar v_F}{|V|\,k_F^2}\right) \tag{11.20}$$

below the Fermi level, then they tend to be bound together in a 'quasi-molecule'. Such a state is called a *Cooper pair*.

The possibility of the existence of such states does not solve the problem of superconductivity. It can be argued that all the electrons in the metal tend to form such pairs, $\mathbf{k}$ with $-\mathbf{k}$, and that this gas of Bose–Einstein quasi-molecules can then undergo a Bose–Einstein condensation. But the pairs are unstable; this is suggestive rather than convincing as a description of the superconducting state.

The most important result is the formula (11.20) for the energy of the pair. We may use a free-electron model, which gives

$$\begin{aligned} \frac{k_F^2}{2\pi^2\hbar v_F} &= \tfrac{1}{2}\mathcal{N}(\mathscr{E}_F) \\ &= \frac{3}{4}\frac{ZN}{\mathscr{E}_F} \end{aligned} \tag{11.21}$$

(in the BCS theory, this quantity is called $N(0)$, the density of states *for one spin* at the Fermi level). Putting (11.6) into (11.20), and assuming that $w \approx k\Theta$, we should find

$$\Delta_0 \approx 2k\Theta \exp\{-2/\mathcal{N}(\mathscr{E}_F)|V|\}$$
$$\approx 2k\Theta \exp\{-8\mathscr{E}_F^2/9|\mathscr{U}|^2\}. \tag{11.22}$$

If we suppose that Δ_0 is some measure of the energy of the superconducting transition and put it equal to kT_c, we find that the transition temperature T_c should be somewhere between 1/10 and 1/10,000 of the Debye temperature, depending on the magnitude of the electron-ion interaction potential $|\mathscr{U}|$. This is indeed what is observed—we can now understand why T_c should be so very different from Θ.

Another effect that follows from this argument is the *isotope effect*. If we consider a change of the mass M of each ion, without any change in the interatomic forces, then we expect the velocity of sound to vary as $M^{-\frac{1}{2}}$. Thus, we expect to find Θ varying in the same way, so that

$$T_c \propto M^{-\frac{1}{2}}, \tag{11.23}$$

which is, approximately, the observed result in many cases.

A further point about Cooper pairs is that they are more strongly bound if the electrons have opposite spins as well as opposite momenta. This is not obvious from the argument that we have just given, but arises in the calculation of the attractive potential (11.2). The argument of § 11.1 assumes that we are dealing with distinguishable particles, i.e. electrons of opposite spin. If both electrons had the same spin, then we should have to invoke the Pauli principle, and use antisymmetrized wave-functions in the calculation of the effective interaction. The inclusion of 'exchange' terms would weaken the attractive force, since we must now try to prevent two electrons coming into the same state or being at the same spot. The singlet state of the Cooper pair is the lowest state of the quasi-molecule.

11.3 The superconducting ground state

Two important features of the superconducting state are suggested by the above result: it is a state that cannot be reached by a finite number of terms in a perturbation expansion, for the function (11.20) is not an analytic function of V near $V = 0$; it is a state in which electrons are somehow paired, with $\mathbf{k}$ associated with $-\mathbf{k}$. The construction of this state was first achieved by Bardeen, Cooper and

Schrieffer. To demonstrate their solution we shall use the procedure of Bogoliubov.

It is very convenient to avoid all discussion of wave-functions, and to set up the whole problem in terms of operators. We need to express the fact that we have N electrons, whose energy is largely kinetic, in the states labelled by the vector $\mathbf{k}$ and the spin index σ, and that they interact through the potential $V(\mathbf{K})$. We define annihilation and creation operators, $b_{\mathbf{k}\sigma}$, $b^*_{\mathbf{k}\sigma}$ which are similar to the operators $a_{\mathbf{q}}$, $a^*_{\mathbf{q}}$ defined in (2.129) and (10.109), except that we are now dealing with fermions. That means that they must obey the *anticommutation relations*

$$[b_{\mathbf{k}\sigma}, b^*_{\mathbf{k}'\sigma'}]_+ \equiv b_{\mathbf{k}\sigma} b^*_{\mathbf{k}'\sigma'} + b^*_{\mathbf{k}'\sigma'} b_{\mathbf{k}\sigma} = \delta_{\mathbf{k}\mathbf{k}'} \delta_{\sigma\sigma'}. \tag{11.24}$$

The result of this definition is that the operator

$$n_{\mathbf{k}\sigma} = b^*_{\mathbf{k}\sigma} b_{\mathbf{k}\sigma} \tag{11.25}$$

measures the occupation number of the state $(\mathbf{k}, \sigma)$, and has eigenvalues 0 and 1, as required by the Pauli principle. When $\mathbf{k}$ and $\mathbf{k}'$, or σ and σ' are not the same in (11.24), the anticommutator reflects the antisymmetry principle, by which the interchange of labels of two states reverses the sign of the wave-function.

The Hamiltonian of our system may be written

$$\mathscr{H} = \sum_{\mathbf{k}\sigma} \xi(\mathbf{k}) b^*_{\mathbf{k}\sigma} b_{\mathbf{k}\sigma} + \sum_{\mathbf{k}_1, \mathbf{k}_2, \mathbf{K}} V(\mathbf{K}) b^*_{\mathbf{k}_2-\mathbf{K},\downarrow} b^*_{\mathbf{k}_1+\mathbf{K},\uparrow} b_{\mathbf{k}_1,\uparrow} b_{\mathbf{k}_2,\downarrow}. \tag{11.26}$$

The first term represents the kinetic energy of the electrons, measured, as in (11.12), from the Fermi level. It is often convenient to assume that this is variable—that the total number of electrons n is not definitely fixed—by introducing the chemical potential, ζ, of the electron gas, and writing the first term

$$\begin{aligned}\mathscr{H}_0 &= \sum_{\mathbf{k}\sigma} n_{\mathbf{k}\sigma} \xi(\mathbf{k}) \\ &= \sum_{\mathbf{k}\sigma} n_{\mathbf{k}\sigma} \{\mathscr{E}(\mathbf{k}) - \zeta\} \\ &= \sum_{\mathbf{k}\sigma} n_{\mathbf{k}\sigma} \mathscr{E}(\mathbf{k}) - n\zeta.\end{aligned} \tag{11.27}$$

The second term in (11.26) represents the scattering defined by the interaction (11.2), whose effect is to 'destroy' the electrons in $\mathbf{k}_1$, and $\mathbf{k}_2$, and then to 're-create' them in $\mathbf{k}_1 + \mathbf{K}$ and $\mathbf{k}_2 - \mathbf{K}$, with the matrix element $V(\mathbf{K})$. We assume here that only opposite spins interact.

Bogoliubov's method is to make a canonical transformation of the set, $b_{\mathbf{k}\sigma}$ and $b^*_{\mathbf{k}\sigma}$, to new annihilation and creation operators having the same commutation relations. In particular, these must show some pairing properties—they must link the state $(\mathbf{k}, \uparrow)$ with the state $(-\mathbf{k}, \downarrow)$, as suggested by the Cooper effect.

For simplicity let us drop the spin indices, and write

$$\left.\begin{aligned} \beta_{\mathbf{k}} \quad & u_{\mathbf{k}} b^*_{\mathbf{k}} - v_{\mathbf{k}} b_{-\mathbf{k}}, \quad & \beta_{\mathbf{k}} &= u_{\mathbf{k}} b_{\mathbf{k}} - v_{\mathbf{k}} b^*_{-\mathbf{k}}, \\ \beta^*_{-\mathbf{k}} &= u_{\mathbf{k}} b^*_{-\mathbf{k}} + v_{\mathbf{k}} b_{\mathbf{k}}, \quad & \beta_{-\mathbf{k}} &= u_{\mathbf{k}} b_{-\mathbf{k}} + v_{\mathbf{k}} b^*_{\mathbf{k}}. \end{aligned}\right\} \tag{11.28}$$

The operators $\beta_{\mathbf{k}}, \beta^*_{\mathbf{k}}$ will have the proper anticommutation relations for fermions if

$$u^2_{\mathbf{k}} + v^2_{\mathbf{k}} = 1 \tag{11.29}$$

The transformation is, indeed, essentially the same as (10.130) for antiferromagnetic spin waves, except that now we have a 'real rotation' in the space of $\mathbf{k}$ and $-\mathbf{k}$, so that

$$u_{\mathbf{k}} = \cos\theta_{\mathbf{k}}, \quad v_{\mathbf{k}} = \sin\theta_{\mathbf{k}} \tag{11.30}$$

satisfies (11.29). This is because we are now dealing with fermions instead of bosons.

When we solve (11.28) for the original operators, and substitute into the Hamiltonian (11.26), we get a variety of different terms containing products of the new operators $\beta_{\mathbf{k}}, \beta^*_{\mathbf{k}}$. Now when any of these terms contains an annihilation operator $b_{\mathbf{k}}$, we can use the anticommutator to shift it to the right, e.g.

$$\beta_{\mathbf{k}} \beta^*_{\mathbf{k}'} = \delta_{\mathbf{k}\mathbf{k}'} - \beta^*_{\mathbf{k}'} \beta_{\mathbf{k}}. \tag{11.31}$$

Suppose that our superconducting state is the 'vacuum', $|0\rangle$, of these new operators. By definition, there are no objects in the vacuum that can be annihilated, so that

$$\beta_{\mathbf{k}} |0\rangle = 0, \tag{11.32}$$

for all $\mathbf{k}$. Such terms in the Hamiltonian do not contribute to the energy of the ground state, $|0\rangle$, which is thus an eigenstate of all this part of $\mathscr{H}$.

But there remain terms which cannot be eliminated in this way—terms which contain only creation operators like $\beta^*_{\mathbf{k}}$. These arise from the kinetic energy part of (11.26), and also from the reduction of the interaction terms, as when $\mathbf{k}$ and $\mathbf{k}'$ in (11.31) happen to be equal. We find that our Hamiltonian can be written

$$\begin{aligned} \mathscr{H} = {} & 2 \sum_{\mathbf{k}} \xi(\mathbf{k}) v^2_{\mathbf{k}} + \sum_{\mathbf{k},\mathbf{K}} \mathscr{V}(\mathbf{K})\, u_{\mathbf{k}} v_{\mathbf{k}} u_{\mathbf{k}+\mathbf{K}} v_{\mathbf{k}+\mathbf{K}} \\ & + \sum_{\mathbf{k}} \{2\xi(\mathbf{k}) u_{\mathbf{k}} v_{\mathbf{k}} + (u^2_{\mathbf{k}} - v^2_{\mathbf{k}}) \sum_{\mathbf{K}} \mathscr{V}(\mathbf{K})\, u_{\mathbf{k}+\mathbf{K}} v_{\mathbf{k}+\mathbf{K}}\} \beta^*_{\mathbf{k}} \beta^*_{-\mathbf{k}}. \end{aligned} \tag{11.33}$$

The first terms will give us the energy of our ground state—if only we can eliminate the creation operators in the final sum.

This is very similar to what we found in (10.132)—and we use the same device. We choose our transformation coefficients $u_{\mathbf{k}}, v_{\mathbf{k}}$ so as to make each 'dangerous' term vanish. We let

$$2\xi(\mathbf{k})\, u_{\mathbf{k}} v_{\mathbf{k}} + (u_{\mathbf{k}}^2 - v_{\mathbf{k}}^2) \sum_{\mathbf{K}} V(\mathbf{K})\, u_{\mathbf{k}+\mathbf{K}} v_{\mathbf{k}+\mathbf{K}} = 0. \tag{11.34}$$

In general, this is a complicated integral equation for the unknown functions $u_{\mathbf{k}}$ and $v_{\mathbf{k}}$, which are subject also to the condition (11.29). But let us make the assumptions of (11.8) and (11.9), simplifying the attractive potential $V(\mathbf{K})$ to a constant, V, within an energy range $\pm w$, and zero beyond. We let the sum over $\mathbf{K}$ in (11.34) be a parameter

$$\Delta_0 = -V \sum_{-w}^{w} u_{\mathbf{k}+\mathbf{K}} v_{\mathbf{k}+\mathbf{K}}, \tag{11.35}$$

and solve (11.34) and (11.29) simultaneously. The result is

$$u_{\mathbf{k}}^2 = \frac{1}{2}\left[1 + \frac{\xi(\mathbf{k})}{\sqrt{\{\Delta_0^2 + \xi^2(\mathbf{k})\}}}\right], \quad v_{\mathbf{k}}^2 = \frac{1}{2}\left[1 - \frac{\xi(\mathbf{k})}{\sqrt{\{\Delta_0^2 + \xi^2(\mathbf{k})\}}}\right], \tag{11.36}$$

which may be substituted back in (11.35) to give

$$\begin{aligned} 1 &= -\tfrac{1}{2} V \sum_{-w}^{w} \{\Delta_0^2 + \xi^2(\mathbf{k})\}^{-\frac{1}{2}} \\ &= -\tfrac{1}{2} \mathcal{N}(\mathscr{E}_F)\, V \ln \frac{2w}{\Delta_0}, \end{aligned} \tag{11.37}$$

by an integration over the range of ξ near the Fermi surface (remembering that only one spin is to be counted). This is exactly the same as (11.19). The parameter Δ_0 is precisely the binding energy of a Cooper pair, as in (11.20) or (11.22).

We can put this back into the remainder of (11.33), from which all the operators have now been removed. We get

$$\mathscr{E}_0 = 2 \sum_{\xi<0} \xi(\mathbf{k}) - \tfrac{1}{2}\Delta_0^2 \sum_{\mathbf{k}} \{\Delta_0^2 + \xi^2(\mathbf{k})\}^{-\frac{1}{2}}. \tag{11.38}$$

Thus, the state $|0\rangle$ is an eigenstate of $\mathscr{H}$—but with lower energy than the full Fermi sphere of electrons, whose energy would just be the sum of the values of $\xi(\mathbf{k})$ for the occupied states, as in (11.27). It can also be shown that this is a minimum of the energy for variation of the expectation number n of the total number of electrons, as required by (11.27).

11.4 Quasi-particles and the energy gap

What is the nature of the superconducting state $|0\rangle$? By definition it is the 'vacuum' of the operators $\beta_{\mathbf{k}}^*$, $\beta_{\mathbf{k}}$. These operators can create and destroy excitations of the vacuum, just as the original operators $b_{\mathbf{k}}^*$ and $b_{\mathbf{k}}$ can 'create' and 'destroy' electrons. What would such an excitation look like? According to (11.28), $\beta_{\mathbf{k}}^*$ has the effect of 'creating' an electron in the state $\mathbf{k}$, with amplitude $u_{\mathbf{k}}$, and at the same time 'destroying' an electron in the state $-\mathbf{k}$, with amplitude $v_{\mathbf{k}}$.

These coefficients are given by (11.36). Suppose that $\mathbf{k}$ is well above the Fermi surface, so that $\xi(\mathbf{k}) \gg \Delta_0$. Then $u_{\mathbf{k}}^2 \approx 1$, and $v_{\mathbf{k}}^2 \approx 0$. Our ex-

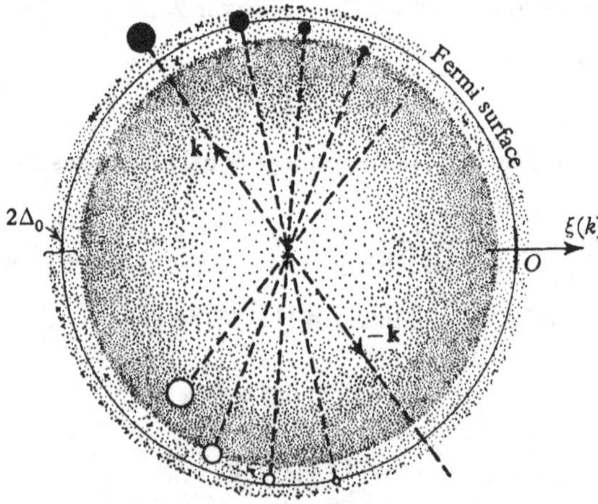

Fig. 203. Quasi-particles.

citation will look like an ordinary electron. But now, as we approach the Fermi level, $\xi(\mathbf{k})$ tends to zero, and the amplitude of $b_{\mathbf{k}}^*$ will decrease, whilst that of $b_{-\mathbf{k}}$ will increase. When $\xi(\mathbf{k})$ is negative, the value of $v_{\mathbf{k}}^2$ will be greater than $\frac{1}{2}$; when we are well below the Fermi level, $v_{\mathbf{k}}^2 \approx 1$ and $u_{\mathbf{k}}^2 \approx 0$. Our excitation now corresponds to the 'destruction' of an electron at $-\mathbf{k}$, i.e. it is a 'hole' in the otherwise occupied states.

The excitations of the superconducting state are thus rather peculiar *quasi-particles* which change from being 'electrons' to being 'holes' as they pass through the Fermi level. In the range of energy around Δ_0 each quasi-particle is a mixture of an electron in $\mathbf{k}$ and a hole in $-\mathbf{k}$. The Cooper pairs, as such, no longer appear, but the correlation between electrons of opposite momentum (and, of course, opposite spin) is evident. In effect, we are insisting that there should be a relationship between the phases of the wave functions in the paired states.

The energy of an excitation can also be calculated. We look for the coefficient of terms containing the operator

$$n'_{\mathbf{k}} = \beta^*_{\mathbf{k}} \beta_{\mathbf{k}} \tag{11.39}$$

in the expansion of (11.26), since this measures the number of quasi-particles that have been excited in this mode. The result is

$$\begin{aligned} \epsilon(\mathbf{k}) &= \xi(\mathbf{k})(u_{\mathbf{k}}^2 - v_{\mathbf{k}}^2) - 2u_{\mathbf{k}} v_{\mathbf{k}} \sum_{\mathbf{K}} V(\mathbf{K})\, u_{\mathbf{k}+\mathbf{K}} v_{\mathbf{k}+\mathbf{K}} \\ &= \sqrt{\{\Delta_0^2 + \xi^2(\mathbf{k})\}} \end{aligned} \tag{11.40}$$

for our simplified interaction, using (11.35) and (11.36).

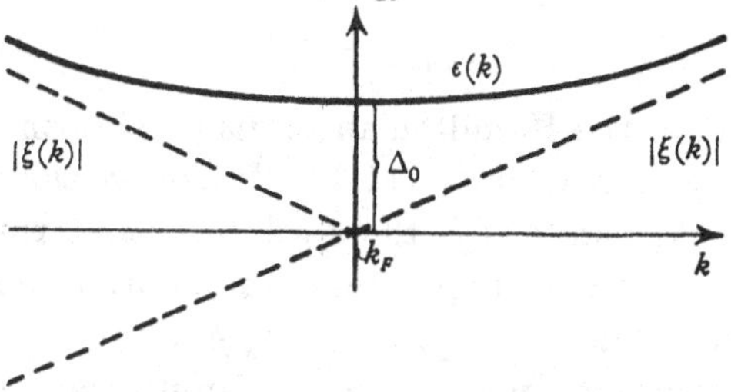

Fig. 204. Energy of quasi-particles.

This is very important, for it demonstrates the existence of an *energy gap* Δ_0 above the superconducting ground state. When $\xi(\mathbf{k}) \gg \Delta_0$, then $\epsilon(\mathbf{k}) \approx \xi(\mathbf{k})$—the energy of the quasi-particle is the energy required to excite an ordinary electron above the Fermi level. But as $\mathbf{k}$ approaches the Fermi surface, where $\xi(\mathbf{k})$ is zero, the energy of the quasi-particle tends to the constant value Δ_0. Then, as $\mathbf{k}$ decreases still further, $\epsilon(\mathbf{k})$ increases again, until it becomes nearly equal to $|\xi(\mathbf{k})|$. This is, of course, the energy required to remove an electron from the state of negative energy $\xi(\mathbf{k})$, just as we expect from our interpretation of the quasi-particle operators.

The superconducting state is thus a 'condensed' state in the sense that a finite energy, Δ_0, is required to produce an excited state of the whole system. In some ways, it is a bit like a semiconductor, where a finite energy gap must be surmounted if one wants to put an electron into the conduction band. This analogy helps to explain, crudely, some special properties of superconductors, for example electromagnetic absorption (cf. § 8.5) and tunnelling (§ 6.8).

11.5 Temperature dependence of the energy gap

The ground state is only appropriate at $T = 0$. At a finite temperature we expect quasi-particles to be excited, according to the usual Fermi–Dirac function

$$f_{\mathbf{k}} = \frac{1}{e^{\epsilon(\mathbf{k})/kT}+1}. \tag{11.41}$$

In the normal state such excitations, above or below the Fermi level, are independent of one another. In the superconducting state they tend to interact, co-operatively, and destroy the energy gap.

Let us suppose that we are dealing with the state $|f_{\mathbf{k}}\rangle$, in which the average number of quasi-particles in the **k**th state is given by (11.41). The effect on this state of a destruction operator $b_{\mathbf{k}}$ is not zero. On the average, we can replace (11.25) by

$$b_{\mathbf{k}}^* b_{\mathbf{k}} |f_{\mathbf{k}}\rangle = f_{\mathbf{k}} |f_{\mathbf{k}}\rangle. \tag{11.42}$$

The reduction of the Hamiltonian to diagonal form, when applied to the state $|0\rangle$, depended upon (11.32). If now we try to do the same thing, applied to the state $|f_{\mathbf{k}}\rangle$, there will be extra terms arising from the reduction of products of operators to standard order as in (11.31), multiplying the 'dangerous' products $\beta_{\mathbf{k}}^* \beta_{-\mathbf{k}}^*$ in (11.33). To remove these 'pair-production' parts of the Hamiltonian, our condition (11.34) must be changed to read

$$2\xi(\mathbf{k})\, u_{\mathbf{k}} v_{\mathbf{k}} + (u_{\mathbf{k}}^2 - v_{\mathbf{k}}^2) \sum_{\mathbf{K}} V(\mathbf{K})\, u_{\mathbf{k}+\mathbf{K}} v_{\mathbf{k}+\mathbf{K}} (1 - 2f_{\mathbf{k}+\mathbf{K}}) = 0 \tag{11.43}$$

(allowing for the two possible spin states associated with each **k**).

This integral equation can be solved, for the simplified interaction V, by replacing Δ_0 in (11.35) and (11.36) by

$$\Delta = -V \sum_{-w}^{w} u_{\mathbf{k}+\mathbf{K}} v_{\mathbf{k}+\mathbf{K}} (1 - 2f_{\mathbf{k}+\mathbf{K}}). \tag{11.44}$$

Instead of (11.37), we then get, using (11.41),

$$\begin{aligned} 1 &= -\tfrac{1}{2}V \sum_{-w}^{w} \frac{1}{\epsilon(\mathbf{k})} \tanh \frac{\epsilon(\mathbf{k})}{2kT} \\ &= -\tfrac{1}{2}\mathcal{N}(\mathscr{E}_F)\, V \int_{-w}^{w} \frac{1}{\epsilon} \tanh \frac{\epsilon}{2kT}\, d\xi, \end{aligned} \tag{11.45}$$

where, as before, ϵ is the energy of an excitation,

$$\epsilon(\mathbf{k}) = \sqrt{\{\Delta^2 + \xi^2(\mathbf{k})\}}. \tag{11.46}$$

The two equations (11.45) and (11.46) yield an implicit relation between Δ and T. This relationship is rather untidy, but the main point is that Δ decreases from Δ_0 at $T = 0$ as T increases. In other words, the energy gap, Δ, decreases as the temperature increases, and closes to zero at a well-defined temperature, T_c, where, by (11.46), $\epsilon(\mathbf{k}) = |\xi(\mathbf{k})|$. Thus, we can find T_c by integrating

$$\frac{2}{\mathcal{N}(\mathscr{E}_F)\,V} = 2\int_0^{w} \frac{1}{\epsilon} \tanh\left(\frac{\epsilon}{2kT_c}\right) d\epsilon, \tag{11.47}$$

whose solution is

$$\Delta_0 \approx 1{\cdot}76kT_c, \tag{11.48}$$

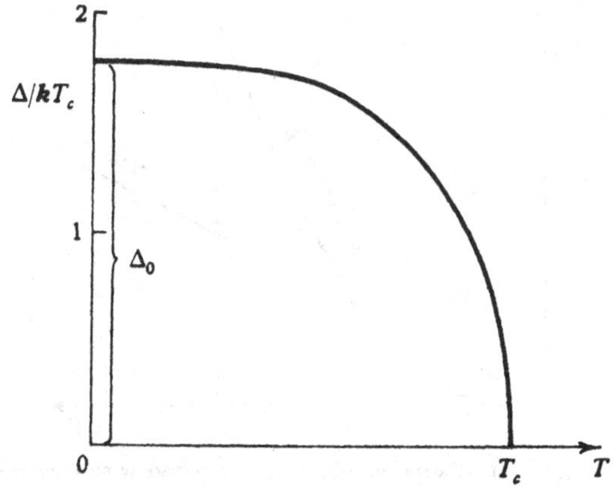

Fig. 205. Variation of energy gap with temperature.

when expressed in terms of (11.22), the energy gap at $T = 0$. Thus, the discussion of § 11.2, where T_c was estimated from the binding energy of a Cooper pair, is justified.

Above T_c there is no solution of the equations. Just below T_c the energy gap rises steeply from zero, varying as

$$\Delta(T) \approx 3{\cdot}2kT_c\left(1 - \frac{T}{T_c}\right)^{\frac{1}{2}}, \tag{11.49}$$

as may be verified by seeking an approximate solution of (11.45) and (11.46) in the neighbourhood of the solution of (11.47). At lower temperatures it increases more slowly, and flattens off to Δ_0 below about $T \sim \frac{1}{2}T_c$.

From a careful calculation of the energy of our system of quasi-particles, allowing for the variation of the energy gap with temperature, one can write down formulae for the specific heat. At the transition temperature itself there is a jump of magnitude

$$C_{es} = 1{\cdot}43\gamma T_c, \tag{11.50}$$

where γ is the coefficient in the formula (4.46) for the electronic specific heat in the normal state. At lower temperatures the energy gap tends to dominate the specific heat, but a simple formula like

$$C_{es} \approx e^{-\Delta(T)/T} \tag{11.51}$$

is not adequate, until Δ settles down to its limiting value Δ_0.

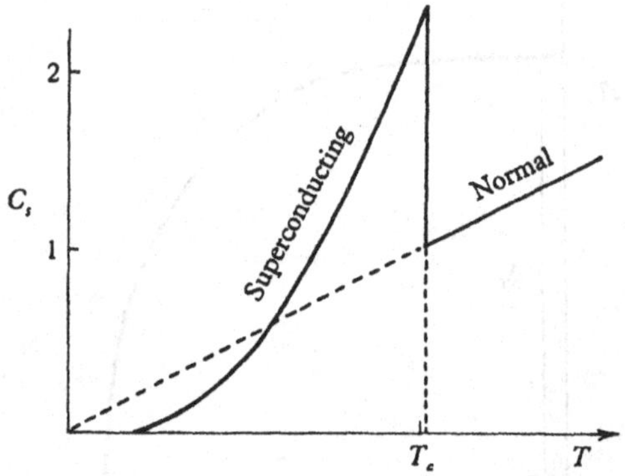

Fig. 206. Specific heat of superconductor, compared with electronic specific heat of normal metal.

11.6 Persistent currents

The most striking feature of the superconducting state is that the metal seems to offer absolutely no resistance to the flow of an electrical current. To understand this in the BCS theory we notice that the energy gap is not tied to the lattice—it is not, as in a semiconductor, fixed to the zone boundaries in reciprocal space—but simply occurs at the Fermi level. The argument that we have used above does not depend upon the Fermi surface being a sphere; it ought to be true for a Fermi surface of arbitrary shape.

As shown in § 7.2 the effect of an electric field is to displace the Fermi distribution by some constant vector relative to the reciprocal lattice. The current **J** corresponds to the displacement

$$\delta\mathbf{k} = \frac{m}{ne\hbar}\mathbf{J} \tag{11.52}$$

of a free electron sphere. The energy gap does not inhibit this shift; it is carried with the Fermi surface to its new position.

To describe the current-carrying state in the language of the BCS theory, we simply assume that the centre of gravity of each electron pair is moved by the amount $\mathbf{\delta k}$. In constructing the quasi-particle operators $\beta_{\mathbf{k}}$, $\beta_{\mathbf{k}}^*$ of (11.28), we now associate the state $(\mathbf{k}+\mathbf{\delta k}, \uparrow)$ with the state $(-\mathbf{k}+\mathbf{\delta k}, \downarrow)$; the rest of the argument follows unchanged.

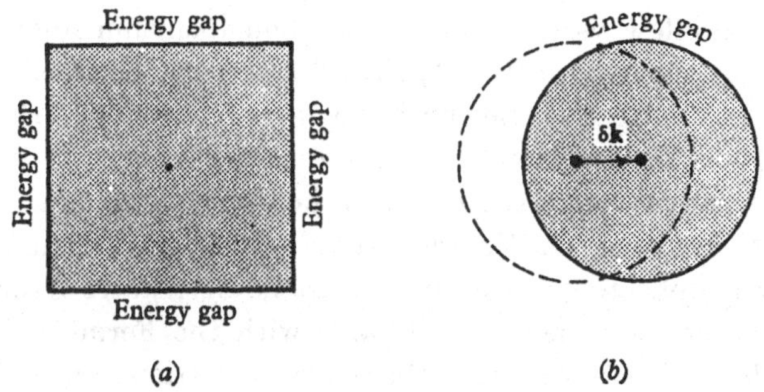

Fig. 207. (*a*) In semiconductor the energy gap is tied to the Brillouin zone. (*b*) In superconductor, the energy gap is carried by the Fermi surface.

The only point to notice is that the total kinetic energy of such a distribution of electrons will have been increased by an amount

$$\delta\mathscr{E} \approx \frac{\hbar^2(\delta k)^2}{2m} \tag{11.53}$$

per electron. If this energy were to exceed the binding energy (11.38) of the superconducting state, then we should not expect the system to remain condensed below the energy gap. This sets a limit to the current that can be carried.

But the most important property of the current-carrying state is that, once established, it is not easily destroyed. The processes that resist a normal current, as discussed in § 7.5, are 'one-electron' transitions, in which a single particle is scattered from one side to the other of the Fermi surface by interaction with a phonon or with an impurity. Such a process in the superconducting state can only proceed by the creation of quasi-particles—by the production of a hole where the electron was, at $(\mathbf{k}+\mathbf{q})$ say, and the materialization of the electron again at $(-\mathbf{k}+\mathbf{q})$. This requires the energy Δ, which is not available from a single phonon. To reduce the current to zero, we need, somehow, to slow down all the electrons at once, making the whole displaced

Fermi surface return smoothly to the origin. There is no simple mechanism for such an unlikely process, so that in a ring of superconducting material a circulating current, once established, may persist almost indefinitely.

The BCS theory seems to depend critically on the existence of well-defined Bloch States $|\mathbf{k},\uparrow\rangle$ and $|-\mathbf{k},\downarrow\rangle$. Yet superconductivity is frequently observed in very disordered and 'dirty' alloys, where the mean free path of an electron for ordinary conduction would be much too short to allow us to treat $\mathbf{k}$ as a good quantum number. The truth is that superconductive 'pairing' can occur between any two states that are related to one another by *time reversal* (§ 3.11). All that is necessary is for the electron, having reversed its spin, to be able to retrace its path through the material, however tortuous and erratic this might have been. The theory of the Fröhlich interaction may then seem very complicated, but provided that the eigenstates of the system can be classified in time-reversed pairs, with the Fermi energy in a region of high density of states, there is no objection in principle to a macroscopically superconducting condensed phase. For the same reason, the reinterpretation of nearly independent electron states as many-body quasi-particle excitations (§ 5.8) is no obstacle to the BCS theory.

If, however, the material contains built-in magnetic fields, as in a ferromagnet, the assumption of time-reversal symmetry is not valid (§ 9.3), for we cannot assume that these fields would also be reversed in the hypothetical process of sending a single electron back along its path. Quite a small concentration of magnetic impurities (§ 10.6) is therefore sufficient to destroy superconductivity. At a slightly lower concentration of impurities, a condition known as *gapless superconductivity* may be observed—superficially superconducting and macroscopically ordered (§ 11.9), but without an actual gap (11.40) in the quasi-particle spectrum.

11.7 The London equation

A superconductor behaves in a peculiar fashion in an external magnetic field. Some of this behaviour can apparently be understood as a consequence of infinite conductivity. We cannot build up a magnetic field inside a superconductor because persistent currents are induced at the surface of the specimen as the field is created, and the interior is screened. But this explanation is not adequate for all the phenomena.

We need to consider the currents induced by a magnetic field with vector-potential $\mathbf{A}(\mathbf{r})$. According to (10.3), this causes a change in the kinetic energy operator; in the state $\psi(\mathbf{r})$ the expectation value of the change is

$$\psi^*\left\{\frac{1}{2m}\left(\frac{\hbar}{i}\nabla-\frac{e}{c}\mathbf{A}\right)^2-\left(-\frac{\hbar^2}{2m}\nabla^2\right)\right\}\psi=-\frac{1}{c}\mathbf{J}(\mathbf{r})\cdot\mathbf{A}(\mathbf{r}), \quad (11.54)$$

where we have equated this local 'kinetic energy density' to the classical energy density associated with currents of density $\mathbf{J}(\mathbf{r})$ moving in the vector potential $\mathbf{A}(\mathbf{r})$. Thus, the current density is the derivative of (11.54) with respect to $\mathbf{A}(\mathbf{r})$, and takes the form

$$\mathbf{J}(\mathbf{r})=\frac{e\hbar}{2m}\frac{1}{i}(\psi^*\nabla\psi-\psi\nabla\psi^*)-\frac{e^2}{mc}\psi^*\psi\mathbf{A}. \quad (11.55)$$

The first term is the well-known formula for the current in the absence of a magnetic field.

In an ordinary metal, in a static field, these two contributions nearly cancel out, leaving a small diamagnetism (cf. § 10.1). In a superconductor, however, the states of the 'wave-function' ψ (which will be discussed in more detail in § 11.9) tend to have a certain 'rigidity'. Suppose, for example, that the imposition of the magnetic field has no effect at all on the wave-function, so that ψ remains equal to ψ_0, its value in zero field. We can be sure that

$$\begin{aligned}\mathbf{J}_0(\mathbf{r})&=\frac{e\hbar}{2mi}(\psi_0^*\nabla\psi_0-\psi_0\nabla\psi_0^*)\\&=0.\end{aligned} \quad (11.56)$$

Thus, in the superconductor, we should be left with

$$\begin{aligned}\mathbf{J}(\mathbf{r})&=-\frac{e^2}{mc}\psi^*(\mathbf{r})\psi(\mathbf{r})\mathbf{A}(\mathbf{r})\\&=-\frac{e^2n}{mc}\mathbf{A}(\mathbf{r}),\end{aligned} \quad (11.57)$$

where n is the local electron density.

This formula, in which the hypothetical 'wave function' no longer appears, is known as the *London equation*; it goes a long way towards explaining many of the properties of superconductors. We know, for example, that the conductivity is infinite, so that there is no electric field inside the specimen. A static magnetic field will be associated with a current $\mathbf{J}$ by the Maxwell equation

$$\nabla\wedge\mathbf{H}=\frac{4\pi}{c}\mathbf{J}. \quad (11.58)$$

But (11.57) can be written

$$\mathbf{H} = \nabla \wedge \mathbf{A}$$

$$= -\frac{mc}{e^2 n} \nabla \wedge \mathbf{J}. \tag{11.59}$$

Combining (11.58) and (11.59) we get

$$\nabla^2 \mathbf{H} = \frac{1}{\lambda^2} \mathbf{H}, \quad \nabla^2 \mathbf{J} = \frac{1}{\lambda^2} \mathbf{J}, \tag{11.60}$$

where

$$\lambda = \sqrt{\frac{mc^2}{4\pi n e^2}}. \tag{11.61}$$

The solution of (11.60), for a magnetic field $\mathbf{H}_0$ just outside and parallel to the surface of the superconductor is

$$\mathbf{H} = \mathbf{H}_0 e^{-z/\lambda}, \tag{11.62}$$

where z is the distance measured into the specimen. Thus, the field falls off rapidly, and penetrates only the distance λ—the so-called *penetration depth*, which is of the order of 5×10^{-6} cm.

This result does not depend on how the field was created. It holds for a specimen that is cooled down, from above the transition temperature, in a magnetic field. When the metal becomes superconducting the magnetic field is expelled. This is the *Meissner effect*, which cannot be explained simply on the basis of infinite conductivity.

The fact that the superconducting state demands time reversal symmetry, and is 'perfectly diamagnetic' suggests that a strong enough magnetic field should be capable of modifying that state. Indeed, bulk superconductivity in metals is destroyed by the imposition of a magnetic field exceeding a critical value H_c. For type I superconductors (see § 11.10), this field can be calculated by thermodynamic arguments from the specific heat curve. The field energy $H_c^2/8\pi$ at the critical field must be the difference in *free energy* between the superconducting and normal states. Suppose, for example, that we assume that the specific heat of the superconducting state is proportional to T^3, whilst the normal state has the ordinary electronic specific heat γT. Then one can show that the critical field should be a function of temperature;

$$H_c = \sqrt{(2\pi)}\, \gamma T_c \{1 - (T/T_c)^2\}, \tag{11.63}$$

where T_c is the critical temperature (11.48). In fact, as noted in (11.51), the specific heat of the superconductor, is a complicated function of temperature, so that this formula is only approximately correct.

It is worth noting that this argument assumes that we can exclude *all* the magnetic field from the specimen. In a very small or thin piece of metal, of thickness comparable with λ, there is penetration of the magnetic field, and the thermodynamic argument is not valid. The superconducting state can then persist to much higher magnetic fields, than would be calculated by (11.63). This is what happens in type II superconductors, where such filaments may be stabilized, even in bulk specimens, by interaction with dislocations, impurities, etc. (see § 11.10).

11.8 The coherence length

The assumption in the derivation of the London equation (11.57) is that the superconducting wave-function is 'rigid', and unaffected by the application of the magnetic field. This is obviously incorrect; we know that a strong enough magnetic field will destroy the superconducting state. We need to calculate the change in the wave-function ψ, in (11.55), produced by the perturbation $\mathbf{A}(\mathbf{r})$, when acting on the superconducting ground state.

This is not difficult to work out inside the framework of the Bogoliubov formalism of § 11.3. Suppose we consider a single Fourier component of the vector potential

$$\mathbf{A}(\mathbf{r}) = \mathbf{A}_{\mathbf{q}}\, e^{i\mathbf{q}\cdot\mathbf{r}}. \tag{11.64}$$

The matrix element of one of these terms, between two ordinary free-electron states like $\exp(i\mathbf{k}\cdot\mathbf{r})$ and $\exp(i\mathbf{k}'\cdot\mathbf{r})$ will be

$$\int e^{-i\mathbf{k}'\cdot\mathbf{r}}\mathbf{A}(\mathbf{r})\, e^{i\mathbf{k}\cdot\mathbf{r}}\, d\mathbf{r} = \mathbf{A}_{\mathbf{q}}\, \delta(\mathbf{k}+\mathbf{q}-\mathbf{k}'). \tag{11.65}$$

Thus, $\mathbf{A}(\mathbf{r})$ will behave like the operator

$$\mathsf{A} = \mathbf{A}_{\mathbf{q}}\, b^*_{\mathbf{k}+\mathbf{q}} b_{\mathbf{k}} \tag{11.66}$$

in the language of the annihilation and creation operators (11.24).

To first order in $\mathbf{A}$, the actual perturbation produced in the Hamiltonian by the magnetic field is the operator

$$-\frac{e\hbar}{2mci}(\nabla\cdot\mathsf{A}+\mathsf{A}\cdot\nabla) = -\frac{e\hbar}{2mc}\sum_{\mathbf{k}}\{(\mathbf{k}+\mathbf{q})+\mathbf{k}\}\cdot\mathbf{A}_{\mathbf{q}}\, b^*_{\mathbf{k}+\mathbf{q}} b_{\mathbf{k}}. \tag{11.67}$$

This is easily obtained by analogy with (11.65) and (11.66); the operator on the right has the same matrix elements as the operator on the left between all 'bare-electron' states.

Now let us apply the transformation (11.28) to this expression. The result, expressed in terms of quasi-particle operators $\beta_{\mathbf{k}}$, $\beta_{\mathbf{k}}^*$, is a bit complicated, but since we are going to apply the perturbation to the superconducting ground state $|0\rangle$ we can use (11.32) to eliminate all terms containing any annihilation operators. We are left with terms containing products of two creation operators; our perturbation becomes

$$-\frac{e\hbar}{2mc}\sum_{\mathbf{k}}(2\mathbf{k}+\mathbf{q})\cdot\mathbf{A}_{\mathbf{q}}(v_{\mathbf{k}+\mathbf{q}}u_{\mathbf{k}}-v_{\mathbf{k}}u_{\mathbf{k}+\mathbf{q}})\beta_{\mathbf{k}+\mathbf{q}}^*\beta_{-\mathbf{k}}^*, \tag{11.68}$$

where $v_{\mathbf{k}}$, $u_{\mathbf{k}}$, etc., are the coefficients in (11.28). The magnetic field creates pairs of quasi-particles, of net momentum $\mathbf{q}$, just as we might expect from our argument (§ 11.6) for the nature of persistent currents.

We use (11.68) as a perturbation, to calculate the change in the ground-state wave-function, and then recalculate the current from (11.55), using this 'non-rigid' state in place of ψ. The result is as follows

$$\mathbf{J}_{\mathbf{q}} = -\frac{e^2n}{mc}\mathbf{A}_{\mathbf{q}}+\frac{e^2}{2m^2c}\sum_{\mathbf{k}}\frac{(2\mathbf{k}+\mathbf{q})\{(2\mathbf{k}+\mathbf{q})\cdot\mathbf{A}_{\mathbf{q}}\}(v_{\mathbf{k}}u_{\mathbf{k}+\mathbf{q}}-v_{\mathbf{k}+\mathbf{q}}u_{\mathbf{k}})}{\epsilon(\mathbf{k})+\epsilon(\mathbf{k}+\mathbf{q})}, \tag{11.69}$$

where the energy $\epsilon(\mathbf{k})+\epsilon(\mathbf{k}+\mathbf{q})$ of the quasi-particle pair is calculated from (11.40).

There are formal difficulties in this calculation—difficulties that often arise in the quantum theory of particles in magnetic fields—about the choice of gauge for the vector potential. We can avoid these difficulties here by assuming that $\mathbf{A}(\mathbf{r})$ is purely transverse, so that $\mathbf{q}\cdot\mathbf{A}_{\mathbf{q}} = 0$. It is convenient to symmetrize the product in $v_{\mathbf{k}}$, etc.; we get

$$\mathbf{J}_{\mathbf{q}} = -\frac{e^2n}{mc}\mathbf{A}_{\mathbf{q}}+\frac{2e^2}{m^2c}\sum_{\mathbf{k}}\frac{\mathbf{k}(\mathbf{k}\cdot\mathbf{A}_{\mathbf{q}})(v_{\mathbf{k}-\frac{1}{2}\mathbf{q}}u_{\mathbf{k}+\frac{1}{2}\mathbf{q}}-v_{\mathbf{k}+\frac{1}{2}\mathbf{q}}u_{\mathbf{k}-\frac{1}{2}\mathbf{q}})}{\epsilon(\mathbf{k}-\frac{1}{2}\mathbf{q})+\epsilon(\mathbf{k}+\frac{1}{2}\mathbf{q})}. \tag{11.70}$$

This sum can be evaluated quite easily from the formulae (11.36) for $u_{\mathbf{k}}$ and $v_{\mathbf{k}}$ in terms of the energy $\xi(\mathbf{k})$. We shall not attempt to follow through this argument in detail; it is elementary geometry. We shall simply write (11.70) in the form

$$\mathbf{J}_{\mathbf{q}} = -\frac{c}{4\pi}\Gamma(\mathbf{q})\,\mathbf{A}_{\mathbf{q}} \tag{11.71}$$

to represent the relation between current and vector-potential; $\Gamma(\mathbf{q})$ is a well-defined function of the wave-number q of the applied magnetic field.

It is obvious enough, from the form of (11.70), that the sum vanishes as $q \to 0$. Thus, for slowly varying magnetic fields, the London formula

(11.57) is valid; the Meissner effect follows, in general. But as q increases the sum in (11.70), which is essentially positive, tends to cancel out the leading term. Indeed, for large values of q we are considering excitations of energy much greater than the energy of the gap, and the quasi-particles behave like ordinary electrons and holes (according as they are above or below the Fermi level). One can then show—it is a good exercise in the use of the theory—that the London term is exactly cancelled. In other words,

$$\Gamma(q) \to 0 \tag{11.72}$$

as q becomes large.

The actual form of $\Gamma(q)$ between these limits is indicated in Fig. 208. It is natural to suppose (and again, it may be verified from (11.70)) that we shall pass from the 'London' régime to the régime indicated by (11.72) when q is large enough to create an excitation, i.e. when q exceeds q_c where

$$\hbar v_F q_c \sim \Delta_0. \tag{11.73}$$

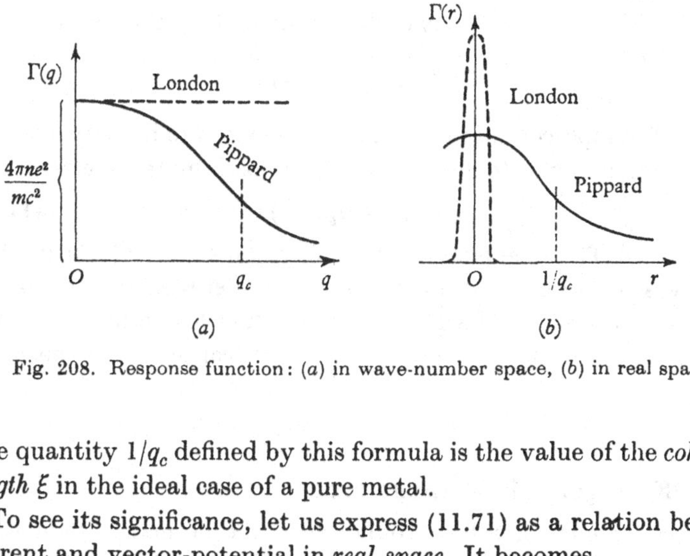

Fig. 208. Response function: (*a*) in wave-number space, (*b*) in real space.

The quantity $1/q_c$ defined by this formula is the value of the *coherence length* ξ in the ideal case of a pure metal.

To see its significance, let us express (11.71) as a relation between current and vector-potential in *real space*. It becomes

$$\mathbf{J}(\mathbf{r}) = \int \Gamma(\mathbf{r}-\mathbf{r}')\,\mathbf{A}(\mathbf{r}')\,d\mathbf{r}', \tag{11.74}$$

where $\Gamma(\mathbf{r})$ is the Fourier transform of $\Gamma(\mathbf{q})$. In the London theory $\Gamma(\mathbf{q})$ is a constant, so $\Gamma(\mathbf{r})$ is a delta-function; the value of $\mathbf{J}(\mathbf{r})$ depends only on the *local* value of $\mathbf{A}(\mathbf{r})$ as in (11.57). But the transform of (11.71) is more complicated. The range of $\Gamma(\mathbf{r})$ will be the coherence length ξ, and we shall have a *non-local* relation between **J** and **A**. It is not unlike the Chambers formula (8.101) used in the theory of the anomalous skin effect.

This result, established phenomenologically by Pippard, is of importance in the detailed analysis of properties such as the anomalous skin effect and the penetration of magnetic fields. In effect, ξ is the 'size' of a Cooper pair; it is the distance over which there must be correlation of the phases of the paired electrons. In a 'dirty' metal where the electron mean free path Λ is less than $1/q_c$, the apparent value of the coherence length ξ is no more than Λ itself. The phase correlation is now, to some extent spoiled, although the pairing itself is not destroyed and the material remains a superconductor.

It is interesting to notice how this coherence is expressed in the mathematical theory. We can see it very clearly in the steps from (11.67) to (11.68). The operator (11.67) causes scattering between single-electron states. In the normal metal we should treat these as independent events, and sum the squares of the separate matrix elements. Thus, in (11.69) we should have just the terms corresponding to

$$(v_{\mathbf{k}} u_{\mathbf{k}+\mathbf{q}})^2 + (u_{\mathbf{k}} v_{\mathbf{k}+\mathbf{q}})^2 \tag{11.75}$$

in the numerator.

But in the superconducting state the operator produces two excitations, of nearly opposite momenta, with a *single* matrix element

$$(v_{\mathbf{k}} u_{\mathbf{k}+\mathbf{q}} - u_{\mathbf{k}} v_{\mathbf{k}+\mathbf{q}}) \tag{11.76}$$

to be squared. This does not give the same result as (11.75); the product term gives rise to correlation, or coherence, between the two quasiparticles produced by the perturbation. But this term is only important for q less than q_c; that is the significance of the coherence length.

11.9 Off-diagonal long range order

A supercurrent may flow round a circuit of many kilometres. Physical conditions at one point of such a circuit are affected by those at other points over enormous distances. The electron system thus exhibits a special form of *long range order*, which we want to characterize mathematically, making due allowance for changes of material, or of temperature, or of magnetic field as we go from one locality to another.

In the macroscopic theory of semiconductors, these local characteristics are summed up in the densities and mobilities of the different types of carrier. Similarly, in the *Gorter–Casimer two-fluid model*, we assume that the supercurrent is carried by a number density n_s of 'superfluid electrons', which depends upon the temperature, etc. If each such electron travels with velocity $\mathbf{v}_s$ and carries charge e^* (not necessarily the ordinary electron charge e), the total supercurrent density must be

$$\mathbf{J}_s = n_s e^* \mathbf{v}_s. \tag{11.77}$$

This model can almost be justified within the BCS theory by identifying the equilibrium concentration of 'normal' electrons,

$$n_n = n - n_s, \tag{11.78}$$

with the concentration of quasi-particle excitations, as given by (11.41). At $T = 0$, there are no quasi-particles, so that $n_s = n$; at the critical temperature, T_c, n_s falls rapidly to zero. We may even approximate thermodynamically to the specific heat anomaly of Fig. 206 by assigning a difference of free energy between the superfluid and 'normal' phases. We might then suppose, for example, that spatial effects such as superconductive tunnelling (§ 11.11) could be described mathematically as variations of the local value of n_s with position.

This description is not, however, adequate. As we saw in § 11.8, the superconducting ground state is characterized by strong correlations in the *wave-functions* of pairs of particles, over distances larger than the coherence length ξ defined by (11.73). Let us go back, therefore, to the London theory of § 11.7, and deliberately introduce *a macroscopic wave-function* $\Psi(\mathbf{r})$ (usually known as the *Ginzburg–Landau order parameter*) for the superfluid component of the electron density. As we learnt in § 11.8, this could well be an appropriate local measure of superfluid order, provided that we were not concerned with properties varying rapidly over distances shorter than the coherence length.

In accordance with the canonical formalism of quantum mechanics, we naturally assume that the local superfluid density is given by

$$n_s(\mathbf{r}) = |\Psi(\mathbf{r})|^2 \tag{11.79}$$

and that

$$\mathbf{J}_s(\mathbf{r}) = \frac{e^*}{2m^*}\left[\Psi^*\left\{\frac{\hbar}{i}\nabla - \frac{e^*}{c}\mathbf{A}\right\}\Psi + \Psi\left\{-\frac{\hbar}{i}\nabla - \frac{e^*}{c}\mathbf{A}\right\}\Psi^*\right] \tag{11.80}$$

defines the supercurrent density, just as in (10.3) and (11.55). But having a suspicion that the supercurrent particles are 'pairs' rather than single electrons, we keep e^* and m^* undefined.

If $\Psi(\mathbf{r})$ were everywhere real, then it could be eliminated in favour of n_s: (11.79) and (11.80) reduce immediately to the London equation (11.57), with e, m and n 'renormalized' to e^*, m^* and n_s. Within the limitations discussed in § 11.8, this provides a good qualitative description of the Meissner effect, including the temperature dependence of the penetration depth implicit in the variation of n_s with T.

But we cannot describe a supercurrent in a homogeneous material unless $\Psi(\mathbf{r})$ is *complex*. The unexpected feature of this phenomenological theory is that the physical situation depends directly on the variation of the *phase* of the order parameter and not simply on its absolute magnitude: in the language of matrix mechanics, we have to deal with *off-diagonal* long range order.

Suppose, for example, that we have a homogeneous material of constant temperature, where $n_s(\mathbf{r})$ would be constant. Writing

$$\Psi(\mathbf{r}) = (n_s)^{\frac{1}{2}} e^{i\chi(\mathbf{r})}, \tag{11.81}$$

we arrive at

$$m^*\mathbf{v}_s(\mathbf{r}) = \hbar\nabla\chi(\mathbf{r}) - (e^*/c)\,\mathbf{A}(\mathbf{r}): \tag{11.82}$$

in the absence of a magnetic field, the phase function $\chi(\mathbf{r})$ plays the part of a velocity potential for the superfluid velocity $\mathbf{v}_s$ already defined by (11.77). But if χ did not vary in space, there would be no supercurrent.

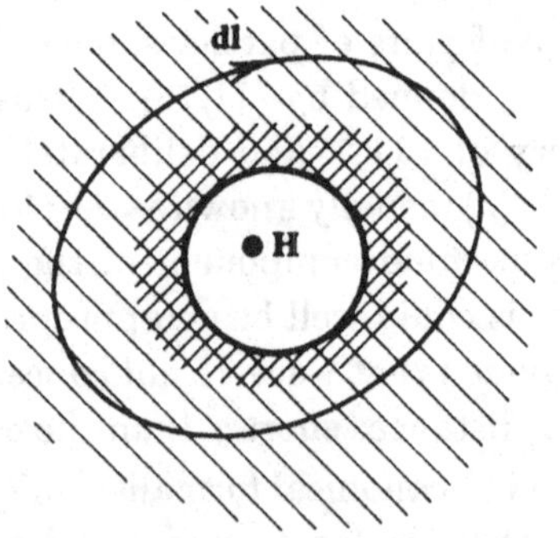

Fig. 209. Path of integration for flux quantization.

Consider now the case of a loop of superconducting material threaded by a magnetic field with vector potential $\mathbf{A}(\mathbf{r})$ (Fig. 209). Let us integrate (11.82) along a path round this circuit, keeping well within the superconducting medium so as to avoid the penetration region where diamagnetic screening currents might occur. In the absence of other sources of E.M.F. in the circuit, $\mathbf{v}_s(\mathbf{r})$ must be zero along this path. The order parameter $\Psi(\mathbf{r})$ itself must be a single-valued function of position within the superconductor, so that its phase angle must come back to the same value, plus some integral multiple of 2π, when we get back to our starting point. By elementary vector calculus we get

$$\Phi = \oint \mathbf{A}\cdot d\mathbf{l} = (\hbar c/e^*)\oint \nabla\chi(\mathbf{r})\cdot d\mathbf{l} = n2\pi\hbar c/e^*, \tag{11.83}$$

where n is an integer. The *fluxoid* Φ—the total magnetic flux through the loop—is *quantized* in units of $2\pi\hbar c/e^*$.

Flux quantization, together with the analogous *quantization of circulation* in liquid ^{4}He, is one of the most striking macroscopic manifestations of quantum mechanics. It is found experimentally that the flux frozen into superconducting systems jumps by units of the quantum

$$\Phi_0 = 2\pi\hbar c/2e = 2{\cdot}07\times 10^{-7}\,\mathrm{G.cm^2}. \tag{11.84}$$

The charge carried by the superfluid particles is exactly that of an electron pair, i.e.

$$e^* = 2e. \tag{11.85}$$

Apart from this renormalization, the basic London hypothesis is thus fully confirmed experimentally.

The pair charge (11.85) is indeed, exactly what we should expect on the basis of the BCS theory. The derivation of equations like (11.80) from first principles (making allowance for the approximation of 'locality') relies upon the *Gor'kov equations*, which are written in the language of Green functions. In this formalism the order parameter $\Psi(\mathbf{r})$ is related to an 'anomalous propagator'

$$F_{\uparrow\downarrow}(\mathbf{r},t;\mathbf{r}',t') = \langle\psi_\uparrow(\mathbf{r},t)\,\psi_\downarrow(\mathbf{r}',t')\rangle \tag{11.86}$$

which measures the average phase coherence of the wave-functions of pairs of electrons of opposite spin at points $\mathbf{r}$ and $\mathbf{r}'$, at times t and t'. As we saw in (11.76), it is characteristic of the superfluid state that this quantity does not vanish, as it would automatically in the 'normal' state of the metal.

11.10 Superconducting junctions

The energy gap in a superconductor may be observed directly by *Giaever tunnelling*. As in § 6.8, we study the current/voltage characteristic of a thin insulating film between, say, a normal metal and a superconductor. Normal electrons can only enter the superconductor if there are vacant quasi-particle states to receive them. At a temperature well below T_c, the current must be zero until the applied bias ϕ equals the energy gap Δ (§ 11.4), and then jumps sharply (Fig. 210). The transmission coefficient $\mathscr{T}$ of such a junction is difficult to calculate, but the variation of current with voltage allows us to check in detail the density of quasi-particle states in the gap region.

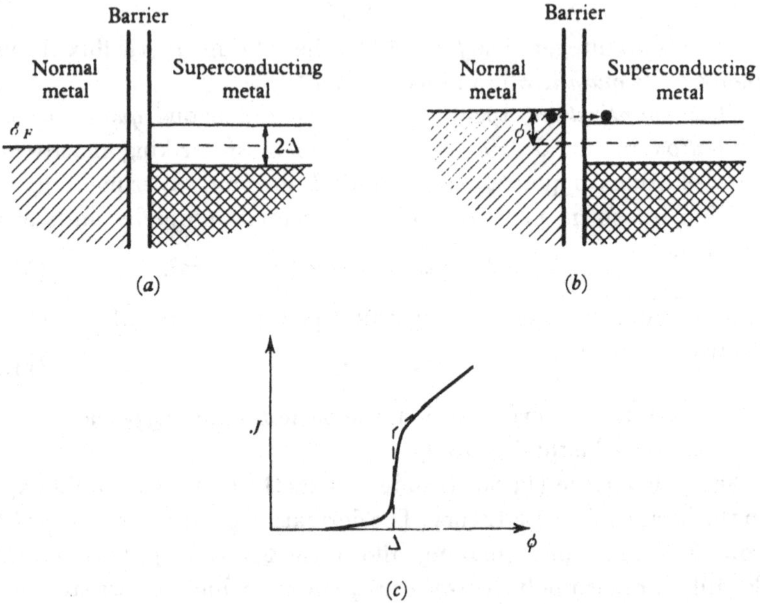

Fig. 210. Quasi-particle tunnelling into a superconductor from a normal metal: (a) at zero bias; (b) for $\phi > \Delta$; (c) current/voltage characteristic, showing effect of temperature.

In practice the transition is not perfectly sharp, for the reasons already discussed in § 6.8. For strongly-coupled superconductors, where the parameter $\mathscr{N}(\mathscr{E}_F)V$ in (11.22) is large, the electron–phonon interaction (11.5) is sufficient to produce a substantial proportion of *phonon-assisted tunnelling* (cf. Fig. 116) of quasi-particles. The fine structure to be observed in the tunnelling current then yields detailed information about the phonon spectrum of the superconductor.

A more striking phenomenon is *coherent tunnelling*, where a supercurrent flows across the junction. This may be observed not only through a thin insulating film between two superconductors but also through quite a thick layer (5000 Å) of a normal metal, which can be made temporarily superconducting by the *proximity effect*. In other words, the strength of the pairing interaction in the true superconductors is sufficient to preserve the phase correlations of the electrons as they pass through the barrier and to re-establish coherence on the other side.

For all such phenomena, the macroscopic order-parameter formalism of § 11.9 is obviously appropriate. How does this function behave at such a boundary; what is the relation between Ψ_L on the left-hand side of the junction and Ψ_R on the right?

Since leakage through a barrier is a time-dependent process, we need to think about the equation of motion of Ψ. Canonical quantum mechanics would say that this involves the energy of a superfluid particle, which we would put at some value μ. We thus write a time-dependent Schrödinger equation for the order parameter, i.e.

$$i\hbar\frac{\partial\Psi}{\partial t} = \mu\Psi. \tag{11.87}$$

In an ordinary supercurrent, of course, μ would be the chemical potential for the superfluid pairs—twice the quantity ζ used as an origin of energy in (11.27)—and hence precisely constant round the whole circuit. Such a system would therefore appear stationary in time, as implied by (11.80), etc.

For transmission through a junction between two identical superconductors, however, we have to consider the coupled equations

$$\left.\begin{aligned} i\hbar\frac{\partial\Psi_L}{\partial t} &= \mu_L\Psi_L+\mathscr{T}\Psi_R, \\ i\hbar\frac{\partial\Psi_R}{\partial t} &= \mu_R\Psi_R+\mathscr{T}\Psi_L, \end{aligned}\right\} \tag{11.88}$$

in which we allow superfluid order to be transferred at the rate $\mathscr{T}$ each way. Note that this coefficient appears here as a *matrix element*, although it was originally defined as a *transition probability* for quasi-particle tunnelling. This is because the superfluid consists of electron *pairs*, whose tunnelling rate would be expected to go as the *square* of the rate for single electrons.

The equations (11.88) have a general time-dependent solution in which the superfluid densities $|\Psi_L|^2$ and $|\Psi_R|^2$ are changing at equal and opposite rates, as if superfluid particles were being transferred from one side of the junction to the other. This we should interpret as a current density (per unit area)

$$J_s = e^*\frac{\partial}{\partial t}|\Psi_R|^2 = \frac{2}{\hbar}\mathscr{T}n_s e^*\sin(\chi_R-\chi_L), \tag{11.89}$$

where the phase difference between Ψ_R and Ψ_L (cf. (11.81)) varies with time according to

$$\frac{\partial}{\partial t}(\chi_R - \chi_L) = -\frac{1}{\hbar}(\mu_R - \mu_L). \qquad (11.90)$$

These equations may, of course, be deduced rigorously from the BCS theory in the Gor'kov formalism, with a tunnelling Hamiltonian to represent the transfer of electrons from one region to the other. Notice now that the supercurrent (11.89) is directly proportional to the quasi-particle tunnelling coefficient $\mathscr{T}$, showing that phase coherence is playing its part once more.

Suppose now that we apply a small steady voltage ϕ (less than the gap width, to avoid quasi-particle excitation) across the junction. This is equivalent to a difference $2e\phi$ in the chemical potential for superfluid particles. The equations (11.89) and (11.90) describe a supercurrent oscillating in time with frequency $2e\phi/\hbar$. This is the A. C. *Josephson effect*, which again provides direct macroscopic evidence of phase coherence in quantum phenomena, and has proved to be a very exact method for measuring the important ratio of atomic constants, $e/\hbar$.

In the presence of a magnetic field, these formulae need the usual modification (10.3) to preserve gauge invariance in the definition of the vector potential. A typical situation is as follows. Suppose we have a relatively large area of junction between two superconductors (Fig. 211). Since the insulating barrier is not itself superconducting, magnetic flux may penetrate through the junction region, parallel to the surfaces of separation. Consider now pairs of points, such as L_1, R_1; L_2, R_2, lying on either side of the junction, deep enough in each case to avoid the penetration regions, but fairly well separated in the direction parallel to the surface. Suppose we know the phase difference happens to be

$$\Delta\chi_1 = \chi(R_1) - \chi(L_1) \qquad (11.91)$$

across $L_1 R_1$; what is the corresponding phase difference $\Delta\chi_2$ across $L_2 R_2$.

Consider a path from L_2 to L_1, along which we integrate both sides of (11.82). Just as in the deduction of (11.83), there is no contribution from $\mathbf{v}_s$, which can have no component parallel to the junction surface except in the penetration region: we get

$$\chi(L_1) - \chi(L_2) = \frac{e^*}{\hbar c}\int_{L_2}^{L_1} \mathbf{A}(\mathbf{r}) \cdot d\mathbf{l}. \qquad (11.92)$$

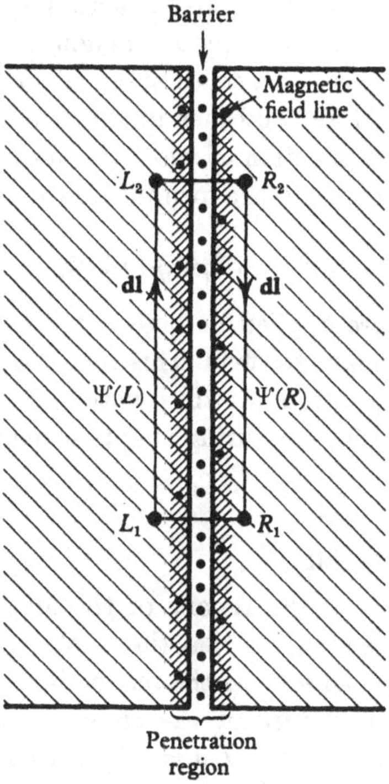

Fig. 211. Josephson junction.

Similarly, on a return journey in the right-hand material,

$$\chi(R_2) - \chi(R_1) = \frac{e^*}{\hbar c} \int_{R_1}^{R_2} \mathbf{A}(\mathbf{r}) \cdot \mathbf{dl}. \tag{11.93}$$

Adding these two equations, and ignoring trivial contributions on the right as we go through the junction, we get

$$\begin{aligned} \Delta\chi_2 &= \Delta\chi_1 + \frac{e^*}{\hbar c} \oint \mathbf{A} \cdot \mathbf{dl} \\ &= \Delta\chi_1 + 2\pi\Phi(1,2)/\Phi_0, \end{aligned} \tag{11.94}$$

where Φ_0 is the unit of quantized flux (11.84), and $\Phi(1,2)$ is the magnetic flux that penetrates the barrier betwen the line $L_1 R_1$ and the line $L_2 R_2$. In other words, the relative phase difference of the order parameter across the boundary varies as one moves about in the junction region.

As a consequence, the total current across a broad junction is a complicated function of the applied magnetic field. We have to evaluate $(\chi_R - \chi_L)$ across each small area, calculate the contribution to the current from (11.89), and then integrate over the whole surface of the junction. For a junction of width W, for example, penetrated by a total flux Φ, one can derive a formula for the maximum *Josephson current*, of the form

$$J \propto |\sin(\pi\Phi/\Phi_0)/(\pi\Phi/\Phi_0)|, \tag{11.95}$$

which can flow at zero applied voltage but which oscillates rapidly as one changes the magnetic field. This D.C. *Josephson effect*, in the more complex configuration of twin junctions in parallel, provides a most sensitive device for measuring magnetic fields.

11.11 Type II material

A detailed analysis of the theory of the Meissner effect (§ 11.7) reveals a paradox. We assumed that the whole specimen would cease to be superconducting as soon as the magnetic field energy exceeded the difference in free energy of the superconducting and normal states—i.e. at the critical field H_c given by (11.63). But this assumption ignores the possibility of constructing an inhomogeneous phase, where the magnetic flux is carried by layers of filaments of normal metal dispersed in superconducting material. Such an *intermediate state* would apparently be thermodynamically stable up to much higher magnetic fields, and would still be electrically superconducting on a macroscopic scale.

Experimentally, something like this state is observed in *type II materials* where the Meissner effect is lost at a lower critical field H_{c1} yet the material exhibits no electrical resistance up to a much higher field H_{c2}. The importance of such materials as windings for electromagnets has focused attention on this phenomenon, which is also of great theoretical interest.

As in the theory of ferromagnetic domains, (§ 10.3) we need a theoretical model of the boundary between normal and superconducting regions in the same metal. This means that the 'superfluid density' n_s of (11.77) or (11.79) must be allowed to vary in space. We need some further equations determining the behaviour of the *amplitude* of the order parameter $\Psi(\mathbf{r})$ as a function of $\mathbf{r}$.

The basic assumption of the *Ginsburg–Landau theory* is that the free energy density in a superconductor containing a magnetic field **H** (measured relative to the normal metal in zero field at the same temperature) is a function of the form

$$g_s - g_n = \alpha|\Psi|^2 + \tfrac{1}{2}\beta|\Psi|^4 + \frac{1}{8\pi}H^2 + \frac{1}{2m^*}\left|-i\hbar\nabla\Psi - \frac{e^*}{c}\mathbf{A}\Psi\right|^2. \quad (11.96)$$

The first two terms merely express the arbitrary assumption that the gain in free energy on going to the superconducting phase is a function of the order parameter Ψ, which can be expanded in powers of the real quantity $|\Psi|^2$. The arbitrary coefficients α and β are themselves dependent on the temperature; at temperatures below T_c, where the superconducting phase is favoured, the coefficient α must become negative; the superfluid concentration in zero magnetic field is then limited by the term in β to the value

$$n_s = |\Psi_0|^2 = -\alpha/\beta. \quad (11.97)$$

For type I material, obeying the Meissner effect, we can bring in the term in H^2, and identify the critical field for the bulk material,

$$H_c^2/8\pi = \alpha^2/2\beta, \quad (11.98)$$

as the point where g_s cannot be reduced below g_n. Again, we can use (11.97) in the London formula (11.61), i.e.

$$\lambda = \left(\frac{m^*c^2}{4\pi n_s e^{*2}}\right)^{\frac{1}{2}}, \quad (11.99)$$

to express the penetration depth in terms of the same parameters whose temperature dependence can thus be estimated.

The final term in (11.96) is also phenomenological, but has firm canonical foundations as a typical quantum mechanical expression for the kinetic energy of a particle in a magnetic field. In the London equations (11.55) and (11.80), it would stand for the kinetic energy of the 'supercurrent', derived from the spatial variation of phase of Ψ as in (11.82). The elegant feature of the Ginsburg–Landau theory is that precisely the same term, without change of coefficients, accounts for the effects of spatial variations of the magnitude of the order parameter—effects that we should attribute to the energy required to break the coherence of the pairing.

Suppose, for example, that we want to go from a superconducting region, where Ψ is given by (11.97), to a normal region where $\Psi = 0$, with the minimum penalty in total free energy. In the absence of a magnetic field, and ignoring the term in β, the modulus of the order parameter must satisfy the partial differential equation

$$(\hbar^2/2m^*)\,\nabla^2|\Psi| - |\alpha|\,\Psi = 0, \tag{11.100}$$

obtained variationally from the free energy integral. This describes a situation in which $|\Psi|$ falls off exponentially, with a characteristic length

$$\xi = \{\hbar^2/2m^*|\alpha|\}^{\frac{1}{2}}. \tag{11.101}$$

In other words, we may connect the coefficient α with the coherence length of § 11.8, a quantity that is approximately equal to the electron mean free path Λ in a typical 'dirty' metal. This description is obviously inadequate to represent in detail the non-local behaviour implicit in the Pippard formulae of § 11.8, but is justifiable from first principles by the Gor'kov method at temperatures near T_c.

The next step in the argument is to study the minimum free energy configuration for a critical boundary between superconducting material (where $|\Psi| = |\Psi_0|$, and $H = 0$) and normal material containing the critical field H_c. The general partial differential equation for Ψ, which is coupled to Maxwell's equations for $\mathbf{H}$ as in (11.58)–(11.60), is rather messy, but the form of the answer is easy enough to work out in two extreme cases.

Suppose, first, that the coherence length ξ is much larger than the penetration depth λ. It is clear from Fig. 212(*a*) that the magnetic field has been excluded from a layer of thickness ξ, which requires an energy

$$\sigma_{\mathrm{I}} \sim \xi(H_c^2/8\pi) \tag{11.102}$$

per unit area of boundary, without any compensatory gain in 'ordering' free energy throughout most of the layer. The surface energy is given approximately by (11.102) and is positive. This is typical of type I superconductivity, where the intermediate state is inhibited by the energy required to make the boundaries between normal and superconducting regions.

On the other hand, if the coherence length ξ is much smaller than the penetration depth λ, the negative free energy of the ordered phase is available throughout most of the region where the magnetic field is being excluded (Fig. 212(*b*)). Roughly speaking, therefore, the surface energy is of the form

$$\sigma_{\mathrm{II}} \sim -\lambda(H_c^2/8\pi). \tag{11.103}$$

More detailed analysis shows that the surface energy changes sign, from positive to negative, when

$$\lambda > \xi/\sqrt{2}. \tag{11.104}$$

This is the basic condition for type II superconductivity; it is obviously favoured by chemical impurities and crystal imperfections, which reduce the coherence length ξ without significantly altering the other macroscopic superconducting parameters such as λ.

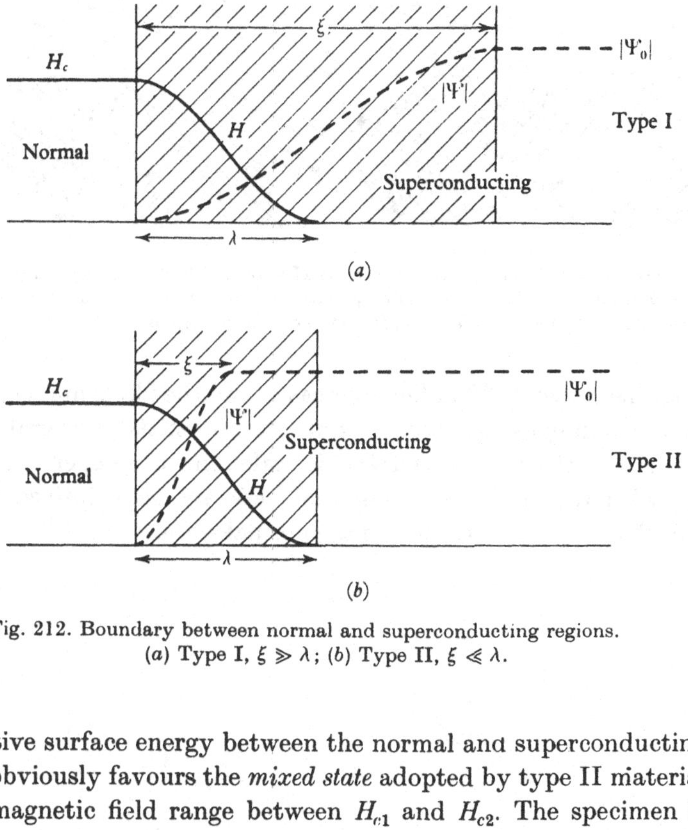

Fig. 212. Boundary between normal and superconducting regions. (a) Type I, $\xi \gg \lambda$; (b) Type II, $\xi \ll \lambda$.

Negative surface energy between the normal and superconducting phases obviously favours the *mixed state* adopted by type II material in the magnetic field range between H_{c1} and H_{c2}. The specimen is penetrated by filaments of magnetic field, each of which has the unit fluxoid strength (11.84). The filament contains a cylindrical core of normal metal of radius ξ, surrounded by a cylinder of radius λ, around which supercurrents must flow to limit the penetration of the magnetic field. Each filament therefore behaves like a vortex line in the superfluid and tends to repel its neighbours. As we increase the magnetic

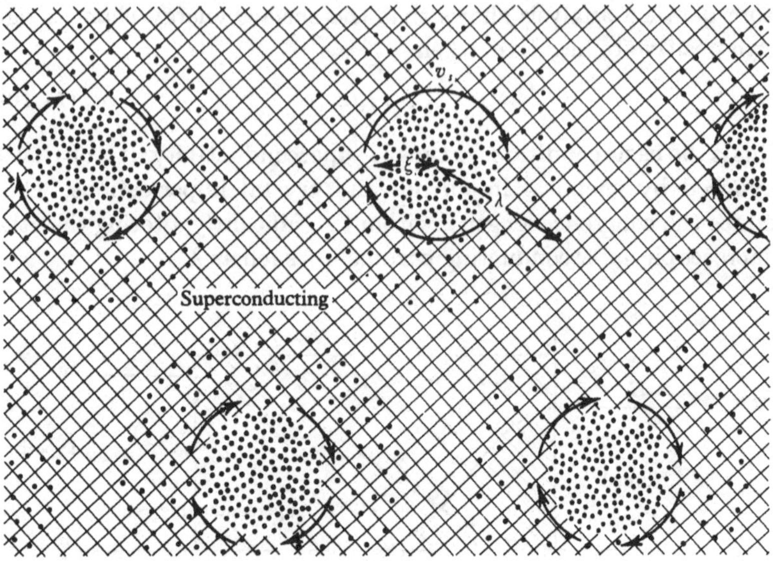

Fig. 213. Array of flux filaments in type II superconductor, showing magnetic field lines concentrated in and around cylindrical cores of normal material. Superfluid vortices flow round the filaments in the penetration regions.

field, the density of flux lines increases, and the filaments 'crystallize' in a two-dimensional lattice. But the interstitial material remains superconducting on a macroscopic scale until the upper critical field H_{c2}, when the core regions of normal metal are forced into contact and the whole specimen reverts to the normal state.

BIBLIOGRAPHY

Chapter 1. Periodic structures

The following are general texts, covering many branches of the subject.

Introductory and descriptive

Slater, J. C. (1951). *The Quantum Theory of Matter.*
Dekker, A. J. (1957). *Solid State Physics.*
Azaroff, L. V. (1960). *Introduction to Solids.*
Cusack, N. (1963). *The Electrical and Magnetic Properties of Solids.*
Rosenberg, H. M. (1963). *Low Temperature Solid State Physics.*
Azaroff, L. V. & Brophy, J. J. (1963). *Electronic Processes in Materials.*
Holden, A. (1965). *The Nature of Solids.*
Zhdanov, G. S. (1965). *Crystal Physics.*
Kittel, C. (1966). *Introduction to Solid State Physics* (3rd edition).
Brown, F. C. (1967). *The Physics of Solids.*
Blakemore, J. S. (1969). *Solid State Physics.*
Cracknell, A. P. (1969). *Crystals and their Structures.*

More advanced

Wilson, A. H. (1953). *The Theory of Metals.*
Peierls, R. E. (1955). *Quantum Theory of Solids.*
Wannier, G. H. (1959). *Elements of Solid State Theory.*
Ziman, J. M. (1960). *Electrons and Phonons.*
Smith, R. A. (1961). *The Wave Mechanics of Crystalline Solids.*
Kittel, C. (1963). *Quantum Theory of Solids.*
Anderson, P. W. (1963). *Concepts in Solids.*
Sachs, M. (1963). *Solid State Theory.*
Weinreich, G. (1965). *Solids: Elementary Theory for Advanced Students.*
Slater, J. C. (1965, 1967). *Quantum Theory of Molecules and Solids:* Vol. 2: *Symmetry and Energy Bands in Solids;* Vol. 3: *Insulators, Semiconductors and Metals.*
Bassani, F., Cagliotti, G. & Ziman, J. M. (eds.) (1968). *Theory of Condensed Matter* (I.A.E.A., Vienna).
Landsberg, P. T. (ed.) (1969). *Solid State Theory.*
Ziman, J. M. (ed.) (1969). *The Physics of Metals. 1. Electrons.*
Harrison, W. A. (1970). *Solid State Theory.*

Chapter 2. Lattice waves

Brillouin, L. (1946). *Wave Propagation in Periodic Structures.*
Bacon, G. E. & Lonsdale, K. (1953). 'Neutron Diffraction.' *Rep. Prog. Phys.* **16**.

BORN, M. & HUANG, K. (1954). *Dynamical Theory of Crystal Lattices.*

DE LAUNAY, J. (1956). 'Theory of Specific Heats and Lattice Vibrations.' *Solid State Phys.* **2.**

SHULL, C. G. & WOLLAN, E. O. (1956). 'Applications of Neutron Diffraction to Solid State Problems.' *Solid State Phys.* **2.**

BOUMAN, J. (1957). 'Theoretical Principles of Structural Research by X-rays.' *Handb. Phys.* **32.**

GUINIER, A. & VON ELLER, G. (1957). 'Les Méthodes Expérimentales des déterminations de structures cristallines par rayons X.' *Handb. Phys.* **32.**

LEIBFRIED, G. & LUDWIG, W. (1958). 'Theory of Anharmonic Effects in Crystals.' *Solid State Phys.* **7.**

SLATER, J. C. (1958). 'Interaction of Waves in Crystals', *Rev. Mod. Phys.* **30.**

KOTHARI, L. S. & SINGWI, K. S. (1959). 'Interaction of Thermal Neutrons with Solids.' *Solid State Phys.* **8.**

COCHRAN, W. (1960). 'Crystal Stability and The Theory of Ferroelectricity.' *Advance Phys.* **9.**

HERBSTEIN, F. H. (1961). 'Methods of Measuring Debye Temperatures, etc.' *Advanc. Phys.* **10.**

BACON, G. E. (1962). *Neutron Diffraction* (2nd edition).

MITRA, S. S. (1962). 'Vibration Spectra of Solids.' *Solid State Phys.* **13.**

FRAUENFELDER, H. (1962). *The Mössbauer Effect.*

COCHRAN, W. (1963). 'Lattice Vibrations.' *Rep. Prog. Phys.* **26.**

JAMES, R. W. (1963). 'The Dynamical Theory of X-ray Diffraction' *Solid State Phys.* **15.**

MARADUDIN, A. A., MONTROLL, E. W. & WEISS, G. H. (1963). 'Theory of Lattice Dynamics in the Harmonic Approximation.' *Solid State Phys.* Suppl. 3.

BAK, T. A. (ed.) (1964). *Phonons and Phonon Interactions.*

WERTHEIM, G. K. (1964). *Mössbauer Effect: Principles and Applications.*

MARADUDIN, A. A. (1966). 'Theoretical and Experimental Aspects of the Effects of Point Defects and Disorder on the Vibrations of Crystals' 1 and 2. *Solid State Phys.* **18, 19.**

STEVENSON, R. W. H. (ed.) (1966). *Phonons in Perfect Lattices and in Lattices with Point Imperfections.*

CHOQUARD, P. (1967). *The Anharmonic Crystal.*

EVANS, R. H. & HAERING, R. R. (eds.) (1969). *Phonons and their Interactions.*

GUYER, R. A. (1969). 'The Physics of Quantum Crystals.' *Solid State Phys.* **23.**

MUNN, R. W. (1969). 'The Thermal Expansion of Axial Metals.' *Adv. in Phys.* **18.**

CHAPTER 3. ELECTRON STATES

REITZ, J. R. (1955). 'Methods of the One-Electron Theory of Solids.' *Solid State Phys.* **1.**

HAM, F. S. (1955). 'The Quantum Defect Method.' *Solid State Phys.* **1.**

SLATER, J. C. (1956). 'The Electronic Structure of Solids.' *Handb. Phys.* **19.**

WOODRUFF, T. O. (1957). 'The Orthogonalized Plane-Wave Method.' *Solid State Phys.* **4.**

CALLAWAY, J. (1958). 'Electron Energy Bands in Solids.' *Solid State Phys.* **7.**

JONES, H. (1960). *The Theory of Brillouin Zones and Electronic States in Crystals.*

HEINE, V. (1960). *Group Theory in Quantum Mechanics.*

PINCHERLE, L. (1961). 'Band Structure Calculations in Solids.' *Rep. Prog. Phys.* **23.**

CALLAWAY, J. (1963). *Energy Band Theory.*

KNOX, R. S. & GOLD, A. (1964). *Symmetry in the Solid State.*

MEIJER, P. H. (ed.) (1964). *Group Theory and Solid State Physics.*

FALICOV, L. M. (1966). *Group Theory and its Physical Applications.*

HARRISON, W. A. (1966). *Pseudopotentials in the Theory of Metals.*

NUSSBAUM, A. (1966). 'Crystal Symmetry, Group Theory and Band Structure Calculation.' *Solid State Phys.* **18.**

LOUCKS, T. (1967). *Augmented Plane Wave Method.*

ALDER, B., FERNBACH, S. & ROTENBERG, M. (eds.) (1968). *Methods in Computational Physics 8: Energy Bands in Solids.*

CRACKNELL, A. P. (1968). *Applied Group Theory.*

COHEN, M. L., HEINE, V. & WEAIRE, D. (1970). Articles on pseudopotentials in *Solid State Phys.* **24.**

KILLINGBECK, J. (1970). 'Group Theory and Topology in Solid State Physics.' *Rep. Progr. Phys.* **33.**

ZIMAN, J. M. (1971). 'The Calculation of Bloch Functions.' *Solid State Phys.* **26.**

CHAPTER 4. STATIC PROPERTIES OF SOLIDS

MOTT, N. F. & GURNEY, R. W. (1940). *Electronic Processes in Ionic Crystals.*

FAN, H. Y. (1955). 'Valence Semiconductors, Germanium and Silicon.' *Solid State Phys.* **1.**

WIGNER, E. P. & SEITZ, F. (1955). 'Qualitative Analysis of the Cohesion in Metals.' *Solid State Phys.* **1.**

KEESOM, P. H. (1956). 'Low Temperature Heat Capacity of Solids.' *Handb. Phys.* **14.**

NYE, J. F. (1957). *Physical Properties of Crystals: their Representation by Tensors and Matrices.*

SMITH, R. A. (1959). *Semiconductors.*

SCANLON, W. W. (1959). 'Polar Semiconductors.' *Solid State Phys.* **9.**

HILSUM, C. & ROSE-INNES, A. C. (1961). *Semiconducting III–V Compounds.*

GSCHNEIDER, K. A. (1964). 'Physical Properties and Interrelationships of Metallic and Semimetallic Elements.' *Solid State Phys.* **16.**

MOTT, N. F. (1964). 'Electrons in Transition Metals.' *Adv. in Phys.* **13.**

TOSI, M. P. (1964). 'Cohesion of Ionic Solids in the Born Model.' *Solid State Phys.* **16.**

MARCH, N. H. (1968). *Liquid Metals.*

PHILLIPS, J. C. (1969). *Covalent Bonding in Crystals, Molecules and Polymers.*

CHAPTER 5. ELECTRON-ELECTRON INTERACTION

PINES, D. (1955). 'Electron Interaction in Metals.' *Solid State Phys.* **1.**

DE WITT, C. & NOZIÈRES, P. (1959). *The Many Body Problem* (Cours donnes a l'École d'Été de Physique Théorique, Les Houches, 1958).

MOTT, N. F. & TWOSE, W. D. (1961). 'The Theory of Impurity Conduction.' *Adv. in Phys.* **10.**

THOULESS, D. J. (1961). *The Quantum Mechanics of Many-Body Systems.*

CAIANIELLO, E. R. (ed.) (1962). *Lectures on the Many-Body Problem.*

PINES, D. (1962). *The Many Body Problem.*

NOZIÈRES, P. (1963). *Le Problème à N Corps.*

PINES, D. (1963). *Elementary Excitations in Solids.*

ZIMAN, J. M. (1964). 'The Method of Neutral Pseudoatoms in the Theory of Metals.' *Adv. in Phys.* **13.**

PINES, D. & NOZIERES, P. (1966). *The Theory of Quantum Liquids. 1. Normal Fermi Liquids.*

MARCH, N. H., YOUNG, W. H. & SAMPANTHAR, S. (1967). *The Many Body Problem in Quantum Mechanics.*

MOTT, N. F. (1967). 'Electrons in Disordered Systems.' *Adv. in Phys.* **16,** 49.

ADLER, D. (1968). 'Insulating & Metallic States in Transition Metal Oxides.' *Solid State Phys.* **21.**

HALPERIN, B. I. & RICE, T. M. (1968). 'The Excitonic State at the Semiconductor Semimetal Transition.' *Solid State Phys.* **21.**

MARCH, N. H. & STODDART, J. C. (1968). 'Localisation of Electrons in Condensed Matter.' *Rep. Prog. Phys.* **31.**

HEDIN, L. & LUNDQVIST, S. (1969). 'Effect of Electron–Electron & Electron–Phonon Interaction on the One Electron States of Solids.' *Solid State Phys.* **23.**

ZIMAN, J. M. (1969). *Elements of Advanced Quantum Theory.*

BOSMAN, A. J. & VAN DAAL, H. J. (1970). 'Small-polaron versus Band Conduction in some Transition Metal Oxides.' *Adv. in Phys.* **19.**

Chapter 6. Dynamics of electrons

Kohn, W. (1957). 'Shallow Impurity States in Silicon and Germanium.' *Solid State Phys.* **5**.

Lewis, H. W. (1958). 'Wave Packets and Transport of Electrons in Metals.' *Solid State Phys.* **7**.

Blount, E. I. (1962). 'Formalisms of Band Theory.' *Solid State Phys.* **13**.

Sham, L. J. & Ziman, J. M. (1963). 'The Electron–Phonon Interaction.' *Solid State Phys.* **15**.

Knox, R. S. (1963). 'Theory of Excitons.' *Solid State Phys. Supp.* **5**.

Markham, J. J. (1966). '*F*-centres in Alkali Halides.' *Solid State Phys. Supp.* **8**.

Brown, E. (1968). 'Aspects of Group Theory in Electron Dynamics.' *Solid State Phys.* **22**.

Fowler, W. B. (ed.) (1968). *Physics of Colour Centres.*

Joshi, S. K. & Rajagopal, A. K. (1968). 'Lattice Dynamics of Metals.' *Solid State Phys.* **22**.

Burstein, E. & Lundqvist, S. (eds.) (1969). *Tunnelling Phenomena in Solids.*

Duke, C. B. (1969). 'Tunnelling in Solids.' *Solid State Phys. Supp.* **10**.

Chapter 7. Transport properties

Jones, H. (1956). 'Theory of Electrical and Thermal Conductivity in Metals.' *Handb. Phys.* **19**.

MacDonald, D. K. C. (1956). 'Electrical Conductivity of Metals and Alloys at Low Temperatures.' *Handb. Phys.* **14**.

Blatt, F. J. (1957). 'Theory of Mobility of Electrons in Solids.' *Solid State Phys.* **4**.

Jan, J. P. (1957). 'Galvanomagnetic and Thermomagnetic Effects in Metals.' *Solid State Phys.* **5**.

Klemens, P. G. (1958). 'Thermal Conductivity and Lattice Vibrational Modes.' *Solid State Phys.* **7**.

Rosenberg, H. M. (1958). 'The Properties of Metals at Low Temperatures.' *Prog. Metal. Phys.* **7**.

Keyes, R. W. (1960). 'The Effects of Elastic Deformation on the Electrical Conductivity of Semiconductors.' *Solid State Phys.* **11**.

Ziman, J. M. (1960). *Electrons and Phonons.*

Drabble, J. R. & Goldsmid, H. J. (1961). *Thermal Conduction in Semiconductors.*

Mendelssohn, K. & Rosenberg, H. M. (1961). 'The Thermal Conductivity of Metals at Low Temperatures.' *Solid State Phys.* **12**.

MacDonald, D. K. C. (1962). *Thermoelectricity: an Introduction to the Principles.*

Beer, A. C. (1963). *Galvanomagnetic Effects in Semiconductors.*

DELVES, R. T. (1965). 'Thermomagnetic Effects in Semiconductors and Semimetals.' *Rep. Progr. Phys.* **28**.

GOLDSMID, H. J. (1965). 'Transport Effects in Semimetals and Narrow-gap Semiconductors.' *Adv. in Phys.* **14**.

CHAPTER 8. OPTICAL PROPERTIES

FRÖHLICH, H. (1949). *The Theory of Dielectrics.*

GIVENS, M. P. (1958). 'Optical Properties of Metals.' *Solid State Phys.* **6**.

MOSS, T. S. (1959). *The Optical Properties of Semi-Conductors.*

PIPPARD, A. B. (1962). 'The Dynamics of Conduction Electrons.' *Low-Temperature Physics* (ed. De Witt, Dreyfus, de Gennes).

SCHULMAN, J. H. & COMPTON, W. D. (1962). *Colour Centres in Solids.*

SCAIFE, B. K. P. (1963). 'Dispersion and Fluctuation in Dielectrics.' *Progr. in Dielectrics*, **5**.

STERN, F. (1963). 'Elementary Theory of the Optical Properties of Solids.' *Solid State Phys.* **15**.

KUPER, C. G. & WHITFIELD, G. D. (1963). *Polarons and Excitons.*

LOUDON, R. (1964). 'The Raman Effect in Crystals.' *Adv. in Phys.* **13**.

EINSPRUCH, N. (1965). 'Ultrasonic Effects in Semiconductors.' *Solid State Phys.* **17**.

MARTIN, D. H. (1965). 'The Study of the Vibrations of Crystal Lattices by far Infra-red Spectroscopy.' *Adv. in Phys.* **14**.

ABELES, F. (ed.) (1966). *Optical Properties and Electronic Structure of Metals and Alloys.*

SPECTOR, H. N. (1966). 'Interaction of Acoustic Waves and Conduction Electrons.' *Solid State Phys.* **19**.

TAUC, J. (ed.) (1966). *The Optical Properties of Solids.*

DANIEL, V. V. (1967). *Dielectric Relaxation.*

APPEL, J. (1968). 'Polarons.' *Solid State Phys.* **21**.

AUSTIN, I. G. & MOTT, N. F. (1969). 'Polarons in Crystalline and Non-crystalline Materials.' *Adv. in Phys.* **18**.

NEWMAN, R. C. (1969). 'Infra-red Absorption due to Localized Modes of Vibration of Impurity Complexes in Ionic and Semiconductor Crystals.' *Adv. in Phys.* **18**.

PICK, R. M. (1970). 'Phonons and Photons in Crystals.' *Adv. in Phys.* **19**.

CHAPTER 9. THE FERMI SURFACE

SHOENBERG, D. (1957). 'The de Haas–van Alphen Effect.' *Prog. Low Temp. Phys.* **2**.

KAHN, A. M. & FREDERIKSE, H. P. R. (1959). 'Oscillatory Behaviour of Magnetic Susceptibility and Electronic Conductivity.' *Solid State Phys.* **9**.

HARRISON, W. A. & WEBB, M. B. (eds.) (1960). *The Fermi Surface.* (This conference Report contains several reviews of different aspects.)

LAX, B. & MAVROIDES, J. G. (1960). 'Cyclotron Resonance.' *Solid State Phys.* **11**.

AZBEL', M. YA. & LIFSHITZ, I. M. (1961). 'Electron Resonances in Metals.' *Prog. Low Temp. Phys.* **3**.

PIPPARD, A. B. (1961). 'Experimental Analysis of Electronic Structure of Metals.' *Rep. Prog. Phys.* **23**.

PIPPARD, A. B. (1962). 'The Dynamics of Conduction Electrons.' *Low-Temperature Physics* (ed. De Witt, Dreyfus, de Gennes).

LIFSHITZ, I. M. & KAGANOV, M. I. (1959, 1962, 1966). 'Some Problems of the Electron Theory of Metals.' *Uspekhi Fiz. Nauk,* **69**, **78**, **87**.

FAWCETT, E. (1964). 'High Field Galvonomagnetic Properties of Metals.' *Adv. in Phys.* **13**.

KUBO, R., MIYAKE, S. J. & HASHITSUME, N. (1965). 'Quantum Theory of Galvanomagnetic Effect at Extremely Strong Magnetic Fields.' *Solid State Phys.* **7**.

KANER, E. A. & GANTMAKHER, V. F. (1968). 'Anomalous Penetration of Electromagnetic Field in a Metal and Radiofrequency Size Effects.' *Uspekhi Fiz. Nauk,* **94**.

KANER, E. A. & SKOVOB, V. G. (1968). 'Electromagnetic Waves in Metals in a Magnetic Field.' *Adv. in Phys.* **17**.

ZYRYANOV, P. S. & GUSEVA, G. I. (1968). 'Quantum Theory of Thermomagnetic Phenomena in Metals and Semiconductors.' *Uspekhi Fiz. Nauk* **95**.

KHAIKIN, M. S. (1968). 'Magnetic Surface Levels.' *Uspekhi Fiz. Nauk* **96**.

BAYNHAM, A. C. & BOARDMAN, A. D. (1970). 'Helicon and Alfvèn Wave Propagation in Non-Magnetic Semiconductors and Semimetals: Active and Passive Waves.' *Adv. in Phys.* **19**.

CHAPTER 10. MAGNETISM

VAN VLECK, J. H. (1932). *The Theory of Electric and Magnetic Susceptibilities.*

NEWELL, G. F. & MONTROLL, E. W. (1953). 'On the Theory of the Ising Model of Ferromagnetism.' *Rev. Mod. Phys.* **25**.

KNIGHT, W. D. (1956). 'Electron Paramagnetism and Nuclear Magnetic Resonance in Metals.' *Solid State Phys.* **2**.

VAN KRANENDONK, J. & VAN VLECK, J. H. (1958). 'Spin Waves.' *Rev. Mod. Phys.* **30**.

DOMB, C. (1960). 'On the Theory of Cooperative Phenomena in Crystals.' *Advanc. Phys.* **9**.

ABRAGAM, A. (1961). *The Principles of Nuclear Magnetism.*

GRIFFITH, J. S. (1961). *The Theory of Transition-Metal Ions.*

ROWLAND, T. J. (1961). 'Nuclear Magnetic Resonance in Metals.' *Prog. Mater. Sci.* **9**.

Siversten, J. M. & Nicholson, M. E. (1961). 'The Structure and Properties of Solid Solutions.' *Prog. Mater. Sci.* **9**.

Herpin, A. (1962). 'Magnetisme.' *Low-Temperature Physics* (ed. De Witt, Dreyfus, de Gennes).

Kittel, C. (1962). 'Magnons.' *Low-Temperature Physics* (ed. De Witt, Dreyfus, de Gennes).

Vonsovskiĭ, S. V. (1962). 'Magnetism and Electrical Properties of Solids.' *Uspekhi Fiz. Nauk*, **76**.

Vonsovskiĭ, S. V. & Izyumov, Yu. A. (1962). 'Electron Theory of Transition Metals.' *Upsekhi Fiz. Nauk*, **77, 78**.

Anderson, P. W. (1963). 'Theory of Magnetic Exchange Interactions: Exchange in Insulators and Semiconductors.' *Solid State Phys.* **14**.

Slichter, C. P. (1963). *Principles of Magnetic Resonance.*

Yafet, Y. (1963). '*g*-Factors and Spin-Lattice Relaxation of Conduction Electrons.' *Solid State Phys.* **14**.

Mattis, D. C. (1965). *The Theory of Magnetism.*

Thompson, E. D. (1965). 'Unified Theory of Ferromagnetism.' *Adv. in Phys.* **14**.

Bailyn, M. (1966). 'Theory of Magnetic Impurities in Simple Metals.' *Adv. in Phys.* **15**.

Rado, G. T. & Suhl, H. (eds.) (1966). *Magnetism*, Vols. 1–4.

Smart, J. S. (1966). *Effective Field Theories of Magnetism.*

Blank, A. Ya & Kaganov, M. I. (1967). 'Ferromagnetic Resonance and Plasma Effects in Metals.' *Uspekhi Fiz. Nauk.* **92**.

Fisher, M. E. (1967). 'The Theory of Equilibrium Critical Phenomena.' *Rep. Prog. Phys.* **30**.

Nagamiya, T. (1967). 'Helical Spin Ordering—1. Theory of Helical Spin Configurations.' *Solid State Phys:* **20**.

Tyablikov, S. V. (1967). *Methods in the Quantum Theory of Magnetism.*

Cooper, B. R. (1968). 'Magnetic Properties of Rare Earth Metals.' *Solid State Phys.* **20**.

Coqblin, B. & Blandin, A. (1968). 'Stabilité des moments localisés dans les métaux.' *Adv. in Phys.* **17**.

Grazhdankina, N. P. (1968). 'Magnetic First Order Phase Transitions.' *Uspekhi Fiz. Nauk*, **96**.

Kittel, C. (1968). 'Indirect Exchange Interactions in Metals.' *Solid State Phys.* **22**.

Heeger, A. J. (1969). 'Localized Moments and non-moments in Metals: the Kondo Effect.' *Solid State Phys.* **23**.

Kondo, J. (1969). 'Theory of Dilute Magnetic Alloys.' *Solid State Phys.* **23**.

Hebborn, J. W. & March, N. H. (1970). 'Orbital and Spin Magnetism and Dielectric Response of Electrons in Metals.' *Adv. in Phys.* **19**.

White, R. M. (1970). *Quantum Theory of Magnetism.*

CHAPTER 11. SUPERCONDUCTIVITY

LONDON, F. (1950). *Superfluids*. Vol. I.

SHOENBERG, D. (1952). *Superconductivity*.

BARDEEN, J. (1956). 'Theory of Superconductivity.' *Handb. Phys.* **15.**

KHALATNIKOV, I. M. & ABRIKOSOV, A. A. (1959). 'The Modern Theory of Superconductivity.' *Advanc. Phys.* **8.**

KUPER, C. G. (1959). 'The Theory of Superconductivity.' *Advanc. Phys.* **8.**

SCHAFROTH, M. R. (1960). 'Theoretical Aspects of Superconductivity.' *Solid State Phys.* **10.**

BARDEEN, J. & SCHRIEFFER, J. R. (1961). 'Recent Developments in Superconductivity.' *Progr. Low Temp. Phys.* **3** (ed. Gorter).

FRÖHLICH, H. (1962). 'Theory of Superconductive State.' *Rep. Prog. Phys.* **24.**

LYNTON, E. A. (1962). *Superconductivity*.

TINKHAM, M. (1962). 'Superconductivity.' *Low-Temperature Physics* (ed. De Witt, Dreyfus, de Gennes).

BLATT, J. M. (1964). *Theory of Superconductivity*.

RICKAYZEN, G. (1965). *Theory of Superconductivity*.

KUPER, C. G. (1968). *An Introduction to the Theory of Superconductivity*.

PARKS, R. D. (ed.) (1969). *Superconductivity* (2 vols.).

INDEX

9 780521 297332

Printed in the United States
By Bookmasters